Optimization Methods in Mobile Communication Systems

Milind Pande • Anand J. Kulkarni
Apoorva S. Shastri
Editors

Optimization Methods in Mobile Communication Systems

A Machine-Generated Literature Overview

Springer

Editors
Milind Pande
Dr. Vishwanath Karad MIT World Peace University
Pune, Maharashtra, India

Anand J. Kulkarni
Institute of Artificial Intelligence
Dr. Vishwanath Karad MIT World Peace University
Pune, Maharashtra, India

Apoorva S. Shastri
Institute of Artificial Intelligence
Dr. Vishwanath Karad MIT World Peace University
Pune, Maharashtra, India

ISBN 978-981-95-1809-8 ISBN 978-981-95-1810-4 (eBook)
https://doi.org/10.1007/978-981-95-1810-4

This Springer imprint is published by the registered company Springer Nature Singapore Pte Ltd.
The registered company address is: 152 Beach Road, #21-01/04 Gateway East, Singapore 189721, Singapore

Preface

The domain of mobile communications is one of the major drivers and contributors to the ever changing and progressive industrialization. It connects people and systems who live in geographically distant locations. Moreover, with the increasing population, communication becomes more challenging. Researchers are working hard to create the fastest and simplest communication methods. This domain prominently involves channel modeling, 3G, 4G, and 5G networks, WiMAX networks, data storage, network latency and lag, wireless broadband, radio network parameters, etc. The optimization methods, techniques, and approaches play an important role in the efficient employment of these areas.

The book provides a machine-generated comprehensive yet classified review of the optimization methods, techniques, and approaches associated with different mobile communications and systems including Wavelet-based CR-OFDM Systems, MIMO Slot Antenna with Low Mutual Coupling for 5G Networks, etc. It also covers mainly the techniques for performance analysis of MIMO systems such as the Hybrid Filtering Technique for MIMO-OFDM Systems and artificial intelligence solutions beyond 5G radio access networks. The complexities associated with the planning and dimensioning of worldwide interoperability for WiMAX Networks have also been discussed in detail. The important aspects of optimization methods for the energy and delay of wireless sensor networks, V2X side link, link-level, and system-level simulation, analysis of macro-diversity, cross-layer resource allocation scheme, as well as several optimization approaches of the multichannel allocation for hybrid overlay/underlay cognitive radio, controller placement problem for wireless SDN, interference mitigation scheme with D2D have also been elaborated. Moreover, certain optimization methods for unstructured, scalable, and 6G networks and the analyses of some protocols related to WSN, viz., energy-efficient routing protocols, Asynchronous Anycast Cross-layer Protocol, Cross-Layer Protocol, and Data Fusion and Cluster-Based Multi-hop Relay-Assisted Routing Protocol have been critically surveyed. The algorithms such as sleep scheduling algorithm (iSleep) and Whale Optimization Algorithm (i-WOA) for clustering in Vehicular Ad Hoc Networks are also reviewed. The reviews are specifically from the elite Springer publications. The articles chosen as well as the reviews are necessarily driven by the keywords recommended by the editor of the book as well as generated by the machine based on the suggestions. The review covers the rationale of the problem being addressed, the associated short literature supporting the contribution, as well as the choice of optimization methods and approaches, the case studies, results, and the significance of the solutions. Such classified review may help the UG and PG students, professors, and researchers from academia as well as industry gain a critical insight at the various systems and aspects of mobile communications systems.

Many thanks to the Editorial Advisors Shrutika Dasari, Chaitra Jawalkoti, Ha-Linh Vu, and Prajana Kate as well as Kamiya Khatter, the Associate Editor of Springer for their unparalleled support in several ways.

Pune, India

Milind Pande
Anand J. Kulkarni
Apoorva S. Shastri

About this Book

Auto-summaries can be generated by either an abstractive or extractive auto-summarization:

An extraction-based summarizer identifies the most important sentences of a text and uses the original sentences to create the summary.

An abstraction-based summarizer creates new text based on deep learning. New phrases are created to summarize the content.

The auto-summaries you will find in this book have been generated via an extractive summarization approach.

Each chapter was carefully edited by Dr. Milind Pande, Dr. Anand J. Kulkarni, and Dr. Apoorva S. Shastri. The editors selected the papers which were then auto-summarized. The editors have not edited the auto-summaries due to the extraction-based approach and have not changed the original sentences. You will find the editors' reviews and guidance on the auto-summaries in their chapter introductions.

In machine-generated books, editors are defined as those who curate the content for the book by selecting the papers to be auto-summarized and by organizing the output into a meaningful order. Next to the thoughtful curation of the papers, editors should guide the readers through the auto-summaries and make transparent why they selected the papers.

The ultimate goal is to provide a current literature review of Springer Nature publications on a given topic in order to support readers in overcoming information overload and to help them dive into a topic faster; to identify interdisciplinary overlaps; and to present papers which might not have been on the readers' radar.

Please note that the selected papers are not used to train an LLM while the auto-summaries are created.

Contents

Chapter 1
Introduction to Mobile Communications

Milind Pande, Anand J. Kulkarni, and Apoorva S. Shastri

Introduction by the Editor

Communication is always a link between people who live in geographically distant locations. With the increasing population, communication becomes more challenging. Researchers are working hard to create the fastest and simplest communication methods. A mobile gadget is one such example. Mobile communication enables the transfer of voice and multimedia data without the use of physical or permanent Links. Mobile communication has become the backbone of society. This chapter addresses the evolution of mobile communication as well as the important approaches. This chapter reviews different mobile communication approaches including Wavelet-based CR-OFDM Systems, MIMO Slot Antenna with Low Mutual Coupling for 5G Networks, PAPR Reduction in OFDM, Spectrum Sharing with L-Band Distance Measuring Equipment, Energy Efficient Cognitive Radio Network (CRN) for 5G Wireless Communication, Latency Minimization, TOA Estimation Using OFDM, Real-time 5G NR signal transmission, Correction of Common Phase Error, UF0MC-IOTA Based Cognitive Radio Transceiver, Transmission of Chaotic-based Encrypted Audio Through OFDM, SM-OFDM Systems, SC-FDMA Communication with Channel Estimation Errors. In addition, MIMO-OFDM channel estimation, GFDM and UFMC Techniques for 5G Wireless Communication, Multilevel Redundant Discrete Wavelet Transform (ML-RDWT),

M. Pande
Dr. Vishwanath Karad MIT World Peace University, Pune, Maharashtra, India
e-mail: milind.pande@mitwpu.edu.in

A. J. Kulkarni · A. S. Shastri (✉)
Institute of Artificial Intelligence, Dr. Vishwanath Karad MIT World Peace University, Pune, Maharashtra, India
e-mail: anand.j.kulkarni@mitwpu.edu.in

M. Pande et al. (eds.), *Optimization Methods in Mobile Communication Systems*, https://doi.org/10.1007/978-981-95-1810-4_1

and Optimal Red Deer Algorithm (ORDA) Centered Approach, Smart antenna with automatic beam have also been discussed.

Communication evolved over several generation from first to fifth generation and future generation that is sixth is on the way. 5G have lowest latency rate and highest connectivity. 5G network radio (5GNR) is develop by third generation partnership project, it designs to consider as a global standard for air interface.

In current 4G Orthogonal frequency division multiplexing (OFDM) has been adopted as it supports less signal interference and improved signal quality. It includes sending data simultaneously into small segments over multiple channels. It uses Fourier transform for modulation and demodulation of signals. It provides robust communication. OFDM have a disadvantage of high peak to average power ratio (PAPR).

$$PAPR = \frac{p_{\max}}{p_{avg}} \tag{1.1}$$

Lower the ratio better is efficiency, this ratio can be reduced by clipping, filtering, coding etc. Single-carrier frequency division multiple access (SC-FDMA) system have low PAPR. It uses frequency-domain equalizers which reduces inter carrier interferences. These equalizers are used at the receiver which helps to improve accuracy of channel by measuring channel behaviour in terms of minimum mean square error (MMSE).

L-band whose spectrum ranges from 960 MHz up to 1164 MHz namely, L-band digital aeronautical communication system type I (LDACSI) channels which are based on OFDM. The (Distance measuring equipment) DME channels occupy a band over the range from 962 MHz up to 1213 MHz. A DME system is a radio navigation system which measures the slant distance between the aircraft and the DME ground station. So, this L-band and DME may share spectrum which reduces signal quality. Solution to this problem is Cognitive Radio Network (CRN). Spectrum sensing is an important process performed by CRN. Cognitive radio detect which communication channels are in use. The transceiver then instantly moves into vacant channels, while avoiding occupied ones. It increases spectrum efficiency by minimizing interference.

Latency is time required to transfer data from source to destination and back to source, also called round trip time for signal to transfer back. Latency can be minimized by advanced network architectures including edge computing, network slicing MIMO usage also by optimizing networks.

Time of arrival (TOA) Estimation: It is time require for signal to get transfer from transmitter to receiver. OFDM can be used to estimate TOA, larger bandwidth gives less TOA and smaller bandwidth gives more value of TOA.

Intercarrier interferences in OFDM can be reduce with the help of universal filtered multicarrier (UFMC) in this techniques filtering is used to eliminates carrier which is out of band which helps to reduce interference with adjacent users in transmission. A major part of communications is audio, and its rate increases dramatically. Securing audio during such communications is very important issue. Though

OFDM have strong resistance for transmission impairments, different time and frequency features makes OFDM prone to intrusions. Chaotic based encryptions work as shield for audio data from illegal access, extinction, modification and use.

Wavelet-based Cognitive radio Orthogonal frequency division multiplexing (CR- OFDM) systems allows selecting channels from available channels which in turn helps in clear signal transmission. It uses wavelet transform for signal processing where wavelet transform outperforms tradition Fourier transform as it extracts required signal information.

Antenna system ensures efficient transmission of signals between mobile devices and network. Antennas at network which provide signals are base station antennas and antenna in mobile devices is called mobile device antenna. There are multiple technologies which can help in this signal transmission like beamforming, Multiple input multiple output (MIMO), frequency bands etc. 4G and 5G has adopted MIMO technology as it supports multiple transmission and multiple receiving signals. MIMO antenna supports efficient data transfer which in turn results in improved quality of signal however mutual coupling in antennas can degrade signal quality hence slots structure is used which helps to keep distance between neighbors antennas and reduces signal interference. Spatial modulation orthogonal frequency division multiplexing (SM-OFDM) is a technique in which antenna coveys extended information to avoid inter-antenna interferences. Generalized frequency division multiplexing (GFDM) It is latest multicarrier technique used in 5G mobile communication it overcomes disadvantage of OFDM Inter Carrier Interference, it also helps in reducing out-of-band effect of radio signals. Universal Filtered Multicarrier (UFMC) is also use in 5G network. It is a group of sub carriers are filtered. UFMC effectively transmit small bursts under very tight response time requirements. But UFMC is very sensitive for the applications where we required loose time synchronization. UFMC performs very well in vehicle-to-vehicle (V2V). Redundant Discrete Wavelet Transform (ML-RDWT), used in 5G to achieve improved spectral performance. Complex down-sampling operation is eliminated. This method employs optimal red deer algorithm (ORDA) which helps to enhance spectral efficiency while reducing bit error rate.

Machine Generated Summaries

Disclaimer: The summaries in this chapter were generated from Springer Nature publications using extractive AI auto-summarization: An extraction-based summarizer aims to identify the most important sentences of a text using an algorithm and uses those original sentences to create the auto-summary (unlike generative AI). As the constituted sentences are machine selected, they may not fully reflect the body of the work, so we strongly advise that the original content is read and cited. The auto generated summaries were curated by the editor to meet Springer Nature publication standards. To cite this content, please refer to the original papers.

Machine generated keywords: ofdm, cfo, offset, antenna, papr, frequency offset, mimo antenna, estimation, cognitive, ber, carrier, ici, phase, synchronization, slot

Effects of Synchronization Errors on Wavelet-Based CR-OFDM Systems in Doubly Selective Fading Channels [1]

This is a machine-generated summary of:

Arjun, K. R.; Surekha, T. P.: Effects of Synchronization Errors on Wavelet-based CR-OFDM Systems in Doubly Selective Fading Channels [1]

Published in: Wireless Personal Communications (2023)

Link to original: https://doi.org/10.1007/s11277-023-10224-7

If you want to cite the papers, please refer to the original.

For technical reasons we could not place the page where the original quote is coming from.

Abstract-Summary

"We present the novel joint effects of Wavelet-based Cognitive Radio OFDM (CR-WOFDM) systems under Synchronization Error in terms of based bit error rate (BER)."

"BER Outputs of WOFDM is compared with FFT based OFDM with Cyclic Prefix (CP-OFDM) systems in a doubly-selective channel by designing a communication system for computer simulation."

"We show the behaviour of wavelets in terms of BER by considering different doubly selective channel Power Delay Profile like, AWGN, FLAT, Pedestrian and Vehicular and channel Doppler models like, Uniform and JAKES."

"We calculate and plot Signal-to-Interference Ratio of WOFDM with Time and Frequency Offset and compared the results with FFT based CP-OFDM."

Introduction

"MCM techniques used in CR systems include Orthogonal Frequency Division Multiplexing (OFDM), Wavelet Packet Modulation (WPM) or Wavelet-Based OFDM (WOFDM)."

"The data from MCM are up converted at a certain carrier frequency by means of a transmit antenna and then transmitted in a channel, which is often a time-dispersive environments [2]."

"During this process of transmission and reception of data through time-disruptive channel, the properties of original data changes and multiple copies of transmitted data arrives at receiver resulting in severe synchronization errors [3]."

"For systems with MCM, there are three types of channel estimation and synchronization methods [4]; Firstly, The blind approach improves spectral efficiency but is less used in most practical applications because for accurate estimation it requires a longer observation window and limits their tracking capability by avoiding the overhead of training sequences."

Literature Review

"For OFDM/FBMC, an improved joint estimation method has been proposed for sampling time-offset; carrier frequency offset, and channel impulse response [5]."

"The closed form formulation of the KI cancellation ratio (KICR) is then calculated for the multi-path block fading channel, taking into consideration imperfect time-frequency synchronization and channel estimation, as well as phase noise."

"Author's theoretical analysis is supported by simulation findings, where the simulation result shows KICR is more vulnerable to phase noise in comparison to time-frequency synchronization and channel estimation errors."

"Using MSE, it was demonstrated that the new method significantly improves time, frequency offset, and channel estimation performance when compared to previously published methods."

"The author proposed a novel algorithm for time and frequency synchronization in uplink OFDM using a Doubly Selective Channel (DSC) and a Basis Expansion Model (BEM) to represent a DSC in [4]."

Synchronization Errors in OFDM and WOFDM

"The term channel characterization refers to the process of determining whether a channel is frequency or time selective."

"Oscillator drifts, as well as motion between transmitter and receiver, can cause channels with time variations within the channel."

"Doubly-selective channels have a response that is both time and frequency selective."

"The ISI attributable to symbols transmitted on the same sub-channel is given in the second term, and the ICI computed over the entire frame is given in the third term."

"The presence of a time synchronization error is denoted as: where, R[n] is the received signal, S[n] is the transmitted signal and w[n] is the channel noise."

"The ISI attributable to symbols transmitted on the same sub-channel is given in the second term, and the ICI computed over the entire frame is given in the third term."

Methodology Used
"In-order to learn the abilities and the impact of Wavelets in multi-carrier modulation it is necessary to analyze the interference in Wavelet based CR OFDM system under the influence of Synchronization errors and doubly selective channels."

"Spectrum sensing block decides the modulation type, configures radio and selects the wavelet basis depending upon channel conditions."

"These data is passed through Analog to Digital Converter to process further."

"A combination of Bit Error Rate (BER) over Signal to Noise Ratio (SNR) doubly-selective channel for different wavelet basis like Daubechies (db), Symlets (Sym), Fejér-Korovkin filters (fk), Bi-Orthogonal (bior) are used and Signal to Interference Ratio (SIR) is calculated for Time and Frequency offset errors."

Results and Discussions
"BER versus SNR and calculation of SIR with respect to Time offset error and Frequency offset error has been used as performance parameters."

"This section provides the simulation results in-terms of BER performance for the provided SNR under varying channel conditions and various Wavelet basis [6]."

"This section contains simulation results in which the signal-to-interference ratio is determined for time and frequency offsets with various Wavelet bases and compared to Cyclic Prefix OFDM."

"The performance of OFDM and WOFDM is almost same under 0ns delay spread or when both transmitter and receiver are stationary."

"The BER performance of WOFDM is worst when there is a delay spread of 42ns and 360ns."

"It is evident from the above result that, OFDM with Cyclic Prefix offers some resistance to the delay spread compared to WOFDM under same channel conditions."

Conclusion
"We have analyzed and compared the performance of various Wavelet Basis in Multi Carrier Modulation Communication systems or simply WOFDM with the FFT based OFDM with Cyclic Prefix jointly under doubly selective fading channel resulting in Time and Frequency Synchronization Errors."

"It is evident from the results that FFT based CP-OFDM shows considerable resistance to Synchronization error under time and frequency offsets, mainly because of the Cyclic Prefix."

"This paper considers the readily available off the shelf wavelets, which shows relatively higher capacity as CP is not required and good provides BER under ideal channel conditions."

"Wavelets specifically designed to match the channel conditions are required, which may show strong resistance to channel impairments due to time or frequency and can majorly contribute to increase the channel capacity."

Dual-Band Millimeter Wave 2 × 2 MIMO Slot Antenna with Low Mutual Coupling for 5G Networks [7]

This is a machine-generated summary of:

Ali, Wael A. E.; Ibrahim, Ahmed A.; Ahmed, Ashraf E.: Dual-Band Millimeter Wave 2 × 2 MIMO Slot Antenna with Low Mutual Coupling for 5G Networks [7]

Published in: Wireless Personal Communications (2023)

Link to original: https://doi.org/10.1007/s11277-023-10267-w

If you want to cite the papers, please refer to the original.

For technical reasons we could not place the page where the original quote is coming from.

Abstract-Summary

"A dual-band 28/38 GHz multiple-input multiple-output (MIMO) two elements antenna for the next 5G network is proposed in this paper."

"The innovative antenna of dual-band works at 28/38 GHz and is achieved by adjusting the stub length and the crescent radius."

"The proposed two ports MIMO achieved a reasonable gain that reaches 5.7 and 6.9 dBi in the 28/38 GHz bands, respectively."

"The proposed MIMO antenna performance is also analyzed and achieved the envelope correlation coefficient, a diversity gain, and channel capacity Loss of about 10^{-4}, 10 dB, and less than 0.3 bit/s/Hz, respectively over the 28/38 GHz bands."

"From the simulated and measured outcomes, it can be confirmed that the suggested MIMO antenna system can be applied to 5G wireless telecommunication applications."

Introduction

"In [8], two and four elements of dual-band 28/38 GHz MIMO antennas are discussed with I-shaped slots in the patch."

"In [9], two elements monopole MIMO antenna worked at 27/39 GHz is introduced."

"In [10] wide band from 22.5 GHz, up to 50 GHz using a two-element slots antenna is proposed."

"A slotted patch with circular polarized two elements 28/38 GHz MIMO antenna is demonstrated in [11]."

"Four ports MIMO antennas with defected ground structures operated from 25.1 and 37.5 GHz are proposed in [12]."

"In [13], a four inverted F elements MIMO antenna operated at 28/37 GHz is discussed."

"The two elements MIMO antenna with substrate integrated waveguide (SIW) cavity-backed operated at four frequency bands is presented in [14]."

"A 2-port MIMO slot antenna operated at 28/38 GHz for 5G network wireless systems with low mutual coupling is introduced."

Antenna Configuration and Design Procedures

"The total area of the antenna (W × L) = 15 × 15 mm^2."

"The single-element slot antenna consists of a rectangular slot with $W_{Gs} = 5.6$ mm and $L_{Gs} = 3.9$ mm etched from the back side (ground plane) and is fed by a microstrip transmission line of width $W_F = 0.8$ mm to achieve the desired 50 Ω impedance."

"There are three steps for antenna design (antenna 1, antenna 2, and the proposed mm-wave antenna 3) to achieve the desired requirements."

"The optimized value of the W_{Gs} equals 5.6 mm for good performance."

The 2-Element MIMO Antenna

"The antenna at 1st port is excited while the second port is ended with 50 Ω and vice versa."

"It can be noticed from the radiation patterns at each frequency that there is a phase difference of 180° when exciting each of the two orthogonal ports which emphasizes the diverse behaviour of the proposed MIMO antenna."

"The antenna has peak gain that varies from 5.5 to 6 dBi at the first band (25.5–30.5 GHz) and from 6 dBi up to 7.5 from the second band (35.5–40 GHz)."

"The simulated results only are provided because of the unavailability of measuring mm-wave antennas (from 28 to 40 GHz) in our labs at this time."

"The proposed antenna has almost dipole-like radiation patterns in co-polarized E and H planes at both 28 and 38 GHz."

MIMO Performance Results and Discussions

"This section introduces the investigation of three important parameters as ECC, DG, and finally, CCL to judge MIMO system performance."

"The ECC calculates the linkage between the MIMO antenna elements."

"The ECC simulated result has a good agreement with the measured result and has low values which means good MIMO antenna performance."

"The CCL(Bit/S/Hz) is the third parameter to evaluate the MIMO antenna behavior."

Conclusion

"The two slot antenna elements built-in highly isolated MIMO antenna with dual-band in mm-wave frequency range have been suggested in this work for 5G wireless application."

"The MIMO antenna has experimental return loss S_{11} less than −10 dB at the first band (25.5–30 GHz) and the second band (35–40 GHz), and the isolation coefficient S_{21} result is lower than −31 within the designed two bands."

"It achieved a reasonable gain that reaches (5.7 and 6.9 dBi) in the 28/38 GHz bands, respectively."

"The aforementioned results confirm the possibility of adopting the suggested innovated MIMO antenna for operation in 5G wireless systems at 28/38 GHz."

Crest Dwindling and Interposing for PAPR Reduction in OFDM Signal with Sustainable Spectral Efficiency [15]

This is a machine-generated summary of:

Geetha, M. N.; Mahadevaswamy, U. B.: Crest Dwindling and Interposing for PAPR Reduction in OFDM Signal with Sustainable Spectral Efficiency [15]

Published in: Wireless Personal Communications (2022)

Link to original: https://doi.org/10.1007/s11277-022-09686-y

If you want to cite the papers, please refer to the original.

For technical reasons we could not place the page where the original quote is coming from.

Abstract-Summary

"They may be demodulated using precise time windowing at the receiver despite the fact that they overlap due to the orthogonality of sub-carriers."

"The main disadvantage of this technique is that it affects the spectral efficiency of distorted signals, which is produced by the Peak-to-Average-Ratio of a transmitter's High Power Amplifier (HPA)."

"An Optimistic Return (OR) model for diminishing the integrated losses owing to signal distortion is included."

"The outcome of this work enhances the performance of spectral efficiency with reduction of the distortion loss, which reduces the PAPR in OFDM signal."

Introduction

"High spectral efficiency, bandwidth range, ability to combat multi-path fading effects, and ability to coexist [16] with other systems that share the same band are all advantages of OFDM."

"High PAPR signals are undesirable because high magnitude spikes induce non-linear distortion in the power amplifier prior to transmission, lowering system performance."

"These bits provide information regarding the PAPR reduction technique's parameters, which will aid in restoring the original OFDM waveform to the receiver."

"High computational complexity techniques necessitate more expensive computational hardware and consume more power, both of which are disadvantageous for mobile devices that rely on batteries [17]. From above discussion there is a great need to develop a better more efficient and very low-complex technique for minimizes the OFDM waveform's PAPR with reduction of signal distortion levels, spectral efficiency, and computational complexity in wireless communication."

Related Works

"Sandoval and others [18] proposed the identification and evaluation of efficiency optimization in industrial, public and strategic applications of PAPR reduction strategies."

"Unlike clipping-filtering technology, this paper proposes a new method based on the fast iterative shrinkage-thresholding algorithm (FISTA) to solve the optimal peak-concealed signal."

"DelMarco and others [19] design new multi-component companders to minimize peak-to-average power ratio (PAPR) in multiplexing orthogonal frequency division signals using a restricted optimization approach."

"By employing Offset Quadrature Amplitude Modulation, this approach is employed to give optimal solutions in high computational complexity for different synchronisation parameters such as CFO, timing error, and PAPR (OQAM)."

"Sarowa and others [20] portrays most of the traditional and contemporary PAPR reduction strategies through a comparative investigation aided by simulative analysis and foreshadowing future research possibilities in wavelet-based OFDM to build more efficient systems."

Crest Dwindling and Interposing Model

"A major challenging in wireless communication is the reduction of the peak-to-average power ratio (PAPR) of signals with lessening the spectral efficiency and the level of signal distortion in the orthogonal frequency-division multiplexing (OFDM) system."

"With IVR output and PAPR reduction efficiency, this analysis indicates that CDI technique can be an excellent scheme for existing and future technology like 5G. The problem of high peak to average power ratio (PAPR) occurs when the Inverse Fast Fourier Transformation (IFFT) is performed to provide signals such as Orthogonal Frequency Division Multiplexing (OFDM) with the in-band multi-carrier property, which gives them robustness against multi-path fading and leads to low cost receivers compared to single carrier signals, i.e. Multiplexing of the Frequency Division (FDM) signals used in traditional mobile communications (GSM) systems."

"From above mentioned techniques, proposed methodologies highly reduce the PAPR in OFDM with improvements the performance of spectral efficiency and thus it proficiently reduces the losses due to signal distortion."

Result and Discussion

"In OFDM the main consideration is reduction of PAPR with improving the spectral efficiency and decreases the losses due to signal distortion."

"We analyzed the CCDF to proposed and existing methods."

"When CCDF is 0.8, 0.7, 0.6, 0.5, 0.4, 0.3, 0.2, 0.1, 0 then the value of proposed method is 2.51, 2.53, 2.57, 2.78, 2.89, 2.99, 3.15, 3.17, 6.25 and existing method is 8.92, 8.75, 8.23, 8.01, 9.71, 9.93, 10.21, 10.35 and 15.81 respectively."

"It clearly shown the distortion level of proposed method is considerably reduced than exiting works."

"From above mentioned experimental results described that the greater efficiency of proposed system in reduction and maintaining of major significant issues in OFDM such as PAPR, losses due to signal distortion and Spectral efficiency."

Conclusion

"Because an OFDM network uses a large number of subcarriers to transport the modulated symbols, the OFDM signals have a high peak-to-average power ratio (PAPR)."

"In order to reduce issue in OFDM signals, Crest Dwindling and Interposing technique is used in the proposed work."

"The outcomes of proposed work efficiently attains the PAPR reduction and losses due to signal distortion with great improvements of spectral efficiency performance in OFDM."

Orthogonal Coded Spread Spectrum Digital Beamforming-Based 5G Receiver [21]

This is a machine-generated summary of:

Biswas, Ardhendu Shekhar; Sil, Sanjib; Bera, Rabindranath; Mitra, Monojit: Orthogonal Coded Spread Spectrum Digital Beamforming-Based 5G Receiver [21]

Published in: Arabian Journal for Science and Engineering (2022)

Link to original: https://doi.org/10.1007/s13369-022-07022-x

If you want to cite the papers, please refer to the original.

For technical reasons we could not place the page where the original quote is coming from.

Abstract-Summary

"5G receiver design at acceptably low cost poses a significant challenge, particularly where the 5G new radio (NR) use cases enhanced mobile broadband (eMBB), massive machine type communication (mMTC) or ultra-reliable low latency communication need to be satisfied."

"One of the key design parameters that has to be looked at for successful implementation of these use cases is the maintenance of high signal-to-noise ratio (SNR)."

"For high SNR requirements, implementation of phased array antennas for suitable beamforming in radio access networks has been preferred by 5G systems designers."

"The authors have proposed an orthogonal coded spread spectrum-based digital beam forming (SSDBF) technique, which can be a better choice for 5G designers."

Introduction

"Through implementation of suitable antenna beamforming techniques, the available SNR at the receiver can be improved to a significant extent."

"Spread spectrum-based digital beamforming technique as used in radar applications implemented using several antenna elements is suitably cheap [22–24]."

"We present our proposed work which is on a modified spread spectrum digital beamforming (SSDBF) technique that delivers a very narrow sharp beam to achieve high SNR at the receiver."

"In conventional digital beamforming, consisting of 64 antenna elements, 64 low-noise amplifiers (LNA), 64 down converters (DC) and 64 analog-to-digital converters (ADC) are required."

"The major novelty of the present work therefore lies in the significant reduction of CSWAP requirements with performance equivalent to the significantly more expensive conventional digital beamforming (CDBF) systems."

Literature Survey

"In [23], a millimeter wave spread spectrum digital beamforming scheme is proposed for narrow beamforming where large aperture, two-dimensional phased array and branch-line coupler with bi-phase modulators are used."

"In another paper [24], for cost and performance optimization a code division multiplexing technique is used to implement spread spectrum digital beamforming radar."

"The authors have used direct sequence spread spectrum technique and orthogonal gold codes to achieve very narrow beamforming."

"A co-operative beamforming solution with spread spectrum technique and radio frequency integrated circuit technology is proposed in [25]."

"In [26–30], different kinds of path loss models for 5G wireless communication at millimeter wave frequencies are presented."

"It has been observed that beamforming has emerged as an important technology for effective implementation of 5G communication systems."

"Optimized design of 5G beamforming receiver has been presented in [31], with high-resolution beamforming achieved using phase and beam control."

Proposed Spread Spectrum Digital Beamforming Scheme

"Further, the modulated 28GHz carrier signal is fed to an analog beamformer block which generates a beam using 4 × 4 rectangular antenna array for onward transmission."

"We have considered a 8 × 8 rectangular array antenna with the proposed SSDBF scheme to generate a narrow beam at the receiving section of the system."

"The antenna array and the SSDBF scheme are enclosed in the block marked as 'SSDBF with Receiving Antenna'."

"The output from the SSDBF block is then carrier demodulated to obtain the 5G baseband signal."

"A rectangular array of M receiving antenna elements is considered."

"The M multiplied outputs are then added to generate a multiplexed signal."

"At each summer output the received signals of each antenna elements are available with respective coding gain and at a lower carrier frequency of 1 GHz."

Results and Discussions

"The achieved BERs at 200 and 300 m distances are relatively at lower SNR values in comparison with 150 m. The same argument is applicable for the throughput performance also."

"The achievement of high throughput with nominal change in SNR is achieved due to the application of LDPC decoder in the pre-CRC validation stage, which significantly improves BER performance, leading to sharp rise in throughput achieved with marginal increase of SNR above minimum value."

"The BER with SNR variation is measured before and after the decoder of the 5G NR baseband receiver in the UMi scenario at a distance of 100 m. It should be noted that low density parity check (LDPC) coder and decoder are used for channel coding and decoding."

"The throughput is measured considering 10 MHz system BW and at a distance of 100 m in UMi scenario."

Conclusion

"The authors have suggested a cost-effective antenna array circuit design solution applicable at the front end of a 5G receiver system."

"This optimum circuit design enhances the receiver's SNR."

"The authors have designed, simulated and successfully tested a SSDBF-based 5G receiver."

"The test results such as the variation of BER versus SNR in comparison with CDBF systems are found to be almost equal which qualifies the SSDBF scheme to be used in terms of reduced CSWAP requirement, which is an important and significant feature of the present work contributing to its novelty."

Multi-objective Optimization with FEC Polar Code for Bandwidth Efficient Mobile Network Using Response Surface Methodology [32]

This is a machine-generated summary of:

Jadhav, Makarand; Dongre, Ganesh; Mahalle, Parikshit: Multi-objective Optimization with FEC Polar Code for Bandwidth Efficient Mobile Network Using Response Surface Methodology [32]

Published in: Wireless Personal Communications (2022)

Link to original: https://doi.org/10.1007/s11277-022-09688-w

If you want to cite the papers, please refer to the original.

For technical reasons we could not place the page where the original quote is coming from.

Abstract-Summary

"This paper presents a bandwidth-efficient polar code-based Orthogonal Frequency Division Multiplexing system to allocate optimized bandwidth, and power resources amongst the users."

"The available network bandwidth is limited to facilitate and deliver an array of services to the legions."

"There is a necessity to increase the network capacity within the same accessible bandwidth to meet such demanding service requirements."

"A fixed Peak to Average Power Ratio of 14 dB and zero carrier frequency offset is considered."

"The intrinsic second-order mathematical regression model is developed for the entire system."

"The number of users is doubled keeping available bandwidth at a Bit Error Rate of 10^{-4}."

"The best possible result has shown Effective Isotropic Radiated Power of 26.99 dB with 5.78 dB Signal to Noise Ratio which is well acceptable for human body radiation hazards."

Introduction

"Further, better error performance, acceptable computational complexity, and code latency based on block length using forward error correcting polar code are present challenges."

"The implementation of a polar FEC code-based system has the bottleneck of block length size, error probability, code rate, and latency in terms of computational complexity [33]."

"The modeling as well as the design of the network, alleviate the effects of required Effective Isotropic Radiated Power (EIRP) and Signal to Noise Ratio (SNR) for a mobile unit at the receiver side are more challenging."

"The receiver's performance is assessed in terms of a required SNR at the input of the receiver assuming a Hatta propagation channel for achieving the required Bit Error Rate (BER) performance for voice application."

"An efficient polar code-based Orthogonal Frequency Division Multiplexing (OFDM) system is designed by adopting a 16 PSK modulation technique along with a well-optimized non-iterative Euclidian-based decoder which is also presented in this paper."

Literature Review

"In [34] low complexity Split-Reduced SCL polar decoder is proposed."

"In [35] Arikan's Bhattacharya parameter method is proposed for polar code construction that reduces the complexity."

"In [36] the joint SC decoding is proposed for Bit-Interleaved Polar Coded Modulation (BIPCM)."

"The proposed systems memory requirement, as well as decoding complexity, is low as compared to Belief Propagation (BP) as well as list polar decoders."

"SC, List SC (LSC), and Low Complexity LSC (LCLSC) polar decoding algorithms are presented in [37]."

"The stack polar decoder presented in [38] has very low time complexity as compared to the SC list algorithm."

"The simplified SC polar decoder presented in [39] has low decoding latency and algorithm complexity."

"In [40] improved SC polar decoder is presented that uses inner block codes."

"In [41] low complexity SC polar decoder is presented."

Model Development for Simulation and Experimentation

"The generated code-words are converted from serial to parallel stream."

"Step 3: Collect stored bits for polar code mapping to generate code-words."

"Step 4: Generated code-words are stored in matrix form."

"Step 5: MPSK modulation is carried on generated code-words."

"Step 6: The resultant serial bit stream is converted into a parallel bitstream."

"Step 10: At the receiver, the reverse procedure to that of the OFDM generator is done by removing cyclic prefix bits from the received signal."

"Step 11: The serial bit stream is converted to the parallel bitstream."

"The bits are converted from parallel bitstream to serial bit stream."

"Polar decoder using Euclidian distance reconstructs block of data streams again Step 13: Evaluate BER by comparing transmitted and recovered data bits for different values of SNR."

"The throughput and packet dropping ratio is considered as a performance metric to record QoS parameter results."

Design of Experiments (DOE) Using RSM Method

"An RSM is used for the optimization of the response in terms of BER and latency."

"There is various other Design of Experiments (DOE) methods for optimization."

"The basic aim of designed experiments is to determine optimal conditions of EIRP in achieving targeted BER of 1×10^{-4}."

"A validation experiment is carried out for the predicted best settings to ratify the results of optimization."

"This has provided optimal network performance which enhances network density and performance to obtain reliable voice connectivity."

"This optimization technique helps to determine the best optimal settings for multiple responses."

"In this design, an optimization is performed to determine the input variable settings that result in a minimum requirement of EIRP from 27 to 30 dBm."

Results and Discussion

"At the acceptable value of BER, E_b/N_0 obtained from the simulation results achieves an SNRi of 21.66 dB. Consider that a network is operating at an EIRP of 10 dBm to cover the distance of 1 km."

"Main effect analysis concludes that for achieving minimum EIRP, FFT points should be at a higher level of 250, distance should be at a minimum of 1 km, and BER should be achievable at 7.5×10^{-5} and NF at 3 dB. However, in actual practice, we cannot directly use all the above conditions at the same time as they are interrelated to each other."

"Further, assuming that distance and FFT points are selected at the same required EIRP, the NF of the device should be in the range of 3–5 dB as displayed in the graph of Distance vs NF."

Multi-objective Optimization

"The optimized distance for global solutions is 1.5932 km."

"The multi-objective optimization aims at optimizing more than one output simultaneously by comparing the desirability of some outputs."

"In this paper, two outputs such as E_b/N_0 and EIRP are considered for optimization."

"At these settings EIRP of 26.99 dB (Target 27 dB) with a minimum E_b/N_0 of 5.784 dB is achieved."

Development of Regression Model and Model Validation

"Information, the validation of the simulation and regression model is verified."

"The transmission rate is doubled with 8 PSK modulation."

"The transmission rate is doubled with 8 PSK modulation."

"Further, the required E_b/N_0 is 3 dB at 2–3 km for EIRP of 26 dBm."

Conclusions

"RSM is used as an optimization technique to balance output responses in achieving targeted BER of 1×10^{-4}."

"Further, an effort in minimizing the amount of power transmitted from the user device in the range of 27–30 dBm is presented in this paper."

"The amount of maximum EIRP that can be transmitted from the mobile unit is limited to 20 dBm."

"The efficient designing of polar encoder and decoder demands minimum SNR value and is achieved with 16 PSK modulation techniques."

"It has been observed from the simulation results that a higher modulation technique demands more transmitted power from mobile handsets."

"The experimentation is limited to up to 16 PSK modulation techniques."

"Further, saving in EIRP of 3.05 dBm is achieved at BER of 2×10^{-5} by adopting 16 QAM modulation technique."

Underlay Spectrum Sharing with L-Band Distance Measuring Equipment for Aeronautical Communications [42]

This is a machine-generated summary of:

Abd-Elaty, Emad; Zekry, Abdelhalim; El-Agooz, Salah; Helaly, Anwar M.: Underlay Spectrum Sharing with L-Band Distance Measuring Equipment for Aeronautical Communications [42]

Published in: Wireless Personal Communications (2022)

Link to original: https://doi.org/10.1007/s11277-022-10045-0

If you want to cite the papers, please refer to the original.

For technical reasons we could not place the page where the original quote is coming from.

Abstract-Summary

"The previous work depends on both DME interference detection and mitigation techniques."

"A DME interference mitigation technique is proposed to improve the performance of the LDACS type I (LDACSI) system without requiring the detection process."

"The proposed technique depends on the underlay spectrum sharing for LDACSI signal."

"The underlay spectrum sharing concept is performed using spread spectrum technique."

"The spread spectrum technique is applied based on the idea of multiplexing the in-phase and the quadrature components of LDACSI signal by two Walsh orthogonal spreading codes in which they are transmitted over the quadrature branch."

"The spread LDACSI signal is friendly coexist on the top of DME interference signal."

Introduction

"Several techniques have been proposed to mitigate the DME impulsive interference in ATM systems [43, 44]."

"Despite simplicity, there is a common drawback of these techniques which is the nonlinear harmonic distortions impressed to the desired OFDM-LDACSI signals due to its direct clipping/blanking with the existence of the DME interference signal."

"In these techniques, a frequency-selective pulse blanking unit, a blanking mitigation unit, and an iterative ICI cancellation are jointly operated to mitigate the possible DME interference."

"The previous DME interference detection is based on the correlative spectrum sensing technique which can be performed using a Correlative interference in the time domain to obtain the DME signal level inlayed in the received LDACSI [45]."

"A DME interference mitigation technique is proposed to improve the performance of the LDACSI system without requiring the detection process."

DME Scenario and Interference Problem

"The DME channels occupy a band over the range from 962 MHz up to 1213 MHz with 252 stations, 126 channels assigned for X mode and 126 channels assigned for Y mode."

"The aircraft interrogates the DME ground station by transmitting a series of pulse-pairs which is called interrogations."

"The period between pulse pairs is 12/36 μs depending on the operating DME X/Y mode, respectively."

"In each DME cell, a DME ground station can detect the strongest 2700 ppps (90 aircrafts in the tracking mode and 18 aircrafts in the searching mode) and re-transmits (replies) them to the aircrafts' after a constant time delay of 50 or 56 μs depending on the operating X or Y mode, respectively."

"The concern of the LDACSI system is the interference from the co-located DME station (station 1) only."

The Proposed LDACSI Underlay Spectrum Sharing System

"The discrete-time domain CP OFDM symbol is applied to a proposed underlay spectrum sharing spread spectrum transmission and then up-converted."

"The band pass up converted output signal wirelessly radiated through the aeronautical channel with an AWGN."

"This section presents the spreading of the LDACSI signal on the top of DME interference at the transmitter side."

"This is because of the splitting of I and Q components results in decreasing the rate to half and the chip rate of each I and Q components after the spreading resulting in increasing the rate to the double."

"Decreasing the rate to the half before the spreading process is equivalent to increasing the rate to the double after spreading."

Complexity Analysis

"The computational complexity of the proposed LDACSI underlay spectrum sharing system is analyzed in terms of complex multiplications (CM)."

"Since, the LDACSI underlay system uses the spread spectrum technique."

"It is noted that the maximum CM required in the proposed underlay spectrum sharing technique is 2880 which is smaller than the minimum CM required in a previous work of [46] which is 11,398."

Simulation Results and Analysis

"In our MATLAB simulation, 25 Monte Carlo iterations is used."

"The Monte-Carlo simulation is performed for accurate BER calculation."

"Throughout the simulation setup, QPSK scheme is employed."

"Throughout the simulation setup, the quadrature phase shift keying (QPSK) scheme are employed."

"The mitigation technique overcome the interference provided by the DME transmission."

Conclusions

"The proposed underlay spectrum sharing is used on the top of DME interfering signal."

"The spread LDACSI signal is friendly coexist on the top of DME interference signal."

"The simulation results shows that the proposed underlay spectrum sharing for LDACSI system provides lower bit error rate and less complexity compared to the previous work."

Carrier Frequency Offset (CFO) Synchronization and Peak Average Power Ratio (PAPR) Minimization for Energy Efficient Cognitive Radio Network (CRN) for 5G Wireless Communication [47]

This is a machine-generated summary of:

Balachander, T.; Krishnan, M. B. Mukesh: Carrier Frequency Offset (CFO) Synchronization and Peak Average Power Ratio (PAPR) Minimization for Energy Efficient Cognitive Radio Network (CRN) for 5G Wireless Communication [47]

Published in: Wireless Personal Communications (2021)

Link to original: https://doi.org/10.1007/s11277-021-08726-3

Copyright of the summarized publication:

If you want to cite the papers, please refer to the original.

For technical reasons we could not place the page where the original quote is coming from.

Abstract-Summary

"The Cognitive Radio Network (CRN) plays significant role in mobile technology for energy efficient wireless 5G communication."

"The Non Orthogonal Frequency Multiple (NOMA) communication technology is applied for energy efficient CRN wireless communication."

"The CRN is an important research area for 5G wireless communication that permit unlicensed Secondary Users (SU) to access the spectrum by achieving energy efficiency."

"The energy efficiency of SU can be optimized by providing Carrier Frequency Offset (CFO) synchronization and Peak Average Power Ratio (PAPR) minimization in a cooperative spectrum sensing of the CRN."

"The synchronization and error correction techniques are proposed to improve the energy efficiency of CRN for improving 5G wireless communication."

"The experimental results prove that proposed methodology attains energy efficiency by improving synchronization and minimizing errors."

Introduction

"Computational results reveal that the improved cognition interconnected wireless network is capable of achieving channel cooperation while also increasing spectrum efficiency levels [48]."

"Because of the hypothetical influence of the wavelet packet multi-carrier with output quadrature phase shift encoding system, most boundary estimation methods for wireless communication technique called orthogonal frequency division multiplexing (OFDM) systems [49] are not particularly relevant to FBMC-OQAM systems."

"In the FBMC-OQAM network, a new linear regression algorithm for fading data transmission is recommended."

"The next-generation data transmission is designed to provide increased spectrum utilization, network connectivity, spectral efficiency, and huge device connectivity."

"This proposed methodology is utilized for OQAM/NOMA networks with PAPR reduction and CFO synchronization to meet the need that OQAM/NOMA performance always outperforms OQAM/NOMA capacity for each paired user."

The State of Art for Spectrum Energy Efficient Models

"Zhang and others [50] has proposed energy efficiency by improving the channel capacity of CRN."

"The CRN is implemented based on MIMO-OFDM wireless communication for improving secondary user's energy efficiency."

"The suggested method provides visible improvement in the PAPR performing of the FBMC/OQAM signaling in comparison with other existing techniques without any impact on Spectral efficiency, according to simulation solution derived from different computer models."

"By exploiting the efficiency of the CRN, an improvement of the number of secondary users (SUs) in a cooperation energy detection is proposed."

"A new inter objective function is proposed that enhances the ergodic capacity and eliminates the average signal strength of the secondary user network while reducing the average interruption power implemented on the primary user transmitter in order to increase energy savings."

Energy Efficient CRN NOMA Model

"The signal X[k] that is emitted can be expressed as given below: W[k] is the NtU channel estimation matrix to lessen the IUI, and S[k] is the U1 frequency domain matrix, where is the TX power normalization factor, which is computed as NtK. The transfer function signal xt can be described as, while the spatial frequency Impedance matching of the tth transmitter is signified as Xt = [Xt [0], Xt [1], Xt [N1]]T. where FHN is the N-point frequency to time domain conversion Inverse fast Fourier transform (IFFT) vector, that can be denoted as The PAPR of the NOMA signal for the tth receiver is denoted by, From the above equations, by integrating DFT distributing, accommodating the situation of the phase factors, and a method described that cooperates with the specification of OQAM, an innovative method to reduce PAPR in based on OQAM is introduced."

Results and Discussion

"The channel estimation of wireless communication techniques are compared with OFDM, HUFMC/OQAM (Hybrid Universal Frequency Multi-Carrier/and OQAM/NOMA."

"NOMA intends to facilitate CRN supplementary customers using the same spatial-frequency as OFDM by using different energies, resulting in increased spatial multiplexing."

"The OQAM/NOMA for 5G wireless communication reaches maximum energy efficiency as 65 Mbits/J/s while compare to other schemes such of UFMC, HUFMC/QAM, OFDM-QAM and OFDM-BPSK."

"For CFO, this means SNR depletion of SNRDCFO = 0.2 dB. This demonstrates the benefits of the sophisticated estimation methods selected for the CRN NOMA's energy efficiency synchronization."

"Despite the fact that it provides certain alternatives for preserving efficiency and spectrum requirement, like cyclic prefix based (CP) and discrete Fourier transform spreading based (DFT) systems, the PAPR is still as high as 13 dB. When the throughput is expanded, this value increases."

Conclusion and Future Work

"The CRN NOMA is implemented for energy efficient wireless 5G communication."

"The novel methodology is developed for cooperative spectrum sharing for CRN NOMA by utilizing energy efficiency that can sense the secondary users (SU)."

"The energy efficiency of SU could be improved by implementing CFO mismatch error correction and reducing PAPR error."

"The NOMA system is integrated for sending data by the SU with proposed modulation."

EEHC Approach for Latency Minimization in 3D Network Architecture Using 5G+ with UAVs [51]

This is a machine-generated summary of:

Sivarambabu, P. V.; Malarvezhi, P.; Dayana, R.; Vadivukkarasi, K.; Rao, P. Rama Koteswara: EEHC Approach for Latency Minimization in 3D Network Architecture Using 5G+ with UAVs [51]

Published in: Wireless Personal Communications (2021)

Link to original: https://doi.org/10.1007/s11277-021-08931-0

If you want to cite the papers, please refer to the original.

For technical reasons we could not place the page where the original quote is coming from.

Abstract-Summary

"This octahedron cell architecture in the combination of Unmanned Aerial Vehicles (UAV—mostly known as DRONES in many parts of the world) helps develop 5G networks with the instance of Millimeter Wave (mm-Wave) approach in communication systems."

"The proposed method is the Energy Efficient Hierarchical Clustering Approach formulated to minimize the power usage of users (UE of the drone) and trace latency for different conditions during transmission."

"This approach helps in optimizing the latency of the system, Signal-Interference Noise Ratio (SINR) based communication spectral efficiency, and cumulative distribution function concerning SINR results in communication channel strength in sharing the information without dropping."

"The proposed approach initially develops energy infliction-based clusters and decides the optimal routing path for communication that allows 50% minimization in average latency compared with existing schemes."

Introduction

"The key reason is that the drone communication unit under UAV is not easy to control."

"The base station got down in a cluster, and the UE's became active to help other users, especially in terms of latency, but the primary issue is the energy consumption by the drone systems [52, 53]."

"The research moves on three different conditions for understanding the communication process, which helps in reducing the energy and power usage, decreased latency based on distance and energy consumption bases the drone selection for communication."

"of drones) and transmission bandwidth under different approaches (cell association approach, SINR based approach and Energy Efficient Hierarchical Clustering Approach), iteration number was calculated, based on packet size Backhaul latency, computational Latency and Transmission Latency was calculated and additional latency to sampling time was also calculated."

System Model

"To operate the drone in an unmanned region 'n' energy consumption and latency in data transmission are the essential factors [54–56]."

"The first phase in this scenario being; to locate drone-UE's and factor of frequency re-use scheme for improving the QoS, which will be further verified concerning the location."

"Commencing to transmission section from a distance-based location, the SINR of the system was given by: Here σ^2 is the noise power of the total 3D cell architecture."

"Q = 8, Where less interference exists and all N drones can be observed and allocated with different bandwidth depends on the usage and enhances the systems QoS. Nevertheless, to prove this based on edge and face connection between neighbouring 3D cells obtained and considered from [57]."

"The drone-UE's because during the network access time latencies between the base stations and remote servers hosting for controlling scheduling."

Proposed Methodology

"The proposed methodology Energy Efficient hierarchical Clustering Approach (EEHCA) formulated for pruning the power usage of users (UE of the drone) and traces latency for different conditions during transmission."

"This approach helps in optimizing the latency of the system and calculated Backhaul latency, computational Latency and Transmission Latency based on packet size."

"The proposed approach initially develops an energy utilization based on clusters and decides the optimal routing path for communication."

"Energy is the major constraint for transmission and the reception of the signal will be calculated and the optimized values will be sorted based on their availability."

"Using the latency of the system, the node optimization was implemented using reallocation of drones for data transmission based on connectivity issues."

Simulation Results

"At long last, this strategy to adapt to the assessment of gadget UE spatial dissipating is as in a state of harmony with the accompanying."

"The spot of a drone is totalled every T second, by at that point, degree the rehash of drone UE's to use it for 3-D cell association for the timeframe of the resulting T seconds is done."

"Being a final product of the very certainty that, in this strategy, alongside SINR, the effect of forestalling up at the transmission, backhaul, and computational latencies is comparable."

"At outstanding, the whole slowness significantly is fundamentally founded on the 3D portable pieces which are incapable of the gadget UEs' stream."

"Along the lines of the one, an assessment bungle inside the framework UEs' dispersing brings a deviation from the optimal phone sections."

"By diminishing T, the instrument can grow a significantly extra prominent aggregated dispersing assessment and, subsequently, a low slowness."

Conclusion

"The application of energy utilization was applied on the network, thus helping in identifying the best drone selection with energy minimization, but this selection depends entirely on the SINR of the system-based Rayleigh fading coefficient."

"The latency optimization was performed on the system, which helped UAV based drone systems in the communication of secured networks, mainly in terms of military application."

"The latency minimization ultimately depends on selecting the drone and finds 50% more efficient in selecting drones for communication."

"The proposed model works on frequency reuse mode, but the work can be carried on frequency offset-based channel optimization for latency minimization in the future."

Enhanced TOA Estimation Using OFDM over Wide-Band Transmission Based on a Simulated Model [58]

This is a machine-generated summary of:

Obeidat, Huthaifa A.; Ahmad, Imran; Rawashdeh, Mohammad R.; Abdullah, Ali A.; Shuaieb, Wafa S.; Obeidat, Omar A.; Abdallhameed, Raed A.: Enhanced TOA Estimation Using OFDM over Wide-Band Transmission Based on a Simulated Model [58]

Published in: Wireless Personal Communications (2021)
Link to original: https://doi.org/10.1007/s11277-021-09297-z

If you want to cite the papers, please refer to the original.

For technical reasons we could not place the page where the original quote is coming from.

Abstract-Summary

"The study investigates the effect of using various spectrum bandwidths and a different number of carriers on localization accuracy."

"In the case of using large BW with a large number of subcarriers, the algorithm was effective to reduce localization errors compared to the conventional TOA technique."

"The performance degrades and becomes similar to the conventional TOA technique while using a small BW and a low number of subcarriers."

Introduction

"For reliable signal transmission, the OFDM technique requires precise timing and accurate frequency synchronization [59, 60]."

"Authors in [61] used the OFDM-Time of Arrival (TOA) algorithm for position estimation of a Long Term Evolution (LTE) signal."

"In [62], a simple OFDM-TOA-based algorithm is proposed to estimate the Direct Path (DP)."

"This paper introduces a coded OFDM system to be used in the application of localization systems and enhance the TOA estimation over the wideband transmission."

"A fractional spectrum of the lower Ultra-Wideband (UWB) was utilized to mitigate the effects of indoor channel propagations due to the high number of multipath, in which several carriers are spread over the bandwidth based on the impulse response of the channels using Wireless-InSite (WI) and then suitable estimations of time arrivals are applied, followed by a numerical technique to estimate the position of the hidden objects."

Mathematical Model of OFDM for Time of Arrival Estimation

"The idea is to use simple lateration localization, based on TOA; however, time is estimated based on analyzing the OFDM signal."

"We first extracted the arrival rays' corresponding information including phase, time delay, and received signal strength."

"We then mapped the high-resolution time with the output of the IFFT, after that we choose the time with the highest value of IFFT as the TOA, which is the second task of the code, in other words, the TOA is estimated based on the mapping between the discrete-time domain samples with the highest output of the IFFT."

"For an electromagnetic wave, the elapsed time and travel distance are related by the speed of light, given the TOA, the distance between the transmitter and receiver is estimated."

Simulation Model

"WI is a site-specific radio propagation software that provides user-requested outputs for different indoor and outdoor environments and has the ability to provide efficient and accurate predictions of channel propagation characteristics like received signal strength, propagation paths, complex impulse response, delay spread, and time of arrival/departure, etc. [63]."

"The obtained channel response is to be used with the OFDM multi-subcarrier system for our proposed localization method."

"Transmitter and receivers' antennas used were omnidirectional and the operating frequency was 5 GHz with 2 GHz bandwidth."

"By convolving the channel impulse response with the OFDM multi-subcarriers spread over the bandwidth, the proposed TOA is estimated by taking the inverse fast Fourier Transform (IFFT) of the convolution."

Results and Discussion

"LE using the proposed algorithm was reduced compared to the conventional TOA approach."

"LE was 0.354 m which is 2 m less than the conventional TOA LE."

"The LE was 0.6946 m which is 5.5 m less than the conventional TOA localization error."

"Most LEs of the proposed algorithm are less than the minimum conventional TOA LE except for the 16 carriers case."

Conclusions

"This paper describes the OFDM technique for wireless communications and desirous advantages of OFDM to improve the localization of targets, in order to provide accurate mobile station (MS) positioning within indoor environments, and provided an overview of different delimiting factors that affect the performance and capability of OFDM."

"The localization of objects within the indoor environment over a wideband spectrum adopting multi-carrier has been presented."

"The method was based on the TOA for various spectrum bandwidths and the number of carriers covering the LOS and the NLOS positions of the transmitter's positions."

"The method adopted the impulse response of the channel to estimate the transmitter position."

Real-Time High-Bandwidth mm-Wave 5G NR Signal Transmission with Analog Radio-over-Fiber Fronthaul over Multi-Core Fiber [64]

This is a machine-generated summary of:

Rommel, Simon; Grivas, Evangelos; Cimoli, Bruno; Dodane, Delphin; Morales, Alvaro; Pikasis, Evangelos; Bourderionnet, Jerome; Feugnet, Gilles; Barros Carvalho, Juliana; Katsikis, Michail; Ntontin, Konstantinos; Kritharidis, Dimitrios; Spaleniak, Izabela; Mitchell, Paul; Dubov, Mykhaylo; Tafur Monroy, Idelfonso: Real-time high-bandwidth mm-wave 5G NR signal transmission with analog radio-over-fiber fronthaul over multi-core fiber [64]

Published in: EURASIP Journal on Wireless Communications and Networking (2021)

Link to original: https://doi.org/10.1186/s13638-021-01914-6

If you want to cite the papers, please refer to the original.

For technical reasons we could not place the page where the original quote is coming from.

Abstract-Summary

"This article presents an experimental demonstration of a high-capacity millimeter-wave 5G NR signal transmission with analog radio-over-fiber (ARoF) fronthaul over multi-core fiber and full real-time processing."

"The demonstration validates the core of the blueSPACE fronthaul architecture which combines ARoF fronthaul with space division multiplexing in the optical distribution network to alleviate the fronthaul capacity bottleneck and maintain a centralized radio access network with fully centralized signal processing."

"A space division multiplexing based radio access network with multi-core fibre allows parallel transport of the uplink ARoF signal and reference local oscillator at the same wavelength over separate cores."

Introduction
"The resulting load on optical distribution networks (ODNs) from the fronthaul required for a highly densified network deployment with the use of very large bandwidths and advanced radio techniques such as beamforming or multiple-input multiple-output (MIMO) transmission however poses major challenges and creates an important bottleneck."

"As the increased spatial control over the radio frequency (RF) signal afforded by MIMO transmission or beamforming is required to achieve the capacity and energy efficiency targets for 5G networks, substantial research activity is focused on improved fronthaul solutions."

"A successful demonstration of ARoF fronthaul for high-bandwidth mm-wave 5G NR signals using full real-time processing in an ARoF baseband unit (BBU) and multi-core fiber (MCF) transport was demonstrated and discussed in a paper at EuCNC 2020 [65]."

"The demonstrated ARoF fronthaul link achieves real-time transmission of 800 MHz wide extended 5G NR orthogonal frequency division multiplexing (OFDM) signals over 10 km of 7-core MCF and 9 m mm-wave wireless at 26.5 GHz, i.e., in the 3GPP FR2 n258 band [66]."

Analog Fronthaul with Optical Beamforming for mm-Wave 5G
"In current 4G LTE networks where bandwidths and frequencies are limited, this segment is covered by transmitting digitized in-phase and quadrature (IQ) samples of the target radio waveform using traditional BB optical links, detecting the signal at the RU and performing digital-to-analog conversion there (and vice versa for uplink (UL) direction)."

"If the RF waveform is transported or optical heterodyning is used for IF to mm-wave upconversion, ARoF further allows centralizing all RF oscillators and simplifies the mm-wave generation process [67]."

"When it comes to the use of optical beamforming in ARoF fronthaul as in the blueSPACE architecture, the placement of the OBFN has significant consequences for the fronthaul transport and may pose specific requirements to the ODN."

"While this configuration allows centralization of the DL OBFN and thus maintains a simpler RU, it does require signals to remain temporally aligned across transport in the ODN."

High-Capacity ARoF Fronthaul Demonstration Experimental Setup
"The blueSPACE BBU mainly consists of five modules, (1) the digital data interface acting as data source/sink for the DSP chains, (2) the cyclic prefix (CP) OFDM transmitter side DSP datapath, (3) the CP-OFDM receiver side DSP datapath, (4) the analog data interface towards DAC and ADC, and (5) the embedded central processing unit (CPU)."

"This module receives plain digital data, originating from either of the sources in the digital interface, and exports data samples that describe the analog BB signals towards the analog interface which drives the DACs."

"This module receives data samples of the incoming analog signals, as encoded by the ADCs, from the analog interface and exports the decoded digital data to the digital interface, destined for either the backhaul interface or the built-in BERT for bit error rate (BER) measurements."

Real-Time Analog Fronthaul Demonstration Results and Discussion
"The optical two-tone generation and ARoF transmitter setup has a fundamental impact on the performance of the overall link, as it defines the targeted RF frequency and largely determines the levels of phase noise in the final RF signal generated through the optical heterodyning."

"The received IF signals further shows a somewhat reduced CSR which, together with the shape of the noise around the signal, suggests the presence of nonlinearities in the link, likely due to saturation of either the PA or LNA."

"The received IF signal appears slightly tilted, with powers of the higher frequency subcarriers somewhat lower than for the lower frequency subcarriers."

"The BER of the received signal is evaluated continuously by the ARoF BBU and retrieved repeatedly after short time intervals for each measured optical power."

"The system demonstrates transmission at a data-rate of 1.4 Gbit/s after 10 km MCF and 9 m mm-wave wireless using fully real-time processing in the ARoF BBU for both signal generation and analysis."

Conclusions
"This article presented the blueSPACE fronthaul architecture with ARoF transport over SDM and optical beamforming and showed a demonstration of a high-bandwidth mm-wave 5G NR signal transmission with ARoF fronthaul over MCF."

"The blueSPACE architecture introduces the combination of ARoF with SDM in the ODN to solve the fronthaul capacity bottleneck and uses optical beamforming to seamlessly enable multi-beam transmission and achieve spatial control over the RF signal."

"The demonstrated fronthaul link used the blueSPACE ARoF BBU for real-time signal processing of 800 MHz extended 5G NR OFDM signals, which are upconverted to a tunable IF between 2.25 and 5.5 GHz."

"The demonstrated real-time fronthaul link highlights the functionality of the blueSPACE ARoF BBU and validates the ARoF fronthaul concept at the heart of the blueSPACE architecture."

Efficient Cross-Correlation Algorithm for Correction of Common Phase Error Employing Preamble for Orthogonal Frequency Division Multiplexing (OFDM) Receivers [68]

This is a machine-generated summary of:

Zhang, Lingfei; Ma, Jun; Wu, Ziqin; Wang, Yiming; Liu, Chen; Habib, Kashif: Efficient Cross-Correlation Algorithm for Correction of Common Phase Error Employing Preamble for Orthogonal Frequency Division Multiplexing (OFDM) Receivers [68]

Published in: Wireless Personal Communications (2021)
Link to original: https://doi.org/10.1007/s11277-020-08030-6

If you want to cite the papers, please refer to the original.

For technical reasons we could not place the page where the original quote is coming from.

Abstract-Summary

"We propose a cross-correlation algorithm for correction of common phase error (CPE) in orthogonal frequency division multiplexing (OFDM) systems with high implementation efficiency."

"Offset compensation algorithm must be adopted in the OFDM system."

"The phase offset is traditionally estimated and compensated by introducing pilots in each OFDM symbol."

"The BER performance of the traditional algorithm PCP is poor because it is limited by the quantity of the pilot constraints of the wireless communication standard."

"The proposed algorithm takes advantage of preamble sequences to remove the limitation of the pilots and improve the BER performance."

"Numerical analyses show that the proposed algorithm presents better estimation performance in additive white Gaussian noise (AWGN) channel."

"The mean square error (MSE) is reduced to 10^{-3}, which is lower than that of the traditional algorithm (10^{-2}) at 10 dB signal-to-noise ratio (SNR)."

Introduction

"Ref. [69] shows a time domain compensation algorithm of phase noise in low complexity for OFDM system."

"It estimates the phase noise using the Karhunen–Loeve, and chooses the basic elements for least squares (LS) phase noise estimation using data-directed algorithm, which can enhance the performance of OFDM system."

"In [70], a time domain optimization algorithm for joint phase noise and channel estimation is presented to use majorization–minimization technique."

"Compared with the LS-based scheme, an algorithm combining phase noise correction and channel noise variance estimation for OFDM in [71] can surmount the over-fitting phenomenon, thus improves performance."

"A frequency-domain phase noise estimation algorithm is developed for an OFDM system."

"A pilot subcarrier allocation algorithm in a non-iterative manner is proposed for OFDM system, which enables phase noise estimation efficiently, low-cost and low-complexity [72]."

System Model and Problem Formulation

"The algorithm in [73] is denoted by phase correction with pilots (PCP)."

"These pilots could be away from each other when placed in OFDM systems."

"The space of the pilots is 14 in each symbol with the length of 64 in IEEE802.11 a/n/ac standard when the bandwidth is 20 MHz."

"The PCP algorithm estimation performance may degrade if the number of the pilot data is fewer."

Common Phase Error Correction Algorithm

"The TPC algorithm computes the correlation value of s_n and r_n to obtain the phase shift information of the symbol."

"Generate the preamble containing two equal training sequences with length N and send it from the transmitter."

"Since the FFT module is necessary in OFDM systems, CPE correction can be executed in the frequency domain."

"Generate the preamble containing two uniform training sequences with the length of N and send it from the transmitter."

"Compared with the TPC algorithm, only the FFT and the gap controller modules are added to the proposed FPC algorithm."

"The FFT module is necessary in OFDM systems, and the gap controller module is nothing but a selector."

Performance Analysis and Discussion

"The simulation results show that the TPC and FPC algorithms can correct the CPE well, and these two algorithms exhibit satisfactory estimation performance."

"These two figures show the QPSK and 16QAM BER performance of the TPC and FPC algorithms after quantized and the other compared quantized results."

"The simulation results show that the MSE performance of the proposed TPC/FPC algorithms were reduced to 10^{-3} when gap is 1, both of which are better than that the 10^{-2} of the traditional PCP algorithm in the same SNR state at 10 dB. Hence, the TPC and FPC algorithms have better estimation performance and can be used in OFDM systems."

"The TPC and FPC algorithms can achieve high accuracy of estimation and reduce the complexity of implementation."

"We compare the performance of the proposed TPC/FPC algorithm utilizing preamble with the performance of the conventional PCP algorithm employing pilot in [74]."

Conclusions

"Simulation results show that MSE is reduced to 10^{-3}, while that of the traditional PCP algorithm is 10^{-2} at 10 dB SNR."

"The proposed algorithms can be easily implemented with low complexity of hardware for practical applications as well as improve the estimation accuracy and simplify the implementation compared with the traditional algorithms."

"Experimental results verify the accuracy and effectiveness of the proposed algorithm."

UF0MC-IOTA Based Cognitive Radio Transceiver [75]

This is a machine-generated summary of:

Dayana, R.; Malarvezhi, P.; Vadivukkarasi, K.; Kumar, R.: UF0MC-IOTA Based Cognitive Radio Transceiver [75]

Published in: Wireless Personal Communications (2020)

Link to original: https://doi.org/10.1007/s11277-020-07467-z

If you want to cite the papers, please refer to the original.

For technical reasons we could not place the page where the original quote is coming from.

Abstract-Summary

"Implementation of cognitive radio (CR) is one of the promising technologies for the effective utilization of spectrum."

"Sensing is the critical component of cognitive radio."

"Encouraged by the element of sensing is critical for consistently defining spectral opportunity and succeeding opportunity in spectrum access for the fifth generation (5G) wireless networks."

"This article focuses on the performance of UFMC-IOTA based CR transceiver sensing challenges, downlink capacity of cognitive radio network (CRN)."

Introduction

"The roles of cognitive radio are spectrum sensing, sharing, controlling and spectrum agility."

"One of the most important element or key aspect in CRN is spectrum sensing."

"CRN is capable of automatically detecting and allocating the unused spectrum for the applications on demand basis."

"It can sense the spectrum based on multidimensional like frequency, time, geographical space, code and angle etc. Unused spectrum also called as white space or spectrum-hole."

"Through suitable sensing methodology, the CRN sensing the underutilized or unutilized spectrum i.e., white spaces and it allocate the same to the SU without causing interference to the PU."

"Sensing peripheral radio spectrum exactly and rapidly plays a major role in cognitive radio."

"Conventional method of sensing is energy detection based sensing which cannot differentiate users at low signal-to-noise-ratio (SNR)."

"The matched filter based sensing requires the prior knowledge about primary user."

Existing Systems

"Generation (5G) cellular network has been allocated new spectrum and a wider spectral bandwidth per radio channel."

"The 5G systems will necessitate much more grouped spectrum to facilitate flexible bandwidth scaling and extension [76]."

"The following two sub-sections explain the 5G waveform candidate for higher spectral efficiency and rapid spectrum sensing methodology for effective utilization of spectrum."

"Among the various 5G waveform candidate UFMC is one of the competent multi-carrier system."

"The tender of this UFMC waveform will promise the gratification of the challenging out-of band emission (OOB) requirements."

"It avoids the carrier frequency offset by applying sub-band and sub-carrier basis filtering [77]."

"Because of the non-orthogonal property of sub-carriers in UFMC, no synchronization is required."

"5G network requires wide flexibility in order to refuge wide range of spectrum."

"The power spectral components of periodic signals are measured from the radio frequency signals [78]."

System Model

"From the social perception, 5G networks have the potential to develop the cellular network access in rural areas."

"By sensing white spaces of unused spectrum can be automatically allocated for the different applications using proposed UFMC-IOTA based CR transceiver."

"The SNR level is poor in rural area, cyclo-stationary feature detection spectrum sensing technology senses the white spaces exactly."

Proposed UFMC-IOTA Transceiver

"Time domain symbols are estimated from the baseband signal with the help of windowing technique."

"CFD spectrum sensing technique is operated by the FFT digital signal processor."

"Processor output is used for to find autocorrelation of the signal."

"Based on the detected signal the following hypothesis is taken."

CFD Sensing Algorithm

"Step 1 Acquire the FFT Signal from the receiver Step 2 Calculate Auto correlation and Spectral Correlation of received signal Step 3 Calculate the Decision Threshold."

"Optimum selection of threshold 'λ' is the important parameter to decide the spectrum."

"If not go to step 1 Step 5 Do the SU need to be allocated?"

"If No, return to step 1 Step 7 If PU re-appear, then vacate the SU without interference to the PU and accommodate the respective PU."

Simulation Results

"The proposed UFMC-IOTA system throughput analysis in CRN environment is simulated by using CFD spectrum sensing technique."

"For the same sensing time, other multicarrier modulation waveforms like FBMC, UFMC, OFDM-IOTA and conventional OFDM achieves the normalized throughput of 0.92, 0.86, 0.81 and 0.79 respectively."

"Whereas other multi-carrier modulation systems like FBMC, OFDM-IOTA and Conventional UFMC achieves detection probability of 0.38, 0.20 and 0.13 respectively."

"The proposed UFMC-IOTA system gives 37%, 30% and 12% improvement in probability of detection on comparison with conventional UFMC, OFDM-IOTA and FBMC system respectively."

"The proposed system gives 53%, 36% and 18% improvement in probability of detection on comparison with conventional UFMC, OFDM-IOTA and FBMC system respectively."

"The proposed UFMC-IOTA system provides 47%, 34% and 27% improvement in probability of detection on comparison with conventional UFMC, OFDM-IOTA and FBMC system respectively."

Conclusion

"Proposed Universal Filtered Multicarrier-Isotropic Orthogonal Transform Algorithm based cognitive radio transceiver is used to trounce the challenges in spectrum sensing."

"It is proved that the proposed UFMC based CR transceiver outperforms in spectrum sensing than the existing sensing methods."

"To better ways of developing the already allocated spectrum band accessing through superior 5G waveform candidate such as UFMC-IOTA based CR."

Transmission of Chaotic-Based Encrypted Audio Through OFDM [79]

This is a machine-generated summary of:

El-Zoghdy, S. F.; El-sayed, Hala S.; Faragallah, Osama S.: Transmission of Chaotic-based Encrypted Audio Through OFDM [79]

Published in: Wireless Personal Communications (2020)

Link to original: https://doi.org/10.1007/s11277-020-07187-4

If you want to cite the papers, please refer to the original.

For technical reasons we could not place the page where the original quote is coming from.

Abstract-Summary

"This paper proposes a communication model for encrypting and transmitting audio over OFDM systems."

"The 2D Baker, Logistic and Standard chaotic encryption schemes are utilized in the suggested model to encrypt and decrypt the transmitted audio."

"A simulation model for the proposed communication system is built using various encryption/decryption quality indicators to assess and compare its performance."

"The results demonstrate a good performance for the suggested communication model using all of the three studied chaotic encryption schemes."

"It also shows that the proposed communication mode is immune against noise distortion."

"One more noticed point from the result is that, the performance of the 2D Standard Map chaotic map in the proposed communication model is superior to that of the other two studied chaotic encryption schemes in terms of Spectral Distortion, Log Likelihood Ratio, Correlation Coefficient, Segmental Signal-to-Noise Ratio (SNRseg) and SNR encryption/decryption quality metrics."

Introduction

"In [80], the authors studied the effects of carrier frequency offset (CFO) restitution and the guard-interval techniques on the act of DCT-OFDM, DWT-OFDM and FFT-OFDM communication models using various encryption schemes over distinct wireless communication channels."

"During their experiments, they utilized the DES, ADES, RC6 and chaotic 2D Baker map encryption techniques in encrypting images that are transmitted over various OFDM versions and compare the performance of such encryption techniques."

"This research paper merely investigates the decryption quality of various chaotic maps for encrypting and modulating digital audio using FFT-OFDM to be utilized in wireless communication systems."

"Various audio encryption/decryption performance indicators like Spectral Distortion, Log Likelihood Ratio, Correlation Coefficient, SNR and Segmental SNR are utilized in evaluating and comparing the performance of the suggested OFDM audio encryption techniques for wireless communication systems."

FFT-OFDM Communication Model

"At the sender side, the source 1D audio is reformatted to 2D audio."

"Generating encrypted 2D audio follows the same method of generating the encrypted image in [80, 81]."

"CP or ZP is eliminated from the delivered signal at the receiving side."

"The decryption process is conducted as in [80–82] by: where, A^c represents the received corrupted encrypted 2D audio, D represents utilized decryption technique, KC represents the decryption key."

"It could be equal or distinct from ciphering key K, and A^r represents the retrieved 2D audio which is then reformatted to 1D audio to reconstruct the source audio."

Utilized Encryption Techniques

"The proposed communication model for transmitting encrypted speech through OFDM system utilizes chaotic 2D Baker, Logistic and Standard maps for the encryption/decryption processes."

"Suppose that $B(n_1, \ldots, n_k)$ is the discretized map and the secret key S_{key} is represented by the vector $[n_1, \ldots, n_k]$."

"According to the chaotic 2D Baker map, the values of elements at position (q, z) are shifted to the position: where $N_i \leq q < N_i + n_i$, $0 \leq z < N$ and $N_1 = 0$."

"This transposing is conducted as: (1) The given $N \times N$ matrix to be randomized is partitioned into K rectangles."

"Each rectangle is of width n_i and has N elements. (2) Starting by the upper left rectangle then the lower ones, every rectangle elements' are reordered and mapped to a row in the attained permuted $N \times N$ matrix. (3) The scan in every rectangle starts from undermost left element to topmost elements."

Transmission of Encrypted Audio over OFDM

"Apply inverse fast Fourier transform (IFFT) to the encrypted 2D audio. (8) Pass the encrypted 2D audio through the S/P converter. (9) Insert the guard interval to eliminate any remaining ISI after multicarrier modulation process. (10) Send the resulted symbols serially via the wideband channel to the receiver."

"Pass them through P/S converter. (4) Decrypt the transmitted 2D audio using any of 2D Baker Map or 2D Logistic Map or 2D Standard Map chaotic techniques. (5) Inverse clipping by the addition of 2 to the values less than -1 in the obtained decrypted 2D audio. (6) Mask decrypted 2D audio with the same chaotic technique used in decryption process. (7) Subtract obtained mask form decrypted 2D audio. (8) Reshape the retrieved 2D audio to its corresponding 1D format. (9) Assemble and rebuild the audio."

Performance Assessment Keys

"To assess a cryptosystem's encryption quality, the CC is computed between corresponding samples in original and encrypted signals."

"Since CC is computed between source signal s and its corresponding encrypted signal y, the low the CC value is, the best the encryption quality."

"The SD encryption/decryption quality indicator is performed on the source and encrypted audio frequency spectra."

"The lower the SD value among source and encrypted audio is, the best the encryption quality."

"SNR may be considered as the suitable encryption/decryption quality indicator for evaluating the distortion in waveform-based audio processing systems which regenerate the input waveform."

"SNRseg may be considered as a good indicator the audio waveform quality."

Experimental Results and Discussions

"In this simulation, the SD between plain audio and decrypted audio utilizing 2D Baker, 2D Logistic and 2D Standard chaotic maps with OFDM system at different SNR is computed."

"In this simulation, the SNR quality indicator is evaluated between SNR of original audio and that of decrypted audio signals using 2D Baker, 2D Logistic and 2D Standard chaotic maps with OFDM system at different SNR."

"This simulation, the SNRseg quality index is gauged between the original audio and that of decrypted audio signals using 2D Baker, 2D Logistic and 2D Standard chaotic maps with OFDM system at different SNR."

"The LLR key performance index between the original and decrypted voice signals is evaluated using 2D Baker, 2D Logistic and 2D Standard chaotic maps with OFDM system at different SNR is computed."

Conclusion

"This paper merely investigates the decryption quality of various chaotic maps for encrypting and modulating digital audio using FFT-OFDM."

"Various encryption/decryption quality indices are used in evaluating and comparing the suggested communication system with 2D Baker, Logistic and Standard chaotic encryption schemes."

"The results reveal that the proposed communication model behaves well in transmitting encrypted audio and its immunity against noise distortion is high."

"It also show that, in the proposed communication system, the performance of 2D standard chaotic map always outperforms that of 2D Baker and 2D Logistic chaotic maps in terms of encryption/decryption quality metrics SNR, SNRseg, CC, SD and LLR."

On Effects of Frequency Offset on Simulink Model of OFDM [83]

This is a machine-generated summary of:

Kaur, Navdeep; Kumar, Naresh: On Effects of Frequency Offset on Simulink Model of OFDM [83]

Published in: Wireless Personal Communications (2020)

Link to original: https://doi.org/10.1007/s11277-020-07685-5

If you want to cite the papers, please refer to the original.

For technical reasons we could not place the page where the original quote is coming from.

Abstract-Summary

"OFDM is a multicarrier transmission technique, widely preferred due to advantages like high data rate, less bandwidth consumption and channel robustness."

"IEEE 802.11a standard uses conventional OFDM which uses IFFT for generating orthogonal carriers and multiplexing."

"Due to extra bandwidth consumption by cyclic prefix in conventional OFDM, DWT has now emerged as new alternative over FFT."

"We investigated the offset effect on conventional and DWT-based OFDM under different communication channels (AWGN and Rayleigh)."

"Effect of different modulation techniques and wavelets on DWT-OFDM model is also analysed."

Introduction

"Other than being bandwidth efficient, research has also shown that wavelet based OFDM systems have less bit error rate (BER) and better orthogonality than conventional OFDM."

"There are some problems faced by OFDM which are inter carrier interference (ICI) and peak to average power ratio (PAPR) [84]."

"Phase noise also effect the OFDM system leading to frequency offset which occurs due to fluctuation in transmitter and receiver oscillators."

"This paper studies the effect of frequency offset on FFT and DWT-OFDM using different wavelets and modulation schemes in Simulink."

"Evaluation of wavelet based OFDM and conventional OFDM is presented in [85], where ICI self cancellation scheme was implemented."

"Comparison of ICI self cancellation based techniques for wavelet based OFDM was analysed uner different communication channels [86]."

"In the present paper, effects of carrier frequency offset is analysed in OFDM (conventional as well as wavelet based OFDM) using Simulink models."

Simulink Models

"Random integer generator block is used to generate data."

"Symbols generated from modulation block are fed to IFFT block to generate 64 sub-carriers."

"At the receiver end, FFT block is used to convert spectra to time domain and then demodulation is performed."

"In FFT-OFDM, cyclic prefixing is done using matrix concatenate block and multipath selector."

"The low frequency components contain most of the information while smallest details are contained in high frequency components, so, zero padding is done to feed zeroes to high pass filter and data symbols are fed to low pass filter of IDWT block."

"At the receiver end, output from high pass filter of DWT block can be neglected."

Simulation and Results

"Two values of phase noise has been taken for comparison that is −70 and −100 dBc/Hz."

"Results show that for less phase noise of −100 dBc/Hz, Bit error rate is less."

"Phase noise is taken as −100 dBc/Hz and frequency offset as 0.02."

"QAM is more sensitive to phase noise and frequency offset at high SNR than PSK but as QAM is more spectrally efficient, it is widely used in broadband communications."

Conclusion
"Simulink model for OFDM using FFT and DWT transform has been created which is more near to hardware implementation."

"DWT-OFDM model shows better performance in terms of BER than conventional model."

"The effect of offsets on FFT and DWT-OFDM systems has been analysed."

"Results show that 64-QAM is more prone to offsets at high SNR than BPSK and DWT-OFDM has performed better than conventional OFDM in the simulink model."

Future Scope
"Simulation model can be further utilised to implement and analyse various ICI cancellation and equalisation techniques."

"Effect of multiple CFO on Wavelet based OFDM can also be analysed."

Physical Layer Encryption for Wireless OFDM Communication Systems [87]

This is a machine-generated summary of:

Jacovic, Marko; Juretus, Kyle; Kandasamy, Nagarajan; Savidis, Ioannis; Dandekar, Kapil R.: Physical Layer Encryption for Wireless OFDM Communication Systems [87]

Published in: Journal of Hardware and Systems Security (2020)

Link to original: https://doi.org/10.1007/s41635-020-00097-8

If you want to cite the papers, please refer to the original.

For technical reasons we could not place the page where the original quote is coming from.

Abstract-Summary
"Devices must be secure and reliable to support the emergence of large scale heterogeneous networks."

"Physical layer security leverages existing components of the physical or PHY layer to provide a low-complexity solution appropriate for wireless devices."

"This work presents a PHY layer encryption technique based on frequency induction for Orthogonal Frequency Division Multiplexing (OFDM) signals to increase security against eavesdroppers."

"The secure transceiver consists of a key to frequency shift mapper, encryption module, and modified synchronizer for decryption."

"Both simulation and hardware evaluation results demonstrate that the proposed system is capable of providing secure communication from an eavesdropper with no decrease in performance as compared with the baseline case of a standard OFDM transceiver."

"The techniques developed in this paper provide greater security to OFDM-based wireless communication systems."

Introduction

"Secret keys are generated using properties of the wireless channel to help secure a communication link."

"This work explores PHY layer encryption of Orthogonal Frequency Division Multiplexing (OFDM) signals, which is a widely adopted communication technique [88]."

"A method is developed to secure wireless OFDM communication links by inducing frequency offsets to the payload portion of the packet."

"The frequency synchronization requirements of OFDM are leveraged to secure communications using PHY layer encryption."

"An algorithm is developed that uses the generated key along with the measured mean CFO value to induce frequency shifts during wireless communication—which are applied at different granularities to the payload of an OFDM packet."

"This is the first technique to artificially induce frequency shifts at the PHY layer for encryption during wireless communication, which provides a method of low-complexity to secure OFDM-based wireless communication links."

OFDM System Design

"The parallel streams of frequency domain symbols undergo an N-point Inverse Fast Fourier Transform (IFFT) operation that results in time-domain values."

"A cyclic prefix is appended to the time-domain parallel data by repeating a set number of values from the end of the IFFT to the beginning of the output, which reduces the effects of inter-symbol interference (ISI)."

"The baseband OFDM signal is converted to the analog domain and is up-converted to a carrier frequency for transmission."

"Timing and frequency synchronization is implemented by applying a joint algorithm that utilizes the received preamble and known reference signals."

"A serial-to-parallel conversion is performed after synchronization to form the same number of streams that were utilized by the transmitter, while the samples used for synchronization are removed."

"Channel estimation is performed by dividing the received frequency domain training data with the original reference values known to the receiver."

System and Threat Models

"Alice and Bob possess matching keys that are unique to the key used by Eve."

"The following assumptions are made regarding the eavesdropper Eve: Eve knows the signal structure, reference signals, as well as the encryption and decryption algorithms used by Alice and Bob."

"Eve is separated from both Alice and Bob by a distance of at least half of the signal wavelength."

"Eve does not interfere with the communication between Alice and Bob or manipulate the physical environment."

Key Generation and Management Policy

"Alice and Bob exchange packets during a probing period, extract channel estimates and CFO measurements, reconcile the PHY layer features with each other, and create unique and reciprocal secret keys."

"Alice and Bob exchange time indices of the channel estimate threshold crossings and distances from the quantization boundaries of the CFO."

"The exchange allows both Alice and Bob to reconcile the extracted bit sequences and CFO estimates."

"Eve observes packet exchanges between Alice and Bob during the reconciliation stage and develops her own key using the same agreement algorithm."

"The channel estimates that form the basis for her generated key differ, which results in a different extracted bit sequence for Eve."

"Once a key bit is generated, the bit is stored in a shift register that is updated periodically with new channel estimate and CFO information by applying the management protocol previously developed by Chacko and others [89]."

Encryption at Transmitter

"The 8-bit sub-key is used to modify the mean CFO value and produce a single frequency shift."

"The mean CFO observed from a window of samples changes over time, and the length of the secret key is much larger than the sub-key to facilitate multiple frequency shifts per packet."

"The shift mapper requires the Boolean key bitstream, the reconciled mean CFO value in 32-bit floating-point format, and a corresponding "key valid" signal."

"The shift duration and enable signals are utilized with three 12-bit free running counters to control the operation of the RAM and provide the properly repeated and aligned shift sequence."

"Integer frequency induction requires minimal complexity to implement as the operation is simply a circular shift in the frequency domain."

"Fractional frequency induction in the frequency domain requires upsampling, a larger FFT, scaling of the output, and complex control signals."

Decryption at Receiver

"The frequency acquisition range, normalized to the sub-carrier spacing, was limited to the fractional CFO, which is a practical choice for an indoor Wi-Fi setting."

"The execution of coarse timing and CFO estimation are primarily based on the contributions described in [90], in which a modified correlation is applied to leverage a repetitive preamble structure."

"If the true CFO of a signal is 3.7 sub-carrier spacings, the fractional estimate is −0.3 with integer 4 since the search space is limited to even sub-carriers."

"A payload valid signal from the coarse timing module is utilized to apply the additional security-based frequency shift to only the payload portion of the packet."

"The modified synchronizer of the receiver outputs the updated normalized frequency shifts with data type S32_23 to the CFO correction module, which compensates for the inherent CFO of the packet and decrypts the payload."

Performance Evaluation
"Although Eve implements the decryption method applied by Bob, the differences in key value and mean CFO result in deviations in the calculated frequency shifts from the values used by Alice during encryption."

"The encryption module, baseline synchronizer, and decryption-modified synchronizer are synthesized on a Xilinx Virtex-7 FPGA VC707 Evaluation board to characterize the additional resources utilized for each component."

"The timing and power requirements of the encryption and decryption modules were analyzed through implementation on a Xilinx Artix-7 AC701 FPGA board."

"The power consumption attributed to the encryption technique is 22 mW. The overhead in power is primarily a result of the resources utilized to implement the BRAM and CORDIC."

"The decryption module does not result in a timing overhead, and requires 11mW of dynamic power for the FPGA resources utilized to implement the BRAM."

Related Work
"Both methods require perfect channel knowledge and consider reference signals of the preamble without discussion of pilot tone manipulation from within the payload."

"An attacker synchronized to the victim transceiver is able to recover the signal using pilot tone based channel estimation techniques with interpolation methods."

"A measured threshold for channel gain was applied to generate pilot symbols that represented whether a sub-carrier contained legitimate information or dummy data throughout the transmission of the signal."

"The method also assumed perfect channel knowledge at the transmitter, did not consider pilot tones within the payload of the signal, and required a large overhead in transmission bandwidth as an equal number of data bearing and dummy sub-carriers were used."

"A PHY layer encryption technique and the corresponding hardware components are described that induce frequency offsets to the payload portion of a packet to secure signals for wireless communication."

Conclusion
"The implementation of the techniques on a Xilinx Virtex-7 FPGA VC707 board resulted in an increase of 1.65% in the number of LUTs, 0.8% in the number of FFs, and less than 0.4% in the number of BRAM and multipliers required at the transmitter."

"Implementation of the decryption technique at the receiver required minimal additional FPGA resources, with the utilized LUTs, FFs, BRAM, and multipliers each increasing by less than 0.4%."

"The techniques were implemented on a low-power Xilinx Artix-7 AC701 FPGA Board to validate suitability for OFDM-based applications."

"An increase of 3.73% in the number of LUTs, 1.82% in the number of FFs, 1.10% in the number of BRAMs, and 1.35% in the number of multipliers is required at the transmitter."

"The technique is, therefore, suitable for a wide range of applications that require wireless communication."

Interoperability in Heterogeneous Wireless Networks Using FIS-ENN Vertical Handover Model [91]

This is a machine-generated summary of:

Rajinikanth, E.; Jayashri, S.: Interoperability in Heterogeneous Wireless Networks Using FIS-ENN Vertical Handover Model [91]

Published in: Wireless Personal Communications (2019)

Link to original: https://doi.org/10.1007/s11277-019-06406-x

If you want to cite the papers, please refer to the original.

For technical reasons we could not place the page where the original quote is coming from.

Abstract-Summary

"Wireless network system aims to provide dedicated links to the users for improving the quality of services."

"Vertical handover (VH) mechanism provides uninterrupted service to the users in a multiple wireless network region by selecting the best available network."

"To improve the QoS of VH model the decision phase needs efficient metrics for providing excellent service."

"A fuzzy based operating model is most suitable for decision making and to implement a seamless handover, neural network is used in the proposed research work."

Introduction

"Various approaches are considered in the handover execution process to provide uninterrupted services to the users."

"Hard handover process operated based on the connection establishment between the target network and the current network."

"If the handover process is initiated and the user device is allowed to connect only one network at a time and to establish a new connection the previous link must be released."

"Such process defines hard handover and it has the limitation in connection establishment."

"Based on the threshold value system discovery process initiates the current network only if the network doesn't able to provide service to the ongoing connection and this reduces the QoS of the network."

"Handover decision is the second factor in which decision making process is performed based on the available wireless access networks."

"A handover execution is the last phase in which procedure for signalling messages and establishing connection to the user from one network to another network."

Related Works

"Research work [92] focuses about the handover process using the cellular mobile communication modules such as mobile system and base station."

"Different versions of radio protocol is described and presented in the research work for signalling process between the NB and UTRAN."

"Early handover preparation is discussed in this research work for defining the authoritative transmission of the handover even in weak signal areas."

"An average detected signal power based research work is addressed in literature [93] to developing a simple handover module based on the detected signal power."

"The last part of the survey work concludes with the research work [94] which discusses the time delay and prediction system of handover in LTE based on location tracking."

"Work user direction is predicted using the previous location as reference and selecting the target NB from UTRAN NB for handover process."

Proposed Work

"The proposed model uses fuzzy as decision making tool along with the feed forward artificial neural network."

"A fuzzy multi criteria vertical handover algorithm is mainly based on the logic controls defined in the fuzzy system."

"A heuristic multilayer feed forward artificial network is used for developing vertical handover decision model."

"The network features are used to identify the handover decision based on the cost, transmission, range and capacity."

"The algorithm is based on the fuzzy inference system and Elman neural network."

"Fuzzy considers the bandwidth and velocity for the predicted number of users and the input parameters makes handover based on the rules."

"The following describes the algorithm for Elman neural network based handover model based on calculating the weights and the error representation values."

"Based on the algorithm the necessity of the handover requirements is identified and passes on the decision module."

Experimental Results and Discussion

"The proposed Fuzzy interference system Elman Neural Network decision model defined in three different stages based on their weights depending on the output value and threshold value."

"These three different stages helps to predict the error and weight estimation in the proposed Fuzzy interference system Elman Neural Network decision model."

"Based on the observation the proposed model has less throughput value when compared to ANN model and artificial neural network vertical handover decision model."

"The ratio is smaller for proposed Fuzzy interference system Elman Neural Network model compared to the ANN model and artificial neural network vertical handover decision model."

"The proposed Fuzzy interference system Elman Neural Network decision model have lesser packet loss rate even in the three different condition."

"The proposed Fuzzy interference system Elman Neural Network decision model have high throughput and lesser packet loss rate in three different conditions compared to existing artificial neural network and vertical handover decision model."

Conclusion

"The proposed Fuzzy interference system Elman Neural Network decision model vertical Handover model overcomes the existing model issues and it is useful in wide ranging of applications."

"The proposed Fuzzy interference system Elman Neural Network decision model of VHD management has less handover latency, end to end delay and packet loss when compared to the ANN module."

"This proposed Fuzzy interference system Elman Neural Network decision model achieves high throughput and that leads to maximize the users in the network without any loss and interference."

"Fuzzy interference system Elman Neural Network decision model concentrate on the routing overhead to lead the successful network for the effective communication in the artificial neural network."

Optimization of FBMC Waveform by Designing NPR Prototype Filter with Improved Stopband Suppression [95]

This is a machine-generated summary of:

Hua, Jingyu; Wen, Jiangang; Wang, Anding; Xu, Zhijiang; Li, Feng: Optimization of FBMC Waveform by Designing NPR Prototype Filter with Improved Stopband Suppression [95]

Published in: Mobile Networks and Applications (2019)

Link to original: https://doi.org/10.1007/s11036-019-01342-3

If you want to cite the papers, please refer to the original.

For technical reasons we could not place the page where the original quote is coming from.

Abstract-Summary

"This paper investigates a design method of nearly perfect reconstruction (NPR) prototype filter to effectively achieve both the filter flexibility and the FBMC frequency selectivity, in which the Nyquist condition can be linearly constrained by using the filter autocorrelation coefficients, and then, the spectral factorization aimed at minimum stopband energy is applied to retrieve the final prototype filter."

"Such advantages finally result in bit-error-rate (BER) superiority in FBMC simulations, which definitely confirms the effectiveness of the proposed design method for NPR prototype filter."

"We also study the coefficient quantization and length reduction of the proposed NPR prototype filter."

"The quantized NPR filter can preserve its advantage on sidelobe suppression as well as BER."

Introduction

"To guarantee the perfect reconstruction (PR) of FBMC signal, the generalized Nyquist condition along both time and frequency axises should be satisfied by FBMC prototype filter [96]."

"Another popular filter was the PHYDYAS filter, which was designed by frequency sampling method in discrete time and had been employed for its superior sidelobe falling rate [97]."

"The IOTA filter and the PHYDYAS filter could not effectively relax Nyquist condition for better sidelobe suppression."

"Different from the conventional minimum phase factorization [98, 99], the proposed factorization can maintain the optimized performance after factorization, making the prototype filter qualified for FBMC application."

"In the BER tests, the designed NPR filter outperforms the PHYDYAS and IOTA filters under scenarios of additive white Gaussian noise (AWGN) and carrier-frequency offset (CFO), which demonstrates the advantage of the proposed method for NPR prototype filter."

Preliminaries

"The generalized Nyquist condition requires the FBMC prototype filter make regular zero crossings both along the time and frequency axises."

"Because the interference originated form NPR filter bank is rather small in contrast to the residual interference due to transmission channel, and the NPR filter can enhance FBMC frequency selectivity at the controllable cost of generalized Nyquist condition [100]."

"The Nyquist condition below is usually relaxed in the design of FBMC prototype filter."

"Due to the relaxation of Nyquist condition, the designed prototype filter causes time-domain sampling interference when applied in filter banks."

"It can be formulated as where G (ω, g) is the frequency response of length-N FBMC prototype filter with $g = [g(0), g(1), \ldots, g(N-1)]^T$."

"To reduce the out-of-band emission, the FBMC prototype filter with narrow band has an increased length, which is K times of the symbol period."

Design of the NPR Prototype Filter

"An spectral factorization different from that of [99] is employed to produce the prototype filter g(n)."

"In this spectral factorization, the minimum stopband energy rather than conventional minimum phase is taken as the factorization criteria, which can reduce the difficulty of zero grouping and maintain the optimized performances of the NPR prototype filter after factorization."

"(d) The reciprocal pairs could be separated, resulting in nonlinear-phase prototype filter along with improved performance, such as the performance of stopband energy."

"Of spectral factorization, the conjugate zero pairs and the reciprocal zero pairs are halved into two groups."

"Figure, the number of reciprocal pairs is far less than the densely distributed conjugate pairs on unit circle, which is actually related to the small value of K. Therefore, the MSBE factorization is effective in the FBMC prototype filter design."

Simulation and Analysis

"When CFO is added into FBMC simulation, the adjacent orthogonality is destroyed, then the advantage attributed by suppressed sidelobe is reduced and the BER gaps between the PHYDYAS filter and the designed NPR filters are diminished."

"According to above results, it is reasonable to relax the Nyquist condition for better stopband suppression in the design of NPR prototype filter, and it can be efficiently implemented by the proposed method."

"The proposed NPR filter still produces smaller BER than the PHYDYAS filter in case of filter quantization."

"Besides, in the AWGN case, the NPR filter $g1_F$ with $K = 3$ can obtain better BER results than the PHYDYAS filter with $K = 4$, which is actually traded from ISI sacrifice."

"It is clear that the proposed method can flexibly exploit NPR relaxation to improve filter performance and BER performance in FBMC systems."

Conclusions

"We have achieved a flexible design of FBMC waveform, i.e., the corresponding NPR prototype filter."

"The stopband suppression is strengthened through the optimization with relaxed Nyquist, in which the autocorrelation coefficients are employed to convert the original nonconvex constraints to linear ones, and a spectral factorization aimed at minimum stopband energy is used to retrieve the NPR filter."

"From the performance comparisons and BER tests, the designed NPR filters are demonstrated being superior to the widely-used PHYDYAS filter and IOTA filter, which will benefit the practical FBMC applications."

Effective Scheduling Policies to Optimize Radio Resources Between NR-gNodeB and Device to Device Systems in 5G [101]

This is a machine-generated summary of:

Priyadharsini, Naveena A.; Selvi, S. Tamil: Effective Scheduling Policies to Optimize Radio Resources between NR-gNodeB and Device to Device Systems in 5G [101]

Published in: Wireless Personal Communications (2019)

Link to original: https://doi.org/10.1007/s11277-019-06604-7

If you want to cite the papers, please refer to the original.

For technical reasons we could not place the page where the original quote is coming from.

Abstract-Summary

"Technology behind every communication systems, especially commercial applications requires periodic modifications to satisfy the subscriber's desires."

"Since radio resources are limited and expensive, Device to device communication underlying New Radio (NR-gNodeB) access network is encouraged in fifth generation (5G) systems."

"The research hotspot is to find ideal solution for resource allocation and interference management."

"Further the intention is to mitigate interference in such system so that functionalities of scheduling schemes such as Greedy, Round Robin and Proportional fair algorithms are analyzed for efficient resource allocation."

"Coded GFDM with proportional fair scheduler results in high throughput and reduced latency with no compromise in its fairness."

"The proposed scheme could be a wise choice for reliable communication in smart cities."

Introduction

"It would play a vital role in various fields such as home automation, smart transportation, online transactions, security systems, climate monitoring systems, cellular Communication systems etc. However there is still a dramatic need for reliable communication with very high data rate, low latency, and increased throughput, multiple device connectivity and enhanced device to device communication, new methodology should be followed."

"They noticed that as the density of users were increased, the path loss and blockage probability of mm wave communication links were increased which directly reduces the system sum rate."

"Their result explored that the combination of frequency reuse and proper scheduling policies by intelligent packing of cellular transmission to users reached high rate in D2D communication."

"Attracted by the prior work, we are tempted to develop a novel method to mitigate the effects of interference and trigger the data throughput towards upward with reduced latency in D2D communication underlying cellular communication for 5G systems."

System Architecture
"Both the NR-gNodeB access users and the D2D users are expected to share the same radio spectrum."

"There are a large number of radio access devices within a cell."

"They are designed to operate at a specific frequency range."

"Enormous number of high gain, high frequency, high directional antenna can be packed into the device."

"It may result in inclusion of different interferences like co-channel interference, adjacent channel interference, ISI, ICI, inter cell interference etc In order to mitigate the effects of such interferences, an optimization of radio resources and transmission power is required."

Proposed System
"LDPC is a linear error correcting code, which is a method for transmitting information over noisy transmission channel."

"In GFDM, the out of band radiation of the transmitted signal is controlled by an adjustable pulse shaping filter which is applied to each and every subcarrier."

"In GFDM, a block of data symbol is transmitted per subcarrier and each subcarrier is circularly convolved with a pulse shaping filter."

"The receiving elements capture the entire signal which contains the mixture of transmitted signal and the channel noise."

"To suppress the ICI due to non-orthogonality of the subcarriers and also to recover the transmitted data, linear receivers such as Matched filter (MF), Zero forcing equalizer (ZF) and Minimum Mean Square Error equalizer (MMSE) are considered."

"The channel with greater SINR is considered for major contribution in the sum of received signal."

"The received signal is transmitted through FDE."

Network Model
"Every cellular user is permitted to share the same sub carrier with different power levels."

"K cellular users are multiplexed through the same channels and are isolated by their power levels."

"The received SINR of cellular user k is given as, From Shannon's channel capacity theorem the data rate achieved by the cellular user k is The performance of the multiple access schemes is analyzed based on energy efficiency and spectral efficiency of the network."

"Energy efficiency is the ratio of data rate to the overall power consumed by the transmitter."

Optimization of Radio Resources

"The steps involved in Greedy scheduling policy are, Step 1 Calculate number of subcarriers to be allocated to each user."

"Case (3) In proportional fair scheduling algorithm, the channel allocation to end user is based on the following features."

"They are (1) information on channel with best quality in terms of SINR, (2) data rate, (3) average throughput of each user on every resource block in the past window length t_c and (4) fairness about the system."

"In time slot M, the proportional fair scheduling policy gives priority to user k in the mth TTI, based on quality of channel in terms of data rate."

"Within the past window t_c, average throughput of each user on every resource block is calculated as, The Proportional fair scheduling algorithm treats all the radio resource blocks independently and updates the system in every time slots."

Experimental Results and Discussion

"OFDMA results in smaller spectral efficiency than NOMA because a part of sub carriers are dedicated permanently to users with poor channel conditions."

"The rest of our discussion is about optimization of radio resources by scheduling mechanisms such as Greedy, round Robin and Proportional Fair algorithms to mitigate interference in the system and also to improve overall system throughput."

"Performance of scheduling policies is investigated in terms of parameters like Path loss, Latency, Throughput."

"Scheduling is based on fairness, throughput and channel quality."

"Proportional fair scheduling is employed in such a way that, the channel with high SINR has less attenuation factor, by which path loss is reduced."

"Among all the three algorithms, Proportional Fair scheduler results in reduced path loss for increased distance."

"We witnessed that, for all the three scheduling schemes, the throughput peaks up as the number of user pair increases."

Conclusion

"This paper reveals an optimistic way for D2D in 5G NR access network."

"SER performance of coded GFDM with different receiver structures is analyzed mathematically and experimentally and concludes GFDM system with MMSE-FDE equalizer results in almost error free transmission."

"NOMA fits perfectly for the proposed system with better spectral efficiency."

"We therefore applied resource optimization mechanism to resolve various interference persist in the system."

A Robust Transceiver for Narrow-Band Communications in Presence of Carrier Frequency Offset [102]

This is a machine-generated summary of:

Amerimehr, Mohammad H.; Rezaee, Mohsen; Dalaei, Hamid Reza: A Robust Transceiver for Narrow-Band Communications in Presence of Carrier Frequency Offset [102]

Published in: Wireless Personal Communications (2019)
Link to original: https://doi.org/10.1007/s11277-019-06873-2

If you want to cite the papers, please refer to the original.

For technical reasons we could not place the page where the original quote is coming from.

Abstract-Summary

"We address narrow-band communications in presence of large and time-varying carrier frequency offset (CFO) at the receiver."

"A robust transceiver system is proposed to dynamically estimate and compensate the CFO."

"The proposed system is particulary attractive for low data-rate communications (e.g., voice service) where the CFO is significantly larger than the data bandwidth."

"The proposed method provides accurate CFO estimation with affordable complexity."

"We also devise a robust frame synchronization scheme to provide accurate timing in presence of CFO."

"The proposed transceiver design is implemented on FPGA and real-time experiments are conducted on a Xilinx ZC706 evaluation base board along with AD-FMCOMMS3 daughterboard."

Introduction

"In wireless communications, carrier frequency offset (CFO) arises from two important factors, frequency mismatch in the transmitter and receiver oscillators and the Doppler effect due to movement of the transmitter or receiver."

"Non-data aided maximum likelihood (ML) carrier synchronization is a method that jointly estimates CFO and its changing rate by observing the received symbols over an interval [103, 104]."

"Although the carrier recovery block would be able to mainly estimate and compensate the CFO, the residual frequency offset still introduces phase ambiguity."

"The two-step carrier recovery method gives accurate CFO estimation with affordable complexity. (3) The two-step frame synchronization is able to detect signal in low signal to noise ratio (SNR) regime and achieves acceptable performance in presence of large CFO.4.We employ a non-coherent demodulation scheme i.e., differential quadrature phase shift keying (DQPSK), which is robust to the phase drift caused by residual frequency offset."

System Description

"We consider a transmission system where the received signal has a large and time varying CFO compared to the transmitted signal."

"Due to the time-varying CFO at the receiver, the CFO estimation is performed frequently."

"We assume that between two successive estimations, the CFO changes linearly with the maximum possible changing rate."

"Although we use this model to simulate the received signal for the worst case, in practice the changing rate can also be time-varying and is upper bounded by the maximum changing rate considered here."

"If the maximum changing rate is tolerated by the demodulator until the next CFO estimation, it is ensured that any changing rate below this maximum value is also handled by the demodulator."

"The above model is sufficient to validate the proposed estimation method for practical time varying CFO."

Transmitter Structure

"The data is first encoded using a concatenation of convolutional and Reed–Solomon encoders."

"The Reed–solomon encoder computes the parity bits while sending the data bits to the next block."

"The Reed–Solomon encoder introduces no delay at the transmitter."

"The encoded data is then mapped to constellation points according to DQPSK modulation."

"Each frame consists of N data symbols, a sync preamble which is utilized at the receiver for fine timing synchronization and a set of training sequences which are utilized for frame acquisition as well as fine frequency synchronization."

Receiver Structure

"This lets us to use the same training sequence (used in frame detection) for the purpose of fine CFO estimation."

"Since the coarse STO estimation is based on the correlation of the received signal with its delayed version, it is robust to CFO (residual CFO from coarse CFO estimation)."

"Once we have found the approximate boundaries of the training sequence, fine CFO estimation can be performed by exploiting the training sequences in the frame."

"The residual frequency offset after the fine CFO estimation must be small enough to perform fine STO estimation."

"Lower residual frequency offset can be achieved by using larger number of samples which requires longer training sequences."

"The number of samples per symbol is represented by N. We use a differential method to reduce the sensitivity of the correlation output to residual frequency offset as well as constant phase offset."

Experimental Results

"We evaluate our proposed frequency and time synchronization methods using software simulations."

"The digital output of the FPGA is converted to analog signal and then mixed to the intermediate frequency (IF)."

"The IF signal is then mixed to a low-IF frequency using the local oscillator."

"Further processing is done in the FPGA to estimate and compensate the CFO and demodulate the signal."

"In order to analyze the system performance, we first verify the CFO estimation capability of the system by Matlab simulation."

"The resulting signal is used as a variable local oscillator and is mixed with the transmitted signal to simulate the variable CFO effect."

"The peak value of the correlation value computed in each method is plotted over residual frequency offset remaining after fine CFO compensation."

"The residual CFO at the receiver slightly increases the EVM values compared to the case where signal is only deteriorated by noise."

Conclusion

"We proposed a narrowband robust transceiver system in presence of CFO."

"We implemented the proposed system on FPGA hardware."

"The simulation and implementation results verified the CFO estimation capability of the system."

A Carrier Selection Method Based on Single RF Chain SM-OFDM Systems [105]

This is a machine-generated summary of:

Fan, Zhuyun; Jin, Jiyu; Jin, Guiyue; Mou, Jun: A Carrier Selection Method Based on Single RF Chain SM-OFDM Systems [105]

Published in: Mobile Networks and Applications (2019)

Link to original: https://doi.org/10.1007/s11036-019-01296-6

If you want to cite the papers, please refer to the original.

For technical reasons we could not place the page where the original quote is coming from.

Abstract-Summary

"Orthogonal Frequency Division Multiplexing (OFDM) applied in SM can improve the system performance in frequency selective fading."

"Multiple radio frequency (RF) chains used in SM-OFDM system are required equal to the number of transmitting antennas, which greatly increases the energy loss of the system."

"A carrier selection method based on single RF chain SM-OFDM systems is proposed, a carrier grouping method for carrier selection and precoding are considered to reduce system complexity."

"Compared with the conventional SM-OFDM system, only single RF chain is used which avoids the inter-antennas interference and improves the energy efficiency."

Introduction

"The combination of SM and OFDM technology (SM-OFDM) effectively improves the practicability of the system in frequency selective fading channels."

"In Rayleigh and Rician fading Channel environments, the performance of SM-OFDM in frequency selective fading channels is analyzed."

"A novel symbol-by-symbol-aided iterative detector is proposed to mitigate the effect of the carrier frequency offset in SM-OFDM systems [106]."

"The iterative channel estimation problem is explored for the SM-OFDM systems to improve system performance [107]."

"The number of radio frequency (RF) chains is required equal to the number of transmit-antennas in SM-OFDM system, which violates the original intention of SM to improve energy efficiency."

"In order to improve the energy efficiency, a carrier selection method based on single RF chain SM-OFDM system is proposed in this paper."

"Compared with the conventional SM-OFDM system, the proposed SM-OFDM system can significantly improve the energy efficiency due to the single RF chain."

System Model

"The basic idea of SM-OFDM as follows, the input bit stream is mapped to an information symbol and the corresponding antenna number by SM, the bits used for antenna selection as implicit information are transmitted with information symbol simultaneously."

"Q (k) is a binary matrix of size $n \times N_c$, where n represents the number of bits/symbol/carrier and N_c is the number of OFDM carriers."

"The first two bits of each column in Q (k) are used to select antenna, and the third bit is modulated by BPSK."

"At the receiver, the reception matrix at t slot can be expressed as: where S (t) is the transmitting signal by OFDM modulator, N (t) is the additive white Gaussian noise (AWGN) which follows CN (0, 1), * is the time-domain convolution."

"Not only the signal but also the active antenna number can be obtained from Y (k)."

The Proposed Method

"There are $C^{4/2}{}_4 = 6$ different combination cases and $\lfloor \log_2 6 \rfloor = 2$ bits are used to select carrier (i.e. the first four cases are took), where $\lfloor \cdot \rfloor$ represents the maximum integer less than it."

"Although the carrier selection method mentioned above improves the energy efficiency of the system, it still has some shortcomings."

"The method of carrier selection is relatively simple, and algorithm complexity will rise with the increase of carriers."

"The carriers are divided into G groups, and half of the carriers are selected, the carrier combination in each group is $C^{Nc/2G}{}_{Nc/G}$."

"That four carriers are divided into two groups, there are $2 \times C^{4/4}{}_{4/2} = 4$ different combination cases less than carrier selection method, it decreases the error of decision which combination used."

"If the input is complex, the number of additions required at the transmitter is In carrier selection system mentioned above, only one IFFT is needed."

Simulation

"Consider a Long Term Evolution (LTE) channel model, the analysis of the simulations is carried out in this section."

"LTE Extended Vehicular A (EVA) model is used in a bandwidth B of 20 MHz, BPSK, QPSK are employed."

"This is because the error rate of the conventional SM-OFDM system depends on signal detection and transmit-antenna number decision, while the single RF chain system depends on signal detection and carrier number decision, which only needs to detect the carrier energy."

"It can be noticed that the more groupings, the better performance of the system."

"Because of randomness, the number of carriers per antenna is not uniformly distributed in the conventional SM-OFDM system, but the PAPR is still higher than single RF chain system with half carriers."

Conclusion

"This paper presents a single RF chain SM-OFDM system to improve the energy efficiency."

"Carrier grouping is utilized for carrier selection, which can mitigate ICI and signal decision deviation of the system."

"Compare to conventional SM-OFDM system, the proposed single RF chain system simplifies the design of transmitter and improves the system performance."

A New Method to Enhance SC-FDMA Communication with Channel Estimation Errors [108]

This is a machine-generated summary of:

Abd El-Rahman, Amal F.; Abd El-Samie, Fathi E.; Khamis, Salah Eldeen A.; Napoleon, Sameh A.: A New Method to Enhance SC-FDMA Communication with Channel Estimation Errors [108]

Published in: Wireless Personal Communications (2019)

Link to original: https://doi.org/10.1007/s11277-018-5818-y

If you want to cite the papers, please refer to the original.

For technical reasons we could not place the page where the original quote is coming from.

Abstract-Summary

"This paper presents a comprehensive study of the SC-FDMA system with PAPR reduction using clipping and filtering in the presence of channel estimation errors for two different versions of SC-FDMA adopting the FFT and the DCT."

"The effect of peak power reduction and the channel estimation error on the system bit error rate (BER) is investigated."

Introduction

"This system avoids ISI and ICI, but its disadvantage is that it has a large PAPR, and we can think in reducing this PAPR using SC-FDMA."

"The single carrier nature of SC-FDMA produces low PAPR [109], increases the data rate, and reduces the BER leading to an increase in signal-to-noise ratio (SNR)."

"It has been proved in [110, 111] that the large PAPR can push the power amplifiers (PA) at the transmitter into saturation."

"One of the best solutions for the PAPR problem is to utilize a transmit PA with a large enough linearity range and a suitable back-off operating point."

"We extend the application of the clipping and filtering technique to the SC-FDMA system for PAPR reduction."

"We investigate the effect of channel equalization in the presence of channel estimation errors and PAPR reduction to develop an integrated framework for the operation of the SC-FDMA system in a real scenario."

SC-FDMA System

"For high-data-rate wireless communications, multi-user (MU) transmission can be achieved through orthogonal frequency division multiple access (OFDMA) and/or SC-FDMA."

"The SC-FDMA has a discrete Fourier transform (DFT) module and an inverse discreet Fourier transform (IDFT) module in its transmitter and receiving ends."

"Signals are modulated and encoded, thereafter converted into multipath parallel signals through a serial-to-parallel transformation."

"Multipath signals become a serial signal after parallel-to-serial transformation and the M-point IDFT, then the cyclic prefix (CP) is added to the signal."

Fast Fourier Transform Based SC-FDMA

"After M-point inverse fast Fourier transform (IFFT), the signal can be expressed as follows: At the transmitter, the data is modulated through the modulator and then FFT is applied."

"The sub-carrier mapping has two methods localized and distributed mapping methods."

"Each user's data is transmitted with consecutive sub-carriers, while in the distributed mapping method transmission; the user's data is transmitted with distributed sub-carriers [112]."

"At the end of transmission, the data is converted from parallel to serial and converted from digital to analog to be suitable for RF transmission."

"We apply on each symbol the sub-carrier de-mapping."

Discrete Cosine Transform Based SC-FDMA (DCT-SC-FDMA)
"The discrete cosine transform (DCT) is a real-valued transform as it uses real arithmetic operations [112]."

"The DCT of a sampled signal x (α), can be written as: β (k) can be expressed as follows [113, 114]: The discrete time signal can be reconstruction via the inverse discrete cosine transform (IDCT) using the following relation [115, 116]:The DCT can be used instead of the FFT to facilitate the operation based on real arithmetics."

"At the transmitter side, the modulated symbols are combined into blocks, each including N symbols, and an N-point DCT is performed."

"At the receiver, frequency-domain equalization is applied."

Channel Estimation and Equalization
"We use equalization at the receiver to eliminate the effect of the channel impulse response."

"The aim of the equalizer is to reduce ISI to permit restoration of the transmit signals [117]."

"Zero forcing is a step-wise equalizer algorithm used in communication systems which applies the inverse of the channel frequency response to the received signal to restore the signal after the channel [118]."

"The MMSE equalizer does not amplify the received signal just multiplys by the inverse of the channel."

"Oversampling by taking a longer time IFFT can reduce the in-band distortion effect as a portion of the noise is reshaped outside the signal band, and it can be rejected later by filtering [119]."

"The BER is used as the evaluation metric after PAPR reduction and channel estimation and equalization."

Conclusion
"This paper presented a unified framework for SC-FDMA based on PAPR reduction and channel estimation and equalization."

"We have investigated the power reduction in SC-FDMA system and its effect on the system BER performance."

"We have investigated two different versions of OFDM with channel estimation and equalization."

"Simulation results have revealed that through clipping and filtering, we can maintain a good BER performance in the SC-FDMA scenario by applying a reliable channel estimation and equalization approach."

Sparse Massive MIMO-OFDM Channel Estimation Based on Compressed Sensing over Frequency Offset Environment [120]

This is a machine-generated summary of:

Li, Guang; Li, Tianyun; Xu, Mankun; Zha, Xiong; Xie, Yunfei: Sparse massive MIMO-OFDM channel estimation based on compressed sensing over frequency offset environment [120]

Published in: EURASIP Journal on Advances in Signal Processing (2019)
Link to original: https://doi.org/10.1186/s13634-019-0627-3

If you want to cite the papers, please refer to the original.

For technical reasons we could not place the page where the original quote is coming from.

Abstract-Summary

"In massive MIMO-OFDM systems, channel estimation is a significant module which can be utilized to eliminate multipath interference."

"In realistic communication systems, carrier frequency offset (CFO), which often exists in receive end, will deteriorate the performance of channel estimation."

"Of this paper, to reduce system complexity and correctly compensate CFO, we propose an improved OFDM frame structure."

"We theoretically analyze the catastrophic influence of CFO on conventional PN-sequence-based compressed sensing (CS) channel estimation scheme."

"As our solution, based on the improved OFDM frame structure, a novel massive MIMO-OFDM channel estimation method under CFO environment is proposed."

"Using the PCS information as a priori information, we improve the CS reconstruction scheme to estimate the accurate channel."

"The major advantage of our scheme is its anti-CFO ability and independence to channel sparsity level."

Introduction

"Precise channel estimation is a significant module to maintain the performance of MIMO-OFDM [121, 122]."

"Compared with orthogonal frequency division multiplexing (OFDM), accurate channel estimation is more challenging in massive MIMO-OFDM systems."

"In actual MIMO-OFDM communication systems, accurate channel estimation is obtained by the aid of training data."

"In order to take advantage of the channel sparsity, we introduced the compressed sensing (CS) theory to estimate channel in MIMO-OFDM systems [123]."

"In [124], authors have proposed a CS-based channel estimation method for MIMO-OFDM by simultaneously exploiting the time-domain pseudo-noise (PN) sequence and frequency-domain pilots."

"We discuss the channel estimation of massive MIMO-OFDM systems over frequency offset environment."

"We eliminate the influence of CFO on channel estimation via exploiting differential correlation algorithm to obtain the PCS information under CFO environment."

System Model

"We consider a $N_t \times N_r$ massive MIMO-OFDM system which consists of N_t transmit antennas and N_r receive antennas."

"Due to the multipath property of wireless channel, the time-domain CIR from the n_tth transmit antenna to the n_rth receive antenna is given as where $\tau_{L-1} > \cdots > \tau_0$ represents the paths' delay, α_i is the tap coefficient of the ith path, g (·) is the shaping pulse, and T_s represents the sampling period [125]."

"Since the pilot-based CS channel estimation is discussed in this work, we suppose that N_p pilots are inserted at each OFDM symbol."

The Proposed OFDM Frame Structure

"Since the paths' delay stays unchanged in an OFDM frame duration, the PN sequence can be employed to precisely obtain the PCS information for each OFDM symbol; hence, the superior performance of CS-based channel estimation can be guaranteed."

"For the purpose to estimate the channel between different transmit-receive antenna pairs, PN sequence inserted in different antennas should not be the same."

"Under the multipath channel, the cross-correlation of different PN sequences in the receive end will affect the detection of the correlation peak."

"In the proposed OFDM frame structure, the PN sequence location at different antennas is orthogonal."

PCS Information Acquisition and the CFO Effect on It

"The PCS information acquisition is based on the autocorrelation property of the PN sequence."

"When there is the known PN sequence in the received signal r (n), the cross-correlation between r (n) and c (n) can be expressed as where s is the shift in the sampling period [126]."

"Due to the excellent anti-noise performance of the PN sequence, the peak of R_{rc} (s) will appear at the starting position of the PN sequence (s = 0)."

"That the CFO is f_c subcarrier spacing, the received PN sequence including CFO effect can be expressed as And the CFO-effected R_{rc} (s) is given by where $\Delta\varepsilon_1$ and $\Delta\varepsilon_2$ are the interference terms related to ε_1 and ε_2, respectively, and sinc(·) is the abbreviation of Singer function."

CFO Compensation

"The traditional PN-based CFO compensation scheme is coherent autocorrelation algorithm."

"After the frame synchronization is completed, the conjugated PN sequence generated by the receive c^* (n) is multiplied by the received PN sequence r (n) to get z (n) Then, compute the autocorrelation function of z (n) and the estimated CFO is given by [127] Literatures [128] shows that the CFO variance of coherent autocorrelation algorithm under AWGN channel is With the multipath channel interference, its variance could deteriorate even further."

"D is the differential delay, circshift (·) denotes cyclic shift, and K is an integer related to D. That is to say, the differential operation of the PN sequence is still a PN sequence which has great autocorrelation property [129]."

"We here choose the highest correlation peak to estimate the CFO [130]."

Improved CS Reconstruction Algorithm

"Greedy iteration CS reconstruction schemes are proved to be feasible in OFDM system channel estimation."

"For most conventional greedy iteration CS reconstruction schemes, the channel sparsity level has to be transmitted to receive end as a priori information [131], which is unachievable in normal actual systems."

"While the sparsity adaptive greedy iteration CS reconstruction schemes suffer from low accuracy and high computational complexity [132]."

"Unlike conventional greedy iteration CS reconstruction schemes such as OMP and CoSaMP, where the number of iterations is the value of channel sparsity level [133], the proposed scheme stops iteration only when the residual is close to 0 ($\|r_t\|_2 < \varepsilon$)."

Results and Discussion

"We simulate the proposed CS-based channel estimation scheme under CFO environment."

"When SNR ≥ 15 dB, for SARFT-8 channel, the MSE of 0.05, 0.1, and 0.2 normalized CFO are about 2×10^{-4}, 8×10^{-5}, and 2×10^{-5}, respectively."

"For ITU-VB channel, the MSE of 0.05, 0.1, and 0.2 normalized CFO are about 4×10^{-4}, 3×10^{-4}, and 8×10^{-5}, respectively."

"Even when SNR = 0 dB, for both SARFT-8 and ITU-VB channel, the MSE of 0.05, 0.1, and 0.2 normalized CFO are still lower than 2×10^{-3}, 3×10^{-4}, and 10^{-4}, respectively."

"It can be seen that the MSE performance under SARFT-8 channel is better than that of ITU-VB channel."

"From the above results, the MSE and BER performances of the proposed scheme are better than other schemes."

Conclusions

"This paper discusses sparse massive MIMO-OFDM channel estimation based on compressed sensing over frequency offset environment."

"Based on the proposed OFDM frame structure, the CFO is estimated by utilizing the proposed differential correlation algorithm, so its influence on channel estimation can be eliminated."

"Based on the obtained PCS information, the CS reconstruction scheme is exploited to estimate the channel."

"It does not need to know the channel sparsity level as a priori information; therefore, it is more practicable than other CS-based channel estimation schemes."

Comparative Study on Implementation Performance Analysis of Simulink Models of Cognitive Radio Based GFDM and UFMC Techniques for 5G Wireless Communication [134]

This is a machine-generated summary of:

Gupta, Manisha; Kang, Ankur Singh; Sharma, Vishal: Comparative Study on Implementation Performance Analysis of Simulink Models of Cognitive Radio Based GFDM and UFMC Techniques for 5G Wireless Communication [134]

Published in: Wireless Personal Communications (2020)

Link to original: https://doi.org/10.1007/s11277-020-07561-2

If you want to cite the papers, please refer to the original.

For technical reasons we could not place the page where the original quote is coming from.

Abstract-Summary

"We studied different reduction methods for peak to average power ratio (PAPR) and bit error rate (BER) in generalized frequency division multiplexing (GFDM) and universal filtered multicarrier (UFMC) as well as the relation between UFMC and GFDM on the basis of PAPR and BER for different windowing techniques in cognitive radio for future wireless communication systems."

"With the help of MATLAB as well as Simulink model, we have discussed all the conclusions related to GFDM and UFMC parameters in a Comparative manner."

"The comparative analysis done between UFMC and GFDM schemes in terms of robustness against multipath channels shows that the simulation results obtained have a deep impact on study of performance analysis of multicarrier modulation techniques implemented through Simulink models in a ubiquitous and pervasive scenario."

Introduction to 5G

"The 5G technology provides high resolution to users and large bandwidth."

"The uploading and downloading speed of 5G technologies is very fast because of larger bandwidth."

"The main aim of 5G technology is high capacity and 10 Gbps network speed."

"5G based equipments are cheaper with lower power consumption and lower latency than 4G. 5G is working in the growth of many industries in entertainment, agriculture, IT and manufacturing industries.5G is mainly based on UFMC and GFDM multicarrier modulation techniques which reduce the level of peak to average power as well as bit error rate for input data as compared to the other wireless technology generations [135]."

Literature Review

"The author also described the routing of signal, protocol layering architecture, paging message system and radio channel structure etc. Salkintis and others 136 proposed the combination of wireless local area network and cellular networks which is capable for providing omnipresent data services and very high data rates."

"Choi and Xu [137] proposed that 3G is based on CDMA 2000 and because of it, third generation wireless technology provides mobile internet and multimedia services."

"Krenik [138] proposed that fourth generation provides downlink data rates, low latency, effective use of spectrum and low-cost design."

"Datta and others [139] proposed that in cognitive radio if obligatory user doesn't need for White Space Device's (WSD) operation for opportunistic users use in the same spectrum then Generalized Frequency Division Multiplexing multicarrier modulation technique makes it possible because GFDM provides extremely low out-of-band radiation which makes it a good choice for the PHY layer of cognitive radio."

A Brief Overview: PAPR, BER and Window Techniques

"PAPR is the ratio of peak power to the average power of a signal expressed in dB. The RF communication system contains a high power amplifier at the transmitter side to produce adequate power of transmission."

"PAPR reduction is necessary to make the receiver error prone because it makes the system poorly efficient to power."

"To make this measurement easier, it is possible to define BER in terms of a probability of error [140]. In a communication system, BER gets affected by the transmission channel noise, interference, distortion, bit synchronization problems, attenuation, wireless multipath fading etc. The BER can be reduced by using a strong signal, a slow and powerful modulation scheme and by applying channel coding schemes."

"It is essential to stabilize all the available factors to get an acceptable bit error rate as it is used to determine the proper communication path between transmitter and receiver [140]."

Role of GFDM and UFMC Techniques in Wireless Communication

"Rectangular pulse shaping that is used in OFDM produces extensive spectral leakage to the adjacent frequency bands but by using pulse shaping filters in GFDM, inter-carrier interference is reduced and performance of the system gets improved."

"D_{MF} can be represented as, where A represents matrix (b) Zero Forcing Receivers This is simply multiplication of inverse of the matrix A and received signal which is expressed as [141], UFMC is one of the 5G waveform candidates which is also based on filtering method like filtered bank multicarrier."

"Different filters like Dolph-Chebyshev filter, hamming, hanning etc. can used in UFMC instead of cyclic pulse to decrease the out of band radiation to improve the suitability of spectrum and as the length of filter is lesser than FBMC, hence it is a good choice for short burst communication."

PAPR Reduction Techniques for UFMC and GFDM

"There are some techniques like selective mapping, reduced clipping filtering, tone reservation, Iteration method etc. which are used to reduce the PAPR as well as BER values in GFDM and UFMC."

"Clipping reduced the PAPR value but at the cost of out of band radiation and higher BER."

"By enlarging the small signal, there is no effect on peak value, hence, average power is increased and peak to average ratio is reduced."

"Random filter mapping method is used to reduce peak to average power ratio mainly by decreasing the likelihood of the same phase multicarrier signal."

"A linear precoder which is used to decrease the variance of instantaneous power is decreased so that value of PAPR for GFDM signals can be reduced but it maintains the average value of IP constant."

BER Reduction Techniques for GFDM

"Clipped method is mainly used to reduce the peak to average power ratio but with the use of iterative reception, BER of the GFDM signal is also reduced by affecting PAPR of the system."

"It also reduces the self-interference effect and gives better results for BER performance in AWGN and frequency-selective channels."

"This is a frequency domain based Gabor transform method in which Discrete Gabor Transform (DGT), based on frequency domain reduces the process of recovery of the GFDM signal."

"By analyzing the interference caused by the frequency-domain DGT, the channel with high coherence and a small roll-off factor of the synthesis window can lead to small interference to the received signal."

Simulink Models for UFMC and GFDM

"The recent work has its focus on the performance of universal filtered multicarrier based cognitive radio and generalised frequency division multiplexing based cognitive radio under different filtering methods to reduce the peak to average power ratio and bit error rate."

"Cognitive radio can work on many frequency bands at a time for many users."

"It reduces the inter carrier interference by using filtering on sub band basis instead of cyclic pulse because of non-orthogonal sub-carriers even that it also suffers from PAPR and BER problem."

"GFDM is based on block transmission in which each block is divided into subsymbols and subcarriers in frequency and time domain, so as to allocate the resource partitions to the input signals."

"To protect from false users, there is a need to sense all the signals carefully, to avoid the transmission of other cognitive radio signal when a GFDM signal is already there in the frequency band under consideration."

Methodology of GFDM Simulink Model

"Up sampling is used to increase the number of points in a digital signal to match the sampling rate with some other signal and down sampling is usually used to conserve memory and signal processing time in case of oversampling."

"Signals are up sampled when signals continue to move through synthesis filter bank while down sampled as they pass through a series of analysis filter bank."

"Digital up converter is used to convert digital baseband samples into digital IF samples."

"Digital to analog converter is used after this block in order to convert digital IF samples into analog IF signal."

"The Digital Up Converter consists of following modules namely Interpolation filter, Digital Mixer, Digital Local Oscillator."

"While Digital down converter is used to convert digital IF samples into digital baseband samples."

"The Digital Down Converter consists of following modules, namely Digital Mixer, Digital Local Oscillator, Low Pass Filter."

BER Results for UFMC Simulink Model

"From figure, it is clear that raised cosine filter gives better result with decreasing trend in BER."

"According to it, PAPR value for GFDM is very large in comparison to PAPR value for UFMC and among these values hanning window (yellow) is showing the best results because value of PAPR is very small for hamming in UFMC."

"UFMC with hanning window can be used to reduce the value of peak to average power ratio in next generation wireless communication under cognitive radio because higher PAPR is the main drawback for orthogonal frequency division multiplexing, that's why OFDM is not able to work better with future wireless technology, hence, we recommend the use of UFMC technique instead of GFDM."

"Fluctuations shows the level of noise in signal and UFMC gives better results for BER reduction."

Conclusion

"Value of PAPR and BER changes with the change in windowing techniques."

"Both UFMC and GFDM have a much lower adjacent channel leakage ratio, even when it operates without time or frequency synchronization to the primary system."

"According to our conclusion, UFMC with Hanning Window improve the spectrum sensing ability of cognitive radio by reducing PAPR and BER values so that it can become favourable choice for 5G communication."

"Our study also proposed that Hamming windowing is more effective in improving the PAPR and BER of the system than Dolph- chebyshev, Kaiser and Hamming Window."

Impact of Study and Future Scope of Work

"In the earlier work done, both GFDM and UFMC have been implemented and parameterized to fit the sampling and framing of the LTE standard."

"LTE algorithm and physical/MAC layer development, aids reference verification and testing which will enable the generation of test waveforms."

"One can make and reuse the test bench for verifying the design prototypes and all other implementations complying with LTE standards."

"The work can be extended to standard compliant propagation channel models."

Computational Design, Analysis and Characterization of Beetle Shaped High Isolation Multiple-Input-Multiple-Output Reconfigurable Monopole-Antenna with Dual Band Filters for Wireless Applications [142]

This is a machine-generated summary of:

Sharma, Manish; Singh, Sukhbir; Varma, Ranjit: Computational Design, Analysis and Characterization of Beetle Shaped High Isolation Multiple-Input-Multiple-Output Reconfigurable Monopole-Antenna with Dual Band Filters for Wireless Applications [142]

Published in: Wireless Personal Communications (2021)

Link to original: https://doi.org/10.1007/s11277-021-08248-y

If you want to cite the papers, please refer to the original.

For technical reasons we could not place the page where the original quote is coming from.

Abstract-Summary

"Article, a beetle shape Multiple-Input-Multiple-Output (MIMO) antenna including dual notched band characteristics are presented for different wireless band applications (UWB, X-Band and Ku Band) which are also experimentally and computationally investigated."

"The proposed antenna offers measured isolation between two ports less than −20 dB in entire operating bandwidth (2.57–19.15 GHz)."

"The wider impedance bandwidth of the antenna suggest that it is not only capable of working in UWB range but also finds applications in X- and Ku-Band applications allocated for Satellite and RADAR."

"An antenna is also reconfigured using PIN diodes to control each notched bands independently."

"Proposed MIMO antenna also offers better diversity performance in terms of ECC < 0.004, DG > 9.95 dB, TARC < -40 dB and CCL < 0.4 b/s/Hz in entire operating bandwidth."

"The antenna also offers a gain of 3.11–4.92 dBi and desired radiation pattern making it a suitable candidate for different wireless applications."

Introduction

"A transparent fork-shaped radiating patch is capable of covering bandwidth 3–13.6 GHz and addition of two L-shaped strips in patch and one in-ground with the implementation of PIN diodes reconfigures antennas in four different working bands [143]."

"Compact MIMO antenna with dimension 22 × 26 × 0.8 mm^3 UWB antenna is dual band-notched which is obtained by loading trident shape strips on feed line [144]."

"A reconfigurable antenna blocking interfering band (5.0–6.0 GHz) is reported which consists of a triangular shape radiating patch and a triangular slot where PIN diode is embedded."

"A reconfigurable antenna consisting C-shaped patch, two symmetrical notches with a parasitic patch works as UWB antenna when the diode is in OFF state and ON state works in two dual bands separated by small band-frequency [145]."

Design Methodology of Proposed Antenna

"Iteration 1 shows hexagonal-type radiating patch with partial ground plane providing resonant bandwidth of 4.25–15.96 GHz."

"Two orthogonal radiating elements provide higher isolation between them with the rectangular ground beneath them."

"As per the Shannon-Hartley, Channel Capacity of MIMO system is given by where C is channel capacity (b/s/Hz), M is number of radiating elements (M = 1) in case of Single–Input–Single–Output (SISO), B is Bandwidth of channel (GHz), S is averaged received signal power in watts, N is average power of the noise in watts."

"When number of radiating elements increases more than 1 (M > 1), Multiple-Input-Multiple-Output configuration is formed which not only ensures increase in channel capacity, but also reduces multiple path fading affects which is an advantage on SISO system."

Parametric of Key parameters, Surface Current Density Distribution and Reconfigurable Characteristics

"Rotated L-shaped stub and C-type electromagnetic coupling slit near feedline provide bandstop filters for WiMAX and DSS notched bands."

"MIMO antenna covering super wideband bandwidth applications finds WiMAX and X-Band Downlink System interfering band which is removed by using an L-shaped slot and C-shaped parasitic element near the feed line of a corresponding quarter wavelength of the interfering band."

"Length of the stub or slot is calculated by the following formula [146] where f_r is the center frequency of corresponding notched band and ε_{reff} is calculated by."

"When the physical length of the slot/stub is varied, the bandwidth of the notched band along with the center frequency also shifts towards lower or higher frequency side."

"For the second band-notched interfering band, physical length L_2 of the C-shaped parasitic element is varied from L_2 = 1.00 mm to L_2 = 3.00 mm and corresponding bandwidth is shifted from 7.33–8.11 to 6.80–7.25 GHz."

Equivalent Circuit Model Analysis

"To further study the mechanism of the proposed designed antenna (without and with filters), the equivalent RF circuit theory model is applied based on input impedance obtained, and hence evaluated result is compared with simulated and measured results [147, 148]."

"For WiMAX notched band, a high impedance of 109 Ω is noted at 3.49 GHz with imaginary impedance varying from positive to negative."

"In the case of the DSS notched band, real impedance value is 7.48 Ω at 7.45 GHZ, and imaginary impedance swings from negative to the positive value of impedance resulting in a series RLC circuit."

MIMO Performance of Proposed Antenna

"Diversity performance of the proposed MIMO antenna is evaluated by calculating the Envelope Correlation Coefficient (ECC), Directive Gain (DG), Total Active Reflection Coefficient (TARC), and Channel Capacity Loss (CCL) [149]."

"Directive Gain (dB) of the proposed MIMO antenna configuration is calculated by."

"Values of DG > 9.95 dB in the operating band of interest and for proposed MIMO antenna, simulated and measured values are greater than 9.95 dB. It can be judged that when 2 × 2 MIMO antenna configuration is working, adjacent radiating elements radiations may affect each other and thereby reduces the overall bandwidth and efficiency."

"MIMO efficiency of the proposed antenna is thus calculated by TARC which provides more meaningful in terms of mutual coupling."

"TARC < 0 dB and for proposed MIMO antenna configuration, simulated and measured values are <−30 dB in the entire operating band."

Results and Discussion

"It can be concluded that gain and radiation efficiency falls at notched bands but maintains permissible variation of 3.08–4.88 dBi for Gain and 74–86% for radiation efficiency."

"For the working frequencies at 4.50 GHz, 8.0 GHz and 12.0 GHz antenna offers omnidirectional pattern in H-plane and dipole like pattern in E-plane."

"At a much higher frequency at 17.0 GHz, radiation pattern in both planes deteriorates due to splitting of radiation lobes."

Comparison with Other State-of-the-Art Designs

"Proposed antenna is compared with size, bandwidth, reconfigurability, gain, radiation efficiency and diversity performance."

"Comparison with other published work suggests that antenna shows very good diversity performance when compared with other."

"Proposed antenna offers maximum radiation efficiency of 88%."

Conclusions

"We have reported enhanced bandwidth, compact, high isolation, and dual notched filter super wideband MIMO reconfigurable antenna."

"The proposed MIMO antenna covers a bandwidth of 16.58 GHz (2.57–19.15 GHz) and high isolation by placing radiating elements orthogonally which achieves better isolation more than −25 dB. Interfering bands are mitigated by etching rotated L-shaped slots on the patch and placing pair of C-type ECT slits near the feedline."

"The diversity performance of the MIMO antenna is characterized by calculating ECC, DG, TARC, and CCL."

"The average gain of the antenna varies between 3.11 and 4.92 dBi with maximum radiation efficiency of 88% and stable radiation pattern suggest that the proposed antenna is suitable for UWB/X/Ku band applications."

Multilevel Redundant Discrete Wavelet Transform (ML-RDWT) and Optimal Red Deer Algorithm (ORDA) Centred Approach to Mitigate the Effect of ICI, BER and CIR in a MIMO-OFDM System [150]

This is a machine-generated summary of:

Nagarajan, K.; Sophia, S.: Multilevel Redundant Discrete Wavelet Transform (ML-RDWT) and Optimal Red Deer Algorithm (ORDA) Centred Approach to Mitigate the Effect of ICI, BER and CIR in a MIMO-OFDM System [150]

Published in: Wireless Personal Communications (2021)

Link to original: https://doi.org/10.1007/s11277-021-09066-y

If you want to cite the papers, please refer to the original.

For technical reasons we could not place the page where the original quote is coming from.

Abstract-Summary

"In this concern, the multiple-input multiple-output orthogonal frequency division multiplexing (MIMO-OFDM) scheme is used on a large scale to achieve UHDR transmission with reduced inter-symbol interference (ISI) and inter-carrier interference (ICI)."

"Discrete wavelet transform-based OFDM (DWT-OFDM) provides better orthogonality due to presence of orthogonal wavelets, which mitigates the effects caused by ISI and ICI."

"For solving this problem, a multilevel redundant discrete wavelet transform (ML-RDWT) is used instead of DWT to achieve improved spectral performance."

"From the simulation outcomes, it is clearly viewed that effects caused by ICI, ISI and BER are mitigated by improving the performance of CIR."

"This algorithm enhances the spectral efficiency by achieving high CIR with reduced BER, ISI and ICI."

"The BER in the proposed MIMO-ML-RDWT-OFDM-ORDA method is 68%, 76%, 38% and 75%, which is very low when compared to the BER in the existing techniques like MIMO-DWT-OFDM-RDA, MIMO-RNS-OFDM-PNMA, MIMO-OFDM-BMA and MIMO-OFDM-ICIMA."

"The ISI in the proposed method is 94%, 91%, 95% low when compared to the ISI in the existing techniques."

"The ICI in the proposed work is 71%, 57%, 73% and 86% low when compared to the ICI in the existing techniques."

"The general performance of the proposed MIMO-ML-RDWT-OFDM-ORDA method is improved in an efficient way with less complexity, error rate and processing delay."

Introduction

"A multilevel discrete wavelet transform is implemented, which is an improved version of RDWT [151] for enhancing the spectral performance of MIMO-OFDM system in an efficient way by mitigating the effects caused with inter-symbol interference (ISI), Inter-carrier Interference (ICI) and Bit Error Rate (BER) with increased carrier-to-interference power ratio (CIR)."

"The growing demand for fifth-generation wireless networks with diminished ISI and ICI for efficient data transmission is possible with the use of MIMO-OFDM."

"The overall performance of proposed MIMO-ML-RDWT-OFDM system may be improved in an efficient way with high spectral efficiency."

"In third section, a detailed description about the proposed Multilevel Redundant Discrete Wavelet Transform (ML-RDWT) in a MIMO-OFDM framework is presented with optimal red deer (ORDA) algorithm to optimize the weight parameters for achieving enhanced spectral efficiency by improving CIR performance in a MIMO-OFDM System."

Related Work: A Brief Review

"The simulation outcomes demonstrate their ICI mitigation system may accomplish an amount of equal service obtained in case without ICI while the speed is approximately 300 km/h. Hao et al. (2016) in [152] has introduced a low complexity ICI mitigation system for MIMO-OFDM systems in assumption of linear channels that vary over time."

"Paek et al. (2019) in [153] has introduced a performance improvement system with coordinated multi-point (CoMP) through spatial phase coding (SPC) depends on MIMO-OFDM on heterogeneous network system (HetNet)."

"Pham et al. (2016) in [154] have introduced a repetitive structure of channel estimation and data detection for MIMO-OFDM systems through an inappropriate cyclic prefix (CP) and restricted number of pilot subcarriers."

"Hussein et al. (2019) in [155] has introduced an innovative fully generalized spatial index (FGSI), light-emitting diode (LED) modulation system of MIMO-OFDM optical system."

Proposed Methodology

"For enhancing the spectral efficiency of MIMO-OFDM system, an innovative method of Multilevel Redundant Discrete Wavelet Transform (ML-RDWT) in a MIMO-OFDM framework is proposed with ICI cancellation by improving the CIR with reduced BER and Inter-symbol-interference (ISI)."

"The disadvantages in discrete wavelet transform (DWT) in a MIMO-OFDM framework can be overcome by the use of improved version of redundant discrete wavelet transform (RDWT) called Multi-level redundant discrete wavelet transform (ML-RDWT)."

"Some of the disadvantages faced by a MIMO-OFDM framework due to the use of DWT are: shift-variance deficiency, poor directional selectivity and reduced spectral efficiency."

"In the proposed MIMO-ML-RDWT-OFDM-ORDA framework, the cyclic prefix (CP) block is not added in the transmitter section."

"The weight parameters are efficiently optimized in the proposed approach by using an optimal red deer algorithm (ORDA), an improved version of red deer algorithm (RDA)."

Simulation Results

"The proposed method achieves high data rate of about 25%, 68%, 15% and 45% when compared to the above mentioned existing techniques."

"The average delay in proposed system is about 84%, 88%, 80% and 90%, which is very low when compared to the average delay in the existing methods."

"The proposed method possesses less free space path loss, when compared to the other existing approaches."

"The proposed method possesses high signal intensity while compared with other existing systems."

"The proposed method exhibits ISI of 94%, 91%, 95% and 86%, which is minimal while compared with existing techniques."

"The BER in the proposed method is 68%, 76%, 38% and 75% minimal while compared with BER in the existing techniques."

"The ICI in the proposed method is 71%, 57%, 73% and 86% less while compared with ICI in the existing techniques."

Conclusion

"The proposed scheme MIM0-ML-RDWT-OFDM-ORDA scheme is highly suitable for use during high-speed data transmission in mobile communication systems with improved spectral efficiency."

"From the simulation results, it is clearly identified that our proposed MIM0-ML-RDWT-OFDM-ORDA method possesses low BER of 68%, 76%, 38% and 75% when compared to the existing methods like MIMO-DWT-OFDM-RDA, MIMO-RNS-OFDM-PNMA, MIMO-OFDM-BMA and MIMO-OFDM-ICIMA."

"The proposed MIM0-ML-RDWT-OFDM-ORDA method possesses low ISI of 94%, 91%, 95% and 85% when compared to the existing methods described."

"The proposed MIM0-ML-RDWT-OFDM-ORDA method possesses low ICI of 71%, 57%, 74% and 86% when compared to the existing methods mentioned above."

Smart Antenna with Automatic Beam Switching for Mobile Communication [156]

This is a machine-generated summary of:

Ma, Yuchen; Wang, Junhong; Li, Yujian; Chen, Meie; Li, Zheng; Zhang, Zhan: Smart antenna with automatic beam switching for mobile communication [156]

Published in: EURASIP Journal on Wireless Communications and Networking (2020)

Link to original: https://doi.org/10.1186/s13638-020-01792-4

If you want to cite the papers, please refer to the original.

For technical reasons we could not place the page where the original quote is coming from.

Abstract-Summary

"A new kind of smart antenna capable of automatically switching its main beam to track a moving target is presented."

"The antenna, which is suitable for mobile communication in long straight spaces, such as railways or highways, integrates a sensing element, signal processing element, and radiating element."

"To verify the efficiency of the proposed antenna, a prototype is fabricated and implemented in a real corridor scenario."

"The results show that this kind of smart antenna is capable of improving mobile communication quality and decreasing energy consumption."

Introduction

"In [157], a smart antenna with beam switching capability that can realize beam switching between 0, 90, 180, and 270° by means of diodes is presented."

"In [158], a smart antenna with beam and polarization switching capabilities in the azimuth plane is proposed."

"In [159], a two-dimensional scanning smart antenna for cognitive radio is proposed."

"In [160], a neural network-based smart antenna is proposed for multiple source tracking."

"Although these antennas have potential applications in smart mobile communication systems, external control systems must be used, and the antennas themselves do not have the function of automatic sensing and adjusting their characteristics according to environmental information."

"The smart antenna proposed in this paper itself has a target distance sensing function, the capability to process sensed target data, and the ability to automatically make judgments and switch its beams."

Method and Experiment

"The radiating element transmits and receives the RF signal to and from the desired directions: two Yagi-Uda antennas and one patch antenna are integrated to construct the radiating element."

"Although the proposed antenna is capable of tracking multiple targets simultaneously by exciting all or several radiating elements, for simplicity, this paper focuses on the case where only one target is tracked, and only one antenna is excited at a time to illustrate the efficiency of the smart antenna."

"When the antenna is in operation, the sensors are active, and when sensor 1 receives a signal from a vehicle, for example, the left Yagi-Uda antenna (sub-element 1) radiates a wave to zone 1."

"When the vehicle moves to the presupposed position (closer to the antenna), the signal processing element generates a signal and sends it to the switch to automatically realize an excitation handover between the left Yagi-Uda antenna and the middle patch antenna, and the radiation area is changed from zone 1 to zone 2."

Results and Discussion

"The measured gains of sub-elements 1–3 at 5.61 GHz are 5.34 dBm, 2.24 dBm, and 4.94 dBm, respectively, and their normalized cross polarization levels are −34.18 dB, −40.91 dB and −29.54 dB. The radiation performance of the proposed antenna was measured in a real scenario of a long straight corridor in a large building."

"The received level along the corridor was measured by an Agilent FieldFox RF Analyzer N9912A. The proposed antenna was suspended at a height of 1.9 m at the midpoint of the corridor in the horizontal direction."

Conclusion

"This paper presents a new kind of smart antenna with the capability of automatic beam switching and low complexity."

"The antenna consists of sensing, signal processing, and radiating elements."

"In the radiating element, two Yagi-Uda antennas for endfire radiation and one patch antenna for broadside radiation are designed on a Rogers 6010 substrate with the same polarization and overlapping working frequency band."

"The results show that the communication quality can be improved by the smart antenna, which could be used for vehicles to facilitate future mobile communication."

Study of Indoor Radio Coverage Performance of Dual Technology Co-existing MIMO Antenna Platform for Low Power Wireless Base Station [161]

This is a machine-generated summary of:

Patro, Saroj K. P.; Mishra, Rabindra K.; Panda, A. K.: Study of Indoor Radio Coverage Performance of Dual Technology Co-Existing MIMO Antenna Platform for Low Power Wireless Base Station [161]

Published in: Radioelectronics and Communications Systems (2020)

Link to original: https://doi.org/10.3103/s0735272720020028

If you want to cite the papers, please refer to the original.

For technical reasons we could not place the page where the original quote is coming from.

Abstract-Summary

"We present the radiated mode indoor radio coverage performance study of a co-existing dual technology, dual 4 × 4 Multiple Input Multiple Output (MIMO) antenna platform."

"One of the 4 × 4 MIMO antenna pairs operates in the 2.5–2.7 GHz licensed Long Term Evolution (LTE) band while the other MIMO antenna pair operates in the 2.4 GHz unlicensed Wireless Fidelity (Wi-Fi) Industrial Scientific Medical (ISM) band."

"This radio performance study is carried out to analyze, predict and verify the impact of multi-radio co-existence on the radiated mode indoor coverage performance in an Over-The-Air (OTA) scenario."

"The study involves OTA radio propagation modeling, prediction and verification both for a near field and far field scenario."

Introduction

"In [162], details of the RF co-existence aspects are covered between Long Term Evolution-Advanced (LTE-A) and Wi-Fi radios in terms of the Receive Signal Strength Indication (RSSI) and other software aspects."

"In [163], the authors analyze the effect of co-existing radio system performance aspects such as receive signal strength, channel quality indicators, system throughput."

"These studies mainly cover the radio coverage aspects of either one of the standalone technologies or in a controlled environment."

"As evident from the above mentioned literature reviews, there is a lack of adequate studies related to the radio coverage aspects of indoor low power BS, arising from the impact of the co-existing multi technology Multiple Input Multiple Output (MIMO) antennas."

"Our co-existing radio system includes a 4 × 4 dual technology (i.e., LTE and Wi-Fi) high isolation MIMO antenna platform."

Target Radio Performance Metrics and MIMO Antenna Platform
"We capture the target radio performance metrics, their requirements under co-existing multi-antenna radio interference scenario, and antenna platform used in this investigation."

"We analyze the radiated RSSI level in the co-existing OTA indoor coverage scenario as a key radio performance metrics."

"Following frequency bands are used for the multi-antenna platform: (1) LTE radio frontend and antenna frequency range: 2.5–2.7 GHz and; (2) Wi-Fi radio frontend and antenna frequency range: 2.4–2.5 GHz."

"The RSSI level of interfering radio (i.e., aggressor) is analyzed in the receive band of the victim receive frequency band (i.e., 2.5–2.6 GHz LTE frequency) while the aggressor wireless transmitter (i.e., 2.4–2.5 GHz Wi-Fi) antennas are operating in close vicinity."

"The elevation plane patterns are of much importance in our study as the downward tilt behavior of these antennas are used effectively for the indoor radio coverage prediction."

OTA Propagation Study for Indoor Coverage Scenario
"The first step of our radio performance study involves theoretical prediction of RSSI in terms of OTA link budget analysis and using a relevant indoor propagation channel model."

"For the link budget of scenario-2, we notice that for the far field case, the RSSI level varies depending on the antenna gain in different elevation directions and having a reasonable spread in the entire indoor propagation environment."

"The second step of our radio performance study involves the modeling of the indoor enterprise scenario through a 3D propagation analysis tool to predict the RSSI."

"Other contemporary literature has partly covered the co-existence radio antenna aspects such as conducted, OTA and single wireless technology and in comparison, our study is captures the multi-radio technology based MIMO co-existence coverage aspects in a much detailed manner covering from theoretical analysis, 3D modeling and actual indoor propagation verification."

Conclusions
"We captured the near field and far field based radio co-existence coverage aspects of a dual-technology multi-antenna MIMO platform in an OTA scenario."

"While the popular literature have so far reported in parts, some or other methods of co-existence performance analysis and measurements, this study tries to take a very comprehensive approach by putting together the radio-antenna co-existence aspects in a co-located environment and also provides directions for robust theoretical, modeling and practical verification related details."

"Similar type of studies and investigations are not reported so far, for the modern low power wireless base station with dual technology, multi-antenna MIMO structures."

"One could also use the methods detailed here for the OTA radio performance aspects of any next generation low power modern wireless BS systems."

Bibliography

1. Arjun, K. R., & Surekha, T. P. (2023). Effects of synchronization errors on wavelet-based CR-OFDM systems in doubly selective fading channels. *Wireless Personal Communications, 129*, 2037–2054. https://doi.org/10.1007/s11277-023-10224-7
2. Chin, W. L., Kao, C. W., Chen, H. H., & Liao, T. L. (2013). Iterative synchronization-assisted detection of OFDM signals in cognitive radio systems. *IEEE Transactions on Vehicular Technology, 63*(4), 1633–1644.
3. Lin, T. T., & Hwang, F. H. (2022). Design of a blind estimation technique of carrier frequency offset for a universal-filtered multi-carrier system over Rayleigh fading. *IEEE Wireless Communications Letters, 11*(5), 1027–1031. https://doi.org/10.1109/LWC.2022.3153309
4. Abdzadeh-Ziabari, H., Zhu, W. P., & Swamy, M. (2019). Timing and frequency synchronization and doubly selective channel estimation for OFDMA uplink. *IEEE Transactions on Circuits and Systems II: Express Briefs, 67*(1), 62–66.
5. Baghaki, A., & Champagne, B. (2018). Joint frequency offset, time offset, and channel estimation for OFDM/OQAM systems. *EURASIP Journal on Advances in Signal Processing, 2018*(1), 1–19.
6. Kumar, S., Chaudhari, M. S., Gupta, R., & Majhi, S. (2020). Multiple CFOs estimation and implementation of SC-FDMA uplink system using oversampling and iterative method. *IEEE Transactions on Vehicular Technology, 69*(6), 6254–6263.
7. Ali, W. A. E., Ibrahim, A. A., & Ahmed, A. E. (2023). Dual-band millimeter wave 2 × 2 MIMO slot antenna with low mutual coupling for 5G networks. *Wireless Personal Communications, 129*, 2959–2976. https://doi.org/10.1007/s11277-023-10267-w
8. Marzouk, H. M., Ahmed, M. I., & Shaalan, A. H. A. (2019). Novel dual-band 28/38 GHz MIMO antennas for 5G mobile applications. *Progress In Electromagnetics Research, 93*, 103–117.
9. Ali, W., Das, S., Medkour, H., & Lakrit, S. (2020). Planar dual-band 27/39 GHz millimeter-wave MIMO antenna for 5G applications. *Microsystem Technologies, 27*, 283–292.
10. Saad, A. A. R., & Mohamed, H. A. (2019). Printed millimeter-wave MIMO-based slot antenna arrays for 5G networks. *AEU-International Journal of Electronics and Communications, 99*, 59–69.
11. Aliakbari, H., AbdolaliAbdipour, A. C., Masotti, D., Mirzavand, R., & Mousavi, P. (2017). ANN-based design of a versatile millimetre-wave slotted patch multi-antenna configuration for 5G scenarios. *IET Microwaves, Antennas & Propagation, 11*(9), 1288–1295.
12. Aghoutane, B., El Ghzaoui, M., Das, S., Ali, W., & El Faylali, H. (2022). A dual wideband high gain 2×2 multiple-input-multiple-output monopole antenna with an end-launch connector model for 5G millimeter-wave mobile applications. *International Journal of RF and Microwave Computer-Aided Engineering, 32*(5), e23088.
13. Ikram, M., Wang, Y., Sharawi, M. S., & Abbosh, A. (2018). Dual band circular MIMO antenna system for 5G wireless devices. In *2018 IEEE International Symposium on Antennas and Propagation & USNC/URSI National Radio Science Meeting* (pp. 247–248). IEEE.
14. Okan, T. (2020). Design and analysis of a quad-band substrate-integrated-waveguide cavity backed slot antenna for 5G applications. *International Journal of RF and Microwave Computer-Aided Engineering, 30*(7), e22236.

15. Geetha, M. N., & Mahadevaswamy, U. B. (2022). Crest dwindling and interposing for PAPR reduction in OFDM signal with sustainable spectral efficiency. *Wireless Personal Communications, 125*, 2799–2818. https://doi.org/10.1007/s11277-022-09686-y
16. IEEE Standard for Air Interface for Broadband Wireless Access Systems. (2018). IEEE Standard 802.16 2017 (Revision of IEEE Standard 802.16–2012) (pp. 1–2726).
17. Jayalath, D. S., & Tellambura, C. (2005). SLM and PTS peak-power reduction of OFDM signals without side information. *IEEE Transactions on Wireless Communications, 4*(5), 2006–2013.
18. Sandoval, F., Poitau, G., & Gagnon, F. (2017). Hybrid peak-to-average power ratio reduction techniques: Review and performance comparison. *IEEE Access, 5*, 27145–27161.
19. DelMarco, S. P. (2017). A constrained optimization approach to compander design for OFDM PAPR reduction. *IEEE Transactions on Broadcasting, 64*(2), 307–318.
20. Sarowa, S., et al. (2020). Evolution of PAPR reduction techniques a wavelet based OFDM approach. *Wireless Personal Communications, 115*(2), 1565–1588.
21. Biswas, A. S., Sil, S., Bera, R., & Mitra, M. (2023). Orthogonal coded spread spectrum digital beamforming-based 5G receiver. *Arabian Journal for Science and Engineering, 48*, 5757–5769. https://doi.org/10.1007/s13369-022-07022-x
22. Bergamo, M. A. (2010). Spread spectrum digital beamforming (SSDBF) radar. In *2010 IEEE International Symposium on Phased Array Systems and Technology* (pp. 665–672). https://doi.org/10.1109/ARRAY.2010.5613293
23. Bergamo, M. A., & Weedon, W. H. (2013). Spread spectrum digital beamforming virtual array and scalability to millimeter wavelength frequencies MMW-SSDBF virtual array. In *2013 IEEE International Symposium on Phased Array Systems and Technology* (pp. 649–652). https://doi.org/10.1109/ARRAY.2013.6731906
24. Tamang, N. D., Sur, S. N., Bera, S., & Bera, R. (2017). Assessment of digital beamforming radar (DBF) and spread spectrum digital beamforming radar (SSDBF). In *2017 14th IEEE India Council International Conference (INDICON)* (pp. 1–5). https://doi.org/10.1109/INDICON.2017.8487745
25. Kalis, A., Kanatas, A. G., & Efthymoglou, G. P. (2010). A co-operative beamforming solution for eliminating multi-hop communications in wireless sensor networks. *IEEE Journal on Selected Areas in Communications, 28*(7), 1055–1062. https://doi.org/10.1109/JSAC.2010.100910
26. Sun, S., et al. (2016). Propagation path loss models for 5G urban micro- and macro-cellular scenarios. In *2016 IEEE 83rd Vehicular Technology Conference (VTC Spring)* (pp. 1–6). https://doi.org/10.1109/VTCSpring.2016.7504435
27. Sun, S., et al. (2016). Investigation of prediction accuracy, sensitivity, and parameter stability of large-scale propagation path loss models for 5G wireless communications. *IEEE Transactions on Vehicular Technology, 65*(5), 2843–2860. https://doi.org/10.1109/TVT.2016.2543139
28. MacCartney, G. R., & Rappaport, T. S. (2017). Rural macrocell path loss models for millimeter wave wireless communications. *IEEE Journal on Selected Areas in Communications, 35*(7), 1663–1677. https://doi.org/10.1109/JSAC.2017.2699359
29. Sulyman, A. I., Alwarafy, A., MacCartney, G. R., Rappaport, T. S., & Alsanie, A. (2016). Directional radio propagation path loss models for millimeter-wave wireless networks in the 28-, 60-, and 73-GHz bands. *IEEE Transactions on Wireless Communications, 15*(10), 6939–6947. https://doi.org/10.1109/TWC.2016.2594067
30. Cheng, H., Ma, S., Lee, H., & Cho, M. (2021). Millimeter wave path loss modeling for 5G communications using deep learning with dilated convolution and attention. *IEEE Access, 9*, 62867–62879. https://doi.org/10.1109/ACCESS.2021.3070711
31. Kim, D.-C., Chi, Y.-E., Park, J., Minz, L., & Park, S.-O. (2022). High-resolution digital beamforming receiver using DDS-PLL signal generator for 5G mobile communication. *IEEE Transactions on Antennas and Propagation, 70*(2), 1428–1439. https://doi.org/10.1109/TAP.2021.3111334

32. Jadhav, M., Dongre, G., & Mahalle, P. (2022). Multi-objective optimization with FEC polar code for bandwidth efficient mobile network using response surface methodology. *Wireless Personal Communications, 125*, 2833–2863. https://doi.org/10.1007/s11277-022-09688-w
33. Jagannatham, A. K. (2017). *Principles of wireless communication systems theory and practice* (2nd ed.). McGraw Hill Education.
34. Zhang, Z., Zhang, L., Wang, X., et al. (2016). A split-reduced successive cancellation list decoder for polar codes. *IEEE Journal on Selected Areas in Communications, 34*(2), 292–302.
35. Taranalli, V., Uchikawa, H., & Siegel, P. (2016). Channel models for multi-level cell flash memories based on empirical error analysis. *IEEE Transactions on Communications, 64*(8), 3169–3181.
36. Tian, K., Liu, R., & Wang, R. (2016). Joint successive cancellation decoding for bit-interleaved polar coded modulation. *IEEE Communications Letters, 20*(2), 224–227.
37. Cao, C., Fei, Z., Yuan, J., et al. (2014). Low complexity list successive cancellation decoding of polar codes. *IET Communications, 8*(17), 3145–3149.
38. Niu, K., & Chen, K. (2012). Stack decoding of polar codes. *Electronics Letters, 48*(12), 695–697.
39. Yazdi, A., & Kschischang, F. (2011). A simplified successive cancellation decoder for polar codes. *IEEE Communications Letters, 15*(12), 1378–1380.
40. Seidl, M., & Huber, J. (2010). Improving successive cancellation decoding of polar codes by usage of inner block codes. In *IEEE 6th International Symposium Turbo Codes and Iterative Information Processing* (pp. 103–106). IEEE.
41. Hussami, N., Korada, S., & Urbanke, R. (2009). Performance polar codes for channel and source coding. In *IEEE International Symposium on Information Theory* (pp. 1488–1492). IEEE.
42. Abd-Elaty, E., Zekry, A., El-Agooz, S., & Helaly, A. M. (2023). Underlay spectrum sharing with L-band distance measuring equipment for aeronautical communications. *Wireless Personal Communications, 128*, 2363–2377. https://doi.org/10.1007/s11277-022-10045-0
43. Zhidkov, S. V. (2003). Impulsive noise suppression in OFDM based communication systems. *IEEE Transactions on Consumer Electronics, 49*(4), 944–948.
44. Epple, U., Shutin, D., & Schnell, M. (2012). Mitigation of impulsive frequency-selective interference in OFDM based systems. *IEEE Wireless Communications Letters, 1*(5), 484–487.
45. Xiao, Y., Xie, J., Yang, J., & Zhang, T. (2016). A time-domain correlative interference mitigation in LDACS1. In *2016 Integrated Communications Navigation and Surveillance (ICNS)*, Herndon, VA (pp. 2D3-1–2D3-9).
46. Hirschbeck, M., & Huber, J. B. (2016). OFDM receiver structure for a subband selective mitigation of time-variant interference. *IEEE Transactions on Vehicular Technology, 65*(10), 8130–8144.
47. Balachander, T., & Mukesh Krishnan, M. B. (2022). Carrier frequency offset (CFO) synchronization and peak average power ratio (PAPR) minimization for energy efficient cognitive radio network (CRN) for 5G wireless communication. *Wireless Personal Communications, 127*, 1847–1867. https://doi.org/10.1007/s11277-021-08726-3
48. Shah, H. A., & Koo, I. (2018). A novel physical layer security scheme in OFDM-based cognitive radio networks. *IEEE Access, 6*, 29486–29498. https://doi.org/10.1109/ACCESS.2018.2842826
49. Sivakumar, V., & Rekha, D. (2018). Node scheduling problem in underwater acoustic sensor network using genetic algorithm. *Personal and Ubiquitous Computing, 22*(5–6), 951–959. https://doi.org/10.1007/s00779-018-1136-3
50. Zhang, P., Ge, W., Zhang, Y., Gao, M., & Zhang, G. (2020). Energy efficiency and user capacity optimization of cognitive MIMO systems via the SCMA-based nonorthogonal time slot allocation. *Symmetry, 12*(7), 1136. https://doi.org/10.3390/sym12071136
51. Sivarambabu, P. V., Malarvezhi, P., Dayana, R., Vadivukkarasi, K., & Rama Koteswara Rao, P. (2022). EEHC Approach for latency minimization in 3D network architecture using 5G+ with UAVs. *Wireless Personal Communications, 122*, 897–914. https://doi.org/10.1007/s11277-021-08931-0

52. Kalantari, E., Bor-Yaliniz, I., Yongacoglu, A., & Yanikomeroglu, H. (2017). User association and bandwidth allocation for terrestrial and aerial base stations with backhaul considerations. In *Proceedings IEEE Annual International Symposium on Personal Indoor and Mobile Radio Communications (PIMRC).*
53. Alam, S., & Haas, Z. J. (2006). Coverage and connectivity in three-dimensional networks. In *Proceedings of Annual International Conference on Mobile Computing and Networking.*
54. Lagum, F., Bor-Yaliniz, I., & Yanikomeroglu, H. (2018). Strategic densification with UAV-BSs in cellular networks. *IEEE Wireless Communications Letters Early Access, 7*(3), 384–387.
55. Crippa, G., Jimenez, C., & Pratelli, A. (2009). Optimum and equilibrium in a transport problem with queue penalization effect. *Advances in Calculus of Variations, 2*(3), 207–246.
56. Saifur, R. S., Azmir, A., Yeongi, C., Kyoung-Jae, L., & Han-Shin, J. (2021). Cognitive UAV-aided URLLC and mMTC services: Analyzing energy efficiency and latency. *IEEE Access, 9*(5011), 5021.
57. Mozaffari, M., TalebZadehKasgari, A. T. Z., Saad, W., Bennis, M., & Debbah, M. (2019). Beyond 5G with UAVs: Foundations of a 3D wireless cellular network. *IEEE Transactions on Wireless Communications, 18*(1), 357–372. https://doi.org/10.1109/TWC.2018.2879940
58. Obeidat, H. A., Ahmad, I., Rawashdeh, M. R., Abdullah, A. A., Shuaieb, W. S., Obeidat, O. A., & Abdallhameed, R. A. (2022). Enhanced TOA estimation using OFDM over wide-band transmission based on a simulated model. *Wireless Personal Communications, 123*, 3449–3461. https://doi.org/10.1007/s11277-021-09297-z
59. Leria, V. J., & Lohan, E. -S. (2012). Timing-based location estimation for OFDM signals with application in LTE, WLAN and WIMAX. In *2012 International Conference on Localization and GNSS* (pp. 1–5).
60. Mensing, C. (2013). *Location determination in OFDM based mobile radio systems.* Technische Universität München.
61. Wang, P., & Morton, Y. (2020). Performance comparison of time-of-arrival estimation techniques for LTE signals in realistic multipath propagation channels. *Journal of the Institute of Navigation, 67*(4), 691–712.
62. Babich, F., Noschese, M., Marshall, C., & Driusso, M. (2017). A simple method for ToA estimation in OFDM systems. In *2017 European Navigation Conference (ENC)* (pp. 305–310).
63. Remcom. (2017). *Wireless InSite reference manual, 3.1.0.* Remcom.
64. Rommel, S., Grivas, E., Cimoli, B., Dodane, D., Morales, A., Pikasis, E., Bourderionnet, J., Feugnet, G., Barros Carvalho, J., Katsikis, M., Ntontin, K., Kritharidis, D., Spaleniak, I., Mitchell, P., Dubov, M., & Tafur Monroy, I. (2021). Real-time high-bandwidth mm-wave 5G NR signal transmission with analog radio-over-fiber fronthaul over multi-core fiber. *EURASIP Journal on Wireless Communications and Networking, 2021*, 43. https://doi.org/10.1186/s13638-021-01914-6
65. Rommel, S., Cimoli, B., Grivas, E., Dodane, D., Morales, A., Pikasis, E., Bourderionnet, J., Feugnet, G., Barros Carvalho, J., Katsikis, M., Ntontin, K., Kritharidis, D., Spaleniak, I., Mitchell, P., Dubov, M., & Tafur Monroy, I. (2020). Real-time demonstration of ARoF fronthaul for high-bandwidth mm-wave 5G NR signal transmission over multi-core fiber. In *European Conference on Networks and Communications (EuCNC)*, Dubrovnik, Croatia (p. 205–208). https://doi.org/10.1109/EuCNC48522.2020.9200921
66. 3GPP TS 38.104, NR: Base Station (BS) radio transmission and reception. 3GPP. 3GPP Rel. 15, V15.7.0 (2019)
67. Hinrichs, M., del Rosal, L. F., Kottke, C., & Jungnickel, V. (2017). Analog vs. next-generation digital fronthaul: How to minimize optical bandwidth utilization. In *International Conference on Optical Network Design and Modelling (ONDM).* https://doi.org/10.23919/ondm.2017.7958539
68. Zhang, L., Ma, J., Wu, Z., Wang, Y., Liu, C., & Habib, K. (2021). Efficient cross-correlation algorithm for correction of common phase error employing preamble for orthogonal

frequency division multiplexing (OFDM) receivers. *Wireless Personal Communications, 118*, 535–549. https://doi.org/10.1007/s11277-020-08030-6
69. Leshem, A., & Yemini, M. (2017). Phase noise compensation for OFDM systems. *IEEE Transactions on Signal Processing, 65*(21), 5675–5686.
70. Wang, Z., Babu, P., & Palomar, D. P. (2017). Effective low-complexity optimization methods for joint phase noise and channel estimation in OFDM. *IEEE Transactions on Signal Processing, 65*(12), 3247–3260.
71. Ngebani, I., Li, Y., Xia, X., et al. (2014). Analysis and compensation of phase noise in vector OFDM systems. *IEEE Transactions on Signal Processing, 62*(23), 6143–6157.
72. Syrjälä, V., Levanen, T., Ihalainen, T., et al. (2019). Pilot allocation and computationally efficient non-iterative estimation of phase noise in OFDM. *IEEE Wireless Communications Letters, 8*(2), 640–643.
73. Garcia Armada, A. (2001). Understanding the effects of phase noise in orthogonal frequency division multiplexing (OFDM). *IEEE Transactions on Broadcasting, 47*(2), 153–159.
74. Liu, T., Chung, W., Yuan, S., et al. (2015). ICI self-cancellation with cosine windowing in OFDM transmitters over fast time-varying channels. *IEEE Transactions on Wireless Communications, 14*(7), 3559–3570.
75. Dayana, R., Malarvezhi, P., Vadivukkarasi, K., & Kumar, R. (2020). UF0MC-IOTA based cognitive radio transceiver. *Wireless Personal Communications, 114*, 2105–2119. https://doi.org/10.1007/s11277-020-07467-z
76. Liu, Y., et al. (2017). Waveform design for 5G networks: Analysis and comparison. *IEEE Access, 5*, 19282–19292.
77. Zhang, L., Ijaz, A., Xiao, P., Wang, K., Qiao, D., & Imran, M. A. (2019). Optimal filter length and zero padding length design for universal filtered multi-carrier (UFMC) system. *IEEE Access, 7*, 21687–21701.
78. Shen, J., & Alsusa, E. (2013). Joint cycle frequencies and lags utilization in cyclostationary feature spectrum sensing. *IEEE Transactions on Signal Processing, 61*(21), 5337–5346.
79. El-Zoghdy, S. F., El-sayed, H. S., & Faragallah, O. S. (2020). Transmission of chaotic-based encrypted audio through OFDM. *Wireless Personal Communications, 113*, 241–261. https://doi.org/10.1007/s11277-020-07187-4
80. Eldokany, I., El-Rabaie, E. M., Elhalafawy, S., Zein Eldin, M., et al. (2015). Efficient transmission of encrypted images with OFDM in the presence of carrier frequency offset. *Wireless Personal Communications, 84*, 475–521.
81. Serag Eldin, S. M. (2014). Optimized OFDM transmission of encrypted image over fading channel. *Sensing and Imaging, 15*(1), 1–14.
82. Henrik, S., & Luders, C. (2005). *Theory and applications of OFDM and CDMA*. Wiley.
83. Kaur, N., & Kumar, N. (2020). On effects of frequency offset on Simulink model of OFDM. *Wireless Personal Communications, 115*(3), 2349–2361. https://doi.org/10.1007/s11277-020-07685-5
84. Veena, M. B., & Swamy, M. N. S. (2011). Performance analysis of DWT based OFDM over FFT based OFDM and implementing on FPGA. *International Journal of VLSI Design & Communication Systems (VLSICS), 2*(3), 119–130.
85. Kumar, N., & Sohi, B. S. (2016). Evaluation of conventional and wavelet based OFDM system for ICI cancellation. *Wireless Personal Communication, 91*(3), 1435–1446. https://doi.org/10.1007/s11277-016-3537-9
86. Kaur, N., & Kumar, N. (2019). Comparative analysis of ICI self cancellation techniques for wavelet OFDM under different channels in Simulink. *Wireless Personal Communication, 105*, 1513–1525. https://doi.org/10.1007/s11277-019-06157-9
87. Jacovic, M., Juretus, K., Kandasamy, N., Savidis, I., & Dandekar, K. R. (2020). Physical layer encryption for wireless OFDM communication systems. *Journal of Hardware and Systems Security, 4*, 230–245. https://doi.org/10.1007/s41635-020-00097-8
88. Proakis, J. G., & Salehi, M. (2007). *Digital communications* (5th ed.). McGraw-Hill.

89. Chacko, J., Juretus, K., Jacovic, M., Sahin, C., Kandasamy, N., Savidis, I., & Dandekar, K. R. (2019). Securing wireless communication via hardware-based packet obfuscation. *Journal of Hardware and Systems Security, 3*, 261–272. https://doi.org/10.1007/s41635-019-00070-0
90. Schmidl, T. M., & Cox, D. C. (1997). Robust frequency and timing synchronization for OFDM. *IEEE Transactions on Communications, 45*(12), 1613–1621.
91. Rajinikanth, E., & Jayashri, S. (2019). Interoperability in heterogeneous wireless networks using FIS-ENN vertical handover model. *Wireless Personal Communications, 108*, 345–361. https://doi.org/10.1007/s11277-019-06406-x
92. Chinnappan, A., & Balasubramanian, R. (2016). Complexity–consistency trade-off in multi-attribute decision making for vertical handover in heterogeneous wireless networks. *IET Networks, 5*(1), 13–21.
93. Zhu, X., Li, M., Xia, W., & Zhu, H. (2016). A novel handoff algorithm for hierarchical cellular networks. *IEEE Journals & Magazines, China Communications, 13*(8), 136–147.
94. Viji, R., & Rathipriya, R. (2015). The network selection using for seamless connectivity of heterogeneous networks based on MADM methods. *Indian Journal of Engineering, 12*(30), 380–387.
95. Hua, J., Wen, J., Wang, A., Xu, Z., & Li, F. (2021). Optimization of FBMC waveform by designing NPR prototype filter with improved stopband suppression. *Mobile Networks and Applications, 26*, 1206–1216. https://doi.org/10.1007/s11036-019-01342-3
96. Farhang-Boroujeny, B. (2011). OFDM versus filter bank multicarrier. *IEEE Signal Processing Magazine, 28*(3), 92–112.
97. Aminjavaheri, A., Farhang, A., Doyle, L. E., & Farhang-Boroujeny, B. (2017). Prototype filter design for FBMC in massive MIMO channels. In *Proceedings of ICC 2017* (Paris, France) (pp. 1–6).
98. Lai, X., & Lin, Z. (2014). Optimal design of constrained fir filters without phase response specifications. *IEEE Transactions on Signal Processing, 62*(17), 4532–4546.
99. Hua, J., Wen, J., Lu, W., Li, F., & Jiang, B. (2014). Design and application of nearly Nyquist and SR-nyquist FIR filter based on linear programming and spectrum factorization. In *Proceedings of ICIEA 2014* (Hangzhou, China) (pp. 64–67).
100. Germany, FEH, France AC. (2015). Final 5GNOW transceiver and frame structure concept D3.3. 5GNOW report. Retrieved from https://www.is-wireless.com/fp7-5gnow/.
101. Priyadharsini, N. A., & Tamil Selvi, S. (2019). Effective scheduling policies to optimize radio resources between NR-gNodeB and device to device systems in 5G. *Wireless Personal Communications, 109*, 1071–1093. https://doi.org/10.1007/s11277-019-06604-7
102. Amerimehr, M. H., Rezaee, M., & Dalaei, H. R. (2020). A robust transceiver for narrow-band communications in presence of carrier frequency offset. *Wireless Personal Communications, 111*, 565–580. https://doi.org/10.1007/s11277-019-06873-2
103. D'Andrea, A. N., & Mengali, U. (1992). Performance of a frequency detector based on the maximum likelihood principle. In *Global telecommunications conference* (vol. 1, pp. 340–344).
104. Ah-Thew, G. (2001). *Doppler compensation for LEO satellite communication systems*. UMI Dissertation Services.
105. Fan, Z., Jin, J., Jin, G., & Mou, J. (2021). A carrier selection method based on single RF chain SM-OFDM systems. *Mobile Networks and Applications, 26*, 1823–1830. https://doi.org/10.1007/s11036-019-01296-6
106. Yan, Z., Yue, X., Dong, B., Ping, Y., & Wei, X. (2017). Efficient detection of spatial modulation OFDM systems with multiple carrier frequency offsets. *IEEE Communications Letters, 21*(2), 426–429.
107. Acar, Y., Leblebici, M. M., Dogan, H., & Panayirci, E. (2016). Data detection based iterative channel estimation for coded SM-OFDM systems. In *Black Sea Conference on Communications & Networking*, Sochi, Russia.

108. Abd El-Rahman, A. F., Abd El-Samie, F. E., Khamis, S. E. A., & Napoleon, S. A. (2019). A new method to enhance SC-FDMA communication with channel estimation errors. *Wireless Personal Communications, 104*, 527–542. https://doi.org/10.1007/s11277-018-5818-y
109. Saraswathi, M., & Babu, S. P. K. (2015). A novel investigation on BER measurement of SC-FDMA system with combined Tomlinson–Harashima precoding and Reed Solomon coding. *Indian Journal of Science and Technology, 8*(26), 23–24.
110. Bisht, M., & Joshi, A. (2015). Various techniques to reduce PAPR in OFDM systems: A survey. *International Journal of Signal Processing, Image Processing and Pattern Recognition, 8*(11), 195–206.
111. Mowla, M. M., Ali, M. Y., & Aoni, R. A. (2014). Performance comparison of two clipping based filtering methods for PAPR reduction in OFDM signal. *International Journal of Mobile Network Communications and Telematics, 4*(1), 23–24.
112. Abd El-Samie, F. E., Al-Kamali, F. S., Alnahari, A. Y., & Dessouky, M. I. (2010). *SC-FDMA mobile communications*. CRC Press.
113. Van Nee, R., & Prasad, R. (2000). *OFDM for wireless multimedia communications*. Artech House.
114. Falconer, D., et al. (2002). Frequency domain equalization for single-carrier broadband wireless systems. *IEEE Communications Magazine, 40*, 58–66.
115. Zhu, X., & Murch, R. D. (2002). Novel frequency domain equalization architectures for a single-carrier wireless MIMO system. In *Proceedings of the IEEE VTC* (pp. 874–878).
116. Oppenheim, A. V., Schafer, R. W., & Buck, J. R. (1999). *Discrete-time signal processing* (2nd ed.). Prentice Hall.
117. Heiskala, J., & Terry, J. (2002). *OFDM wireless LANs: A theoretical and practical guide*. Sams Publishing.
118. Ochiai, H., & Imai, H. (2000). On the clipping for peak power reduction of OFDM signals. In *Proceedings of the IEEE Global Communications Conference (GLOBECOM)* (pp. 731–735).
119. Rahmatallah, Y., & Mohan, S. (2013). Peak-to-average power ratio reduction in OFDM system: A survey and taxonomy. *IEEE Communication Surveys and Tutorials, 15*(5), 1567–1592.
120. Li, G., Li, T., Xu, M., Zha, X., & Xie, Y. (2019). Sparse massive MIMO-OFDM channel estimation based on compressed sensing over frequency offset environment. *EURASIP Journal on Advances in Signal Processing, 2019*, 31. https://doi.org/10.1186/s13634-019-0627-3
121. Song, W. G., & Lim, J. T. (2006). Channel estimation and signal detection for MIMO-OFDM with time varying channels. *IEEE Communications Letters, 10*(7), 540–542.
122. Lin, X., Wu, S., Jiang, C., et al. (2018). Estimation of broadband multiuser millimeter-wave massive MIMO-OFDM channels by exploiting their sparse structure. *IEEE Transactions on Wireless Communications, 17*(6), 3959–3973.
123. Bajwa, W. U., Haupt, J., Sayeed, A. M., et al. (2010). Compressed channel sensing: A new approach to estimating sparse multipath channels. *Proceedings of the IEEE, 98*(6), 1058–1076.
124. Ding, W., Yang, F., Dai, W., et al. (2015). Time-frequency joint sparse channel estimation for MIMO-OFDM systems. *IEEE Communications Letters, 19*(1), 58–61.
125. Li, Z., & Xia, X. G. (2008). An Alamouti coded OFDM transmission for cooperative systems robust to both timing errors and frequency offsets. *IEEE Transactions on Wireless Communications, 7*(5), 1839–1844.
126. Liang, L., Wang, J., & Song, J. (2013). An improved spectrum sensing method for DTMB system based on PN autocorrelation. *IEICE Transactions on Communications, 96*(6), 1559–1565.
127. Wu, J., Chen, Y., Zeng, X., et al. (2007). Robust timing and frequency synchronization scheme for DTMB system. *IEEE Transactions on Consumer Electronics, 53*(4), 1348–1352.
128. Rezki, M., Kale, I., & Sabel, L. (2002). Reduced complexity feedforward frequency offset estimator for AWGN communication channels. In *International Conference on Digital Signal Processing* (vol. 2, pp. 1313–1316).

129. Ma, X., Yang, F., Liu, S., et al. (2018). Sparse channel estimation for MIMO-OFDM systems in high-mobility situations. *IEEE Transactions on Vehicular Technology, 67*(7), 6113–6124.
130. Wei, J., & Liu, Y. (2010). Carrier frequency offset estimation using PN sequence iteration in OFDM systems. In *Proceedings of Second International Conference on Networks Security, Wireless Communications and Trusted Computing* (pp. 405–409).
131. Taubock, G, & Hlawatsch, F. (2008). A compressed sensing technique for OFDM channel estimation in mobile environments: exploiting channel sparsity for reducing pilots. In *Proceedings of the IEEE International Conference on Acoustics, Speech and Signal Processing* (pp. 2885–2888).
132. Jayanthi, P. N., & Ravishankar, S. (2017). Sparse channel estimation for MIMO-OFDM systems using compressed sensing. In *Proceedings of the IEEE International Conference on Recent Trends in Electronics* (pp. 1060–1064).
133. Bi, X., Chen, X. D., Zhang, Y., et al. (2013). Variable step size stagewise adaptive matching pursuit algorithm for image compressed sensing. In *Proceedings of the IEEE International Conference on Signal Processing*.
134. Gupta, M., Kang, A. S., & Sharma, V. (2022). Comparative study on implementation performance analysis of Simulink models of cognitive radio based GFDM and UFMC techniques for 5G wireless communication. *Wireless Personal Communications, 126*, 135–165. https://doi.org/10.1007/s11277-020-07561-2
135. Fagbohun, O. O. (2014). A comparative studies on 3G, 4G and 5G wireless technology. *IOSR Journal of Electronics and Communication Engineering, 9*(3), 88–94.
136. Salkintzis, A. K., Fors, C., & Pazhyannur, R. (2002). WLAN-GPRS integration for next-generation mobile data networks. *IEEE Wireless Communications, 9*(5), 112–124.
137. Choi, M. G., & Xu, Y. (2000). A new multimedia network architecture using 3G CDMA2000. In *Vehicular Technology Conference Fall 2000. IEEE VTS Fall VTC2000. 52nd Vehicular Technology Conference* (Vol. 6, pp. 2937–2944).
138. Krenik, B. (2008) 4G wireless technology: When will it happen? What does it offer? In *IEEE Asian Solid-State Circuits Conference* (pp. 141–144).
139. Datta, R., Michailow, N., Lentmaier, M., & Fettweis, G. (2012). GFDM interference cancellation for flexible cognitive radio PHY design. In *IEEE Vehicular Technology Conference (VTC Fall)* (pp. 1–5).
140. www.electronicsnotes.com. Retrieved 2016.
141. www.rfwireless-world.com. Retrieved 2016.
142. Sharma, M., Singh, S., & Varma, R. (2021). Computational design, analysis and characterization of beetle shaped high isolation multiple-input-multiple-output reconfigurable monopole-antenna with dual band filters for wireless applications. *Wireless Personal Communications, 119*, 1029–1049. https://doi.org/10.1007/s11277-021-08248-y
143. Borhani Kakhki, M., & Rezaei, P. (2017). Reconfigurable microstrip slot antenna with DGS for UWB applications. *International Journal of Microwave and Wireless Technologies, 9*(7), 1517–1522.
144. Jetti, C. R., & Nandanavanam, V. R. (2018). Trident-shape strip loaded dual band-notched UWB MIMO antenna for portable device applications. *International Journal of Electronics and Communications, 83*, 11–21.
145. Pandey, G. P., Kanaujia, B. K., Gupta, S. K., & Gautam, A. K. (2004). A novel C shape antenna with switchable wideband frequency notch. *Wireless Personal Communications, 80*(2), 471–482.
146. Sharma, M., Awasthi, Y. K., & Singh, H. (2019). Compact multiband planar monopole antenna for bluetooth, LTE, and reconfigurable UWB applications including X-band and Ku-Band wireless applications. *International Journal of RF and Microwave Computer-Aided Engineering, 29*(6), 1–11.
147. Rahman, M., Khan, W. T., & Imran, M. (2018). Penta-notched UWB antenna with sharp frequency edge selectivity using combination of SSR, CSRR and DGS. *AEU International Journal of Electronics and Communications, 93*, 116–122.

148. Pandey, G. K., Singh, H. S., & Bharti, P. K. (2015). Design and analysis of multiband notched pitcher-shaped UWB antenna. *International Journal of RF and Microwave Computer-Aided Engineering, 25*(9), 795–806.
149. Mathur, R., & Dwari, S. (2018). Compact Cpw-Fed ultrawideband mimo antenna using hexagonal ring monopole antenna elements. *AEU-International Journal of Electronics and Communications, 93*, 1–6.
150. Nagarajan, K., & Sophia, S. (2022). Multilevel redundant discrete wavelet transform (ML-RDWT) and optimal red deer algorithm (ORDA) centred approach to mitigate the effect of ICI, BER and CIR in a MIMO-OFDM system. *Wireless Personal Communications, 122*, 3347–3370. https://doi.org/10.1007/s11277-021-09066-y
151. Sharief, A. H., & Sairam, M. S. (2019). Performance analysis of MIMO-RDWT-OFDM system with optimal genetic algorithm. *AEU-International Journal of Electronics and Communications, 111*, 152912.
152. Hao, J., Wang, J., & Pan, C. (2016). Low complexity ICI mitigation for MIMO-OFDM in time-varying channels. *IEEE Transactions on Broadcasting, 62*(3), 727–735.
153. Paek, M. J., Kim, W. C., Kim, M. Y., & Song, H. K. (2019). Spatial phase coding with CoMP for performance enhancement based on MIMO-OFDM in HetNet system. *IEEE Access, 7*, 62240–62250.
154. Pham, T., Le-Ngoc, T., Woodward, G. K., & Martin, P. A. (2016). Channel estimation and data detection for insufficient cyclic prefix MIMO-OFDM. *IEEE Transactions on Vehicular Technology, 66*(6), 4756–4768.
155. Hussein, H. S., & Hagag, M. (2019). Optical MIMO-OFDM with fully generalized index-spatial LED modulation. *IEEE Communications Letters, 23*(9), 1556–1559.
156. Ma, Y., Wang, J., Li, Y., Chen, M., Li, Z., & Zhang, Z. (2020). Smart antenna with automatic beam switching for mobile communication. *EURASIP Journal on Wireless Communications and Networking, 2020*, 179. https://doi.org/10.1186/s13638-020-01792-4
157. Gu, C., et al. (2015). Compact smart antenna with electronic beam-switching and reconfigurable polarizations. *IEEE Transactions on Antennas and Propagation, 63*(12), 5325–5333. https://doi.org/10.1109/TAP.2015.2490239
158. Lai, M.-I., Wu, T.-Y., Hsieh, J.-C., Wang, C.-H., & Jeng, S.-K. (2007). A miniature, planar, and switched-beam smart antenna employing a four- element slot antenna array for digital home applications. In *2007 IEEE Antennas and Propagation Society International Symposium*, Honolulu, HI (pp. 3376–3379). https://doi.org/10.1109/APS.2007.4396261
159. Guntupalli, A. B., & Wu, K. (2013). 60 GHz circularly-polarized smart antenna system for high throughput two-dimensional scan cognitive radio. In *2013 IEEE MTT-S International Microwave Symposium Digest (MTT)*, Seattle (pp. 1–3). https://doi.org/10.1109/MWSYM.2013.6697662.
160. El Zooghby, A. H., Christodoulou, C. G., & Georgiopoulos, M. (2000). A neural network-based smart antenna for multiple source tracking. *IEEE Transactions on Antennas and Propagation, 48*(5), 768–776. https://doi.org/10.1109/8.855496
161. Patro, S. K. P., Mishra, R. K., & Panda, A. K. (2020). Study of indoor radio coverage performance of dual technology co-existing MIMO antenna platform for low power wireless base station. *Radioelectronics and Communications Systems, 63*, 66–76. https://doi.org/10.3103/s0735272720020028
162. Hu, Z., Susitaival, R., Chen, Z., Fu, I.-K., Dayal, P., & Baghel, S. K. (2012). Interference avoidance for in-device coexistence in 3GPP LTE-advanced: Challenges and solutions. *IEEE Communications Magazine, 50*(11), 60. https://doi.org/10.1109/mcom.2012.6353683
163. Micheli, D., Barazzetta, M., Carlini, C., Riccardo Diamanti, V. M., & Primiani, F. M. (2016). Testing of the carrier aggregation mode for a live LTE base station in reverberation chamber. *IEEE Transactions on Vehicular Technology, 66*(4), 3024. https://doi.org/10.1109/tvt.2016.2587662

Chapter 2
Mobile Communication Channel Modelling

Milind Pande, Anand J. Kulkarni, and Apoorva S. Shastri

Introduction by the Editor

In the case of mobile communication, the channel that we consider for the purpose of transmission is wireless in nature because when the signal is broadcast there is no direct connection set up between the transmitter and receiver. The signal's properties change as the signal gets influenced by numerous propagation effects. Hence a direct line of sight between the transmitter and receiver may not always be achievable. The fluctuations in channel strength across time and frequency are distinguishing characteristics of the mobile wireless channel. The algorithms and techniques used for channel modelling are covered in this chapter. The chapter discusses the statistical analysis of 5G/6G millimeter wave channels, multiuser OFDMA networks, estimating the frequency of signals with white Gaussian noise and narrowband interference, joint map detection for OFDM, time domain channel estimation for MIMO-FBMC/OQAM systems, wireless channel estimation and beamforming, G2A communication channel modelling. Additionally, it includes analysis of ultra-wideband channels, studying dynamic channel characteristics, measuring, and modelling Wi-MAX channels, vehicular channels, and their implications on communication systems design, and utilizing machine learning for wireless communication channel modelling.

In 5G new radio, spectrum from 30 to 300 MHz is referred to as millimeter waves (mm waves). It has advantage of high data rates, low latency however mm

M. Pande
Dr. Vishwanath Karad MIT World Peace University, Pune, Maharashtra, India
e-mail: milind.pande@mitwpu.edu.in

A. J. Kulkarni · A. S. Shastri (✉)
Institute of Artificial Intelligence, Dr. Vishwanath Karad MIT World Peace University, Pune, Maharashtra, India
e-mail: anand.j.kulkarni@mitwpu.edu.in

M. Pande et al. (eds.), *Optimization Methods in Mobile Communication Systems*, https://doi.org/10.1007/978-981-95-1810-4_2

waves experience higher propagation loss. Gaussian noise is a random noise of equal intensity for all frequencies. Transmission channel can be classified as outdoor and indoor channel. An indoor channel can be modelled for small coverage area resulting into user terminals can move in certain range hence it also called static or quasi static channel. In outdoor channel parameters changes with respect to time. Channel gain that is signal strength is controlled by doppler spectrum which is range of frequency which continuously changes with the change in location of transmitter and receiver and controlled by continuously changing parameter.

Ray tracing (RT) model is a deterministic channel models can properly characterize the propagation phenomena. RT has been applied in indoor and urban environments from early 1990s but interest toward RT simulation platforms has recently got increased as they provide more realistic results than stochastic models at mm Wave frequencies. Shoot and bouncing ray (SBR) method modeled the 60 GHz indoor channel compared the small-scale properties of the channel with the measurements performed using a uniform virtual array. By using data of power delay profile (PDP) and power angular spectrum (PAS) at location RT model can also be used for image data in outdoor microcellular and courtyard environment at 15 GHz, 28 GHz and 60 GHz. Ray tracing model is validated by simulation method which require measuring setup. This setup contains may equipment. Vector network an analyzer (VNA) is use for measuring data in which simulation results are depends. It operates at center frequency of 3.5 GHz. Multiplier and mixers for up and down conversion of signals, 70 MHz oscillator, 18.81 GHz phase locked oscillator (PLO) filters and antennas. The Rx and Tx are connected to VNA through coaxial cables of length 1 m and 20 m, respectively. The receiver antenna is placed on the positioner, and the direction of the Rx antenna is set by the PC installed with LabVIEW program. The measurement data are collected at PC and that are connected to VNA via General Purpose Interface Bus (GPIB) cable.

The deterministic approach of RT utilizes the geometry of the environment, and the location of the Tx and Rx for finding different propagation paths. In case of image theory method, for finding a propagation path with single reflection, a mirror image of the Tx is created with respect to each line of the radio environment model. This helps to calculate prorogation path that radio waves take after surface reflection. Ray tracing tool finds path along with other propagation paths between the Tx and the Rx based on interactions like reflections, diffractions and their combinations. In simulations parameters like antenna heights, transmission power, frequency of operation, direction of Tx and Rx antennas are used. Antenna configurations can be named as Tx1-Config1 and Tx2-Config1, it means the receiver antenna was pointed towards the config 1 direction from where signal strength is strong. Root Mean Square (RMS) between the measured and simulated received power level is around 2.6 dB and 2.9 dB in case of Tx1-Config1 and Tx2-Config1, respectively. At receiver position, measurements are taken at 9 points on grid. Average of these points are considered for calculating delay spread (DS) for receiving signals which is 1 ns and 4 ns for Tx1-Config1 and Tx1-Config2, respectively. HPBW (Half Power Beamwidth) is angle over which the antenna's signal strength is half of its maximum value. RMS and DS are direct function of antenna HPBW, i.e., narrow the HPBW of the antenna the smaller is the RMS-DS.

Vehicular channels are of two types of vehicles to vehicle (V2V) and vehicle to infrastructure (V2I). Two technologies are based on IEEE 802.11p and LTE standards. Intelligent Transport system (ITS) allows vehicle to exchange information. ITS requires setting up Vehicular ad hoc network (VANET). This network supports communication of V2V, V2I and V2P (Vehicle to pedestrian).

WiMAX use microwave for data transfer. It supports long distance communication almost in kilometers. It uses Orthogonal Frequency Division Multiplexing (OFDM). It uses Orthogonal Frequency Division Multiplexing). It is multiplexing method in which signal bandwidth is divided into parallel subcarriers or narrow strips of bandwidth. OFDM uses subcarrier that is orthogonal to each other to reduce signal interference.

MIMO-FBMC/OQAM systems are used to enhance the performance of channels. Multiple-input multiple-output (MIMO) helps to provide wide bandwidth for data transfer, filter bank multicarrier (FBMC) filter signals which in turn helps to avoid signal interference. Offset Quadrature Amplitude Modulation (OOQM) helps in time frequency localization of signal in all to improve spectral efficiency. In wireless communication system channel modelling is crucial milestone. Tradition statistical method for analyzing large amount of data is very difficult task for channel estimation, Machine learning model works efficiently in analyzing such data for channel estimation and modelling which is crucial requirement of new 5G network. Ground-to-air (G2A) communication systems use directional antennas to provide coverage for aircraft. This technique helps in minimize interference from moving objects near ground station and improves signal quality.

Machine Generated Summaries

Disclaimer: The summaries in this chapter were generated from Springer Nature publications using extractive AI auto-summarization: An extraction-based summarizer aims to identify the most important sentences of a text using an algorithm and uses those original sentences to create the auto-summary (unlike generative AI). As the constituted sentences are machine selected, they may not fully reflect the body of the work, so we strongly advise that the original content is read and cited. The auto generated summaries were curated by the editor to meet Springer Nature publication standards. To cite this content, please refer to the original papers.

Machine generated keywords: estimation, channel model, indoor, propagation, channel characteristic, ghz, model, channel estimation, angle, outdoor, cfo, estimate, delay spread, random, multipath

A Study about Signal Variation with Minor Receiver Displacement in a Meeting Room at 60 GHz: Measurements and Simulations [1]

This is a machine-generated summary of:

Sheikh, Muhammad Usman; Ruttik, Kalle; Jäntti, Riku; Hämäläinen, Jyri: A study about signal variation with minor receiver displacement in a meeting room at 60 GHz: measurements and simulations [1]

Published in: EURASIP Journal on Wireless Communications and Networking (2021)

Link to original: https://doi.org/10.1186/s13638-021-01965-9

If you want to cite the papers, please refer to the original.

For technical reasons we could not place the page where the original quote is coming from.

Abstract-Summary

"The first target of this article is to highlight the benefits of an in-house built three-dimensional RT tool at 60 GHz and shows the effectiveness of simulations in predicting different characteristics of the channel."

"To validate the simulation results against the measurements, two different transmitter (Tx) positions and antenna types along with ten receiver (Rx) positions are considered in a typical conference room."

"After validating the simulation results with the measurement data, in the second part of this work, the impact of a small change, i.e., 20 cm in the receiver position, is studied."

"The channel characteristics are considered with respect to the direct orientation (DO), i.e., the Rx antenna is directed toward the strongest incoming path."

"Different antenna configurations at the Tx and Rx side are applied to highlight the role of antenna properties on the considered channel characteristics."

"In the second system configuration the impact of different antenna half power beamwidth on different considered channel characteristics is highlighted through acquired simulation results."

"The validation of results shows the RMS error of only 2–3 dB between the measured and simulated received power levels for different Tx configurations in the direction of DO."

"Results indicate that only a small change of the Rx position may result a large difference in the received power level even in the presence of line-of-sight between the Tx and Rx."

"It is found that the STD of received power level across the room increases with the decrease in HPBW of the antenna."

Introduction

"In that task, the deterministic channel models, e.g., ray tracing (RT) models, are valuable since they are able to properly characterize the propagation phenomena."

"RT has been applied in indoor and urban environments from early 1990s but interest toward RT simulation platforms has recently got increased as they provide more realistic results than stochastic models at mmWave frequencies."

"In [2], authors modeled the 60 GHz indoor channel by shoot and bouncing ray (SBR) method and compared the small-scale properties of the channel with the measurements performed using a uniform virtual array."

"In references [3, 4], authors investigated and validated the results of the RT simulations based on image theory with the measurement data in an outdoor micro-cellular and courtyard environment at 15 GHz, 28 GHz and 60 GHz and provided PDP and power angular spectrum (PAS) for different locations."

Measurement Setup and Scenario

"The measurement setup consists of different elements, e.g., vector network analyzer (VNA) operating at the center frequency of 3.5 GHz, multipliers and mixers for up and down conversion of signals, 70 MHz oscillator, 18.83 GHz phase locked oscillator (PLO), filters and antennas."

"The Rx and Tx are connected to VNA through coaxial cables of length 1 m and 20 m, respectively."

"The receiver antenna is placed on the positioner/motor, and the direction of the Rx antenna is set by the PC installed with LabVIEW program."

"The measurement data are collected at PC and that are connected to VNA via a GPIB cable."

"The main beam of the Tx2 antenna was pointing in the direction of the up-right corner of the room."

Simulation Methodology

"The deterministic approach of RT utilizes the geometry of the environment, and the location of the Tx and Rx for finding different propagation paths."

"In case of IT method, for finding a propagation path with single reflection, a mirror image of the Tx is created with respect to each line of the radio environment model."

"Ray tracing tool finds the LoS path along with other propagation paths between the Tx and the Rx with the given number of interactions, i.e., reflections, diffractions and a mix of reflections and diffractions."

"Other general parameters, e.g., antenna heights, transmission power, frequency of operation, azimuth and downtilt of the Tx and Rx antennas are set in the simulations as were used in measurements."

"We have considered two more special antenna configurations for simulations, as stated here: Tx1-Config2 In this configuration, same Tx antenna position is used as in TX1-Config1."

Results and Discussion

"It is important to highlight here that for Tx1-Config1 and Tx2-Config1 cases the results are reported with DO, that means the antenna of the receiver was oriented in the direction of the strongest incoming path."

"The RMS error between the measured and simulated received power level is around 2.6 dB and 2.9 dB in case of Tx1-Config1 and Tx2-Config1, respectively."

"Considering average value of 9 points grid at each receiver position, the mean value of DS is 1.1 ns and 4 ns for Tx1-Config1 and Tx1-Config2, respectively."

"Positions 9 and 10 lie close to the main beam of the Tx antenna, and at those positions the RMS-DS is the direct function of the antenna HPBW, i.e., narrow the HPBW of the antenna the smaller is the RMS-DS."

Conclusions

"The received power RMS error was only 2.6 dB and 2.9 dB between the measured and simulated results for Tx1-Config1 and Tx2-Config1, respectively."

"As performance indicators, we considered receiver power level, power delay profile, power angular profile, RMS delay spread and RMS angular spread in azimuth plane, assuming the Rx antenna is directly oriented in the direction of the Tx."

"We found that even with LoS a slight displacement of Rx by 20 cm causes an average STD of around 6.5 dB in the received power level in case of directed antennas."

"In the last part of this study, for the transmitter located in the corner of the conference room, the impact of antenna HPBW on different channel characteristics was investigated."

"The acquired simulation results show that the STD of received power level in the conference room increases with the decrease in HPBW of the Tx antenna."

Statistical Analysis of 5G/6G Millimeter Wave Channels for Different Scenarios [5]

This is a machine-generated summary of:

Bedda-Zekri, A.; Ajgou, R.: Statistical Analysis of 5G/6G Millimeter Wave Channels for Different Scenarios [5]

Published in: Journal of Communications Technology and Electronics (2022)
Link to original: https://doi.org/10.1134/s1064226922070063

If you want to cite the papers, please refer to the original.
For technical reasons we could not place the page where the original quote is coming from.

Abstract-Summary
"In 5G NR (New Radio), the wide ranges of radio spectrum, from 30 to 300 GHz and 0.1 to 10 THz, referred to as millimeter waves (mm Waves) and sub-terahertz respectively, remain unemployed."

"Concern, this paper provides an in-depth statistical study of millimeter wave propagation candidates for 5G/6G systems (28, 38, 60, 73, 100, and 120 GHz) based on different scenarios for outdoor Urban Microcell (UMi) channels with various scenarios especially the geometrical parameters (Co-Pol/X-Pol, SISO/SIMO/MIMO), and also the environmental conditions (LOS/NLOS, rain rate, temperature and humidity for an urban Microcell)."

"The results indicated that the 38 and 73 GHz channels are quite tolerant to the impact of deteriorating environmental conditions and geometrical parameter variations, while the 60, 100 and 120 GHz channels are more affected."

"In terms of path loss (PL) and path loss exponent (PLE), the omnidirectional power delay profile (OPDP) showed improved performance over the directional power delay profile (DPDP) for all channels."

Introduction
"Compared with the existing studies, our key contributions in this work can be summarized as follows: (1) An in-depth study of the propagation of different frequency bands using various scenarios was carried out, especially the geometric parameters (Co-Pol/X-Pol, SISO/SIMO/MIMO). (2) The effect of Multi User MIMO (MU–MIMO) on the mmWaves channel characteristics mainly the impact of switching from 1U-MIMO to 2U-MIMO and 2U-MIMO to 4U-MIMO are discussed where in the literature the most researchers have not investigated this study. (3) The effect of switching from LOS to NLOS. (4) Focusing on the analysis of the propagation of frequency bands around 100 GHz (sub-terahertz) that are candidates for 6G systems, taking into account the potential scenarios with additional analysis and details."

"Section I reviews several related works in which we present recent works studying propagations at mmWaves bands and enclose this section with a comparison with the contribution of this paper in terms of mmWave/terahertz bands, scenarios and reported parameters."

Related Works

"Rappaport and others in [6], developed another simulator named SIRCIM (Simulation of Indoor Radio Channel Impulse Response Models) for radio channel impulse response statistical models for the analysis and design of wireless communication systems."

"Due to the scope of millimeter-wave channel modeling and the requirement to operate in frequency bands greater than 6 GHz for wireless communication such as 5G/6G, many groups of researches have begun to share their expertise to create mmWaves channel models and establish a performance evaluation platform."

"Besides, the authors in [7], carried out a NYUSIM simulation of spatial channel modeling characteristics for 5G mmWaves at carrier frequencies (28, 38, 60, and 73 GHz)."

"In [8], the impact of weather conditions on the channel model characteristics of millimeter wave band frequencies (60 and 73 GHz) considering the rainy season and winter season in Bangladesh is discussed."

Path Loss Models

"The Radio propagation models have a significant impact on decisions made in the field of wireless communications."

"The 5G/6G wireless communications offers tremendous promise, to the point where users can already expect to have access to ultra-high data rates from anywhere and at any time."

"5G wireless communications can be severely affected by weather conditions, path loss (PL), attenuation due to atmospheric conditions, reflection loss, building materials, blocking of obstacles and scattering [9, 10]."

"It is necessary to use higher frequencies at 6 GHz and above 100 GHz for 5G/6G technologies."

"To optimize network efficiency and get more details on some candidate frequency bands for future mobile communication technology."

Simulation Scenarios, Analysis and Discussion

"In the case of LOS scenario, for DPDP, the 28 GHz channel has lower PL and PLE 102.8 dB, 2.1 respectively, whereas 73 and 120 GHz have the highest values (116.1 and 116.3 dB) and vice versa for the received power."

"In the case of NLOS situation, for DPDP/OPDP, the 38 and 73 GHz channels have a lower PL and PLE (120.5, 119.7 dB for PL and 2.8, 2.5 for PLE), whereas for 60 GHz, it has the two highest values (144.9 dB, 3.9) and conversely for the received power."

"In the case of 50 mm/h rain rate setting, the 38 GHz channel has the lower PL (107.5 dB), while the 120 GHz channel shows the higher one (116.1 dB) and conversely for the received power, the 100 GHz channel holds the lower PLE (1.8), while 28 GHz has the highest value (2.5)."

Conclusions

"The results, it can be concluded that the parameters of statistical channel modeling are dependent on the geometric parameters (Co-Pol/X-Pol, SISO/SIMO/MIMO) and environmental conditions (LOS/NLOS, rain rate, temperature and humidity)."

"Observation suggests that mmWave channel with carrier frequencies 28, 60, 100, and 120 GHz channels are more affected by the switching from LOS to NLOS."

"Besides, in terms of antenna configurations (SISO/SIMO/MIMO), the results indicated that 120 GHz is more affected, while 28 GHz is less affected than the other channels, except that 38 GHz is more favorable to change this factor."

"For rain, temperature and humidity factors, it was concluded that the attenuations of 60, 100 and 120 GHz are the most important, while 28 and 73 GHz channels are less affected, except that 38 channel is the most capable to withstand the change of these factors."

Low Complexity Carrier Frequency Offset and Channel Estimation for Multiuser OFDMA Networks [11]

This is a machine-generated summary of:

Liang, Yao-Jen; Chen, Kwang-Cheng: Low Complexity Carrier Frequency Offset and Channel Estimation for Multiuser OFDMA Networks [11]

Published in: Wireless Personal Communications (2021)

Link to original: https://doi.org/10.1007/s11277-021-08558-1

If you want to cite the papers, please refer to the original.

For technical reasons we could not place the page where the original quote is coming from.

Abstract-Summary

"We therefore propose a joint carrier frequency offset (CFO) and channel estimation scheme by leveraging the characteristic of transmission signals in the OFDMA uplink systems."

"By modeling the system in a Hadamard product form, we develop an iterative algorithm wherein a closed-form estimate of CFO and channel is applied in each updating iteration."

"Different from the traditional expectation-maximization type methods, which update only a single CFO and channel of one user while keeping the other users' CFOs and channels constant in each iteration, our proposed scheme updates the CFOs and channels of all active users simultaneously."

"The proposed CFO and channel estimator enjoys faster convergence and lower complexity."

Introduction
"CFO estimation in the OFDMA uplink is closely related to the particular subcarrier assignment scheme (CAS) adopted in the system."

"In [12], an ESPRIT-based CFO estimation method is proposed by utilizing rotational invariance techniques with a lower computational complexity."

"Sparse blind CFO estimation methods for interleaved OFDMA uplink are proposed by exploiting the sparsity embedded in the OFDMA data [13, 14]."

"Iterative CFO estimation schemes for tile-based OFDMA uplink are proposed with low computational complexity and fast convergence in [15, 16]."

"Reference [17] leverages Kalman and Particle filtering and proposes two joint CFO and doubly selective channel estimators for MIMO-OFDMA uplinks based on basis expansion modeling (BEM)."

"Based on the simulation results, the estimation performance of the proposed joint CFO and channel estimator can be greatly improved and approaches to the Cramér-Rao Lower Bounds."

System Model
"Each OFDM uplink packet contains one training symbol for frequency synchronization and channel estimation followed by several data symbols."

"III.A. It is noted that pilot sequences with equal power, equal space, and orthogonality are proposed to be optimal for channel estimation of MIMO OFDM systems [18]."

"We extend the orthogonality to cascaded-orthogonality for joint CFO and channel estimation of OFDMA systems."

Proposed Algorithm
"We first analyze some important properties of cascaded orthogonal matrix, and then derive the joint CFO and channel estimation scheme based on the properties."

"Examining the issue of joint CFO and channel estimation in both a coordinated OFDM downlink and an OFDMA uplink system, we find that the requirement of cascaded orthogonality in pilots is automatically satisfied in OFDMA uplink due to the exclusive assignment of subcarriers among different users."

"CFO/channel estimation in OFDMA uplink can be considered as a special case of that in coordinated OFDM systems."

"The procedure for the joint CFO and channel estimation of coordinated OFDM systems [19] can thus be adopted into OFDMA uplink."

"It is sensitive of the number of OFDMA subcarriers, N, to the estimation accuracy in Step 1."

"The computational complexity of [12] is also included for a benchmark, since the ESPRIT-based estimator in [12] provides the accurate CFO estimates without channel estimation, satisfying a well-known low computational complexity."

Simulation Results
"If without further description, the initial CFOs of all active users for AH, LSAGE, and the proposed PFCE are independently drawn from a uniform distribution within the range (−0.5, 0.5)."

"Note: LSAGE and AH both employ sequentially interference cancelation, the order of users' estimation is very important, because the previously estimated CFO and channel of other users would affect current user's estimation."

"While the proposed PFCE utilizes PIC, the issue of the order of users' estimation does not exist."

"From the figure, the AH, LSAGE, and proposed PFCE method all converges to the averaged CRLB after 4 iterations."

"It is worthy noting that our proposed method converges faster than the other benchmark methods since it updates the estimates of all K users in parallel."

Conclusion

"We investigate the joint CFO and channel estimation in OFDMA uplink systems."

"A system model leveraging the Hadamard products is derived to facilitate the joint estimator."

"By leveraging the inherent cascaded orthogonality, we obtain an iterative estimation/compendation algorithm with approximated closed-form solutions at each iteration."

Computationally Efficient Algorithm for Estimating the Frequency of Signals at the Output of a Channel with White Gaussian Noise and Narrowband Interference [20]

This is a machine-generated summary of:

Kazakov, L. N.; Lukyanov, I. V.; Shakhtarin, B. I.: Computationally Efficient Algorithm for Estimating the Frequency of Signals at the Output of a Channel with White Gaussian Noise and Narrowband Interference [20]

Published in: Journal of Communications Technology and Electronics (2019)

Link to original: https://doi.org/10.1134/s106422691910005x

If you want to cite the papers, please refer to the original.

For technical reasons we could not place the page where the original quote is coming from.

Abstract-Summary

"Abstract—A computationally efficient algorithm is proposed for estimating the frequency of a quasi-harmonic signal in the presence of narrowband interference based on iterative calculation of the autocorrelation sequence."

Formulation of the Problem

"Estimate (2) is done over the sampling However, the window technique increases the equivalent noise bandwidth, and as a result, the variance of the estimate increases."

"A disadvantage of this approach is its high computational complexity, due to the need to preliminarily estimate the frequencies of mixture components (1) and numerical solution of (6)."

"A practical compromise is the choice of a window function with fixed parameters and a small equivalent noise band (e.g., the Bartlett window [21]) and subsequent calculation of the ML estimate from (2)."

"Calculation of the ML estimates comes with high computational costs, because it is necessary to calculate a large number of spectral samples with a small frequency step."

Nikiforov Algorithm

"An efficient computational algorithm for estimating frequency under AWGN conditions was proposed by Nikiforov [22, 23]."

"The algorithm is based on an increase in the signal-to-noise ratio (SNR) using an iterative procedure for calculating the autocorrelation sequence (ACS) followed by calculation of the autoregressive (AR) frequency estimate [24]."

"To reduce the error, it is necessary to increase the order of the AR model to the number of complex exponents present in the signal."

"This requires preliminary estimation of the number of complex exponents in process (1) and a significant increase in computational resources, which casts doubt on the efficiency of the AR estimation as a whole."

"The increase in the number of iterations p increases the useful signal/narrowband interference ratio in accordance with the expressions According to (10), a decrease is possible only when $A_0 > A_m$."

An Autoregressive Frequency Estimation Algorithm Immune to Narrowband Interference

"To suppress narrowband interference, it is possible to use a bandpass filter tuned to the useful signal band."

"Such a procedure can be effectively integrated into the estimation algorithm: it multiplies the PSD signal by a rectangular window: where k_0 and k_1 are the x coordinates of the zeros of the main SPD lobe."

"This approach makes it possible not only to increase the immunity of the algorithm, but also to reduce the number of addition and multiplication operations, as well as the required amount of memory, since it is necessary to calculate spectral samples only within the main SPD lobe."

"Such a procedure is effectively carried out by discrete Fourier transform (DFT) or the Goertzel algorithm; there is no need to pad the initial sampling with zeros."

Computational Complexity of the Algorithm

"We introduce the coefficients of the reduction in computational operations:where M_{1mult}, M_{1add} are the number of multiplication and addition operations of the Nikiforov algorithm [23]; M_{2mult}, M_{2add} of algorithm [25]; M_{3mult}, M_{3add} of the proposed algorithm."

"For N = 1024 and p = 3, the number of multiplication and addition operations is reduced by about 2.2 and 3.1 times compared to the original Nikiforov algorithm, and for N = 32 768, by 2.8 and 4.4 times, respectively."

"With small sample sizes (up to 32 samples), the proposed algorithm loses to the Nikiforov algorithm in computational efficiency."

Simulation Results

"The observed "threshold effect," below which there is a discrepancy with the RKB, is due to the excess levels of spectral noise samples above the spectral level of the useful signal, resulting in an error in determining index I_0."

"These results demonstrate that the suppression of narrowband interference using the Nikiforov algorithm is possible only with a number of iterations for calculating the ACS greater than four, while the proposed algorithm provides immunity to such interference already with two iterations."

"The proposed algorithm is free of this disadvantage."

Conclusions

"Combining the filtering procedure with the frequency estimate made it possible to reduce the number of multiplication and addition operations (by 2.8 and 4.4 times, respectively, with a sampling size of N = 32 768). (2) The proposed algorithm yields an estimate for the frequency in AWGN conditions with variance deviating from the Rao–Kramer boundary by no more than 0.2 and 0.4 dB with four and three iterations for calculating the ACS, respectively, up to region of the threshold effect."

"When estimating the signal frequency in the presence of powerful narrowband interference, the variance of the estimate using the Bartlett window deviates from the theoretical limit by no more than 0.52, 0.74, and 1.3 dB with four, three, and two iterations for calculating the ACS in the SNR range, respectively, up to the region of the threshold effect."

Joint MAP Detection for OFDM in Presence of Phase Noise from Free Running and Phase Locked Loop Oscillator [26]

This is a machine-generated summary of:

Shrivastav, Kamayani; Yadav, R. P.; Jain, K. C.: Joint MAP Detection for OFDM in Presence of Phase Noise from Free Running and Phase Locked Loop Oscillator [26]

Published in: Wireless Personal Communications (2019)

Link to original: https://doi.org/10.1007/s11277-019-06579-5

If you want to cite the papers, please refer to the original.

For technical reasons we could not place the page where the original quote is coming from.

Abstract-Summary

"This paper addresses a statistically optimal joint maximum a posteriori (MAP) data detection and phase noise (PHN) estimation algorithm for iterative orthogonal frequency division multiplexing (OFDM) receivers, used for high speed and high spectral efficient wireless communication systems."

"With reduced computational complexity, the proposed optimization incorporates the statistics of PHN spectral components in prior."

"The condition of small PHN is relaxed in the proposed algorithm to achieve near optimal performance."

"Simulation results demonstrate that the proposed optimization achieves "NO PHN" performance for both the PHN models even under multipath channel fading."

Introduction

"The symbol error rate (SER) performance of the system using conventional OFDM receiver gets degraded because of the isolated approach of data detection and PHN estimation [27–29]."

"A reliable data detection of transmitted symbols needs the joint PHN estimation [30–36]."

"A new joint MAP cost function for the data detection and PHN estimation in frequency domain is formulated in this paper."

"The SER for the proposed joint MAP detection is compared with the SER of an OFDM receiver without PHN."

"For the optimal data detection the joint cost function is reduced globally for both data and PHN."

Phase Noise Modelling

"This work addresses the phase synchronization of receiver oscillator with the assumption of perfect time synchronisation in the beginning of every OFDM symbol."

"The steep decrease in the power spectral density (PSD) of PHN outside the loop bandwidth enables the sufficient characterization of the whole process with some lower order elements only."

OFDM System Modelling

"In the packet transmission a packet is formed of many contiguous OFDM symbols."

"Initial symbols of each packet are some full pilot symbols followed by several data symbols."

"In data transmission phase the channel is already been estimated and data and pilot subcarriers are multiplexed."

"After this signal is RF modulated and asserted on the channel."

Multidimensional Circulatory Gradient Descent Optimization

"For the optimum search we can use either exhaustive grid search or less complex random search method."

"In high dimensional spaces the random search method produces effective result for faster convergence than exhaustive grid search method which is potentially expensive and suffers from the "curse of dimensionality" [37]."

Computational Complexity

"For designing an OFDM receiver, computational complexity is a major issue to be addressed."

"Precisely for PHN estimation, the frequency domain estimation is more efficient than time domain because it clears nearly 100% of interference with the estimation of only few lower order PHN spectral components."

"With only additions and subtractions as its major operations, few multiplications and without any division, the implemented random search method [37] becomes non-significant in concern of computational complexity."

Simulation Results

"This section demonstrates the performance of proposed joint MAP detection with simulation results."

"The PHN effect is estimated and then jointly optimized with data detection by using proposed joint MAP detection."

"The performance of proposed joint MAP detection algorithm in presence of applicable receiver PHN model and channel type is simulated now."

"This happens because the inaccurate detection of symbols led to less accurate PHN spectral components estimation and this error propagation at higher SNR values deteriorate the SER performance, which is compensated in the proposed algorithm with joint MAP optimization before next iteration."

Conclusion

"The proposed detector uses MAP criterion to estimate the PHN in joint with data detection."

"Further the estimation is optimized with the proposed multidimensional circulatory gradient descent algorithm to achieve near optimal data detection performance."

"This improves over the cost function optimization and joint MAP detection."

"Simulation results show that for both AWGN and multipath channel model, the proposed joint MAP detection achieves near "No PHN" SER performance for both the PHN models."

Time Domain Channel Estimation for MIMO-FBMC/OQAM Systems [38]

This is a machine-generated summary of:

Singh, Prem; Vasudevan, K.: Time Domain Channel Estimation for MIMO-FBMC/OQAM Systems [38]

Published in: Wireless Personal Communications (2019)

Link to original: https://doi.org/10.1007/s11277-019-06515-7

If you want to cite the papers, please refer to the original.

For technical reasons we could not place the page where the original quote is coming from.

Abstract-Summary

"A general preamble-based time domain channel estimation model for multiple-input multiple-output (MIMO)-filter bank multicarrier (FBMC) system based on offset quadrature amplitude modulation (OQAM) is derived."

"To the conventional frequency domain interference approximation method (IAM)-based channel estimation model, the derived time domain model does not require symbol time to be sufficiently longer than the maximum channel delay spread, viz."

"The derived time domain model does not require the channel to be frequency flat."

"Based on the time domain model, the weighted least squares (WLS) and minimum mean square error (MMSE) channel estimators for MIMO-FBMC/OQAM systems are investigated."

Introduction

"Another variant of the IAM-based channel estimation for MIMO-FBMC/OQAM systems with overhead reduced preambles, has been discussed in [39]."

"Like SISO-FBMC/OQAM systems, most of the channel estimation techniques for MIMO-FBMC/OQAM systems are also based on the IAM model."

"We derive a general time domain channel estimation model for MIMO-FBMC/OQAM systems."

"To estimate the channel in MIMO-FBMC/OQAM systems based on the time domain model, the key contributions of this paper are as follows. (1) The minimum mean square error (MMSE) and weighted least square (WLS) estimators are proposed; (2) we derive the Cramer–Rao lower bound (CRLB) for the proposed estimators; and (3) it is demonstrated via numerically evaluated results that the proposed estimators not only achieve their respective CRLB bounds, but also significantly outperform the existing IAM and time domain LS estimator."

MIMO-FBMC/OQAM System Model

"For ease of mathematical analysis, we define the difference between OFDM and FBMC/OQAM lies in the choice of the prototype pulse p[k]."

"In classical OFDM, p[k] is the rectangular pulse of duration T, whereas in FBMC/OQAM, the duration of the pulse p[k] is greater than T (usually, an integer multiple of T)."

"The increased duration of the pulse p[k] in FBMC/OQAM allows each subcarrier to be shaped by a well frequency–time (FT) localized prototype filter, which is different from the traditional rectangular pulse in classical OFDM with its Sinc-shaped spectrum."

Frequency Domain Channel Estimation Model

"We revisit the conventional IAM-based CFR estimation for MIMO-FBMC/OQAM systems."

"The IAM model is based on the assumption that the symbol time is sufficiently longer than the maximum channel delay spread."

"Since, the IAM model estimates the CFR, it is a frequency domain channel estimation method."

"Under the assumption that the symbol time is sufficiently longer than the maximum channel delay spread, the prototype pulse p[k] has low variations in time over the channel length and the following condition holds approximately true [40] i.e. However, when the above condition does not hold, the frequency domain method suffers from severe performance degradation."

Time Domain Channel Estimation Model

"We derive a general time domain channel estimation model for MIMO-FBMC/OQAM systems."

"Unlike the frequency domain model, this model does not impose any constraints on the length of the symbol time."

"The BCRLB for this estimator is given as"

Numerical Results and Discussion

"The WLS estimator has considerable performance gain over the LS estimator, because the former does not ignore the noise correlation across different sub-carriers."

"Unlike the IAM estimator, the WLS estimator does not require symbol time to be much greater than the maximum channel delay spread."

"The prior knowledge of the channel mean and covariance matrix is indeed the reason for the better performance of the MMSE estimator."

"The proposed WLS and MMSE estimators perform significantly better than the IAM estimator because, unlike the later, these estimators do not require symbol time to be sufficiently greater than the maximum channel delay spread."

"Due to the a priori knowledge of the channel mean and covariance, the MMSE estimator performs better than the WLS estimator."

Conclusions

"The minimum mean square error (MMSE) and weighted least square (WLS) time domain channel estimators have been investigated."

"As compared to the frequency domain interference approximation method (IAM)-based channel estimator, the designed time domain estimators do not require symbol time to be sufficiently greater than the maximum channel delay spread, and their preamble requirement does not increase with the number of transmit antennas."

"Numerical results showed that the proposed time domain channel estimators significantly outperform the existing IAM estimator and achieve their respective bounds."

Analytical and Experimental Investigation of Ultra Wideband Channel Characteristics in the Presence of Door/Window Glass [41]

This is a machine-generated summary of:

Bharadwaj, Richa; Koul, Shiban K.: Analytical and Experimental Investigation of Ultra Wideband Channel Characteristics in the Presence of Door/Window Glass [41]

Published in: Wireless Personal Communications (2019)
Link to original: https://doi.org/10.1007/s11277-019-06753-9

If you want to cite the papers, please refer to the original.

For technical reasons we could not place the page where the original quote is coming from.

Abstract-Summary

"This paper presents ultra-wideband channel characteristics in the presence of float glass slab as obstacle of varied thickness (4 mm, 8 mm, 12 mm, 19 mm) in an indoor environment."

"Two different sets of measurements are performed related to variation in the Tx–Rx antenna position and rotation of the glass slab at angles varying from 0° to 90° to mimic the glass door/window movement."

"Different direct path, partial and total obstructed path situations are observed depending on the position of the Rx antenna, rotation angle of the glass slab and also the position of the Tx–Rx antennas with respect to the glass slab."

"For LOS situations, channel parameters for various glass thickness show similar range of values and are comparable with the scenario when no-glass in present."

Introduction

"Ultra wideband technology is also used to study the channel parameters and detect direct/obstructed path scenarios in indoor environments such as office, home [42, 43] further extended to the study of the channel links in specific environments such as car [44], aircraft cabins [45] and underground mines [46, 47] etc. UWB body centric propagation has also been studied in [48–51] with works related to off-body and on-body communication."

"Indoor propagation characteristics of a UWB signal is analyzed in a high rise apartment for LOS and NLOS scenarios caused due to obstruction from walls and glass/wooden doors [52]."

"To the best of author's knowledge work carried out in this domain is limited and in depth analysis of the channel parameters when the signal passes through glass of varied thickness has not been presented in open literature."

"This paper gives information on the channel characteristics of propagation through glass of various thickness (4 mm, 8 mm, 12 mm and 19 mm) and no glass present, in an indoor environment in the 4–8 GHz frequency range of the UWB spectrum."

Measurement Description

"Float glass sheets which acts as an obstacle is placed between the Tx and Rx antenna."

"The first measurement study analyses the propagation through glass when the glass position is fixed at 0° parallel to the Tx and Rx antennas."

"The initial position of the glass sheet is parallel to both Tx and Rx antennas."

"The five different situations are considered in which the Tx–Rx pair of antennas are present near the edge of the open and hinge end, aligned with the centre of the glass obstacle, leading to different channel links between the Tx and Rx antenna."

"Measurements for both the experimental campaigns are also carried out when no glass obstacle is present to have a comparison between the scenarios when the glass sheet is present between the Tx and Rx antennas."

"The classification of the channel links formed depends on the position of the Rx and Tx antennas and the position/orientation of the glass sheet."

Techniques and Algorithms

"The time of arrival (TOA) techniques are considered for understanding the channel link for the measured data [53]."

"S_{21} parameters (phase and magnitude) are obtained from the measured data is further used to obtain the channel impulse response (CIR) using (Inverse Fourier transform) IFFT."

"The power delay profile (PDP) is the squared magnitude of the impulse response, defined as: By using the impulse response of the measured multi-path channels, the root mean square (RMS) delay spread, received signal strength have been obtained."

"The power delay profile (PDP) gives the intensity of a signal received through a multipath channel as a function of time delay [53–55] where the time delay is the difference in travel time between multipath arrivals."

"The RMS delay spread gives an indication of the multipath components present in the channel [53, 56]."

Results and Analysis

"The major increase in σ_τ is noticed when the propagation takes place near the (open end) edge of the glass sheets (Tx position: C2) and total obstructed scenarios (Tx position: C3–C5) leading to more scattering and multipath components."

"The major scattering and signal distortion leading to multipath effects is for the position C2 of the Tx antenna as the edge (open end) of the glass sheet comes in the way of the signal path between the Tx–Rx antennas."

"For position S3, NLOS scenario can be observed till 60° position of the glass sheet, higher σ_τ is observed as the channel link is obstructed by the glass sheet."

"For LOS links similar P_k values are observed (−65 to −55) for all the glass thickness and comparable with the no-glass present values."

"For glass position at 90°, S1_90° and S3_90° show higher signal amplitude as both links are in LOS situation."

Conclusion
"IR-UWB communication makes it feasible to capture variations in channel parameters of the Tx–Rx link due to change in glass thickness."

"For partial NLOS and NLOS links, there is variation in channel parameter values observed for increase in glass thickness (4 mm → 19 mm) [Mean: (σ_τ) 5 → 12 ns; (P_k) −62 → −68 dB]."

"As observed the standard deviation of the channel parameters for higher glass thickness is more [(σ_τ) 4–6 ns; (P_k) 3.5–5.5 dB] in comparison to the values obtained for lower glass thickness [(σ_τ) 1.5–2.5 ns; (P_k) 1.5–3 dB]."

"The results obtained gives a deeper insight of the channel link characteristics that change in the presence of a glass sheet between the Tx and Rx when in static and rotating scenario through extensive experimental campaign."

Analysis of Dynamic Channel Characteristics Based on a Body Area Network [57]

This is a machine-generated summary of:

Zhou, Lingli; Liao, Wei; Hou, Songyue: Analysis of Dynamic Channel Characteristics Based on a Body Area Network [57]

Published in: Wireless Personal Communications (2021)

Link to original: https://doi.org/10.1007/s11277-021-09035-5

If you want to cite the papers, please refer to the original.

For technical reasons we could not place the page where the original quote is coming from.

Abstract-Summary
"Because static human communication characteristics have been studied more, this paper proposed and verified a dynamic human body communication channel model in the Human Body Communication band, and the dynamic channel characteristics of 35 kinds of human body continuous postures were analyzed."

"The results of using five typical communication links on the dynamic human body model show that the average power delay is following a second order exponential decay."

"To evaluate the safety of human body communication technology on the human body, the frequency domain finite differential analysis method was used to analyze the distribution of the electromagnetic field."

"The absorption rate is the quantification of the electromagnetic energy absorbed by the human body, and the absorptivity of 35 postures demonstrated that human communication technology is not harmful to the human body."

Introduction

"With the rapid development of microelectronic devices and wireless sensors, new microwearable devices that utilize human body communication technology and various sensors have attracted increasing attention [58, 59]."

"Human body communication technology utilizes human biological tissue to communicate in the BANs, which guarantees that communication is secure [60, 61]."

"Although the channel model of human communication has been studied, the dynamic human channel model is seldom studied in HBC."

"Muhammad and others evaluated the secrecy capacity for in-body wireless channels [62].However, these models did not consider dynamic characteristics in the HBC band, and the research considered the channel characteristics of single-person communication."

"It is necessary to study the characteristics of dynamic human communication channels."

"This paper mainly introduced the characteristics of single- and double-dynamic human body communication channels and studied the absorption of electromagnetic energy by the human body."

Computational Methods

"This paper used a numerical model of adult males (33 years old, 168.76 cm in height and 65 kg in weight) to simulate the real human body."

"According to the radiation characteristics of the dipole, it is advantageous for the electromagnetic wave to diffract around the human body if the direction of the dipole transmission and the direction of the receiving electric field component are perpendicular to the human body surface."

"The direction of the electrodes was perpendicular to the human body surface in this paper."

"The FDTD method was used to calculate the distribution of electric and magnetic fields on the surface of the human body model."

"Although the calculation time of the FDTD method is long, the finite element method can simulate complex geometric shapes, and the numerical model more easily changes the posture of the human body."

Channel Characteristic Analyze

"The transmitter terminals send the human body data measured by the sensors to the receiver terminals in the form of electromagnetic waves [63]."

"The path loss caused by the multipath effect produced by the motion human body model is shadow loss."

"Dynamic human body communication should also consider the loss of shadows due to human activity."

"This leads to a shadow area in the human body, and the electromagnetic wave diffraction in the shaded region, causing losses."

"Compared with the large fluctuation of shadow loss in UWB, the change of shadow loss is small in HBC, which indicates that the dynamic human model is more suitable for body surface communication in HBC."

"A dynamic human body model causes a change in the actual communication distance and leads to a multiple-path effect, which leads to the power delay profile (PDP) decaying over time."

Distribute Characteristic

"The first peak of the multipath cluster represents the first path of the signal, counting the arrival time of the first power in different postures."

"The gamma, exponential, Weibull, and lognormal distributions were used to fit the cumulative probability density function of the first path delay."

"The first path delay of the left–right chest link follows the Weibull distribution."

"According to the second-order Akaike information guidelines, the interval delay of the left–right chest link conforms to the Burr distribution."

"The power delay distribution can be characterized by two parameters: the average additional delay and the mean square root delay spread (RMS)."

"Delay spread is a statistical variable, the mean square root delay spread is the square root of the second-order moment of the power delay distribution of the multipath signal, and its calculation formula is as follows."

Handshake Communication

"The motion of the model 1 arm is taken as the starting point to analyze the change in path loss of the arm at different angles."

"When the direction of arm motion is backward, the path loss decreases with the increase of the angle of backward motion."

"This paper believes that the smaller the arm angle is, the greater the influence of other parts of the human body, leading to greater path loss."

"Part of the signal transmission along the arm is through the neck, another part of the signal after the refraction and diffraction in the torso of model 1 is directly to the human body and transmitted through the trunk."

"By comparing the delay spread, it is found that the communication link with the arm as the receiver has better channel characteristics than that with the chest as the receiver."

Safe Inspection

"The specific absorption rate (SAR) is the rate of electromagnetic energy absorbed by the human body per mass unit [64, 65]."

"SAR represents the power of electromagnetic radiation absorbed by the human body per unit mass. where U is the electromagnetic energy and m is the human body mass."

"To evaluate the body area communication transceiver to ensure the safety of local exposure, we calculated the average 10 g quality peak SAR of human tissue."

"A comparison shows that the SAR peak values at 30 MHz did not exceed the exposure limits of the general public, but the SAR peak at 6.85 GHz for the human body of some postures exceeded the exposure limit."

"The SAR of the human body is quite different in the two frequency bands."

Conclusions

"This paper mainly analyzes the channel characteristics of a single dynamic model and a two-person dynamic model in human body regional communication under the HBC frequency band."

"Since it is impossible for the human body to remain stationary all the time, we analyzed the multipath characteristics of dynamic human body communication."

"By comparing the delay spread, it is found that the communication link with the arm as the receiver has better channel characteristics than that with the chest as the receiver."

"Studying the communication of a dynamic human body model can make human body communication more general."

"It is meaningful to analyze the distribution of multipath time delay by statistical dynamic human communication channel characteristics."

"Studying the distribution of electromagnetic fields around the human body gives us a better understanding of human communication technology in terms of human safety."

Measurement and Analysis of Wi-MAX Channel in 3.3 GHz Using Minimal-Ray Fading Model [66]

This is a machine-generated summary of:

Bhise, Suvarna R.; Khot, Uday P.: Measurement and analysis of Wi-MAX channel in 3.3 GHz using minimal-ray fading model [66]

Published in: Wireless Networks (2020)

Link to original: https://doi.org/10.1007/s11276-020-02280-9

If you want to cite the papers, please refer to the original.

For technical reasons we could not place the page where the original quote is coming from.

Abstract-Summary

"This paper presents minimal-ray a sum-of-sinusoids fading model with one direct signal and multipath signal as minimum as three in number, having phase, attenuation, and time of arrival as random variables for random scatterer area."

"At the receiver side, the demodulator has to adjust the phase by trial and error method to recover the original time-variant signal which is a tedious process."

"If the dependency of the random phase and the angle of deviation is proved, then it is easier at the receiver side for the demodulator to adjust the phase directly."

"This work defines the simple relationship between the angle of deviation and random phase which can directly help at the receiver side for adjusting the deviation occurred in the transmitted signal."

Introduction
"It is a challenge to create a random multipath environment in WSS fading simulator with a minimum number of a path (to reduce the number of oscillators)."

"A nonstationary T-junction channel model with randomly distributed scatterers for vehicular communications is proposed in [67], where the receiver does not support the variations of the angle of motion (AOM) of the mobile."

"A fixed scattering area is modeled by a targeted random trajectory approach by considering the different AoM, AoA and variable distances between the mobile receiver (MR) and the transmitter for a non-stationary propagation area."

"Sophisticated SoS models are proposed to efficiently simulate the channel by using the finite number of the scatterer."

"In all the above literature, fading simulators assume independence of random phase and AoA/AoM. Because of this, at the receiver side demodulator has to adjust the phase by trial and error method to recover the original time-variant signal which is a tedious process [68]."

Mathematical Model of Proposed Minimal-Ray Fading Model
"When a signal takes two or more paths from the transmitting antenna to the receiving antenna multipath occurs."

"Multipath fading occurs when the received signal power falls below some arbitrary threshold."

"For every channel, this changes both amplitude and delay difference."

"Even if the user remains perfectly still, the channel will change because of the change in reflectors' position."

"A big difference in the received signal quality occurs when small differences in the arrival time of the different signals happen."

Statistical Characteristics of Minimal-Ray Fading Model
"These characteristics include correlation between field components of real and imaginary parts of the signal, wide-sense stationary (WSS) condition and joint and marginal pdf of time of arrival (ToA) and random phase."

"The geometric shape of the propagation area and distribution of local scatterer within the same area are the aspects of small scale fading (SSF), which are affecting the statistics of the received signal."

"CDSM is modeled for analysis of ToA and random phase [69]."

"For CDSM the behavior of the marginal pdf of ToA is in good agreement with the data available in the literature and random phase is uniformly distributed in the disk model."

"The simple relation between AoD and random phase is derived in this section."

"Because of random phase and attenuation a maximum increase or attenuation occurs in received signal."

"The received signal depends on random phase and attenuation of the multipath component."

Empirical Measurements

"Methodology for measuring path loss and signal strength described in the following section. (1) Measurement procedure and location description The average building height is about 25–30 m (5–7 floors) nearby station so the measurement environment is a typical urban heavily-sized Asian city."

"Four sector antennas are base station BS1, BS2, BS3, BS4."

"Frequency 3316 MHz is latched with selected base station BS1. (2) Calibration The study and validation of received signal is done by considering the Walfish Model in a multipath environment for path loss (PL)where d is the distance in km and f is the frequency in MHz."

Comparison with Other Models

"To avoid the complexity of models, all random variables are not implemented."

"Random attenuation also not implemented in all models."

"In all models, presumption made that AoA is independent on the random phase."

"Minimal-ray model considered all variables are random, not predefined."

"Fading simulators can be made a cost-effective implementation by using less number of oscillators."

"The existing fading simulator's require N number of oscillator, whereas minimal-ray require N number of the phase shifter and attenuator."

"Minimal-ray analysis is enough to simulate the complex environment."

"Two pdfs are modeled and for each distribution goodness of fit used by measuring the similarity between two distributions by finding the maximum difference of cdf's using Kolmogorov–Smirnov (K–S) test."

Conclusions

"Minimal-ray, multipath fading simulator has been developed for measurement and analysis of the Wi-MAX channel in 3316 MHz."

"It shows ToA is a negative exponential and random phase uniformly distributed, and also a dependency of ToA and random phase."

"The design of different fading channel simulators has been compared and it shows that the minimal-ray is having all random variables."

"Because of these random variables, even three-rays produce many amplitudes and phase perturbations result in a complex environment."

"Existing fading simulators require n number of the oscillator, whereas the minimal-ray requires n number of the phase shifter and attenuator."

Wireless Channel Estimation and Beamforming by Using Block Sparse Adaptive Filtering [70]

This is a machine-generated summary of:

Mohanty, Basabadatta; Sahoo, Harish Kumar; Patnaik, Bijayananda: Wireless channel estimation and beamforming by using block sparse adaptive filtering [70]

Published in: Signal, Image and Video Processing (2020)
Link to original: https://doi.org/10.1007/s11760-020-01795-0

If you want to cite the papers, please refer to the original.
For technical reasons we could not place the page where the original quote is coming from.

Abstract-Summary
"Channel estimation normally provides information about indoor and outdoor fading channel statistics."

"The adaptive channel estimation models play an important role to generate the required channel state information (CSI) using the estimated channel coefficient vector."

"The proposed estimation model process the transmitted quadrature amplitude modulation (QAM) data samples in the frequency domain."

"The proposed sparse-FBRLS (Fast BRLS) model has simultaneously addressed the problems of channel estimation and beamforming in case of indoor and outdoor communication."

Introduction
"Different channel estimation models are proposed by researchers to acquire accurate CSI for the indoor and outdoor environments."

"Adaptive techniques are preferred than the non-adaptive methods to estimate CSI due to the adjustment of model coefficients according to time-varying statistics of the wireless channel."

"Channel estimation is performed by a improved least mean square (LMS) algorithm which reduces the noise effect during data transmission better than LMS and normalized LMS (NLMS) [71]."

"The sparse behavior of wireless channels can also be exploited by using adaptive filtering-based estimation model to provide an improved computational complexity [72]."

"Sparse signal processing is proposed by Berraki and others [73] in a beamforming application for efficient angular frequency estimation and interference cancellation for the outdoor spatial channel."

"The impulse response of the channel should be properly estimated in spite of small or large-scale fading condition to design an effective beamformer and improve the SNR of the receiver."

Channel Models for Indoor and Outdoor Communications
"An indoor channel can be modeled for a small coverage area and lower mobility of the user terminal."

"The channel gain is mostly governed by the Doppler spectrum and thus the time-correlated channel variation is modeled with consideration of the speed of the mobile terminal."

"IEEE 802.11 can be used as a standard channel [74] model for the indoor environment."

"A frequency-selective Rayleigh fading channel with time-varying channel gain and Doppler spectrum of 900 Hz with 100 km/h of user mobility is considered to describe the outdoor environment [73]."

Block RLS Algorithm

"The processing time and computational burden in the RLS algorithm increase due to the sample processing of the input data, which can be effectively reduced by converting the input data in the form of blocks."

"This forms a basis to develop the BRLS algorithm, where the block processing to deal with large data samples [75]."

"Let g is the block index and k is the sample index, which is related as: The input blocks in matrix form can be represented as: Frequency-domain block processing can be further implemented with its inherent advantages by the use of Fourier transform and the necessary implementation steps [76] are given below: 1."

"From the same data input blocks, the desired signal of the specified dimension is formulated 3."

Sparse Modeling and Frequency Domain Filtering

"The incorporation of inherent sparsity [73] to the cost function of the FBRLS algorithm using the norm penalty function helps to achieve a bounded error convergence."

"This section describes the implementation steps of the sparse-FBRLS algorithm which can be utilized simultaneously for channel estimation and beamforming."

Simulation Results and Discussion

"The mean square error (MSE), beam pattern, array factor, and channel capacity using all the three algorithms of RLS family (RLS, BRLS, FBRLS) are analyzed and compared with the proposed adaptive model to justify the improvement in system performance of the proposed model."

"The performance of the proposed algorithm is also better for the outdoor environment as compared to the indoor environment as it is observed that the MSE starts decreasing from a value 0.1 onward for indoor channel conditions whereas the value is approximately 0.015 for the outdoor channel."

"It is can be observed that comparatively high signal strength is achieved in the desired direction using the proposed sparse-FBRLS-based model."

"It can be observed from the results that the channel capacity with respect to SNR is increased and achieving higher value using the proposed sparse-FBRLS algorithm."

Conclusion

"The channel coefficients obtained by using sparse adaptive estimation model can accurately describe the time-varying nature of the channel."

"In case of beamformer, only significant coefficients are involved in calculation array factor due to sparse nature of the model."

"The proposed algorithm is quite effective than RLS or BRLS to develop the sparse estimation model."

"The adaptive beamformer using sparse-FBRLS develops an automatic steering mechanism for antenna elements and improves the directive gain of the array."

G2A Communication Channel Modeling and Characterization Using Confocal Prolates [77]

This is a machine-generated summary of:

Mirza, Muhammad-Yasir Masood; Khan, Noor M.: G2A Communication Channel Modeling and Characterization Using Confocal Prolates [77]

Published in: Wireless Personal Communications (2020)

Link to original: https://doi.org/10.1007/s11277-020-07597-4

If you want to cite the papers, please refer to the original.

For technical reasons we could not place the page where the original quote is coming from.

Abstract-Summary

"This paper proposes an appropriate multipath channel model for various ground-to-air (G2A) communication scenarios."

"Most of the research literature supposedly model G2A channels interchangeably in the same manner as that of land mobile communication links."

"This research analyzes G2A multipath channel and thus proposes a geometrically-based physical G2A multipath channel model."

"The proposed channel model clearly justifies the existence of multipath environment in G2A communication due to the existence of aircrafts in the vicinity of the intended aircraft."

"The model gives an insight to analyze multipath multiuser G2A communication in wide-beam and narrow-beam communication link scenarios."

"This model can be used to analyze the performance of high data-rate communication links with high mobile speeds over sparsely distributed multipath channels."

Introduction

"Measurement-based path-loss models can be used to predict an appropriate distance or range of AS to remain in contact with the G. In the above-cited measurement campaigns, the authors in [78–88] observed the behavior of signal propagation path-losses and developed path-loss models for different A2G communication environments."

"The initial work for the representation of air-to-ground (A2G) channels through geometrical channel modeling was proposed in [89], where the authors presented a three-dimensional ellipsoidal geometry based A2G geometric channel model to analyze the temporal and spatial characteristics of A2G communication environment."

"The initial work of characterizing temporal and spatial characteristics of A2G communication environment through geometric-based channel modeling was presented in [89], where, the authors assumed a three-dimensional ellipsoid covering all possible scatters around the ground station while keeping communication units at its focal points."

"2 G2A propagation channel is analyzed by differentiating its attributes from those of A2G communication link and the existence of a multipath environment around the intended aircraft is explored."

Ground-to-Aircraft Multiuser Communication System Model

"If the aircraft is above the communication link then aircrafts lying in the tier of the same separation distance may provide multipath signals with the same propagation delay."

"In ground-to-aircraft (G2A) communication environment, transmitted radio signals undergo reflection, diffraction or scattering mechanisms and arrive at the aircraft receiving unit with random amplitudes and arrival times."

"The scattering aircrafts that will provide multipath signals with the same propagation delay can be presumed on the same prolate spheroid, where AS and G are to be located on its focal points."

"Considering the case of communication link between G and AS, the signals arriving through aircrafts lying on the surface of a defined prolate spheroid with G and AS as its focal points will reach with the same time delay."

Ground-to-Aircraft Channel Model Development

"Intersection points between circular rings and the plane cross-sectioned ellipses are the conceivable locations of scattering aircrafts at which they are assumed to be located during their flight while maintaining the minimum safe separation distance."

"The ellipse formed by the plane cross-section of a prolate spheroid and the circle of separation distance may intersect at either none or two or four points."

"Since, the geometry of the aircraft channel model depends upon multiple intersection ellipses with different major and minor axes and circles of different radii; therefore, the geometrical G2A channel model would thus be a combination of these scenarios depending upon the radius of circle, and the major and minor axes of the intersection ellipse."

"Having the above information in mind, one can visualize the proposed G2A communication system model as a combination of multiple plane cross-sectioned ellipses and concentric circles with varying radii."

G2A Communication Channel Model

"In RADAR systems theory, RADAR Cross-Section (RCS) of scattering object is defined as the ratio between the power density of the radio waves impinged on a scattering object and the power density scattered in the direction of the radar's receiving antenna."

"In situations, when both the transmitter and receiver antennas are placed at different locations by maintaining a considerable distance, the RADARs are known as bistatic radars whereas the RCS is termed as Bistatic RADAR Cross Section (BRCS)."

"The formation of RCS collectively depends upon incident angle of impinging wave, bistatic angle, reflection and absorption properties of the reflecting surface, geometry of the object and the polarization of electromagnetic wave at the reflecting object and the receiver."

"It is the ratio of the reflected power to the incident power intercepted by the aircraft."

Simulation Results and Discussion

"Although, some more scattering aircrafts can be accommodated on the second circular ring; however, the reflected signals coming via those aircrafts will reach with larger delays and negligible power."

"In order to observe the reflection properties of the scattering aircrafts in our proposed G2A propagation scenario, RCS of scattering aircrafts is required."

"Following the geometry of the proposed G2A communication model, incidence angles of radio signals impinged on the surface of aircrafts can be measured."

"In G2A wireless communication environment, a signal transmitted by G may arrive at the receiver of the intended aircraft via various multipath as a result of reflections or scattering from the proximate scattering aircrafts."

"Utilizing the geometry of the proposed G2A communication model and acquiring signal reflection properties of the surface of the aircraft, the power delay profile of the model can be estimated."

Interference Visualization

"The proportion of the transmitted signal power for the intended aircraft would be the same as those for the other proximate aircrafts connected with the same G2A communication link."

"After despreading the received signal at the intended aircraft would thus include a major portion of the desired signal, a portion of Multiple Access Interference (MAI) depending upon the received correlations of the signatures and the noise power."

"The system enhances transmission quality by directing narrow beams towards the intended aircraft while steering nulls towards other proximate aircrafts of no interest that may become a source of interference."

"It is pertinent to mention here that in this scenario, the intended aircraft may receive the interfering signals due to the existence of proximate scattering aircrafts around it; however, these signals will be of negligible power and can thus be neglected."

Conclusion

"G2A multipath channel has been intensively analyzed by thoroughly differentiating its attributes from A2G communication link, and a geometrically-based physical G2A multipath channel model has been proposed."

"This proposed model can be modified for any G2A multipath propagation scenario with appropriate choice of semi major and semi minor axes in accordance with the positions of scattering aircrafts around the intended aircraft."

"The proposed model suitably justifies the existence of multipath scattering environment around the intended aircraft and provides an insight to analyze multiuser G2A communication in the presence of multipath scattering in wide-beam and narrow-beam communication link scenarios."

"The proposed model can not only be used to improve the analytic prediction of signals received at the intended aircraft in the presence of proximate scattering aircrafts but also can be seen as a way to realistically simulate the exact scenario of G2A propagation environment."

Vehicular Channels: Characteristics, Models and Implications on Communication Systems Design [90]

This is a machine-generated summary of:

Cortés, José A.; Aguayo-Torres, Mari Carmen; Cañete, Francisco J.; Gómez, Gerardo; Martos-Naya, Eduardo; Entrambasaguas, J. Tomás: Vehicular Channels: Characteristics, Models and Implications on Communication Systems Design [90]

Published in: Wireless Personal Communications (2019)

Link to original: https://doi.org/10.1007/s11277-019-06269-2

If you want to cite the papers, please refer to the original.

For technical reasons we could not place the page where the original quote is coming from.

Abstract-Summary

"Two technologies, based on the IEEE 802.11p and on the long-term evolution for vehicular communications (LTE-V) standards, have been proposed for this purpose."

"This paper analyzes the relation between the characteristics of vehicular communication channels and the parameters of the referred systems, with particular emphasis on the physical and medium access control layers."

"The primary factors that influence V2V and V2I channels and their main characteristics are firstly described."

"The key parameters of the IEEE 802.11p and LTE-V physical layers are then summarized and its suitability to deal with the time and frequency selectivity of vehicular channels is compared."

"Distortion caused by the time variation of the channel is examined and design challenges related to important aspects like synchronization, multiple access interference and channel estate information estimation are discussed."

Introduction
"Two main technologies have been proposed for this purpose: dedicated short-range communications (DSRC) and cellular V2X (C-V2X)."

"At the physical layer, the precoded version of OFDM used for uplink, known as single carrier frequency division multiple access (SC-FDMA), is also employed in direct V2V LTE-V. C-V2X communications are also included among the services that will be provided by the 5th Generation (5G) cellular systems [91]."

"C-V2X technology has important advantages, such as a physical layer potentially better adapted to the characteristics of the channel [92], greater bandwidth, and the existence of an infrastructure already deployed (which in the case of the DSRC should be likely defrayed by public administrations)."

"V2X communications are carried out over harsh channels with three important characteristics: high vehicle speed, changing environment and low antenna height (except in cellular V2I transmissions)."

Factors Influencing the Channel Characteristics
"The following ones are addressed: the type of communication link (direct between vehicles or through an infrastructure), the kind of scenario (rural, suburban, urban) and the type of vehicle and antenna."

"The surrounding scatterers have considerable influence and the characteristics of the resulting channels significantly differ from the C-V2X ones."

"The composite channel, including the antennas, is referred to as radio channel and is characterized by a non-directional channel response that results from the angle-integration of the former weighted by the gain pattern of the transmitter and receiver antennas [93]."

"The characteristics of the radio channel also depend on the type of vehicle (truck, bus, van or car), mainly because of their different antenna height."

Channel Characteristics
"The simplest characterization of a mobile radio channel is done in terms of the average attenuation as a function of the distance between the transmitter and the receiver."

"It is a four-lane highway in which the transmitter and the receiver are moving in opposite directions at 100 km/h. There are three additional vehicles, which act as moving scatterers, four static discrete scatterers (the one located in (202,7.6) may model an overhead traffic sign) and a large number of (static) diffuse ones."

"The urban and highway environments exhibit the largest delay spread values."

"This same reason also explains that larger delay spread values are obtained both in scenarios with denser traffic, as in the highway one, or NLOS channels."

"Regarding the Doppler spread, since it is associated to the speed of the vehicles, the largest values are obtained in the highway scenario and the lowest in the urban one."

Channel Models
"Even in mobile channels (with a more modest mobility) this model must be complicated to cover non-stationary behavior by creating different setting classes like

"urban", "suburban", "typical", "bad", etc. Second, because the different transmission paths followed by the signal in V2V usually exhibit an important correlation."

"NGSM are the simplest models to implement and their behavioral parameters (e.g. number of echoes, their weights, delays, etc.) are set according to statistical processing of channel estimations from measurement campaigns."

"The work in [94] carries out a characterization in which the echoes of the channel response (apart from the large-scale fading) are identified and the distribution of their amplitudes are modeled with Ricean and Rayleigh functions (for the first echo, corresponding to the LOS component, and for the remaining ones, respectively)."

"The small-scale fading is represented by a TDL model whose low-pass equivalent can be expressed as [85]. where The involved variables are: t, time at which the channel response is observed."

Implications of Vehicular Channels on Link Level Performance

"Since the IEEE 802.11p standard uses a much larger subcarrier spacing than LTE-V, it also requires larger coherence bandwidth (lower channel delay spread) if a single tap equalizer is to be used."

"Long delay spread values, such as those measured over urban scenarios or highways, may yield coherence bandwidth as low as 80 kHz, which is well below the subcarrier bandwidth of IEEE 802.11p technology, while the channel can still be considered flat for LTE-V. Moreover, as LTE-V has a higher number of demodulation reference signals than IEEE 802.11p, the cellular technology is able to obtain a more accurate channel frequency estimation (at the expense of extra overhead)."

"The higher subcarrier spacing of the IEEE 802.11p makes it more resilient than LTE-V to the distortion from large Doppler spread values, or equivalently, its shorter symbol duration reduces ICI as the time-selectivity affects the orthogonality of the carriers."

Conclusions

"The implications that V2V channels features have on communication systems are analyzed in this work."

"The first part of the paper focuses on channel characterization and modeling."

"The physical elements of these high mobility scenarios affecting the channel behavior are discussed and the way they are included in the models proposed in the literature is reviewed."

"A summary of the main channel behavioral parameters, collected from published works, are given for urban, suburban, rural and highway scenarios."

Machine Learning for Wireless Communication Channel Modeling: An Overview [95]

This is a machine-generated summary of:

Aldossari, Saud Mobark; Chen, Kwang-Cheng: Machine Learning for Wireless Communication Channel Modeling: An Overview [95]

Published in: Wireless Personal Communications (2019)
Link to original: https://doi.org/10.1007/s11277-019-06275-4

If you want to cite the papers, please refer to the original.

For technical reasons we could not place the page where the original quote is coming from.

Abstract-Summary

"For highly mobile communications, channel estimation on top of the channel modeling enables high bandwidth physical layer transmission in state-of-the-art mobile communications."

"For the coming 5G and diverse Internet of Things, many challenging application scenarios emerge and more efficient methodology for channel modeling and channel estimation is very much needed."

"Applying machine learning to assist channel modeling and channel estimation has been introduced with evidence of literature survey."

Introduction

"The main focus of this paper is how machine learning will be used in channel modeling."

"The stochastic modeling method is based on the statistical distributions of the channel parameters and can be either narrowband or wideband."

"Fifth generation (5G) with massive MIMO, higher data rate, quick handover, and channel modeling become more and more complex than the traditional stochastic or deterministic methods."

"Machine learning for wireless channel modeling is not only best fitting a model with appropriate parameters but also keeping the model updated to obtain more precise operating of wireless communication systems."

"Using classical estimation method such as SAGE to extract the channel parameters."

"Involving and applying machine learning to extract the CSI and then applying estimation techniques using method such as regression, classification and clustering etc to estimate the channel parameters that will lead to a better performance and lower complexity."

Channel Model

"Fading is identified as the time variation of the received multipath components MPCs power that went through the medium channel effects where it can be categorized into two parts as Small-Scale Fading and Large-Scale Fading."

"Propagation in wireless channel models results in either large or small scale fading base on the variations of the signal."

"Frequency selective fading: Frequency selective fading is due to the time spread of the transmitted symbols within the channel [96]."

"This dispersion is due to the signal bandwidth larger than the coherence bandwidth while frequency selective fading is due to time spread in the time domain."

"If the baseband signal bandwidth is much less than the Doppler spread then effect of Doppler spread is negligible at the receiver and this would be a slow fading channel characteristics."

ML Usage in the Wireless Communication Channel Modeling

"For the purpose of learning and predicting, common methodologies to realize ML are training, testing, and validating wherein training part, learning from the data while the testing method, a trained model is used for predicting such as channel modeling parameters [97]."

"Let (x, y) the measurement data of the wireless channel such CIR is used in train part using back propagation algorithm or Newton algorithm [98] and the DNN algorithms to estimate the weight parameters to build the model channel."

"Zhang [99] went over an interdisciplinary study of using massive channel modeling data and used cluster-nuclei based channel model and matching the propagation signals to machine learning algorithms using measurement data."

"Data mining methods used in Machine learning as other algorithms to build models that predict our needs Authors of [100] believe that the combination of data mining and machine learning methods and huge dataset, channel modeling parameters can be extracted and controlled."

Applications of Machine Learning in Channel Modeling

"Machine learning is needed to examine the channel modeling and focusing in regarding large scale and small-scale fading where fading refers to the time changes of the MPC's power."

"By applying machine learning to the vehicle to vehicle (V2V) or vehicle to infrastructure (V2I) to the wireless propagation channels to measure the parameters such as large scale and small scale fading such as path loss, shadow, delay parameter."

"This section describes some machine learning techniques toward the challenging V2V/V2I channel modeling."

"Applying machine learning to the vehicle to vehicle (V2V) or vehicle to infrastructure (V2I) is still an open subject and how to alternate from the traditional ways of measuring the channel empirical model or the ray-tracing methods."

"Due to the increasing complexity due to a large number of operations and large measurement, researcher tend to perform machine learning to enhances the channel modeling."

"Analysis massive data of channel modeling with advanced machine learning algorithms [101]."

Conclusion

"This paper presents an overview of theoretical framework toward machine learning problems for future wireless communication channel modeling."

"This paper went over some useful methods in machine learning in the field of wireless communication such as regression analysis which were used for data fitting to predict the channel modeling parameters."

"There are so many investments and time from academia and industry have been invested in machine learning of the wireless channel modeling and we still have a long way to overcome the challenges and achieve the precise modeling."

"Involve machine learning to the wireless channel modeling is still on the preliminary phase and such that work will lead and guide scholars to achieve that goal."

"The scientific explanation of using machine learning in wireless communication channel modeling has been shown with evidences and reasons through this journey, while mobility introduces further technological opportunities."

PLC-RF Diversity: Channel Outage Analysis [102]

This is a machine-generated summary of:

Igboamalu, Frank Nonso; Ndjiongue, Alain Richard; Ferreira, Hendrik Christoffel: PLC-RF diversity: channel outage analysis [102]

Published in: Telecommunication Systems (2019)

Link to original: https://doi.org/10.1007/s11235-019-00624-1

If you want to cite the papers, please refer to the original.

For technical reasons we could not place the page where the original quote is coming from.

Abstract-Summary

"Signals from indoor access points in a wireless fidelity suffer from high attenuation as a result of the thick walls and metal structures in the multilevel-storey buildings where they are deployed."

"This is due to the scattering of the retransmitted signals by the thick walls and metal structures."

"A combination of the RF and PLC technologies is proposed for achieving signal transmission in indoor environments taking into account the presence characteristic of multilevel-storey buildings."

Introduction

"The hybrid power line Communication-wireless (PLC-wireless) channel was analyzed for data communication purposes, using statistical modeling, which was presented in [103]."

"A previous study on PLC-wireless diversity combining can be found in [104], which combines the received signals for both narrowband power line communications (NB-PLC) and wireless link along with the diversity combining techniques and then compare them to the conventional combining techniques using closed-form expressions for average bit-error-rate."

"In [105], a solution for combining received signal for diversity combining techniques for hybrid NB-PLC and unlicensed wireless transmission for differential modulation schemes are proposed."

"In all the literatures discussed above, firstly, they all analyzed the performance diversity for PLC-wireless channel using the average BER or closed-form expression for the average BER which is an interesting channel analysis in terms of noise factor."

Description of the System Model

"The obtained coded and modulated signal is treated based on the OFDM principle, then the same signal is sent to both PLC and RF links for cyclic prefix indexation."

"These signals are transmitted using the corresponding channel."

"On the PLC side, a coupling circuit (CC) is exploited to couple the message signal to the channel."

"This constitutes the coupling of the RF signal to the RF channel."

"Another antenna at the receiver side is exploited to receive the message signal from the channel."

"The cyclic prefix on the received signal is removed, and the signal is then sent to the combining module."

"The diversity is achieve in SC, by selecting the signal with the highest SNR between PLC and RF at a particular time interval."

Analysis

"It shows the two coupling circuit connected to the PLC channel."

"The characteristic of the channel depends on the environment (PLC, RF)."

"PLC and RF may have the same fading characteristics which are not dependent on each other statistically."

"The Rayleigh distribution fits well in both PLC and RF channels here because there is no line of sight link between transmitter and receiver when the message is sent through both channels."

Results

"They show that the value of the coupling circuit gain clearly affect the probability of error in the MRC scenario."

"The reason to this is the filtering capacities of the coupling circuit."

"Recall that the analysis is given, considering that both asymmetric channels are Rayleigh distributed, which explains that the line-of-sight between the transmitter and the receiver does not exist."

Hybrid Technology Estimated Implementation Cost

"To analyze the cost implementation of the hybrid technology, it's important to consider the cost of the electronic component for the PLC and RF in general, which have a significate influence in the price of the modem."

"Both the PLC and RF will be controlled by the microcontroller, which will synchronize the communication modules of the two technologies."

"The combination of the PLC and RF components as one module using micro-controller will help to reduce the cost of components."

Conclusion

"This paper aimed to analyze the outage probability of PLC-RF diversity combining."

"Two different techniques are exploited to analyze this probability of failure namely selection combining and maximum ratio combining."

"In each of these cases, the expression of the system SNR is proposed, its PDF and its CDF which led to the system outage probability."

Bibliography

1. Sheikh, M. U., Ruttik, K., Jäntti, R., & Hämäläinen, J. (2021). A study about signal variation with minor receiver displacement in a meeting room at 60 GHz: Measurements and simulations. *EURASIP Journal on Wireless Communications and Networking, 2021*, 81. https://doi.org/10.1186/s13638-021-01965-9
2. Zhou, A., Huang, J., Sun, J., Zhu, Q., Wang, C., & Yang, Y. (2017). 60 GHz channel measurements and Ray tracing modeling in an indoor environment. In *The 9th International Conference on Wireless Communications and Signal Processing (WCSP)* (pp. 1–6).
3. Sheikh, M. U., Säe, J., & Lempiäinen, J. (2018). Multipath propagation analysis of 5G systems at higher frequencies in Courtyard (small cell) environment. In *The IEEE 5G World Forum (5GWF)* (pp. 239–243).
4. Sheikh, M.U., & Lempiäinen, J. (2017). Analysis of multipath propagation for 5G system at higher frequencies in microcellular environment. In *The 13th International Wireless Communications and Mobile Computing Conference (IWCMC)* (pp. 1660–1664).
5. Bedda-Zekri, A., & Ajgou, R. (2022). Statistical analysis of 5G/6G millimeter wave channels for different scenarios. *Journal of Communications Technology and Electronics, 67*, 854–875. https://doi.org/10.1134/s1064226922070063
6. Rappaport, T. S., Seidel, S. Y., & Takamizawa, K. (1991). Statistical channel impulse response models for factory and open plan building radio communicate system design. *IEEE Transactions on Communications, 39*, 794–807. https://doi.org/10.1109/26.87142
7. Hasan, R., Mowla, M. M., Rashid, M. A., Hosain, M. K., & Ahmad, I. (2019). A Statistical Analysis of Channel Modeling for 5G mmWaves Communications. In *International Conference on Electrical, Computer, and Communication Engineering (ECCE)*, Baltimore, Maryland, Sept. 29–Oct. 3, 2019, pp. 1–6. https://doi.org/10.1109/ECACE.2019.8679507
8. Hamid Dutty, H. B., & Mowla, M. M. 2019). Weather impact analysis of mm waves channel modeling for aviation backhaul networks in 5G communications. In *22nd International Conference on Computer and Information Technology*, Tejgaon, Dhaka, Bangladesh, 2019, (ICCIT), pp. 1–6. https://doi.org/10.1109/ICCIT48885.2019.9038447
9. Agrawal, S., & Sharma, K. (2019). 5th generation millimeter wave wireless communication propagation losses dataset for Indian metro cities based on corresponding weather conditions. *Data in Brief, 23*, 103564. https://doi.org/10.1016/j.dib.2018.12.003
10. Rudd, R., Craig, K., Ganley, M., & Hartless, R. (2014). *Building materials and propagation.* Final report; Ofcom, 2604.
11. Liang, Y.-J., & Chen, K.-C. (2021). Low complexity carrier frequency offset and channel estimation for multiuser OFDMA networks. *Wireless Personal Communications, 120*, 2813–2829. https://doi.org/10.1007/s11277-021-08558-1

12. Lee, J., Lee, S., Bang, K.-J., Cha, S., & Hong, D. (2007). Carrier frequency offset estimation using ESPRIT for interleaved OFDMA uplink systems. *IEEE Transactions on Vehicular Technology, 56*(5), 3227–3231.
13. Cheng, P., Chen, Z., de Hoog, F., & Sung, C. K. (2016). Sparse blind carrier-frequency offset estimation for ofdma uplink. *IEEE Transactions on Communications, 64*(12), 5254–5265.
14. Huang, M., Huang, L., Guo, C., Zhang, P., Zhang, J., & Yang, L.-L. (2017). Carrier frequency offset estimation in uplink ofdma systems: An approach relying on sparse recovery. *IEEE Transactions on Vehicular Technology, 66*(10), 9592–9597.
15. Bai, L., & Yin, Q. (2012). Frequency synchronization for the OFDMA uplink based on the tile structure of IEEE 802.16e. *IEEE Transactions on Vehicular Technology, 61*(5), 2348–2353.
16. Zhang, W., & Yin, Q. (2014). Blind carrier frequency offset estimation for tile-based orthogonal frequency division multiple access uplink with multi-antenna receiver. *IET Communications, 8*(8), 1309–1316.
17. Abdzadeh-Ziabari, H., Zhu, W., & Swamy, M. N. S. (2018). Joint carrier frequency offset and doubly selective channel estimation for MIMO-OFDMA uplink with Kalman and particle filtering. *IEEE Transactions on Signal Processing, 66*(15), 4001–4012.
18. Barhumi, I., Leus, G., & Moonen, M. (2003). Optimal training design for mimo ofdm systems in mobile wireless channels. *IEEE Transactions on Signal Processing, 51*(6), 1615–1624.
19. Liang, Y.-J., Stüber, G. L., Chang, J.-F., & Yang, D.-N. (2012). A joint channel and frequency offset estimator for the downlink of coordinated mimo-ofdm systems. *IEEE Transactions on Wireless Communications, 11*(6), 2254–2265.
20. Kazakov, L. N., Lukyanov, I. V., & Shakhtarin, B. I. (2019). Computationally efficient algorithm for estimating the frequency of signals at the output of a channel with white Gaussian noise and narrowband interference. *Journal of Communications Technology and Electronics, 64*, 1360–1366. https://doi.org/10.1134/s106422691910005x
21. Dvorkovich, V. P., & Dvorkovich, A. V. (2014). *Window functions for the harmonious analysis of signals*. Tekhnosfera. [in Russian].
22. Nikiforov, A. A. (2013). *Mekhanizatsiya Stroitel'stva, 11*, 53.
23. Nikiforov, A. A. (2014). *Identification and assessment of information parameters of navigation systems with code separation*. Candidate's dissertation in technical sciences, MGTU im. Baumana.
24. Ostanin, S. A. (2011). *Journal of Radio Electronics, 12*. Retrieved from http://jre.cplire.ru/win/dec11/13/text.pdf.
25. Volkov, V. G., Krivov, Y. N., & Luk'yanov, I. V. (2016). *Journal of Radio Electronics, 10*. Retrieved from http://jre.cplire.ru/jre/oct16/6/pdf.
26. Shrivastav, K., Yadav, R. P., & Jain, K. C. (2019). Joint MAP detection for OFDM in presence of phase noise from free running and phase locked loop oscillator. *Wireless Personal Communications, 109*, 563–577. https://doi.org/10.1007/s11277-019-06579-5
27. Wu, S., & Bar-Ness, Y. (2004). OFDM systems in the presence of phase noise: Consequences and solutions. *IEEE Transactions on Communications, 52*(11), 1988–1996.
28. Petrovic, D., Rave, W., & Fettweis, G. (2007). Effect of phase noise on OFDM systems with and without PLL: Characterization and compensation. *IEEE Transactions on Communications, 55*(8), 1607–1616.
29. Syrjälä, V., Valkama, M., Tchamov, N. N., & Rinne, J. (2009). Phase noise modelling and mitigation techniques in OFDM communications systems. In *Proceedings of the Wireless Telecommunications Symposium* (pp. 1–7).
30. Lin, D. D., & Lim, T. J. (2007). The variational inference approach to joint data detection and phase noise estimation in OFDM. *IEEE Transactions on Signal Processing, 55*(5), 1862–1874.
31. Stefanatos, S., & Katsaggelos, A. K. (2008). Joint data detection and channel tracking for OFDM systems with phase noise. *IEEE Transactions on Signal Processing, 56*(9), 4230–4243.

32. Munier, F., Eriksson, T., & Svensson, A. (2008). An ICI reduction scheme for OFDM system with phase noise over fading channels. *IEEE Transactions on Communications, 56*(12), 1119–1126.
33. Gong, Y., & Hong, X. (2009). OFDM joint data detection and phase noise cancellation for constant modulus modulations. *IEEE Transactions on Signal Processing, 57*(7), 2864–2868.
34. Rabiei, P., Namgoong, W., & Al-Dhahir, N. (2010). A non-iterative technique for phase noise ICI mitigation in packet-based OFDM systems. *IEEE Transactions on Signal Processing, 58*(11), 5945–5950.
35. Salim, O. H., Nasir, A. A., Mehrpouyan, H., et al. (2014). Channel, phase noise, and frequency offset in OFDM systems: Joint estimation, data detection, and hybrid Cramér-Rao lower bound. *IEEE Transactions on Communications, 62*(9), 3311–3325.
36. Negusse, S., Zetterberg, P., & Händel, P. (2015). Phase-noise mitigation in OFDM by best match trajectories. *IEEE Transactions on Communications, 63*(5), 1712–1725.
37. Nelder, J. A., & Mead, R. (1965). A simplex method for function minimization. *The Computer Journal, 7*, 308–313.
38. Singh, P., & Vasudevan, K. (2019). Time domain channel estimation for MIMO-FBMC/OQAM systems. *Wireless Personal Communications, 108*, 2159–2178. https://doi.org/10.1007/s11277-019-06515-7
39. Taheri, S., Ghoraishi, M., & Xiao, P. (2015). Overhead reduced preamble-based channel estimation for MIMO–FBMC systems. In *2015 International Wireless Communications and Mobile Computing Conference (IWCMC)* (pp. 1435–1439). IEEE.
40. Lélé, C., Javaudin, J. P., Legouable, R., Skrzypczak, A., & Siohan, P. (2008). Channel estimation methods for preamble-based OFDM/OQAM modulations. *Transactions on Emerging Telecommunications Technologies, 19*(7), 741–750.
41. Bharadwaj, R., & Koul, S. K. (2019). Analytical and experimental investigation of ultra Wideband Channel characteristics in the presence of door/window glass. *Wireless Personal Communications, 110*, 763–780. https://doi.org/10.1007/s11277-019-06753-9
42. Ye, T., Walsh, M., Haigh, P., Barton, J., & O'Flynn, B. (2011). Experimental impulse radio IEEE 802.15.4a UWB based wireless sensor localization technology: Characterization, reliability and ranging. In *22nd IET Irish Signals and Systems Conference*, Dublin, Ireland (pp. 1–7).
43. Ghassemzadeh, S. S., Greenstein, L. J., Sveinsson, T., Kavcic, A., & Tarokh, V. (2005). UWB delay profile models for residential and commercial indoor environments. *IEEE Transactions Vehicular Technology, 54*(4), 1235–1244.
44. Moghimi, A. R., Tsai, H., Saraydar, C. U., & Tonguz, O. K. (2009). Characterizing intra-car wireless channels. *IEEE Transactions on Vehicular Technology, 58*(9), 5299–5305.
45. Chiu, S., & Michelson, D. G. (2010). Effect of human presence on UWB radiowave propagation within the passenger cabin of a midsize airliner. *IEEE Transactions on Antennas and Propagation, 58*(3), 917–926.
46. Sun, Z., & Akyildiz, I. F. (2010). Channel modeling and analysis for wireless networks in underground mines and road tunnels. *IEEE Transactions on Communications, 58*(6), 1758–1768.
47. Chehri, A., Fortier, P., & Tardif, M. P. (2012). Characterization of the ultra-wideband channel in confined environments with diffracting rough surfaces. *Wireless Personal Communications, 64*(2), 859–877.
48. Ramesh, C., & Vaidehi, V. (2007). Performance analysis of UWB channels for wireless personal area networks. *Wireless Personal Communications, 4*, 169–178.
49. Goulianos, A. A., Brown, T. W. C., Evans, B. G., & Stavrou, S. (2009). Wideband power modeling and time dispersion analysis for UWB indoor off-body communications. *IEEE Transactions on Antennas and Propagation, 57*(7), 2162–2171.
50. Alomainy, A., Sani, A., Rahman, A., Santas, J. G., & Hao, Y. (2009). Transient characteristics of wearable antennas and radio propagation channels for ultra-wideband body-centric wireless communications. *IEEE Transactions on Antennas and Propagation, 57*(4), 875–884.

51. Bharadwaj, R., Parini, C., & Alomainy, A. (2015). Experimental investigation of 3-D human body localization using wearable ultra-wideband antennas. *IEEE Transactions on Antennas and Propagation, 63*(11), 5035–5044.
52. Wong, S. S. M., Lau, F. C. M., & Tse, C. K. (2006). Propagation characteristics of UWB radio in a high-rise apartment. In *2006 8th International Conference Advanced Communication Technology*, Phoenix Park (pp. 5–918).
53. Bharadwaj, R., Swaisaenyakorn, S., Parini, C. G., Batchelor, J., & Alomainy, A. (2014). Localization of wearable ultra-wideband antennas for motion capture applications. *IEEE Antennas Wireless Propagation Letters, 13*, 507–510.
54. Rappaport, T. S. (1996). *Wireless communications principles and practice*. Prentice Hall Inc.
55. Gezici, S., & Poor, H. V. (2009). Position estimation via ultra-wide-band signals. *Proceedings of the IEEE, 97*(2), 386–403.
56. Guvenc, I., & Sahinoglu, Z. (2005). Threshold-based TOA estimation for impulse radio UWB systems. In *Proceeding of IEEE International Conference on Ultra-Wideband*, Zurich, Switzerland (pp. 420–425).
57. Zhou, L., Liao, W., & Hou, S. (2021). Analysis of dynamic channel characteristics based on a body area network. *Wireless Personal Communications, 122*, 2905–2924. https://doi.org/10.1007/s11277-021-09035-5
58. Ramgoolam, K., & Bassoo, V. (2019). Maximising energy efficiency for direct communication links in wireless body area networks. *IET Wireless Sensor Systems, 9*(1), 32–41.
59. Wang, J., Fujiwara, T., Kato, T., et al. (2016). Wearable ECG based on impulse radio type human body communication. *IEEE Transactions on Biomedical Engineering, 63*(9), 1887–1894.
60. Kang, T., Lim, I., Park, K., Kang, S., & Kim, S. (2014). Improving data rate in the human body communications. In *2014 International Conference on Information and Communication Technology Convergence (ICTC)*, Busan (pp. 51–52).
61. Li, M. A., Shen, H. B., & Liu, L. (2013). Analysis of wave propagation on human body based on stratified media model. *Journal of Central South University, 20*(12), 3545–3551.
62. Awan, M. F., Fang, X., Ramzan, M., et al. (2019). Evaluating secrecy capacity for in-body wireless channels. *Entropy, 21*(9), 858.
63. Rout, D. K., & Das, S. (2017). Multiple narrowband interference mitigation using hybrid hermite pulses for body surface to external communications in UWB body area networks. *Wireless Networks, 23*(2), 387–402.
64. Vadivelu, R., Santhakumar, G., Kaviya, P., Menaka, M., Monisha, S., & Balasubramaniam, D. (2021). Human body communication on portable biometric authentication. *Journal of Physics: Conference Series, 1916*(1), 012113.
65. Nath, M., Maity, S., Avlani, S., Weigand, S., & Sen, S. (2021). Inter-body coupling in electro-quasistatic human body communication: Theory and analysis of security and interference properties. *Scientific Reports, 11*(1), 1–15.
66. Bhise, S. R., & Khot, U. P. (2020). Measurement and analysis of Wi-MAX channel in 3.3 GHz using minimal-ray fading model. *Wireless Networks, 26*, 3591–3602. https://doi.org/10.1007/s11276-020-02280-9
67. Borhani, A., & Ptzold, M. (2014). A novel non-stationary channel model utilizing Brownian random paths. *REV Journal on Electronics and Communications, 4*(1–2), 8–15.
68. Rorabaugh, C. B. (2004). *Simulating wireless communication system: Practical models in C++*. Prentice Hall.
69. Bhise, S., & Khot, U. (2019). Random phase and time of arrival statistics for circular disk scattering model. *Springer Wireless Networks, 25*(6), 3579–3585.
70. Mohanty, B., Sahoo, H. K., & Patnaik, B. (2020). Wireless channel estimation and beamforming by using block sparse adaptive filtering. *Signal, Image and Video Processing, 15*, 769–777. https://doi.org/10.1007/s11760-020-01795-0
71. Zhang, Z., Xiao, L., Su, X., Zeng, J., & Xu, X. (2018). A channel estimation method based on the improved LMS algorithm for MIMO-OFDM systems. In *2018 12th International*

Symposium on Medical Information and Communication Technology (ISMICT), Sydney, NSW, pp. 1–5. https://doi.org/10.1109/ismict.2018.8573728
72. Li, Y., Wang, Y., & Jiang, T. (2016). Sparse-aware set-membership NLMS algorithms and their application for sparse channel estimation and echo cancelation. *AEÜ International Journal of Electronics and Communications, 70*(7), 895–902.
73. Berraki, D. E., Armour, S. M. D., & Nix, A. R. (2018). Benefits of the sparsity of mmWave outdoor spatial channels for beamforming and interference cancellation. *Physical Communication, 27*, 170–180.
74. Gao, Q., & Stewart, R. (2011). Improved double angle complex rotation QRD-RLS. In *Proceedings of the 19th ACM/SIGDA International Symposium on Field Programmable Gatearrays-FPGA'* (pp. 79–82).
75. Sahoo, H. K., Mohanty, B., & Patnaik, B. (2017). Block and fast block sparse adaptive filtering for outdoor wireless channel estimation and equalization. *Wireless Personal Communications, 98*(3), 3003–3019.
76. Saxena, P., & Kothari, A. G. (2014). Performance: Analysis of adaptive beamforming algorithms for smart antennas. *IERI Procedia, 10*, 131–137.
77. Mirza, M.-Y. M., & Khan, N. M. (2020). G2A communication channel modeling and characterization using confocal prolates. *Wireless Personal Communications, 115*, 745–787. https://doi.org/10.1007/s11277-020-07597-4
78. Khawaja, W., Guvenc, I., & Matolak, D. (2016). Uwb channel sounding and modeling for uav air-to-ground propagation channels. In *2016 IEEE Global Communications Conference (GLOBECOM)* (pp. 1–7).
79. Matolak, D. W., & Sun, R. (2013). Air-ground channel measurements amp; modeling for uas. In *Proceedings of the 2013 Integrated Communications, Navigation and Surveillance Conference (ICNS)* (pp. 1–9).
80. Matolak, D. W., & Sun, R. (2014). Air-ground channel characterization for unmanned aircraft systems: The hilly suburban environment. In *IEEE 80th Vehicular Technology Conference* (pp. 1–5). IEEE.
81. Matolak, D. W., & Sun, R. (2014) Antenna and frequency diversity in the unmanned aircraft systems bands for the over-sea setting. In *IEEE/AIAA 33rd Digital Avionics Systems Conference (DASC)* (pp. 6A4–1–6A4–10).
82. Matolak, D. W., & Sun, R. (2015) Air-ground channel characterization for unmanned aircraft systems: The near-urban environment. In *2015 IEEE Military Communications Conference* (pp. 1656–1660).
83. Matolak, D. W., & Sun, R. (2017). Air-ground channel characterization for unmanned aircraft systems-part I: Methods, measurements, and models for over-water settings. *IEEE Transactions on Vehicular Technology, 66*(1), 26–44.
84. Matolak, D. W., & Sun, R. (2017). Air-ground channel characterization for unmanned aircraft systems-part III: The suburban and near-urban environments. *IEEE Transactions on Vehicular Technology, 66*(8), 6607–6618.
85. Meng, Y. S., & Lee, Y. H. (2011). Measurements and characterizations of air-to-ground channel over sea surface at c-band with low airborne altitudes. *IEEE Transactions on Vehicular Technology, 60*(4), 1943–1948. https://doi.org/10.1109/TVT.2011.2136364
86. Schneckenburger, N., Jost, T., Shutin, D., Walter, M., Thiasiriphet, T., Schnell, M., et al. (2016). Measurement of the L-band air-to-ground channel for positioning applications. *IEEE Transactions on Aerospace and Electronic Systems, 52*(5), 2281–2297.
87. Sun, R., & Matolak, D. W. (2014). Over-harbor channel modeling with directional ground station antennas for the air-ground channel. In *2014 IEEE Military Communications Conference* (pp. 382–387). https://doi.org/10.1109/MILCOM.2014.69
88. Sun, R., & Matolak, D. W. (2017). Air-ground channel characterization for unmanned aircraft systems-part II: Hilly and mountainous settings. *IEEE Transactions on Vehicular Technology, 66*(3), 1913–1925.

89. Newhall, W. G., & Reed, J. (2002). A geometric air-to-ground radio channel model. In *IEEE Military Communications Conference* (pp. 632–636).
90. Cortés, J. A., Aguayo-Torres, M. C., Cañete, F. J., Gómez, G., Martos-Naya, E., & Entrambasaguas, J. T. (2019). Vehicular channels: Characteristics, models and implications on communication systems design. *Wireless Personal Communications, 106*, 237–260. https://doi.org/10.1007/s11277-019-06269-2
91. 3rd Generation Partnership Project. (2017). *Study on enhancement of 3GPP support for 5G V2X services*. 3GPP TR 22.886, Release 15.
92. Qualcomm. (2018). *V2x technology benchmark testing*. Retrieved October 24, 2018, from https://www.qualcomm.com/media/documents/files/5gaa-v2x-technology-benchmark-testing-dsrc-and-c-v2x.pdf.
93. Steinbauer, M., Molisch, A. F., & Bonekl, E. (2001). The double-directional radio channel. *IEEE Antennas and Propagation Magazine, 43*(4), 51–63.
94. Bernado, L., Zemen, T., Tufvesson, F., Molisch, A. F., & Mecklenbrauker, C. F. (2015). Time- and frequency-varying -factor of non-stationary vehicular channels for safety-relevant scenarios. *IEEE Transactions on Intelligent Transportation Systems, 16*(2), 1007–1017.
95. Aldossari, S. M., & Chen, K.-C. (2019). Machine learning for wireless communication channel modeling: An overview. *Wireless Personal Communications, 106*, 41–70. https://doi.org/10.1007/s11277-019-06275-4
96. Rappaport, T. T. S., DiPierro, S., & Akturan, R. (2011). Analysis and simulation of interference to vehicle-equipped digital receivers from cellular mobile terminals operating in adjacent frequencies. *IEEE Transactions on Vehicular Technology, 60*, 1664–1676.
97. Riihijarvi, J., & Mahonen, P. (2018). Machine learning for performance prediction in mobile cellular networks. *IEEE Computational Intelligence Magazine, 13*(1), 51–60. https://doi.org/10.1109/MCI.2017.2773824
98. Mitchell, T. M. (1997). *Machine learning*. McGraw-Hill.
99. Zhang, J. (2016). The interdisciplinary research of big data and wireless channel: A cluster-nuclei based channel model. *China Communications, 13*(Supplement 2), 14–26. https://doi.org/10.1109/CC.2016.7833457
100. Ma, X., Zhang, J., Zhang, Y., & Ma, Z. (2017). Data scheme-based wireless channel modeling method: Motivation, principle and performance. *Journal of Communications and Information Networks, 2*(3), 4151.
101. He, H., Wen, C.-K., Jin, S., & Li, G. Y. (2018). Deep learning-based channel estimation for beamspace mmWave massive MIMO system. *arxiv*, p. 25.
102. Igboamalu, F. N., Ndjiongue, A. R., & Ferreira, H. C. (2019). PLC-RF diversity: Channel outage analysis. *Telecommunication Systems, 73*, 521–530. https://doi.org/10.1007/s11235-019-00624-1
103. Oliveira, T. R., Picorone, A. A. M., Zeller, C. B., Netto, S. L., & Ribeiro, M. V. (2018). On the statistical characterization of hybrid PLC-wireless channels. *Electric Power Systems Research, 163*, 329–337.
104. Sayed, M., Tsiftsis, T. A., & Al-Dhahir, N. (2017). On the diversity of hybrid narrowband-PLC/wireless communications for smart grids. *IEEE Transactions on Wireless Communications, 16*(7), 4344–4360.
105. Sayed, M., & Al-Dhahir, N. (2016). Differential modulation diversity combining for hybrid narrowband-powerline, wireless smart grid communications. In *Proceedings of the IEEE Global Conference on Signal and Information Processing* (pp. 876–880). December 7–9, Washington, DC, USA.

Chapter 3
Multiple Input Multiple Output Schemes in 3G, 4G, and 5G Networks

Milind Pande, Anand J. Kulkarni, and Apoorva S. Shastri

Introduction by the Editor

Multiple-Input Multiple-Output (MIMO) systems have many applications in Wireless and Mobile Networks. This technique can be used to improve the spectral efficiency of the system and increase the capacity of the channel, avoid the fading of the channel and to avoid interference. In addition, can be used for security as well as tracking and detection. Hence the performance analysis of these systems is of foremost importance. The chapter covers mainly the techniques for performance analysis of MIMO systems such as the Hybrid Filtering Technique for MIMO-OFDM Systems, artificial intelligence solutions beyond 5G radio access networks, 5G-V2X, Spectral Efficiency, and Pilot Contaminated Massive MIMO Systems, SBPA in Crowded Massive MIMO Systems, Algorithms for Automatic Selection of Antennas in a MIMO, Deep Neural Network for Compressive Sensing and Application to Massive MIMO Channel Estimation, MIMO-OFDM System with Diverse Transformation for 5G Applications, Timing Offset Estimation and Correction in OFDM Assisted Massive MIMO Systems, single-cell massive MIMO systems. Various channel estimation methods for MIMO-OFDM and massive MIMO-GFDM systems are discussed, including Taylor-Based Least Square Estimation, Elman Recurrent Neural Network, and Polynomial Expansion-Based MMSE Channel Estimation. Hybrid optimizer-based recurrent neural network long short-term memory, convex combination-based algorithms, and deep learning are

M. Pande
Dr. Vishwanath Karad MIT World Peace University, Pune, Maharashtra, India
e-mail: milind.pande@mitwpu.edu.in

A. J. Kulkarni · A. S. Shastri (✉)
Institute of Artificial Intelligence, Dr. Vishwanath Karad MIT World Peace University, Pune, Maharashtra, India
e-mail: anand.j.kulkarni@mitwpu.edu.in

M. Pande et al. (eds.), *Optimization Methods in Mobile Communication Systems*, https://doi.org/10.1007/978-981-95-1810-4_3

also addressed for channel estimation in 5G systems. Optimized pilot-based channel estimation and large-scale MIMO SC-FDMA uplink systems have also been explored. Additionally, robust beamforming and spatial precoding techniques are being studied.

Orthogonal frequency division multiplexing (OFDM) suffers from the drawback of a high peak-to-average power ratio (PAPR). The signal is clipped directly into the discrete-time domain at the Nyquist sampling rate, in-band distortion affects the high-frequency components of the in-band signal. A hybrid filtering algorithm removes the high-frequency components within the feasible regions, which can efficiently reduce the PAPR of OFDM signals. This can be achieved with two steps in the first step before clipping, the original signal is oversampled at a rate slightly exceeding the Nyquist rate. In the second step, the signal is then processed using a longer inverse fast Fourier transform (IFFT) length to reshape a portion of the noise outside the signal band.

AI/ML techniques have been seen to be promising in many applications in communication networks, and there is no doubt that AI/ML would be an inherent element in 5G systems and beyond. When designing an ML-enabled radio access network (RAN), depending on the locations of data sources, the ML training host, the ML inference host, and the point of actuation, three types of AI architectures can be identified: centralized, completely decentralized, and hybrid.ML algorithms under the hybrid architecture are trained locally by their respective user plane (UP), and the central server in the control plane (CP) is used as a coordinator to orchestrate the different steps of the algorithms and coordinate all the participating nodes during the learning process. In a network with centralized AI architecture, the training node collects data from data sources across the network and trains the model accordingly.

Mobility has been enhanced in 5G technology to support a relative vehicle speed of up to 280 km/h in LTE-based V2X, and for the 5G target, it is expected to reach 500 km/h relative speed. A direct communication interface is needed in two modes of network coverage (in and out) in 5G base V2X to reduce latency and improve network communication between vehicles. Several enhancements, including physical and MAC layer improvements, have been made. In autonomous driving applications, real-time data such as map exchange, video streaming, speeds, and locations are exchanged. Designing the system architecture of 5G for vehicular communications is challenging to support all requirements of vehicles and services. The 5GCAR project is considered as one of the essential 5G projects that propose the 5G-V2X techniques, applications, use cases, and architectures to support the V2X connectivity within 5G. The 5G-V2X design is involving five domains of network management, security, multi-RAT connectivity, edge computing, and network slicing. The 5G-V2X network core is proposed, including the network management using SDN, the slicing technology, the 5G virtualized functions, security, and privacy are considered. The 5GCAR project involves different approaches for the 5G-V2X security check for the V2X messages by applying the security checks at the vehicles while utilizing the network connectivity presence.

To combat the effect of non-orthogonality in generalized frequency division multiplexing (GFDM), interference-free pilots are used in frequency-domain minimum mean square error (MMSE) channel estimation. The direct use of the minimum mean square error (MMSE) channel estimator in massive MIMO systems could result in unacceptable complexity. Polynomial expansion-based estimators reduce the computational complexity in channel estimation for massive MIMO.

In MIMO systems complexity of the signal increases with an increase in antennas. To reduce this complexity automatic antenna selection method is required. Antenna selection algorithm reduces computational complexity and does not require enumeration of all combinations of rows and columns of the initial matrix of the MIMO radio channel. An algorithm based on a minimum trace of the correlation matrix.

Deep neural network exploits the problem of sparse signals in multiple measurement vectors (MMVs). The subspace pursuit (SP) algorithm reconstructs sparse signals. The basic idea of reconstruction sparse signal is to represent this signal as a linear combination of a few columns of a dictionary matrix.

The performance of 5G can be improved by integrating orthogonal frequency division multiple access (OFDMA) and multi-user massive multiple input multiple output (MIMO) technology. Integration increases channel reliability for 5G networks. The performance of multi-user massive MIMO-OFDM will be improved by replacing the conventional Fourier transform (FFT) by discrete wavelet transform (DWT) and fractional Fourier transform (FRFT) in the OFDM physical layer. In the MIMO-OFDM system, the major concern in estimating the channel with high precision lies in maintaining the performance of the system. Taylor-Least Square Error (TLSE) algorithm is based on the traditional Least Square Error (TLSE) algorithm. This algorithm helps to enhance channel estimation with the help of Taylor series.

The Elman recurrent neural network (E-RNN) algorithm estimates the channel in MIMO-OFDM considering reliability and scalability. The E-RNN is based on robust features and remembers information, which is helpful in time-series predictions for the channel estimation, the E-RNN is applied on the receiver side. E-RNN estimates the channel, and the fast Fourier transform (FFT) is applied to these signals. The exploitation of sparseness in adaptive filters upgrades their performance regarding convergence rate, robustness, steady-state performance, and tracking ability. Convex Combination Based Algorithm computationally efficient adaptive algorithm-based channel estimation model. Polynomial expansion is used to approximately compute the matrix inverses in the conventional MMSE estimation and precoding, consequently reducing the cubic computational complexity. The degree of the matrix polynomial can be properly selected to get a required trade-off between complexity and estimation/precoding performance. The polynomial expansion technique can be used to reduce the computational complexity. spatial correlation effects (SCE) and mutual coupling effects (MCE) degrade transmission capability with bit error rate. Beamforming scheme and spatial precoding, optimizing beamforming coefficients and antenna positions for both linear and circular arrays which in turn improve efficiency of transmission of data.

Machine Generated Summaries

Disclaimer: The summaries in this chapter were generated from Springer Nature publications using extractive AI auto-summarization: An extraction-based summarizer aims to identify the most important sentences of a text using an algorithm and uses those original sentences to create the auto-summary (unlike generative AI). As the constituted sentences are machine selected, they may not fully reflect the body of the work, so we strongly advise that the original content is read and cited. The auto generated summaries were curated by the editor to meet Springer Nature publication standards. To cite this content, please refer to the original papers.

Machine generated keywords: mimo, pilot, estimation, channel estimation, massive mimo, mimo system, massive, channel, matrix, csi, mimoofdm, ofdm, sparse, antenna, complexity

Closed-form Analysis of RZF in Multicell Massive MIMO Over Correlated Rician Channel [1]

This is a machine-generated summary of:

Kaur, Harleen; Kansal, Ankush: Closed-form Analysis of RZF in Multicell Massive MIMO Over Correlated Rician Channel [1]

Published in: Wireless Personal Communications (2023)

Link to original: https://doi.org/10.1007/s11277-023-10518-w

If you want to cite the papers, please refer to the original.

For technical reasons we could not place the page where the original quote is coming from.

Abstract-Summary

"Determining correlation matrices while estimating the channel and precoding is vital in providing spatial directivity to the signals."

"This paper considers the performance analysis of the downlink multi-cell multiuser Massive Multiple Input Multiple Output (mMIMO) system with user-specific correlation matrices for the Rician Fading channel."

"Using the Random Matrix Theory (RMT) results, the closed-form for the asymptotic analysis of ergodic downlink rate per user terminal (UT) is derived for the spatially correlated system with Regularized Zero-Forcing (RZF) scheme."

"The numerical results show that the derived expression for ergodic user rate provides a tight approximation for the large-dimensional system and accurate results for small values."

Introduction

"The TDD-based single-cell mMIMO downlink system with an uncorrelated Rayleigh channel environment uses imperfect CSI and calibration errors to design a robust joint combiner and precoder algorithm to calibrate closed-form expression for the achievable sum rate [2]."

"In [3], the Rician-based model is considered to extend the mathematical analysis towards more general conditions by determining the tractable expression for achievable rate in the case of uplink-based single-cell mMIMO using maximum ratio combing (MRC) and ZF receiver."

"The [4] determines the sum rate for the multicell mMIMO downlink system, and [5] provides derivation for achievable rate for uplink and downlink mMIMO system with uncorrelated Rician channel for maximum ratio transmission (MRT) and ZF schemes."

"In [6], with RMT, the expression for ergodic sum rate of downlink-based single-cell mMIMO system is derived for spatially correlated Rician Fading model with RZF technique for perfect CSI."

"In [7], the analytical analysis using ZF, MR, SMMSE, and MMSE techniques is provided but using the spatially correlated Rayleigh fading model for the multicell mMIMO case."

System Model

"The channel considered between the UTs and BSs in different cells is a spatially correlated Rayleigh fading model that is represented as For TDD mode, the system utilizes the pilots transmitted by the UTs in uplink mode to estimate the channel for processing the receive signals and precoding the data for downlink mode."

"The channel's statistical distribution is required to derive the deterministic approximation for the SINR of the system, along with the large dimensions of the mMIMO system, which will finally lead to determining the approximate value of the ergodic rate of the system."

"This paper further extends these results to utilize these fundamental equations for the multi-cell scenario where the channels between the UTs and BSs are spatially correlated with different correlation matrices, not the same for all the channels as considered in [6]."

Determining the Deterministic Equivalents for Closed-Form of the SINR of Multi-cell MIMO System

"It cannot be executed in the same way."

"Using Lemma 5 [8] for correlated vectors."

Results and Discussions

"Allocating UTs to the BS depends upon the most significant channel gain value from all the combinations."

"For UTs and BS existing in the same cell, the model used for assigning channels is the Rician Fading model and for UTs and BSs in separate cells is the Rayleigh fading model."

"By varying the various parameters, such as the number of BS antennas, transmit power, training power, and Rician factor, the accuracy of the derived deterministic equation of SINR for the multiuser multicell mMIMO system is verified."

"It also shows the accuracy of the derived deterministic equivalents, which yields tight approximations to the simulated values."

"The results here also show the accuracy of the derived deterministic SINR equation."

"The above simulated and theoretical results prove that the derived equation for SINR tightly approximates the system behaviour by only using long-term channel statistics."

Conclusion

"Using correlation matrices while estimating the channel and precoding improves the system throughput by providing spatial directivity to the signals."

"This paper uses the spatially correlated Rician Fading model for the downlink multicellular multiuser mMIMO system to determine the closed form for the ergodic user rate for the mmWave range using the RMT and the large system analysis."

"The determined expressions are such that they only depend upon the long-term channel statistics, such as correlation matrices and Rician factors of the users."

"The user-specific correlation matrix and imperfect channel estimation make this system closer to a realistic environment."

"From the derived results, it is proved that considering correlation in the system assists the system in providing much better average user rates."

Performance Analysis of Hybrid Filtering Technique for Reduction of PAPR in Alamouti Coded MIMO-OFDM Systems [9]

This is a machine-generated summary of:

Rajkumarsingh, B.; Rajarai, R. R.; Hosany, M. A.: Performance Analysis of Hybrid Filtering Technique for Reduction of PAPR in Alamouti Coded MIMO-OFDM Systems [9]

Published in: Wireless Personal Communications (2023)

Link to original: https://doi.org/10.1007/s11277-023-10348-w

If you want to cite the papers, please refer to the original.

For technical reasons we could not place the page where the original quote is coming from.

Abstract-Summary

"Such a system suffers from the drawback of a high peak-to-average power ratio (PAPR) when the symbol phases in the OFDM subcarriers line up in such a fashion that constructive formation of a large signal peak in the time domain occurs."

"In-band distortion caused by clipping affects the high-frequency components of the in-band signal because the signal is clipped directly into the discrete-time domain at the Nyquist sampling rate."

"This causes the aliasing phenomenon through which the high frequency component of the OFDM signal's spectrum actually inherits the low frequency identity of the sampled version of the spectrum."

"We propose an additional filtering scheme that has been assigned to our clipping algorithm to remove the high frequency components within the feasible regions (i.e. the constellation areas of the modulated symbols), which can efficiently reduce the PAPR of OFDM signals."

"The performance of the hybrid filtering algorithm is evaluated and compared with other algorithms such as the Partial Transmit Sequence (PTS), Selective mapping, active constellation extension for PAPR reduction by computer simulation in Alamouti Coded MIMO-OFDM systems."

"The hybrid filtering algorithm is found to offer a better PAPR reduction as compared to the other techniques."

"It is shown that at a clipping ratio of 1.8, a PAPR reduction of 9.818 dB, equivalent to a percentage difference of 97.97% with respect to the original signal is achieved, when employing QPSK modulation and 12.42% more when using clipping only."

Introduction

"A selected mapping (SLM) scheme based on a Polar coding technique for PAPR reduction in orthogonal frequency division multiplexing with index modulation (OFDM-IM) systems is proposed in [10]."

"In [11], a companding method using orthogonal frequency division multiplexing with offset quadrature amplitude modulation (OFDM/OQAM or OFDM/OQAM signal was used and a PAPR reduction of approximately 6.5 dB at a CCDF of 10^{-3} was observed with N = 512."

"The motivation of this work is to propose an additional filtering scheme that has been assigned to the clipping algorithm so as to remove the high frequency components within the feasible regions (i.e. the constellation areas of the modulated symbols) and which can efficiently reduce the PAPR of OFDM signals."

"Such analysis is done through adaptation of the following technique: Clipping, Hybrid Clipping and Filtering and compared with Partial Transmit Sequence (PTS), Selective mapping (SLM), active constellation extension (ACE) for PAPR reduction by computer simulation in Alamouti Coded MIMO-OFDM systems."

Methods

"The performance of the hybrid filtering algorithm is evaluated and compared with other algorithms such as the Partial Transmit Sequence (PTS), Selective mapping (SLM), active constellation extension (ACE) for PAPR reduction by computer simulation in Alamouti Coded MIMO-OFDM systems."

"Although clipping is the most basic technique for dealing with high PAPR issues, such a nonlinear process causes in-band signal distortion and out-of-band radiation."

"Because the signal is clipped directly into the discrete-time domain at the Nyquist sampling rate, in-band distortion affects the high-frequency components of the in-band signal."

"The following steps are required to combat the effects of aliasing, prior to clipping, (i) The original signal is oversampled (sampled at a rate slightly exceeding the Nyquist rate), (ii) The signal is then processed using a longer IFFT length to reshape portion of the noise outside the signal band."

Results and Discussions

"It can be examined that the PAPR of the OFDM signals decreases significantly after clipping, and greater CR yields to greater PAPR reduction effect."

"Both the Clipping and Hybrid Clipping-Filtering techniques show the highest values in PAPR reduction compared to the other techniques."

"The PTS scheme showed better performance of PAPR reduction than SLM simply because the effect of distortions occurring in the signal and a loss of subcarrier orthogonality have been greatly suppressed through adoption of the peak value optimisation process and pseudo-random partitioning respectively."

"A relatively lower value of PAPR decrease has been achieved using the ACE technique as compared to Clipping and PTS."

"It is slightly better PAPR reduction than the DE-TR technique with the same type of QPSK modulation and number of subcarriers."

"A better PAPR performance is noted though with the hybrid filtering technique with QPSK and 256 subcarriers."

Conclusion

"The primary technique used was clipping which reduced the PAPR through saturation of the overshooting signal amplitude to a predetermined value."

"The clipper works directly at limiting the signal envelope to the predefined clipping level in case the signal exceeds that level but the hybrid scheme smoothened and reduced the overall peak power regrowth, and thus decreased the PAPR effect further and achieved the largest reduction value."

"In this paper, a hybrid filtering approach is proposed for addressing the PAPR issue in Alamouti coded MIMO-OFDM systems."

"The Hybrid Clipping-Filtering technique shows the highest value in PAPR reduction compared to algorithms such as the Partial Transmit Sequence (PTS), Selective mapping (SLM), active constellation extension (ACE) and other recent improved techniques."

"The PAPR reduction can be compared with other approaches for PAPR reduction such as Tone Injection, Tone Reservation, Peak Windowing, Peak Cancellation and Artificial Bee Colony Algorithm among others, that deserve further investigations on various types of channels."

Channel Estimation Using Hybrid Optimizer Based Recurrent Neural Network Long Short Term Memory for MIMO Communications in 5G Network [12]

This is a machine-generated summary of:

Dash, Lipsa; Thampy, Anand Sreekantan: Channel estimation using hybrid optimizer based recurrent neural network long short term memory for MIMO communications in 5G network [12]

Published in: SN Applied Sciences (2023)

Link to original: https://doi.org/10.1007/s42452-022-05253-z

If you want to cite the papers, please refer to the original.

For technical reasons we could not place the page where the original quote is coming from.

Abstract-Summary

"An optimized Recurrent Neural Network-Long Short-Term Memory (RNN-LSTM) network is presented in this paper for channel estimation."

"History of channel responses of pilot block is collected or estimated using Least Square (LS) channel estimation method."

"Using these collected channel responses, the proposed RNN-LSTM is trained where weight parameters are chosen optimally using hybrid Particle Swarm Optimization (PSO)-Adam optimizer."

"Using the trained PSO-Adam optimizer based RNN-LSTM, the current channel response is predicted or estimated."

"The performance of the proposed channel estimation scheme is analysed by varying pilot sequence length and number of antennas to evaluate the metrics Bit Error Rate (BER) and Mean Square Error (MSE)."

Introduction

"Using millimeter wave (mmWave) in MIMO is one of the major technologies for 5G remote communication frameworks [13–15]."

"In planning a MIMO remote communication framework in 5G, the major complexity encountered is channel estimation."

"Channel state information (CSI) is an inevitable information in remote communications."

"Getting exact CSI is of central significance in ensuring the performance of radio connections in remote communication frameworks."

"To be more explicit, CSI profoundly decides the parameters of physical-layer and methods performed for a radio communication interface in a remote communication framework."

"It is significant to find precise CSI in remote communication frameworks."

"Remote communication frameworks can get precise CSI by performing channel estimation [16, 17]."

"Thusly, such conventional techniques are just reasonable for small-scale channel estimation in remote communication frameworks."

"RNN-LSTM is trained using the estimated channel responses and pilot block."

Elated Works

"Mehrtash Mehrabi and others [18] had used deep neural network for designing the decision directed channel estimation scheme for MIMO space time block coded (STBC) frameworks."

"They had presented a new channel estimation scheme based on compressive sensing."

"Yang Liu and others [19] had proposed decreased dimension decomposition-based channel estimation method for millimetre wave massive MIMO systems."

"The proposed channel estimation scheme only calculated the channel tap part for reducing the computational complexity."

"Besides, to solve the problem of low spectral efficiency, the authors had proposed blind method-based downlink channel gain estimation in massive MIMO system based on Time Division Duplexing."

"Jeya and Amutha [20] had presented an enhanced Semi-Blind Sparse algorithm for solve the complexities of channel estimation in MIMO-OFDM."

"By proposing the proposed scheme, the authors showed better symbol error rate and Peak signal-to-noise ratio (PSNR)."

Channel estimation using hybrid PSO-adam optimizer based RNN-LSTM for MIMO communications in 5 G network

"For predicting or estimating the current channel response, an optimized RNN-LSTM is presented."

"These estimated channel responses of pilot sequences are given as input to the propose RNN-LSTM structure, where the LSTM weight parameters are optimized using hybrid PSO and Adam optimizer to attain the accurate prediction output."

"To estimate CSI from channel matrix H, the number of training pilot sequence should be less than or equal to number of transmitters i.e. $N \leq t$. The received signal can be defined as follows, Based on the knowledge of P and Y, the channel estimator can find the channel matrix H. In this work, Rayleigh fading channel is modelled between the transmitter and receiver."

"These estimated channel responses are given as input to the proposed RNN-LSTM model for predicting the current channel response (H_k) of the pilot sequence."

"These channel responses are given as input to the weight optimized RNN-LSTM model."

Results and Discussions

"The performance of the proposed channel estimation scheme is evaluated in terms of BER and MSE for SNR."

"The evaluation metrics of the proposed PSO-Adam-LSTM is compared with that of the Adam-LSTM, PSO-LSTM and conventional LSTM."

"Besides, the performance of different channel estimation schemes is evaluated under the various pilot lengths and the number of antennas."

"The evaluation metrics of different channel estimation schemes are analysed under various pilot lengths (PL = 128, 136, 152, 160)."

"The evaluation metrics of different channel estimation schemes are analysed by varying numbers of antennas."

"As depicted in the figures, the proposed PSO-Adam-LSTM-based channel estimation scheme outperforms the channel estimation schemes based on Adam-LSTM, PSO-LSTM, and LSTM."

"As depicted in the figure, compared to deep learning-based channel estimation schemes, conventional schemes consume high execution time because of computational complexity."

Conclusion

"To estimate the accurate channel state information or channel response of the pilot data block, an optimized RNN-LSTM is used in this paper."

"The history of channel responses of pilot sequences at time 't-1' has been estimated using the LS estimation scheme."

"These estimated channel responses have been given as input to the proposed hybrid PSO-Adam optimizer-based LSTM in which weight parameters are optimized using PSO and Adam optimizer."

"Channel response at time 't' is estimated or predicted using the optimized LSTM."

"We substantially prove that the proposed estimator demonstrates superior performance over other channel estimation approaches in previous works, owing to better channel learning, resulting in decreased estimation errors."

"Deep learning facilitated channel estimation would also require availability of prototypical datasets publicly, where several pilot types, channel conditions, antenna arrangements and scenarios are considered in a comprehensive manner."

An Architecture and Performance Evaluation Framework for Artificial Intelligence Solutions in Beyond 5G Radio Access Networks [21]

This is a machine-generated summary of:

Koudouridis, Georgios P.; He, Qing; Dán, György: An architecture and performance evaluation framework for artificial intelligence solutions in beyond 5G radio access networks [21]

Published in: EURASIP Journal on Wireless Communications and Networking (2022)

Link to original: https://doi.org/10.1186/s13638-022-02164-w

If you want to cite the papers, please refer to the original.

For technical reasons we could not place the page where the original quote is coming from.

Abstract-Summary

"Network automation requires the development of a network architecture that accommodates multiple solutions based on artificial intelligence (AI) and machine learning (ML)."

"Based on a review of 5G system architecture, the state-of-the-art candidate AI/ML techniques, and the progress of the state of the art, and the on AI/ML for 5G in standards we define an AI architecture and performance evaluation framework for the deployment of the AI/ML solution in B5G networks."

"The framework identifies the logical AI functions, determines their mapping to the B5G radio access network architecture and analyses the associated deployment cost factors in terms of compute, communicate and store costs."

"The framework is evaluated based on a use case scenario for heterogeneous networks where it is shown that the deployment cost profiling is different for the different AI architecture alternatives, and that this cost should be considered for the deployment and selection of the AI/ML solution."

Introduction

"Leveraging recent progress in AI/ML, future radio networks are expected to follow a data-driven paradigm for resource management and for operations, where the level of network automation is increased [22, 23]."

"The natural question that arises is how to integrate AI-based resource management into the architecture of a radio access network, i.e. where should one store the required data and where should the related computations be executed."

"One way of realizing network automation is to have a cloud-based approach in which all intelligence is placed in a central point where data from the network are collected and the computations are executed [24, 25]."

"We propose a framework for comparing three different architectures, in terms of deployment costs and performance metrics, and use the framework comparing the three alternatives for an AI architecture that handles standards-based beyond 5G (B5G) radio access networks."

5G Networks: Architecture and Functions

"One of the fundamental approaches concerning 5G-related 3GPP standards is that the radio access network (RAN) and core network (CN) architectures are described in terms of services and functions."

"The gNB provides 5G NR access to the users by providing NR control plane (CP) and user plane (UP) protocol termination towards the user equipment (UE) side."

"The gNBs are interconnected through the Xn interface [26], while the NG interface [27] allows a NG-RAN consisting of a set of gNBs to be connected to the 5GC (5G core network)."

"The 5G system architecture is defined as service-based, and the interaction between network functions is represented in two ways: (a) a service-based representation, where network functions (e.g. AMF) within the control plane enable other authorized network functions to access their services, and (b) a reference point representation shows the interaction exist between any two network functions (e.g. AMF and SMF)."

AI/ML Techniques and Solutions in B5G Radio Access Networks

"Advances in the combination of AI/ML and wireless communications reside in different aspects of wireless network design and optimization, including channel measurements, modelling and estimation, physical layer optimization and network management and optimization [28]."

"We give an overview of the most prominent machine learning approaches to network optimization including supervised, unsupervised, reinforcement, federated and ensemble learning."

"Traditional supervised learning algorithms include regression models, the k nearest neighbour (KNN) algorithm, support vector machines (SVM) and Bayesian learning, while recent interest in supervised learning focuses primarily on deep neural networks."

"Bayesian learning models can be applied to spectral characteristic learning and estimation in 5G networks."

"To address RRM problems with reduced complexity and cope with the growing variety of scenarios, in [29], the authors propose a lean RRM architecture that consists of one or a few learner(s) that learn RRM policies directly from the data gathered in the network using a single general-purpose learning framework, and a set of distributed actors, which execute RRM policies issued by the learner and repeatedly generate samples of experience."

AI/ML Architecture Performance Framework for RAN

"When designing ML-enabled RANs, depending on the locations of data sources, the ML training host, the ML inference host and the point of actuation, three types of AI architectures can be identified: centralized, completely decentralized and hybrid."

"ML algorithms under the hybrid architecture are trained locally by their respective APs, and the central server in the CP is used as a coordinator to orchestrate the different steps of the algorithms and coordinate all the participating nodes during the learning process."

"To quantitatively evaluate the proposed AI/ML solutions under different AI architectures, apart from the use case-related network performance metrics, a second set of metrics need to be considered."

"In a network with centralized AI architecture, the training node collects data from data sources across the network and trains the model accordingly."

Use Case Analysis

"Following the architecture framework, the uplink scheduling problem is solved by using deep reinforcement learning (DRL) under the centralized, decentralized and the hybrid AI architectures, as described in subsequent subsections."

"In the centralized AI architecture, the training and the inference are located and performed by a central unit where the training data are transmitted from the network devices, i.e. UEs and access nodes, to the central unit."

"In the hybrid AI architecture, the access nodes can exchange information so as to improve the scheduling solutions compared to the decentralized architecture."

"Among the three AI architectures, as expected, the centralized one achieves the best performance, at the price of notable inter-node traffic and a long training time, as well as potential privacy and security issues during data exchange."

"On the opposite, under the decentralized architecture the models could be trained with insignificant data exchange among access nodes, but the performance improvement is lower than that for the centralized architecture."

Conclusions

"AI/ML techniques have been seen to be promising in many applications in communication networks, and it is no doubt that AI/ML would be an inherent element in 5G systems and beyond."

"This path gives birth to the challenging topic on how to integrate AI into the 5G systems such that we could leverage the advantages of ML techniques to optimize and improve the B5G networks."

"We have proposed an AI architecture and performance evaluation framework for the deployment of the AI/ML solution in B5G networks."

"In our future work, we will develop ML algorithms under the hybrid AI architecture and evaluate the performance in demanding heterogeneous network scenarios."

5G-V2X: Standardization, Architecture, Use Cases, Network-Slicing, and Edge-Computing [30]

This is a machine-generated summary of:

Abdel Hakeem, Shimaa A.; Hady, Anar A.; Kim, HyungWon: 5G-V2X: standardization, architecture, use cases, network-slicing, and edge-computing [30]

Published in: Wireless Networks (2020)

Link to original: https://doi.org/10.1007/s11276-020-02419-8

If you want to cite the papers, please refer to the original.

For technical reasons we could not place the page where the original quote is coming from.

Abstract-Summary

"Multiple wireless accessing technologies designed to provide connectivity in vehicular networks such as conventional Wi-Fi, IEEE 802.11p, and cellular communications."

"C-V2X supports two communication modes through a single platform to provide both Wi-Fi and cellular communication."

"LTE-V2X is the current 3GPPRelease 14 standard that has many enhancements to provide the new 3GPPRelease 16 for the new 5G radio generation."

"We propose the Optimizing of 5G with V2X, and analyzing the current V2X standards, introducing the development of 5G, challenges, features, requirements, design, and technologies."

Introduction

"Both safety and other non-safety vehicle applications are presented in intelligent transportation systems (ITS) and the Radio Access Technologies (RATs) used for communication in V2X are Dedicated Short Range Communications (DSRC), which operates in the 5.9 GHz band, and the 3rd Generation Partnership Project (3GPP) has presented Cellular-V2X (C_V2X)—a Long Term Evolution (LTE) based RAT—which also operates in 5.9 GHz band in addition to the licensed carriers of cellular operators and can also operate in the absence of the cellular infrastructure in a distributed manner [31]."

"We continue the work in this paper and summarize the paper contributions as follows: The paper introduces the current V2X technologies It compares the V2X based DSRC technology and V2X based cellular technology It introduces the evolution of 5G New Radio in V2X communication Study the 5G-V2X challenges and issues."

Overview of V2X Technologies

"DSRC was introduced to support vehicular communications with safety and non-safety applications, and thus, IEEE802.11p was in its stack to handle the ad hoc mode and to support traffic safety and prevent accidents in bandwidth spectrum from 5850 to 5925 MHz for both V2V, V2I modes."

"DSRC had many challenges to resolve to support vehicular communications, and these challenges are as follows: Network congestion control IEEE 802.11p includes contention-window techniques and adaptive messages frequency techniques to decrease the congestion, but are not applicable in high-density networks."

"In TS 22.185 [32] the services required by typical V2X applications, which are defined to include: message transfer frequency to be in the range (10–50) Hz, a long communication range to give a longer response time, V2X communication should be supported in and out of coverage of the network, transmission latency less than 100 ms should be supported and up to 20 ms in some cases."

New Radio (NR) V2X: Evolution of C-V2X

"5G targets to support wireless communications in high reliability with ultra-low latency and ultra-high throughput."

"Novel wireless technologies: 5G aims at providing ultra-high efficiency and performance metrics by introducing new use cases."

"Massive IoT or Massive Machine Type Communications (mMTC): applications that require many low cost and low energy devices to communicate small data packets as IoT and smart buildings, homes, and cities could be enhanced with technologies capable of giving them low latency high throughput and speed, multimedia support, location awareness, mobility and updating firmware efficiently over the air."

"Autonomous applications are mostly interested in mmWave technology as an essential part of future 5G technology to offer ultra-high throughput and ultra-low latency, and this can be mainly interesting for the following applications: V2V communications and here it is used to support near direct communications between vehicles as collaborative sensing in use cases of high density."

Key Requirements and Challenges for 5G-V2X

"Mobility has been enhanced in 5G technology to support a relative vehicle speed of up to 280 km/h in LTE based V2X, and for the 5G target, it is expected to reach 500 km/h relative speed [33]."

"For collaborative driving application, in 5G V2X Over-the-air IP communications were examined in a moving vehicle, and the end to end accurate measurements returned less than two μs for each transmitted packet."

"A direct communication interface is needed in two modes of network coverage (in and out) in 5G base V2X to reduce latency and improve network communication between vehicles."

"Several enhancements, including physical and MAC layer improvements, have been made for that enhanced OFDM for channel estimation with less than 1 ms as a latency design target in 5G. In autonomous driving applications, real-time data as map exchanging, video streaming, speeds, locations are exchanged."

"Designing the system architecture of 5G for vehicular communications is challenging to support all requirements of vehicles and services."

5G V2X Use Cases

"UCC1: Cooperative maneuver In this use case, data is shared among vehicles in the vicinity about planned trajectories, driving intentions, and local awareness."

"Goal This use case targets two main goals; the first one is designing a driving assistance system that could extend the local vision of the vehicle to improve safety, and the second goal to make a Local Dynamic Map (LDM) that could help in autonomous and cooperative driving by merging the data of static maps with dynamic data of sensors to help autonomous vehicles through different traffic patterns."

"To realize this, 5GCAR provides the following V2X specification: Offering end-to-end delays under 5 ms Offering ultra-high-reliability near 10^5 Handling the high-density scenarios of connected vehicles Ensuring Positioning precision of 30 cm for cars and 10 cm for pedestrian 5GCAR has outlined the following V2X use case classes: (1) Cooperative maneuver, (2) Cooperative perception, (3) Cooperative safety, (4) Intelligent autonomous navigation, (5) Remote driving."

5G-V2X Proposed Architecture

"The 5GCAR project is considered as one of the essential 5G projects that propose the 5G-V2X techniques, applications, use cases, and architectures to support the V2X connectivity within 5G. In this section, we propose the 5G-V2X design, which considered a result of the collaboration between companies, universities, projects, and V2X chipset makers."

"The 5G-V2X network core is proposed, including the network management using SDN, the slicing technology, the 5G virtualized functions, security, and privacy are considered in [34]."

"The 5G-V2X design is involving five domains of network management, security, multi-RAT connectivity, edge computing, and network slicing."

"The 5GCAR project involves different approaches for the 5G-V2X security check for the V2X messages by applying the security-checks at the vehicles while utilizing the network connectivity presence [35]."

5G-V2X Slicing

"The network slicing, including both the CN and RAN complicate the overall network slice design, we aim at delivering the CN segment functions in the RAN segment to meet the tight constraints of latency and scalability of V2X different applications."

"We introduce the V2X network slicing that can propose different V2X use cases with diverse requirements presented over the 5G network infrastructure."

"The slice that supports the Tele-operated driving use case must guarantee end-to-end high-reliability and ultra-low latency for connectivity between the controlled and the remote vehicle, which is hosted outside the core network (CN)."

"Autonomous driving slicing, that shares several CN functions with different slices, however, requires specific slice configurations over the RAN for V2V connectivity."

"Tele-operated driving slice This slice must ensure the end-to-end high-reliability and ultra-low-latency connectivity among the teleoperation device located outside the core network and the controlled vehicle."

NR V2X Edge Computing

"Edge computing is proposed to manage applications data with hard real-time requirements utilizing edge-computing servers that are close to the UEs that reduces the latency."

"For applications and services with soft-real time requirements or bounded end-to-end latency, the edge servers manage these data if the network response delayed."

"For the non-real-time applications, computing tasks are offloaded to the network cloud for load balancing."

"Another improvement proposed by 5GCAR is relevant to the optimization task execution time, considering the network and computing-edge resources."

"The millimeter-wave access technology utilization can provide valuable results for the offloading of the computation tasks nearby the access vehicles by utilizing the available near edge-cloud servers."

"In 5GCAR, the total latency for tasks offloaded is calculated considering the sending and receiving time of the tasks to the edge cloud and the latency of processing this task at the edge server."

Conclusions

"We proposed a V2X based 5G standard process, different available use cases requirements within the new 5G radio."

"We also describe the 5G-V2X architecture according to the 5GCAR project and the available autonomous V2X classes."

"We served the most recent research papers and project outcomes that describe the 5G-V2X enablers, methodology, concepts, and framework."

"We light the shade on the edge computing capabilities and slicing concept in 5G-V2X according to 3GPP and 5GCAR project."

Polynomial Expansion-Based MMSE Channel Estimation and Precoding for Massive MIMO-GFDM Systems [36]

This is a machine-generated summary of:

Wang, Yanpeng; Fortier, Paul: Polynomial Expansion-Based MMSE Channel Estimation and Precoding for Massive MIMO-GFDM Systems [36]

Published in: Wireless Personal Communications (2022)

Link to original: https://doi.org/10.1007/s11277-022-09943-0

If you want to cite the papers, please refer to the original.

For technical reasons we could not place the page where the original quote is coming from.

Abstract-Summary

"Low-complexity channel estimators and precoders are proposed for massive multiple-input multiple-output generalized frequency division multiplexing (MIMO-GFDM) systems."

"In order to combat the effect of non-orthogonality in GFDM, interference-free pilots are used in frequency-domain minimum mean square error (MMSE) channel estimation."

"Polynomial expansion is used to approximately compute the matrix inverses in the conventional MMSE estimation and precoding, consequently reducing the cubic computational complexity to square order."

"The degree of the matrix polynomial can be properly selected to get a required trade-off between complexity and estimation/precoding performance."

Introduction

"The direct use of the MMSE channel estimator in massive MIMO systems could result in unacceptable complexity."

"Polynomial expansion-based low-complexity channel estimators have been recently proposed and validated for massive MIMO systems [37]."

"Linear precoding methods include zero-forcing (ZF), maximum ratio transmission (MRT) and MMSE, which is shown to be the better choice for massive MIMO systems [38]."

"Motivated by the previous work on polynomial expansion-based detectors, estimators and precoders, in this paper, we consider low-complexity channel estimation and precoding algorithms in massive MIMO-GFDM systems."

"We extend the interference-free pilot pattern in [39] to the context of massive MIMO for channel estimation and approximate the matrix inverses in the MMSE estimator and precoder with the related matrix polynomials."

"Numerical results verify that the proposed channel estimators and precoders achieve near-optimal performance with lower complexity."

System Description

"We consider a massive MIMO system using GFDM modulation."

"For pilot patterns in [40] and [41], the GFDM data act as interference to the pilots at the receiver, which reduces the estimation accuracy."

"An interference-free pilot pattern has been proposed in [39] for GFDM systems."

"We consider a single-cell massive MIMO system that operates in time-division duplex (TDD) mode, where the downlink channel can be obtained from the transpose of the uplink channel."

Polynomial Expansion-Based MMSE Channel Estimation

"Bayesian MMSE estimators provide accurate estimations for MIMO channels when the channel statistics are known at the receiver."

"It is necessary to find low-complexity estimators for massive MIMO channels."

"Polynomial expansion-based estimators [37] have been proposed to reduce the computational complexity in channel estimation for massive MIMO."

"We propose low-complexity MMSE channel estimators for massive MIMO-GFDM systems based on the concept of polynomial expansion."

Polynomial Expansion-Based Precoding

"With the CSI estimated in the previous section, we can generate precoding vectors using appropriate precoding algorithms to process the data before it is transmitted."

"This operation mitigates the effect of the channel on signals at the transmitting end and thus provides reliable data transmission."

"The direct computation of a matrix inverse incurs a cubic computational complexity which is expensive for massive MIMO applications."

"The polynomial expansion technique can be used to reduce the computational complexity."

Complexity Analysis

"The value of V can be properly selected to balance the performance against the complexity."

"We refer readers to [37] for more details on this low-complexity algorithm."

Numerical Results

"The WPE-MMSE estimator reaches a lower MSE level compared to the PE-MMSE estimator because of the introduction and optimization of weights."

"The PE-MMSE and WPE-MMSE estimators show larger MSE values than the MMSE estimator, which is the cost of the polynomial expansion approximation with a finite V. The MSE gaps between polynomial expansion-based estimators and the MMSE estimator decrease with an increasing V. With a sufficiently large V, the polynomial expansion-based estimators will approach the MMSE estimator at the cost of complexity."

"We can see that the MSE of the PE-MMSE and WPE-MMSE estimators decreases as V increases, but there is a noticeable difference between them."

"The WPE-MMSE estimator shows nearly the same MSE as the MMSE estimator while the PE-MMSE estimator exhibits higher values."

Conclusion

"Massive MIMO-GFDM systems can provide flexible waveforms and high spatial resolution at the same time."

"MMSE estimators are optimal when the channel statistics are known at the receiver."

"The cubic computational complexity in the MMSE algorithm is unaffordable for massive MIMO applications."

"We proposed polynomial expansion-based MMSE estimators and precoders for massive MIMO-GFDM systems, which reduce the cubic computational complexity to square order through approximating the matrix inverse with a finite V-degree matrix polynomial."

Spectral Efficiency Optimization of Massive MIMO System Under Channel Varying Conditions [42]

This is a machine-generated summary of:

Panda, Satyasen: Spectral Efficiency Optimization of Massive MIMO System Under Channel Varying Conditions [42]

Published in: Wireless Personal Communications (2020)

Link to original: https://doi.org/10.1007/s11277-020-07924-9

If you want to cite the papers, please refer to the original.

For technical reasons we could not place the page where the original quote is coming from.

Abstract-Summary

"An optimization objective problem is formulated to improve the joint spectral efficiency during both uplink and downlink."

"The primary objective problem is partitioned into two different sub-problems, which are solved by a user defined joint frame allocation and user matching (JFAUM) algorithm based on efficient allocation of pilot resources and matching techniques for user clustering."

"The numerical results prove the improvement in spectral efficiency of massive MIMO system by using the proposed JFAUM algorithm under various practical conditions."

Introduction

"The advanced features of massive MIMO system include high energy efficiency, large spectral efficiency and simultaneous service to large number of user equipments (UEs) [43, 44]."

"The primary architecture of Massive MIMO system involves base station (BS) with large number of antennas, synchronously serving relatively less number of single antenna UEs in the exact frequency band."

"Narasimhan and others [45] discussed message passing detector procedure to work with received matched filter vector for reducing channel aging in large scale MIMO channels."

"It also considered the effects of channel aging in the downlink and multi-cell conditions of Massive MIMO system."

"Conditions, this paper describes optimization procedures to improve the spectral efficiency and the frame utilization of the cell free Massive MIMO system under high speed user mobility environment."

"Through detailed simulations, we discussed the variations of spectral efficiencies with respect to number of users, speed of users and number of antennas of BS."

System Model with Channel Aging

"The communication process involves uplink and downlink of data between BS and UEs."

"The uplink frame sequence contains orthogonal pilot symbols sent by the UEs to the BS for CSI estimation of the channels."

"The downlink frame sequence contains orthogonal pilots sent by BS to the UEs for CSI estimation of the channels."

"The BS sends the precoded data to the UEs after CSI estimation phase."

"Balance should be maintained between the training phase and data communication phase by the UEs for effective frame usage optimization to achieve higher spectral efficiency."

Optimization Procedures

"The sub-problem P2 is concerned to efficient frame utilization and sub-problem P3 is concerned to effective resource utilization by proper choice of UEs and BS antennas."

"Condition 1 describes that each UE can be scheduled to a single BS antenna at maximum."

"Condition 2 describes that each BS antenna can facilitate a number of UEs."

"Conditions 3 and 4 describe the individual matching between UEs and BS antennas."

"To improve the spectral efficiency during uplink and downlink, we consider that each BS antenna assists the UEs with certain preference parameter based on better SINR."

"The conditions of one to many matching are represented as The BS manages the matching operations for improving the spectral efficiency of certain UEs without any decrease in the spectral efficiency of other UEs till the matching process reaches its limit, when no UE can get a BS antenna for matching."

Results and Discussions

"It is found that the per user achievable mean spectral efficiency drops with increasing number of UEs due to increased channel aging effect and interference."

"We observe that the mean spectral efficiency rises with increase in the number of BS antennas because of less interference between UEs."

"The proposed JFAUM algorithm performs the dynamic adjustment of frame and efficient matching of UEs with BS antennas to improve the overall spectral efficiency of the Massive MIMO system."

"In the absence of optimization algorithm, we observe that the spectral efficiency decreases with rise in the number of UEs and speed of UEs because of channel aging and interference."

"In the presence of JFAUM algorithm, the mean spectral efficiency improves even at high speeds because of dynamic adjustment of frame and effective user matching."

Conclusion

"We discussed the performance of cell free Massive MIMO system based on channel aging conditions."

"We derived the spectral efficiency performance with respect to the number of UEs, number of BS antennas and speed of the UEs."

"Following that, we derived the objective problem to optimize the overall spectral efficiency of certain UE based on the uplink and downlink spectral efficiency."

"We demonstrated the improved performance of the Massive MIMO system by the use of optimization algorithm under various conditions."

Estimation of Sparse Channel Using Bayesian Gaussian Mixture and CS-Aided Techniques for Pilot Contaminated Massive MIMO System [46]

This is a machine-generated summary of:

Ravi Babu, T.; Dharma Raj, C.; Adinarayana, V.; Murali Krishna, K.: Estimation of Sparse Channel Using Bayesian Gaussian Mixture and CS-Aided Techniques for Pilot Contaminated Massive MIMO System [46]

Published in: Wireless Personal Communications (2021)

Link to original: https://doi.org/10.1007/s11277-020-07927-6

If you want to cite the papers, please refer to the original.

For technical reasons we could not place the page where the original quote is coming from.

Abstract-Summary

"The pilot contamination becomes a major challenging issue in massive MIMO applications due to erroneous channel state information between neighboring cells."

"For estimating the channel behavior over Sparse Bayesian and Compressed sensing Aided methods are proposed by surviving the Pilot contamination effects by changing the angular to beam domain for sparse channel approximation."

"Simulation results have demonstrated the effectiveness of Bayesian and compressed sensing by compared with the standard estimators in connection with estimation accuracy, especially in pilot contamination."

Introduction

"Pilot contamination affects the effectiveness of channel estimation due to the influence of neighboring cells. [47]."

"The sample length can be received in proportion to match the covariance matrix further, standard channel estimation for small scale MIMO-least square methods are applicable."

"Many researchers have worked on single user MIMO channel estimation by sparse methods of estimating [48, 49]."

"Compressed sensing has been incorporated to estimates short sequence type of sparse channels even with unknown statistics."

"Using CS concepts efficient sparse channel can be achieved by compressive statistics for distribution of non-zero coefficients."

"Distributed compressive technique has been applied by Rao and Lau [50] in studying channel state information so that the pilot overheads are reduced."

"The authors reached the challenges like estimate the channel from the vicinity of pilot contamination under the massive MIMO systems."

Massive MIMO System Model

"The arrangement of Massive MIMO is most considerable and promising technology to deliver great spectral efficiency on target cell and ability to maintain huge number of individual antennas."

"The random pilot signal, channel estimation, channel feedback, beam-forming computations will weaken the effectiveness of massive MIMO [51–53]."

"To extend the system of MIMO model and pilot contamination is analyzed to estimate error quality like antenna correlation coefficient and NMSE on different UEs by CS Aided and Bayesian Gaussian Mixture techniques."

"Ω_{i-1} is orthogonal previous support elements."

"The $s_{i,d}$ and $s_{i,s}$ are dense and sparse vectors respectively."

Error Estimation Methodologies

"From many experimental studies, it has been observed that as M increases, the user channel matrices have proclivity of becoming sparse so that it is inefficient to estimate the entire matrices by using long pilot training symbols."

"Spatial sparsity can be used to have only zero overheads leading to substantial decrease in estimation overhead compared with the traditional LS technique."

"Besides the trace properties can be obtained from Bayesian channel estimation approach improves effectively by allocating the same pilots to spatially separating users in different cells."

"The Sparse Bayesian approximation needs agreeable channel information throughout the statistical characteristics to overcome the channel estimation problem."

"The EM algorithm is utilized to learn the prior properties of the channel."

"In this algorithm, utilizing of central limit theorem for the purpose of approximation with various random variables to addresses as Gaussian."

Discussions and Simulation Results

"The noise level is higher than the channel components; gain of the Bayesian estimator possesses a significant change over the LS and CS-Aided estimators because of the orthogonal pilot sequences."

"The R-LS estimator is used to estimate the performance under orthogonal pilot sequence."

"It also identified that to get more valid MSE results with increasing distance D. In R-LS estimator, the obtained performance of MSE is increasing with SNR states the pilot contamination is severe at D = 1 and D = 2."

"With the Bayesian approach performs far better than CS-aided and R-LS estimator, it improves the MSE at high severity of pilot contamination."

"When T increases, all the MSE estimator's performance is decreased, however, while comparing with the R-LS estimator, the Bayesian approach achieved vital improvement as increasing of T at severe pilot contamination."

Conclusion

"The challenging issue in massive MIMO is pilot contamination; to mitigate this proposed methods are used to estimate the channel parameters of the preferred paths and interference from neighboring cells under the underdetermined system."

"The CS-aided approach is reduced pilot overhead by effective channel separation."

"The better estimation performance is found at higher SNR values, particularly at more burdened pilot contamination cells."

"The comparative results of simulation have proved the effectiveness of Bayesian compressed sensing in connection with estimation accuracy and acceptable estimates in the vicinity of pilot contamination."

Performance of MIMO MC-CDMA System with Relay Using Adaptive Pilot Channel Estimation [54]

This is a machine-generated summary of:

Tamilarasan, N.; Lenin, S. B.; Nithyanandan, L.: Performance of MIMO MC-CDMA System with Relay using Adaptive Pilot Channel Estimation [54]

Published in: Journal of The Institution of Engineers (India): Series B (2022)
Link to original: https://doi.org/10.1007/s40031-022-00773-z

If you want to cite the papers, please refer to the original.

For technical reasons we could not place the page where the original quote is coming from.

Abstract-Summary

"The Multicarrier Code Division Multiple Access (MC-CDMA) along with antenna diversity can reduce the fading effect considerably and provide high-speed data for future mobile communication systems."

"The use of Multi-Input and Multi-Output MC-CDMA mitigates the multipath effect of the transmission impairments caused by the frequency selective nature of the channel."

"The Adaptive Pilot technique is proposed and incorporated, in which, the location and number of pilots in time and frequency domain are dynamically varied with the help of channel estimation itself thus increases spectrum utilization."

"Further to enhance the performance, relaying technique is incorporated into the channel to reduce the Energy per Bits to the Spectral Noise Density (E_b/N_o), increase coverage area, reduce power requirements at mobile stations and will effectively reduce the number of base station."

Introduction

"In non-blind channel estimation technique, part of pilot symbols or training sequence is either multiplexed or superimposed during the transmission of data from the transmitter in the time or frequency domain and is used at the receiver for channel estimation [55–57]."

"In pilot-based technique, estimation accuracy is directly proportional to the pilot symbol, which decreases the data transmission rate as well as spectral efficiency, however it has very low complexity, so it is easy for implementation in mobile where the available computational capacity is limited."

"To overcome the spectrum loss by the pilot signal, the author's proposed blind channel estimation techniques [58–61] which do not use pilot and have better spectral efficiency but requires plenty of information for estimating channel condition also it exhibits a lesser performance in the receiver due to rapid signal variation and phased ambiguities in the demodulated signals under the influence of fading and Doppler's effect."

Transceiver Structure

"During the 1st slot, the Tx send the superimposed signal to the R node and Rx."

"The CSI is required for channel estimation which is obtained from the received pilot signal and the user data symbol is obtained by dividing the received signal by channel response."

"If the channel is in the reference level (good condition), there is no change in pilot symbols."

"If the channel condition is below the reference level, immediately the pilot strength is increased, in order to maintain the quality of the link at the same time user data is temporarily reduced due to the occupation of the pilot signal in the channel."

"The pilot symbols are reduced only when the channel resumes in good condition at the same time reducing pilot strength which results in increasing user data rate in the wireless channel."

MIMO MC-CDMA System Model

"This system considers different Tx and Rx antenna configuration and R node to improve the quality of the communication link and capacity enhancement."

"By using Walsh code function the signal of each carrier is orthogonally spread to keep a chip rate of 1.2288 Mcps."

"During the 1st time slot the Tx sends the modulated signal to the R and Rx and in 2nd slot the Node R retransmits the amplified signal to the Rx in order to overcome the channel impairments."

Simulation Result

"The E_b/N_0 requirement for the BER of 10^{-3} is 15 dB and it is reduced to 13.8 dB by increasing the SCs to 128."

"The E_b/N_0 required to achieve BER = 10^{-3} is 12 dB when the relay has only 1 antenna with MIMO of 16 × 1 × 2, while the E_b/N_0 needed to achieve the same BER is 11 dB when the number of antennas in the relay is increased to 2(16 × 2 × 2) with SCs of 64."

"When the SCs is increased from 64 to 128 for the same system, the required E_b/N_0 for the target BER is 9.8 dB and it is reduced to 8.9 dB by using two antennas in the receiver."

"From the simulation result increasing number of antennas in the receiver as well as relay the E_b/N_0 requirement for target BER is reduced 5.5 dB. The more detailed report is shown in Table."

Conclusion

"This paper discusses the performance of MIMO and Massive MIMO antennas configuration for MC-CDMA with relay using AP channel estimation."

"The proposed system is tested for different SCs, antennas configuration with relay."

"It is also clear that the E_b/N_0 requirement for a given BER is achieved by the increase in number of antennas from 2 × 1 × 2 to 4 × 2 × 4/16 × 2 × 4 and the number of SCs from 64 to 128 SCs."

Spatial Based Pilot Allocation (SBPA) in Crowded Massive MIMO Systems [62]

This is a machine-generated summary of:

Afshar, Abolghasem; Vakili, Vahid Tabataba: Spatial Based Pilot Allocation (SBPA) in Crowded Massive MIMO Systems [62]

Published in: Wireless Personal Communications (2021)

Link to original: https://doi.org/10.1007/s11277-021-08205-9

If you want to cite the papers, please refer to the original.

For technical reasons we could not place the page where the original quote is coming from.

Abstract-Summary

"A new approach of random access and pilot allocation in crowded scenarios of massive multi-input multi-output (MIMO) systems is proposed."

"When the number of users in a cell is more than number of orthogonal pilots, random access to pilots (RAP) is a promising solution to solve the massive access problem of users; though the collision of pilots is a by-product of this method."

"There are several methods for pilot collision resolution such as strongest user collision resolution (SUCR), but they have one major limitation; i.e. they cannot serve users more than the number of orthogonal pilots simultaneously in coherent transmission mode."

"By applying this approach in newly proposed protocol (SBPA), the limitation in number of adopted simultaneous users in crowded scenarios of a massive MIMO system would be resolved."

Introduction

"A set of collided pilots are considered as a graph code, and belief propagation algorithms are applied for collision resolution of random access to pilots and detection of data in [63], which rough orthogonality of user channels and constant channel gain in coherence interval (as results of an ideal asymmetric propagation and channel hardening in massive MIMO systems, respectively) are used in signal processing."

"To solve the access issue in crowded scenarios of massive MIMO, strongest user collision resolution (SUCR) protocol is introduced in [64] that attempts to allocate orthogonal pilots to active users in random access approach and resolve the pilot collision with help of inherent characteristics of the massive MIMO systems bases on power of received pilots."

"Although all of the above-mentioned approaches attempt to provide active pilot allocation with random access process and present innovative methods for collision resolution, they still have not been able to increase total number of admitted users in coherent interval."

Angular Structured Sparsity in Massive MIMO

"Recent research showed that Massive MIMO systems have a sparse structure in angular domain of received antenna array."

"Huge number of the antennas in receiver (high-resolution bin compared to received signal paths) and high-frequency waves are two main reasons for applying this structure."

"Following [60, 65] a massive MIMO channel is modeled by K single-antenna UE and a receiver as base station which deployed by N antennas as uniform linear array (ULA) with."

System Modeling and Problem Definition

"The time-frequency resources are divided into coherence blocks of T channel uses, such that the channel response between each BS and its UEs are constant and flat during coherence interval."

"If the time delay is more than the cyclic prefix, the orthogonality of pilots will be destroyed and to avoid it, it is assumed that there is small cells with a radius of less than 250 m. Since symbol duration is less than the coherence interval, any Doppler spread which impact the orthogonality of OFDM channels can be ignored as well."

"In coherent transmission blocks, due to orthogonality of allocated pilots to admitted users, the solution will be intra-cell pilot contamination-free."

"UE detected spatial feature in each RAP block can be used in several and even tens of coherence intervals for pilot allocation, but the channel state information of each active UE should be estimated in each CI during coherent transmission."

Spatial Based Pilot Allocation (SBPA)

"The spatial signature of every active user in random access blocks can be obtained by sending the primary nonorthogonal pilots which has been provided in pseudo random approach."

"Active UEs send primary pilots (Random Pilot Sequence) to BS for admission: primary pilots act as an identifier of UEs, then there should be one primary pilot for each UE."

"BS allocate available orthogonal pilots in the pool to active UEs as per their spatial feature: in this step, BS knows the spatial feature of each active UEs, then calculates AOA of every transmitted pilot."

"Based on calculated AOA of active UEs, a cell can be divided to several sectors (based on resolution bin) and orthogonal pilot allocation can be done based on divided sectors of a cell."

"Users which are in one group can use same orthogonal pilots in coherent transmission and this is a golden advantage of pilot allocation based on UEs spatial features."

Simulation Results

"To evaluate the performance, normalized MSE of the proposed CS-based detection is considered."

"With increasing number of the antenna in BS, the resolution bin increases too; therefore, the system detects AOA of users more accurately."

"Since there is more active UEs in each sector (there is sector dividing as per potential resolution bin and angular spread of transmitted beams to admit more users with same pilots), the access failure rate, will be increased in crowded cell."

"In practical scenarios of mMTC, IoT and eMBB, all users will not be simultaneously active and there would be UEs activation according to UE activation probability in CI."

Conclusion

"An innovative approach for random access and pilot allocation for TDD multiuser massive MIMO systems is proposed."

"Physical characteristics of ULA and array antenna theory are exploited and a spatial resolution based pilot allocation approach for random access to pilots in crowded scenario of massive MIMO systems is proposed."

"By using detected spatial feature of active users and sectoring the cell based on available resolution bin, spatial resolution based pilot allocation (SRBPA) protocol has been proposed."

"Numerical results shows the impact of changing antenna numbers in BS, length of coherence interval, and number of users in cell on MSE of estimated channel, overhead and access failure of proposed system."

Algorithms for Automatic Selection of Antennas in a MIMO System without Enumeration of Combinations [66]

This is a machine-generated summary of:

Bakulin, M. G.; Kreindelin, V. B.; Pankratov, D. Yu.: Algorithms for Automatic Selection of Antennas in a MIMO System without Enumeration of Combinations [66]

Published in: Journal of Communications Technology and Electronics (2021)

Link to original: https://doi.org/10.1134/s1064226921120032

If you want to cite the papers, please refer to the original.

For technical reasons we could not place the page where the original quote is coming from.

Abstract-Summary

"The most effective antenna selection criteria require an enumeration of all possible antenna combinations."

"The criterion for the maximal throughput implies an algorithm using the Frobenius norm, which does not require an enumeration of all possible combinations of antennas."

"We propose an antenna selection algorithm using the criterion of the minimal trace of the correlation matrix (minimum mean square error, MMSE), which does not require an enumeration of all combinations of antennas and has small losses compared to complete enumeration."

Introduction

"Communication systems with many transmitting and many receiving antennas (multiple-input-multiple-output, MIMO) became widespread due to their high spectral efficiency and noise immunity [67, 68]."

"At an increase in the number of antennas in MIMO systems, the complexity of digital signal processing in such systems significantly increases."

"The technology of automatic selection of antennas for MIMO systems, which allows reduction in the complexity of signal processing and reduce the cost of equipment, is being actively studied."

"In [69], it was shown that the technology of automatic selection of antennas can improve the noise immunity of MIMO systems with space-time coding."

Structural Diagram and a Model of a Mimo System Without and with Automatic Antenna Selection

"The technology of automatic selection of antennas for MIMO systems allows reducing the complexity of such systems due to the fact that the number of antennas used for transmitting and receiving signals is chosen less than the total number of antennas [69, 70]."

"The number of radio frequency circuits (blocks) on the transmitting side is $n < N$ and the number of radio frequency circuits (blocks) on the receiving side is $m < M$. After appropriate processing (coding and digital modulation), the transmitted data are sent to n radio frequency units while n antennas are used for transmission (instead of N antennas in a MIMO system without automatic antenna selection)."

"On the receiving side, m antennas are used (instead of M antennas in a MIMO system without automatic antenna selection) and m radio frequency units from the output of which signals are fed to a digital signal processing unit."

Features of Automatic Selection of Antennas on the Receiving Side

"As has long been known [67], the use of multiple antennas on both the transmitting side (transmit diversity) and the receiving side (receive diversity) provides multiple spatial paths for signal transmission."

"When using diversity reception at the input of RF units (radio frequency blocks), several signal copies, each of which corresponds to its own complex transmission coefficient from matrix H of the MIMO channel, are received."

"There are the following DSP algorithms for receiving signals from various antennas [71]."

"When using the technology of auto-selection of antennas on the receiving side [70], only m of M antennas and m RF units are used."

Hybrid Transmission Method with Antenna Selection by the Hybrid Maximal Ratio Transmission Criterion

"This method requires knowledge of the channel matrix coefficients between all transmit and receive antennas."

"When choosing 1 of 2 rows and 1 of 2 columns in 2×2 matrix H, it is necessary to enumerate $2 \cdot 2 = 4$ combinations."

"We now consider 3×3 matrix H, which has the following form: Matrix H can be represented as column vectors h^j or as row vectors h_i."

"It is easy to see that to select 2 out of 3 rows and 2 out of 3 columns from this matrix, it is necessary to sort out $3 \cdot 3 = 9$ combinations."

Automatic Antenna Selection Based on the Criterion of Maximal Throughput of the Mimo System

"In a MIMO channel with Rayleigh fading in the absence of spatial correlation, the throughput of the MIMO system is known to increase in proportion to min{M, N} [67, 68]."

"By disabling the least efficient channels, it is possible to improve the performance of MIMO systems using automatic antenna selection technology."

"The use of the maximal SNR criterion in space diversity mode (3) is aimed at increasing the maximal noise immunity of the MIMO system with automatic antenna selection."

"In the spatial multiplexing mode, the technology of automatic selection of antennas to increase the spectral efficiency of the MIMO system, it is advisable to use the criterion of maximal throughput [72]."

"The expression for the throughput of the radio channel of the MIMO system with automatic selection of antennas (without allowance for spatially correlated fading) has the following form [68, 73]:where ρ is the SNR at the input of one receiving antenna."

New Automatic Antenna Selection Algorithm Based on the Criterion of the Minimum of the Correlation Matrix Trace

"We propose a new antenna selection algorithm using the criterion for the minimum trace of the correlation matrix, which does not require enumeration of all combinations of rows and columns of the initial matrix of the MIMO radio channel."

"Steps 1–7 are repeated until the dimension of the current channel matrix reaches the required value, i.e., dimension $m \times n$. We consider a particular case of the proposed algorithm for automatic selection of antennas for M = N = 3 and m = n = 2, i.e., it is necessary to select 2 antennas out of 3 on both the transmitting and receiving sides."

"It is easy to see that after the performed transformations, the channel matrix takes the following form: Thus, the first row and first column were removed from initial matrix H and two transmit antennas and two receive antennas were selected."

Analysis of the Computational Complexity of the Proposed New Automatic Antenna Selection Algorithm

"The operation complexity of calculating the correlation matrix of errors of dimension M × M will be determined by the expressions [74]: where N_{mult} is the number of multiplications, and N_{add} is the number of additions."

"In the enumeration algorithm, it is required to calculate correlation matrices with dimension M/2 × M/2."

"The first cycle of exclusion of the transmit and receive antennas requires two calculations of the M × M correlation matrix."

"The next cycle requires two operations to calculate the correlation matrix of a dimension of (M – 1) × (M – 1)."

"This is until the matrix dimension decreases to M/2 × M/2."

Simulation Results

"The performed computer simulation of the indicated algorithms for automatic selection of antennas allows drawing the following conclusions: (1) When using the criterion for minimizing the Frobenius norm, it is necessary to enumerate all possible combinations of rows and columns of the initial matrix of the MIMO channel; however, this does not lead to good performance according to the criterion of the minimal error probability, BER $\sim 10^{-3}$ is achieved only at SNR ~ 25 dB. (2) Other criteria (maximal throughput and minimal trace of the correlation matrix) allow achieving BER = 10^{-3} at SNR values <5 dB, but their implementation requires significant computational costs. (3) The proposed new algorithm approaches in efficiency the algorithms that are optimal in terms of maximal throughput and minimal trace of the correlation matrix with SNR losses of about 1 dB while it has a significantly lower computational complexity (by ~100 times)."

Conclusions

"Optimization criteria for the development of algorithms for auto-selection of antennas in MIMO systems were considered and a new method for automatic selection of antennas with less computational complexity was proposed."

"These criteria can be applied to implement automatic antenna selection algorithms on both the receive side and the transmit side of the MIMO system."

"The proposed algorithm for automatic selection of antennas does not require a complete enumeration of all combinations and, in terms of efficiency, approaches the algorithm that is optimal in terms of the trace minimum of the correlation matrix at a SNR loss of ~1 dB."

Optimality of Linear MIMO Detection for 5G Systems via 1-Opt Local Search [75]

This is a machine-generated summary of:

Khurshid, Kiran; Khan, Adnan Ahmed; Siddiqui, Muhammad Haroon; Hadi, M. Usman; Rashid, Imran; Imran, Muhammad: Optimality of Linear MIMO Detection for 5G Systems via 1-Opt Local Search [75]

Published in: Journal of Electrical Engineering & Technology (2021)

Link to original: https://doi.org/10.1007/s42835-020-00625-0

If you want to cite the papers, please refer to the original.

For technical reasons we could not place the page where the original quote is coming from.

Abstract-Summary

"Vertical Bell-Labs Layered Space Time (V-BLAST) and Sphere Decoder (SD) are non-linear methods used for extracting transmitted data; whereas, Zero-Forcing (ZF) and Minimum Mean Square Error (MMSE) detections are comparatively in complex and effectual linear techniques."

"The heuristic 1-Opt local search method used for solving computationally hard combinatorial optimization problems is applied to the ZF and MMSE detection algorithms."

"Simple MIMO decoding using ZF and MMSE is accomplished to find the estimated symbol then the transmitted symbol is calculated using the heuristic 1-opt approach by means of the estimated symbol."

"Simulation results demonstrate that 1-opt search when applied to the ZF and MMSE displays better bit error rate (BER) performance than the simple ZF and MMSE detectors."

"It is also verified through simulations that the proposed 1-Opt based detectors display better BER performance as compared to the complex nonlinear VBLAST and SD at considerably reduced complexity."

Introduction

"Even modest complexity MIMO decoders could be power hungry."

"With the decreased BER, increased diversity and array gains of MIMO systems comes an increased complexity in receiver design."

"To solve complex MIMO detection problem in an optimal way, a thorough search over the space of all possible symbols is required so that the closet point to the received symbol (in Euclidean sense) can be found out."

"Non-linear MIMO detection algorithms give better BER performance than the ZF and MMSE decoders."

"The BER performance of VBLAST is better than MMSE and ZF decoders; however, at a cost of greater computational complexity."

"We introduce a local search heuristic to solve the NP hard MIMO decoding problem."

"Inspired by the applications of local search heuristics, we holistically combine a VDS inspired local search method termed 1-opt local search with linear MIMO decoders."

MIMO System Model

"The signal after passing through the channel reaches the receiver."

"MIMO system model with number of transmitters N_t and number of receivers N_r has a channel matrix represented by H. Dimensions of the complex channel matrix H are $N_r \times N_t$."

"The coefficients of H represented by [H]i,j signify the transfer function and complex path gains between jth transmitter and the ith receiver."

"Vector y with dimension $N_r \times 1$ signifies complex received symbols and vector x with dimensions Nt × 1 indicates a complex transmitted vector."

"The complex matrix H represents a Rayleigh flat fading channel."

MIMO Decoders

"The most optimal receiver that can detect the transmitted vector x with minimum probability of error is the ML decoder [76]."

"In CO problem, as the problem size increase, the number of workable results also increases exponentially and a simple application of this search approach outcomes in an unaffordable complexity level."

"At a reduced complexity level, these detectors offer optimal solutions."

"Among the linear detectors, ZF receiver is the most popular low-complexity equalizer."

"This complexity involves the calculations of computational complexity of $H^\dagger$. The initiative of nulling in MMSE decoder is to reduce the expected mean square error between the transmitted symbol and the receiver's estimate."

"In SUC all the N_t transmitted symbols are not jointly decoded, the first symbol is decoded by ZF or MMSE estimation whereas remaining data symbols are treated as interference; the decoded symbol subtracts its contribution from received vector offering a reduced order complexing problem having $N_t - 1$ unknowns."

1-Opt Based Detector

"Heuristic methods can be broadly separated into two types: construction and local search approaches."

"The problem associated with these algorithms is that they often return inferior quality solutions as compared to the results of other local search algorithms."

"Local search methods have some initial solution to start with."

"In 1-opt local search, the results are obtained by varying a single component in the existing solution."

"We have applied 1-opt local search to the conventional ZF and MMSE decoders for improving the performance of these linear decoders."

"As near the initial solution estimate other workable solutions exist, so, in the next step 1-opt local search is used to investigate the constellation points about the initial ZF and MMSE solutions."

"1-Opt search explores the solutions by altering a single element in the existing result (symbol)."

Computational Complexity and Performance

"We will now observe the performance and complexity of 1-Opt ZF, 1-Opt MMSE and VBLAST."

"ZF has complexity: In comparison to ML which does an exhaustive search over the whole search space, 1-Opt ZF uses a compact constellation."

"Is to find the computational complexity of ZF/MMSE for initial guess."

"Step is calculation of complexity required in finding neighbours."

"ML detection has the highest computational complexity while 1-Opt MMSE offers almost equal computational complexity as MMSE."

Future Work

"For future work, proposed algorithms might be implemented on Massive MIMO systems where the base station is equipped with hundreds of antennas."

"Massive MIMO systems require computationally less complex decoders, hence, it would be interesting to see the performance of our proposed algorithms with large scale MIMO systems."

"We plan to use the functions proposed in [77] to solve MIMO detection problem in the future."

Conclusion

"ZF, MMSE and 1-Opt supported symbol detection in MIMO systems are comprehensively investigated."

"Proposed 1-Opt based detectors carry an exciting potential to become suitable candidates for real time wireless communication systems."

"When compared with VBLAST, SD and ML detectors, for higher modulation schemes and substantial number of antennas, 1-Opt based decoders are predicted to give near optimal results at reduced complexity levels."

"The lesser complexity level, performance efficiency and ease of implementing the proposed decoders reflect promising options for making them an integral part of the upcoming 5G systems."

Deep Neural Network for Compressive Sensing and Application to Massive MIMO Channel Estimation [78]

This is a machine-generated summary of:

Mohades, Zohreh; Tabataba Vakili, Vahid: Deep Neural Network for Compressive Sensing and Application to Massive MIMO Channel Estimation [78]

Published in: Circuits, Systems, and Signal Processing (2021)

Link to original: https://doi.org/10.1007/s00034-021-01675-z

If you want to cite the papers, please refer to the original.

For technical reasons we could not place the page where the original quote is coming from.

Abstract-Summary

"We consider the problem of sparse signal recovery using a learned dictionary in multiple measurement vectors (MMVs) case."

"Employing deep neural networks, we provide two new greedy algorithms to solve sparse MMV problems."

"In the first algorithm, we create a stacked vector of measurement matrix columns and a new measurement matrix, which can be assumed as the Kronecker product of the primary compressive sampling matrix and a unitary matrix."

"In order to reconstruct sparse vector corresponding to this new set of equations, a four-layer feed-forward neural network is applied."

"In the second algorithm, joint sparse structure of the sparse vectors is considered."

Introduction

"In [79], considering the MMV problem, in which the sparse signals are correlated, a convolutional deep stacking network (CDSN) has been proposed to extract information regarding the correlation of the sparse vectors and based on this information, a greedy algorithm has been presented to reconstruct the sparse signals from the measurement vectors."

"Instead of using conventional CS reconstruction methods, either optimization-based methods or greedy ones, Mousavi and others [80] have shown how to employ deep convolutional networks to reconstruct structured sparse signals, while improving both reconstruction accuracy and runtime."

"In [81] by designing a feature enhancement layer (FEL) and integrating it into the convolutional neural network (CNN), a Wireless Deep Learning Reconstruction Network is proposed to recover signals from compressive measurements transmitted over fading channels."

"The authors in [82] jointly designed the pilot signals and channel estimator for massive MIMO systems, employing a data-driven deep learning algorithm."

Deep Neural Network

"Unlike the feed-forward networks, RNN networks own feedback connections making them capable of referring to the previous states, and hence, they are allowed to process different data sequences."

"The basic idea of reconstruction sparse signal is to represent this signal as a linear combination of a few columns of a dictionary matrix."

"A learned dictionary can represent a given signal with smaller representation error than predefined dictionaries."

"Since the input signals are joint sparse or more specifically have block sparse structure, in this paper we use a block sparse dictionary learning [82] to learn a dictionary which can satisfyingly offer a sparse representation of a massive MIMO channel."

Proposed Methods

"We present our proposed algorithms to reconstruct a sparse signal."

"In [83], BOMP has been proposed to reconstruct block-sparse signals."

"An iterative algorithm is applied for sparse signal recovery."

"In the case of block sparse signals, we generate the training data set as follows."

"In [84], subspace pursuit (SP) algorithm has been proposed to reconstruct sparse signals."

"That for our problem, the input and target data stand for the residual vectors and the corresponding sparse signal vectors, respectively."

Channel Estimation in Massive MIMO System with Proposed Algorithms

"We use our proposed framework for pilot-based channel estimation problem in downlink frequency division duplex (FDD) massive MIMO systems [47]."

"In the conventional downlink channel estimation, the number of required pilots is proportional to the number of antennas."

"For massive MIMO, however, due to the huge number of antennas it is greatly burdensome to estimate the channel using the conventional approach."

"It is shown that a massive MIMO channel benefits sparsity in the angular domain [85], and hence, compressive sensing estimation methods can be utilized for massive MIMO channel estimation."

"We give a more accurate sparse representation of the massive MIMO channel."

Simulation Results

"We compare our results with some well-known traditional MMV reconstruction methods."

"For evaluation of Algorithm I, reconstruction algorithms Simultaneous OMP (SOMP), Group LASSO (G-LASSO), and BOMP have been considered and the algorithms used to evaluate Algorithm II are SOMP, G-LASSO and SP."

"Each of the traditional MMV reconstruction methods owns some properties encouraging us to use them for the sake of comparison."

"SOMP method inherits the properties of OMP method which is one of the fastest traditional reconstruction methods."

"G-LASSO method benefits an extra-norm term relative to the LASSO method which is useful to consider the joint sparsity structure of the signals."

"As observed, the proposed Algorithm I also improves the mentioned reconstruction algorithms."

"By comparing the proposed method with some well-known conventional reconstruction methods, it is seen there is a great desire to adopt such approaches to reconstruct the sparse signals in the case of Sparse Channel Estimation."

Conclusion

"We presented two algorithms to reconstruct sparse vectors for the MMV problems."

"The proposed methods, learn the structure of sparse vectors using DNN."

"Some future works can be more profoundly investigated including, first, considering prior information to the signal and evaluating the performance of the proposed algorithms under such a case, and second, finding the phase transition diagrams which exhibit the number of required measurements to exactly reconstruct sparse signals with different numbers of sparsity orders."

Multiuser Massive MIMO-OFDM System Incorporated with Diverse Transformation for 5G Applications [86]

This is a machine-generated summary of:

Kansal, Lavish; Sharma, Vishal; Singh, Jagjit: Multiuser Massive MIMO-OFDM System Incorporated with Diverse Transformation for 5G Applications [86]

Published in: Wireless Personal Communications (2019)

Link to original: https://doi.org/10.1007/s11277-019-06707-1

If you want to cite the papers, please refer to the original.

For technical reasons we could not place the page where the original quote is coming from.

Abstract-Summary

"The orthogonal frequency division multiple access (OFDMA) technology is being utilised for the uplink and downlink transmission to afford the high spectral efficiency in fading environments."

"The 5G implementation requires additional improvements to meet the futuristic stress."

"This work proposes an innovative solution that combines OFDMA technology with multi-user massive multiple input multiple output (MIMO) technology to meet the required elevated data rates as desired by the growing application needs of 5G. Massive MIMO is capable to fulfil the vision of 5G to realize a huge number of base stations equipped with a large number of terminals to be served in the same time–frequency resource without severe inter-user interference."

Introduction

"The main reason of this efficient integration in 5G is a hybrid combination of orthogonal frequency division multiplexing (OFDM) and multi user massive multiple input multiple output (MIMO) systems."

"This unique merger of OFDM and multi user massive MIMO results in a significant increase in channel capacity and reliability of wireless communication systems without increasing the requirement of the operational bandwidth [87]."

"MIMO technology also has a prominent role to play in improving the system capacity and performance in terms of bit error rate/symbol error rate performance in wireless systems."

"Multi-user MIMO (MU-MIMO) systems hold the promise of elevated data capacity and throughput communications for 5G mobile communication applications."

"Attribute, the FFT is getting replaced by DWT in the existing wireless communication technologies/systems [88, 89]."

"The same methodology can be implemented in MIMO-OFDM systems i.e. 5G mobile communication systems to enhance its performance."

Related Works

"Diverse transforms like FFT, DWT, and discrete cosine transform (DCT) are also being used in OFDM systems to analyse the BER performance over AWGN channel [90]."

"The DWT-OFDM outperforms the FFT-OFDM regarding BER performance and spectral efficiency over Rayleigh fading channel also [91]."

"FRFT appended OFDM has been analysed over diverse wireless channels such as AWGN, Rayleigh, Rician, and Nakagami [92]."

"The BER performance analysis depicts a considerable improvement in comparison to convention FFT based OFDM over AWGN, Rayleigh, Rician and Nakagami channels [93].In the recent past, a significant improvement in the transmission capacity of modern wireless communication systems has been reported by many researchers on using the massive MIMO system [94]."

"The use of MIMO-OFDM transmission in WiMAX systems results in elevated data rate, throughput, resilience against multipath fading and efficient bandwidth usage."

"Based on the study of the research work reported in the literature survey, the following are the research gaps which are required to be addressed:Wavelet transform can be utilised to improve the BER performance in place of Fourier transform in the MIMO-OFDM system."

OFDM Model Description

"Shannon-Fano, Lampel-Ziv, Huffman, etc The source encoding process will reduce the redundancy involved in data representations.2.Channel encoding Channel encoding scheme i.e. FEC is required to detect and correct the errors at the receiver end thereby making the performance of the system more reliable."

"The interleaving dispenses the bits in frequency, time, and time–frequency both domains to achieve minimum burst error at the receiver end.4.Modulation Modulation or symbol mapping in the OFDM is attained through the digital modulation schemes such as M ary phase shift keying (M-PSK) and M ary quadrature amplitude modulation (M-QAM)."

"After the mapping of the modulated symbols on the orthogonal sub-carrier, the parallel to serial conversion is performed to make sure that the data remains in serial form for efficient transmission over the physical channel.6.IFFT/IDWT/IFRFT The IFFT/IDWT/IFRFT will be used to generate the orthogonal subcarriers for the transmission of the data symbols."

MU-Massive MIMO Methodology

"In massive MIMO, it is usually assumed that M ≫ K, for achieving good spatial separation of user signals."

"This is, however, not necessarily a requirement for massive MIMO."

Results Discussions

"The presence of multipath fading result in degradation of the orthogonality of the subcarriers in the conventional FFT based-OFDM."

"The DWT based-OFDM is very robust against the time varying nature of the multipath fading channel."

"The analysis depicts that the DWT based multi-user massive MIMO-OFDM system outperforms the FFT basedmulti-user massive MIMO-OFDM system."

"The performance improvement originates from the fact that the orthogonality between the subcarriers is better maintained in the DWTbased multi-user massive MIMO-OFDM system in comparison to FFT basedmulti-user massive MIMO-OFDM system [95, 96]."

"It is very much evident that the FRFT based multi-user MIMO-OFDM outperforms the FFT based multi-user MIMO-OFDM because it offers much resiliency towards the non-linear behaviour of the multipath fading channel [92–97]."

"It is very much apparent from the previous discussion that the BER performance of DWT/FRFT incorporated multi-user massive MIMO-OFDM system is better than the conventional FFT based multi-user massive MIMO-OFDM system."

Conclusion

"After analysing the SNR vs BER results, it is very much evident that the FRFT based multi-user massive MIMO-OFDM offers an SNR improvement of 8–13 dB in comparison to FFT based multi-user massive MIMO-OFDM for achieving a BER of 10^{-4} in Rayleigh fading channel for diverse modulation levels and varying number of users in the downlink."

"Further, this SNR improvement is elevated to 8–17 dB in DWT based multi-user massive MIMO-OFDM to achieve the same BER of 10^{-4}."

"It can beconcluded that the performance of multi-user massive MIMO-OFDM will be improved by replacing the conventional FFT by diverse transforms such as DWT and FRFT in OFDM physical layer."

Preamble Based Timing Offset Estimation and Correction in OFDM Assisted Massive MIMO Systems in the Presence of Inter-User-Interference [98]

This is a machine-generated summary of:

Singh, Sandeep Kumar; Gandhi, Abhay S.: Preamble Based Timing Offset Estimation and Correction in OFDM Assisted Massive MIMO Systems in the Presence of Inter-User-Interference [98]

Published in: Wireless Personal Communications (2019)

Link to original: https://doi.org/10.1007/s11277-019-06216-1

If you want to cite the papers, please refer to the original.

For technical reasons we could not place the page where the original quote is coming from.

Abstract-Summary

"This paper presents a new preamble design for correct timing synchronization and symbol time offset (STO) estimation in orthogonal frequency division multiplexing (OFDM) based massive multi input multi output (M-MIMO) systems."

"Mean and standard deviation of the evaluated STO and probability of the correct timing synchronization are considered as the performance parameters."

"Considering the effect of fading and IUI, result shows that the proposed preamble gives better performance at low SNR."

Introduction

"Several designs for MIMO-OFDM systems are proposed by the researchers [99–104]."

"Various beamforming techniques have been presented to suppress IUI in M-MIMO systems [105–107]."

"Since M-MIMO assists large number of users, high IUI will be present at low SNR."

"With these motivations, in this paper we present a new preamble design for better timing synchronization at low range of SNRs."

"Proposed design achieves a better correlation to estimate offset at the receiver."

"The main contribution of this paper are summarized as follows: This paper presents a new preamble design using simple PN sequence for better timing synchronization."

"Preamble is designed specifically for low range of SNR in M-MIMO-OFDM systems where most of the previous preamble designs under perform."

"For validating the accuracy and robustness of the proposed design, timing synchronization for Rayleigh fading channel is investigated."

System Model

"This section presents basic overview of MIMO-OFDM followed by detailed explanation of M-MIMO system."

"M-MIMO is considered similar to MIMO system except with much more antennas at the base station."

"It depicts that, with the help of various techniques like precoding or digital beamforming, it can assist tens of users simultaneously."

"With the use of these techniques, data is transmitted in a particular direction for a specific user."

Effect of Synchronization Error

"To combat these errors and to perform correct demodulation of the received data, one must perform correct timing synchronization and detect the exact starting point of the symbol."

"In the case of M-MIMO, data intended for other users acts as interference for the desired user."

"These techniques do not work well when users are too close to each other."

"User 1,2,3 and 5 enjoy IUI free reception, where as, user 4 is interfered by the data of user 6."

"As observed clearly user 4 and 6 are too close and IUI is unavoidable."

Proposed Preamble Design for Correct Timing Synchronization

"At receiver, timing synchronization is performed to estimate STO and detect the start of transmitted frame."

"This is performed by correlating with same preamble generated locally at receiver, thus searching the transmitted preamble."

"The PN sequence with good correlation properties is adopted for preamble design [108]."

"While correlating received signal, peak is found after the complete overlap between transmitted and locally generated preambles."

"Step is to form an array of zeros of length 1 by N. This array size is equal to the size of one OFDM symbol, thus, that of the proposed preamble."

"We insert generated PN sequences in the even index of the zeros array."

"Desired preamble is achieved by applying N-point IFFT."

"Same preamble is likewise generated at the receiver."

"Cross-correlation of this preamble with the received signal gives the required timing metric."

Simulation Results

"For the evaluation of the algorithm mean, standard deviation of the estimated STO and probability of the correct timing synchronization are considered to be the performance parameters."

"It compares the probability of correct timing synchronization of five previous (S&C [109], SAG [99], Luo [102], Liu [103], UPSP [104]) and proposed preamble designs."

“Proposed algorithm gives good performance in low SNR for the single user case.”

“In case of IUI, for every interfering user a random delay is introduced at receiver of the destined user.”

“As the number of interfering users increases, IUI also increases.”

“Difference is nearly 100% with no IUI, whereas for 5 users this difference is about 50%.”

“While with a probability of 0.95, which is acceptable range for practical cases [102, 110], it serves upto 8 users simultaneously.”

“In case of low SNR, using proposed design along with digital beamforming techniques more users can be supported.”

Conclusion

“A preamble based method is proposed for accurate timing synchronization and STO estimation in OFDM based M-MIMO systems at low SNR.”

“In results, start of the symbol has been estimated very accurately for high SNR and estimation is acceptable for low SNR values.”

“While for 8 users, acceptable performance is obtained for practical range of SNR values.”

“Performance of the proposed algorithm has been verified to computer based simulations.”

Improving MIMO Relay Compressed Sensing-Based Channel Estimation and Pilot Allocation [111]

This is a machine-generated summary of:

Akbarpour-Kasgari, Abbas; Ardebilipour, Mehrdad: Improving MIMO relay compressed sensing-based channel estimation and pilot allocation [111]

Published in: EURASIP Journal on Wireless Communications and Networking (2019)

Link to original: https://doi.org/10.1186/s13638-019-1384-6

If you want to cite the papers, please refer to the original.

For technical reasons we could not place the page where the original quote is coming from.

Abstract-Summary

"Multiple-input multiple-output (MIMO)-orthogonal frequency division multiplexing (OFDM) relay networks as a key technology is one the main techniques to maintain the performance and data rate."

"Being influenced by fading channels, the MIMO-OFDM relay networks need some reliable approaches to estimate the channel response."

"We have utilized CS-based approaches to estimate the MIMO-OFDM relay channel."

"This approach benefits from backward correction which distinguishes it from other iterative approaches and helps interestingly in channel estimation accuracy."

"In order to improve the channel estimation, pilot allocation approaches are proposed based on the system model and probability function."

"A cross entropy-based approach is utilized to propose two different approaches called sequential cross entropy self-coherence (SCE-SC) and parallel cross entropy self-coherence."

"It is demonstrated that in FBP-based approach self-coherence is important in mutual coherence."

Introduction

"MIMO-OFDM relay channel estimation has been developed using DCS-based approach and a novel algorithm has been proposed using compressive sampling matching pursuit (CoSaMP) called block-verified CoSaMP (B-vCoSaMP) [112]."

"In DCS-based channel estimation, the trivial pilot pattern is the random pilot allocation; while utilizing restricted isometry property (RIP), the measurement matrix could be designed to develop channel estimation."

"In DCS-based channel estimation, the pilot sequences are translated to the measurement matrix identification and one may optimize the mutual coherence to design appropriate pilot sequence."

"In [113] a suboptimal pilot allocation algorithm is designed based on GA for DCS-based channel estimation."

"He and others [113] and Akbarpour-Kasgari and Ardebilipour [112] generate optimized pilot sequences for MIMO-OFDM compressed channel estimation using a GA and cross entropy (CE)-based approach."

"In order to exploit the FBP-based channel estimation and ζ-norm, we have designed appropriate measurement matrix using pilot symbols."

Methods

"The channel estimation is addressed using CS perspective."

"In order to increase the channel estimation accuracy together with decreasing the spectrum utilization, we have proposed the FBP-based channel estimation and SCE-SC and PCE-SC pilot allocation algorithms."

"To compare the proposed method with the existing ones, we have considered a GA-based approach and B-vCoSaMP-based channel estimation."

System Model

"The channel between transmit-receive pair which is denoted by h_{uv} consists of L resolvable paths."

"These resolvable paths are resulted from L scatterers which are encountered by the signal conveying from the transmitter antenna u to the receiver antenna v. K of these L scatterers are significant scatterers where K<

"The delays of different paths are the same in all the channel ensembles between two terminals."

"The channel coefficient could be represented as where $\tau(I-1)\geq\cdots\geq\tau(1)\geq\tau(0)$ are the respective paths' delay and g(.) is the shaping pulse in continuous domain."

"The shaping pulse is zero outside the the interval $[0,T_g]$, where T_g is the integer multiple of chip time T. Without loss of generality, we assumed that $\tau(i)$ are integer multiples of T. Thus, the number of channel paths, caused by the channel itself and shaping filter is derived by $L = \tau(I - 1)/T + T_g/T + 1$."

Forward-Backward Pursuit Channel Estimation

"FBP which is based on MP algorithm is used here because of its forward selection and backward fixing."

"OMP which is a greedy algorithm is a special case of FBP which the forward selection is present but the backward fixing is absent; consequently, it cannot fix its own mistakes in the previous steps."

"The FBP algorithm constructs the new subspace by adding just one atom to the previous subspace, and in the backward steps, it reconstructs the subspace by omitting bad atoms."

"The proposed FBP algorithm could be compared with its greedy one called OMP, where OMP is the special case of FBP without backward steps to increase the estimation accuracy."

Pilot Allocation for Compressed Channel Estimation

"In FBP, jointly sparse channels are estimated altogether."

"In order to improve the accuracy of estimated channels, it is mandatory to select pilot subcarriers to optimize the estimation metric."

"There is no polynomial time approach to calculate RIP; thus, we have used mutual coherence to design optimal measurement matrix."

"Mutual coherence is defined as where λ_i and λ_j are pilot subcarriers among N available subcarriers."

"Mutual coherence is defined as maximum off-diagonal entries of Gram matrix $G\{\Phi\} = \Phi^H\Phi$ if Φ is orthonormal [112, 114]."

"▮ According to the above, among other jointly sparse estimation, our proposed method and formulation leads to self-coherence while others lead to self-coherence and cross-coherence [112, 115]."

Proposed Pilot Allocation Algorithms

"We will demonstrate two numerical algorithms to optimize the mutual coherence of measurement matrix and defining optimized pilot sequences."

"The pdf in each of the iteration is updated utilizing appropriate population in each generation."

"In each generation, the updated pdf is utilized to generate the new individuals."

"Each generation is constructed from (I)individuals which are referred to each of the sample pilot sequence."

"Elite individuals in each generation are J individuals with best fitness function."

"Since the generation of individuals consists of sampling from the probability function, this method is more robust than other mutation-based approaches in trapping to local minima."

"In each generation, the pdf vector is utilized to generate I individuals."

"Iteratively the pdf is updated based on the initial generation."

"The proposed method will decrease the local minima trapping more than the SCE-SC approach and increase the accuracy of the resulted steady state point."

Numerical Results

"We demonstrate the performance of the proposed algorithms in channel estimation and pilot allocations called FBP, SCE-SC, and PCE-SC by simulation results."

"We will consider the applicability of the proposed algorithm and the obtained pilot sequences utilizing two main characteristics of channel estimator called NMSE and BER."

"The BER gap between PCE-SC pilot placement using 45 pilots and the ideal case is approximately 1 dB which represented the superiority of the proposed approach in terms of bandwidth efficiency."

"NMSE of the proposed method is better than a GA-based approach approximately 3 dB. Moreover, concerning BER, this superiority is almost 1.5 dB. Evidently, by utilizing $N_p = 45$ optimized pilots on each antenna, the performance is almost 1 dB away from the optimal performance which could be compensated using other facilities."

Results and Discussion

"Channel estimation of AF MIMO relay is considered and CS-based approaches is utilized due to their spectral efficiency and accuracy improvement."

"We have proposed FBP-based channel estimation algorithm for forward selection and backward elimination."

"Two pilot allocation algorithms are proposed based on the cross entropy where the number of multiplications are decreased rather than other approaches since they utilize cross-coherence and self-coherence, simultaneously."

"PCE-SC and SCE-SC algorithms are proposed for pilot generation and channel estimation accuracy improvement."

Study of Single-Cell Massive MIMO Systems with Channel Aging and Prediction [116]

This is a machine-generated summary of:

Younas, Talha; Li, Jiandong; Tulu, Muluneh Mekonnen; Munir, Hafiz Mudassir; Liaqat, Mahrukh: Study of single-cell massive MIMO systems with channel aging and prediction [116]

Published in: Wireless Networks (2019)

Link to original: https://doi.org/10.1007/s11276-019-02030-6

If you want to cite the papers, please refer to the original.

For technical reasons we could not place the page where the original quote is coming from.

Abstract-Summary

"We study and investigate common impairment named as channel aging caused by the relative movement between BS antenna and UTs."

"We study some commonly used channel predictors."

"We apply Wiener predictor and ARMA predictor to mitigate the effects of channel aging."

"We provide signal-to-interference-plus-noise ratios for each channel predictor and demonstrate their overall performance."

Introduction

"In some prior works of massive MIMO, sum of achievable rates have been investigated by including the effects of pilot contamination in multi-cell scenarios by calculating signal-interference-plus-noise (SINR) expressions and applying different precoding techniques [117–122]."

"Channel aging is one of the important impairments for massive MIMO systems and needs to be addressed before the practical deployment of massive MIMO."

"We incorporate channel aging in massive MIMO scenario during uplink (UL) and downlink (DL) transmission."

"This paper provides a complete single-cell massive MIMO modeling with channel aging and prediction."

"We numerically apply channel aging effects and the benefits of prediction for single-cell scenario with reasonable parameters."

"We also present a brief overview of some commonly used channel predictors which have been applied in prior works to mitigate channel aging."

"Our simulation results suggest that aged channel effects the over all system performance."

System Model

"We assume a single-cell scenario in which a single BS equipped with large number of antennas, N, communicates with U user terminals UTs."

"We also assume that all UTs are equipped with single-antenna for the brevity of our analysis."

"In this phase BS communicates from its N antennas to U single antenna UTs."

"The antenna arrays at BS receive the signal in the form of pilot sequences from UTs in UL and send the pre-coding vector for each UTs in the process of DL."

"In single-cell scenario it is relatively easy to estimate channel because the effect of pilotcontamination is not considered."

"Pilot contamination arises when UTs from neighboring cells send their set of pilot sequences to BS at the same time and some of that information sneaks down to the neighboring cells corrupting the channel estimate at BS."

Phenomena of Channel Aging and Prediction

"There is a delay between these two stages due to time variation of channel."

"The delay is actually one of the very important properties of real-time channels."

"We need a time-variant model to analyse the impact of channel aging."

"It depends upon several parameters such as velocity of UTs and medium of channel propagation [123]."

"The channel predictors for AR(1) are easy to design because less parameters need to be optimized."

"One of the approaches being used to mitigate the effects of channel aging is known as channel prediction."

"Different channel predictors have been used in prior works."

"One approach that has been used for channel prediction in prior works is known as Polynomial fitting."

"It exploits the derivative property of mobile radio channel over time."

"It predicts the channel as polynomial of time [124]."

"The variation of wireless channel over time is represented by ACF."

Achievable Rates

"By using the concept of uplink-downlink duality from [125, 126] and applying in our massive MIMO system, we assume thatThis can be done by choosing efficient power control constraints."

"The total transmission power is same, but is divided differently over UTs."

Results and Discussion

"In first case the achievable rate has been calculated without the effect of channel aging and prediction, in second case we apply effects of channel aging and use Wiener predictor to calculate achievable rates and in third case we apply effects of channel aging and use ARMA predictor to calculate achievable rates."

"We observe that increasing the number of antennas helps in getting good results of achievable rates with or without channel predictors."

"We see that the performance of Wiener predictor is better than ARMA predictor."

"We observe that with same number of antennas and other parameters the fluctuations in Wiener predictor are less than that of ARMA predictor and also Wiener predictor gives good peak achievable rates and goes down with less ripples so again it performs better than ARMA predictor."

"Over all we can say that Wiener predictor performs better than ARMA predictor under this scenario."

Conclusion

"We proposed a framework to include some practical effects like channel aging to large scale MIMO systems."

"We studied in the detail the phenomena of channel aging."

"We also studied some channel predictors in detail."

"We observe that Wiener predictor performs better than ARMA predictor in all aspects but it is more complex and is difficult to apply practically."

Optical Wireless Multiple-Input Multiple-Output System Based on Avalanche Photodiode Receiver [127]

This is a machine-generated summary of:

Du, Hao; Xu, Guoning: Optical wireless multiple-input multiple-output system based on avalanche photodiode receiver [127]

Published in: Annals of Telecommunications (2019)

Link to original: https://doi.org/10.1007/s12243-019-00736-4

If you want to cite the papers, please refer to the original.

For technical reasons we could not place the page where the original quote is coming from.

Abstract-Summary

"With the rising demand for high data-transmission rates, optical wireless communication (OWC) is considered one of the particularly appropriate solutions."

"An avalanche photodiode (APD)-based optical wireless configuration is used to approximate the receiver output."

"This paper also presents bit error rate (BER) evaluation statistics compared with the theoretical results for optical wireless multiple-input multiple-output (MIMO) communication systems."

Introduction

"Optical wireless communication (OWC) has been considered a strong candidate in applications that require high-speed free-space communications [128–131]."

"Its highly recognized advantages include cost-effectiveness, fast bandwidth access, high-quality data transmission, no interference with radio frequency (RF) communication systems, and high security [132]."

"OWC has attracted increasing interest and shown promising results for both outdoor and indoor systems [133, 134]."

"Visible-light LEDs, infrared LEDs, and lasers are widely used to transmit data in OWC systems."

"In [135], a 400-Mbps bi-directional line-of-sight (LOS) configuration was demonstrated, based on visible-light/infrared (IR) LEDs."

"Sugiura and others proposed a visible light communication (VLC) multiple-input multiple-output (MIMO) system that allows parallel transmission and reception of symbol streams in a practical LOS-channel scenario [136]."

"An APD-based OWC MIMO system configuration is introduced, and the numerical analysis is discussed."

Optical MIMO

"Overcoming this limitation motivates the study of FSO MIMO systems with an APD receiver array."

"The key idea is that FSO MIMO techniques can improve performance in a manner similar to RF MIMO systems [137], in this case via spatial diversity [138, 139], while APDs help approach shot-noise-limited operation."

"To address both challenges, consideration is given to the use of non-coherent optical arrays, analogous to the use of antenna-array technology for microwave systems, as means of combatting fading."

"To the additive complex Gaussian noise for RF, signal-dependent shot noise becomes the limiting factor for the optical array."

Noise in APDs

"The exact form of the statistics of the random gain introduced by APDs is known [140, 141] but too complicated to produce tractable analyses."

"An approximate analysis of the APD receiver statistics, the WMC approach [142], yields more useable results."

"The primary electrons produced in the APD follow a Poisson distribution for a fixed light intensity with parameter λ_1 when a "1" is sent and λ_0 when a "0" is sent."

SISO Systems

"Subsequent calculations with a more realistic value of 10 dB employed means that when using a mean value λ of primary electrons, $\lambda_1 = \lambda + \lambda_d$ and $\lambda_0 = \varepsilon\lambda + \lambda_d$, with ε equal to the inverse of the numerical value of the extinction ratio."

"To make the approximation, the means and variances of the WMC distribution APD are taken to be the means and variances of Gaussian distributions."

"The Bayesian decision rule is commonly employed in digital communication systems [143] and is used to investigate the costs of information lost when a one is interpreted as a zero and vice versa."

"This approach leads to a ratio of the conditional probabilities for the received signal r as follows:where P(0) and P(1) are the probabilities that the transmitter transmit a "0" and a "1," respectively, and the respective costs are C0 and C1."

OW Diversity

"Taking the transmitted signals here as s_1 for one and s_0 for zero, the conditional probabilities become The same arguments as in the preceding section may be applied to the ratio of the variances and thus, for this case, become Continuing with the same method produces a threshold: The actual values of s_1 and s_0 must be scaled by M so that the optical power of the diversity system is the same as that of the original SISO link, i.e., $s_1 = T_1/M$ and $s_0 = T_0/M$. T_1 and T_0 are the signals in transmitter s_1 and s_0, respectively."

"An approximation to the expectation may be found by considering a sequence of random numbers U_1, U_2, …, U_N drawn from the uniform distribution Although this method does not compete in terms of efficiency with deterministic quadrature rules, offering slow convergence and needing many samples, it does come into its own for multi-dimensional integrals since the number of samples required to achieve a given accuracy is independent of the dimension of the integral."

Conclusions

"Looking further into long-term and sustainable applications, optical wireless MIMO communication is a highly promising option to present a viable supplemental technology to radio systems."

"The WMC model is employed to analyze the SISO configuration using a large deviations approach."

"To progress further, simulation is implemented via IS and successfully compared with the analytical results."

"This paper illustrates the feasibility of using diversity schemes for APD-based OWC systems."

"In the presence of strong fading, the SISO approach is rendered virtually useless, whereas diversity offers acceptable BER performance, and the generated 4 × 4 MIMO system BER performance is significantly better than that of the SISO and other MIMO systems when the mean transmitted energy is higher than − 165dBJ."

Performance Analysis of Spatial Multiplexing MIMO-MFSK Based on Energy Detection for Fast-Fading Environments [144]

This is a machine-generated summary of:

Li, Shuaijun; Qiu, Hongbing; Zheng, Lin; Yang, Chao: Performance analysis of spatial multiplexing MIMO-MFSK based on energy detection for fast-fading environments [144]

Published in: EURASIP Journal on Wireless Communications and Networking (2022)

Link to original: https://doi.org/10.1186/s13638-022-02116-4

If you want to cite the papers, please refer to the original.

For technical reasons we could not place the page where the original quote is coming from.

Abstract-Summary

"Noncoherent multiple-input multiple-output (MIMO) has received attention in recent years, since it is less influenced by factors such as phase fluctuations and has fewer requirements for channel estimation and synchronization."

"Spatial multiplexing MIMO-MFSK based on energy detection is different from conventional noncoherent MIMO in that it can achieve higher spatial multiplexing gain in the independent distribution of channel fading statistics."

"With partial real CSI replacing instantaneous CSI, the system performance is inferior to conventional coherent MIMO under the influence of noise."

"By energy detection, the noise of the MIMO-MFSK system conforms to the generalized gamma distribution."

"We investigate the signal-dependent noise problem of minimum Euclidean distance detection."

"Numerical results show that the SER formula fits well with the simulation results under the condition of a high signal-to-noise ratio."

Introduction

"Noncoherent MIMO with differential detection could also reduce the dependence on the channel information."

"Due to the lack of CSI, conventional noncoherent MIMO systems are unable to meet the requirements in fast-fading environments in terms of detection performance, channel utilization efficiency, and resistance to fast time-varying fading."

"Goldsmith proposed a noncoherent SIMO scheme that does not use the iCSI but rather only the knowledge of the channel statistics, a transmitter that modulates information only in the amplitude of the symbols, and a receiver that measures only the average received energy across the antennas [145–147]. [148] generalizes the constellation design in [145] for arbitrary numbers of users and channel conditions."

"MIMO based on energy detection is mainly considered from the perspective of an infinite number of receiving antennas and mostly uses amplitude modulation with limited modulation space."

"There is a lack of research on spatial multiplexing techniques for noncoherent MIMO in fast time-varying channels without CSI or with partial CSI."

Method

"The system is equated to a real linear model by a nonlinear treatment of the square-law energy detection."

"We analyze the probability distribution of the noise after energy detection and the theoretical performance of the system."

"Following that we present a suboptimal signal detection method and discuss the signal-dependent noise problem."

Channel Environment

"Considering a far-field environment where the antenna array size is much smaller than the propagation distance, it can be deemed that all concurrent transmission channels of the system experience approximately the same Doppler shift."

"Even though the Doppler shift for each space division path is slightly different due to different direction angles, the effect is not very obvious."

"It has the characteristics of anti-Doppler frequency offset and phase offset in fast-fading environments."

"The Doppler shift will also cause the signal frequency offset, which can be ignored under the condition that the Doppler frequency deviation is far less than the symbol rate."

System Model

"After V-BLAST encoding, the transmitted data stream is modulated with MFSK, upconverted, and sent in parallel through multiple transmitting antennas."

"At the receiving end, the transmitted signal is received in a noncoherent manner."

Detection and Performance Analysis

"We analyzed the spatial multiplexing MIMO-MFSK system and derived the equivalent system model."

"The MIMO-MFSK system is equivalent to a linear model."

"Following that we perform a theoretical analysis of the detection performance."

"Since the Bessel function is nonlinear, the implementation complexity of the receiver is high."

"Approximate processing of the Bessel function at high SNR can simplify the optimal decision rule."

Results and Discussion

"The theoretical average SER is compared with the numerical results, and the simulation also demonstrates the accuracy of the received signal distribution by energy detection."

"The isometric constellation design of the transmitting signal can also reduce the overlap regions between neighboring distributions and improve the accuracy of the minimum Euclidean distance detection."

"As the number of antennas increases, SER performance improves."

"As the number of antennas increases, the reliability of detection will be enhanced."

"As the number of receiving antennas increases while the transmitting antenna remains the same, the performance of the system improves."

"The receiving antenna remains the same, and increasing the transmitting antenna will deteriorate performance."

"If the diversity gain at the receiver side remains constant and the number of transmitting antennas increases, the mutual interference under spatial multiplexing will increase, and the system performance will deteriorate."

Conclusion

"This paper presents a spatial multiplexing MIMO-MFSK system based on energy detection with a limited number of transceiver antennas, further extending the modulation space to mutually orthogonal amplitude and frequency dimensions."

"The energy detection technique can help avoid phase synchronization problems, reduce the complexity of the training sequence, and eliminate pilot pollution."

"The MIMO-MFSK system is equivalent to a real linear model through nonlinear processing, which reduces channel estimation and signal detection complexity."

"As the primary metric for judging system performance, the theoretical expression for the SER is given, and we analyze the problem of signal-dependent noise that affects the detection performance of the system."

An Efficient MAPSO Model for Interference Cancellation and Optimal Channel Estimation in MIMO-OFDM System [149]

This is a machine-generated summary of:

Venkateswarlu, Chittetti; Rao, Nandanavanam Venkateswara: An Efficient MAPSO Model for Interference Cancellation and Optimal Channel Estimation in MIMO-OFDM System [149]

Published in: Wireless Personal Communications (2022)

Link to original: https://doi.org/10.1007/s11277-022-09955-w

If you want to cite the papers, please refer to the original.

For technical reasons we could not place the page where the original quote is coming from.

Abstract-Summary

"Numerous researches are performed based on estimating the channel to attain optimal output."

"To decrease the bit error rate and to enhance the system performances, this article aims in proposing the mutable weight-based adaptive particle swarm optimization (MAPSO) based MIMO-OFDM approach for the optimal channel estimation."

"This approach consists of four phases: OFDM with MIMO, STBC based OFDM coding, signal recognition, and MC-CDMA for the evaluation of optimal channel estimation."

"The MAPSO algorithm is used in the MC-CDMA approach for tuning the parameters at every level of channel estimation to achieve optimal value."

Introduction

"To this, the MIMO-OFDM uses the frequency division duplex (FDD) mode of operation because of its decreased interference and enhanced coverage [150, 151]."

"In the current massive MIMO systems, the amplifier leads to signal distortion and so it is better to utilize amplifiers with non-linear and high power efficiency [152]."

"The high data rate is obtained with high spectral effectiveness when the MIMO system is merged with the orthogonal frequency division multiplexing."

"The system with the OFDM modulation with MIMO is accepted by various applications like WiMAX, high-speed LAN standards, digital audio broadcasting (DAB), and several other application regions of MIMO-OFDM."

"The MIMO-OFDM is affected by the high fluctuations in the envelope of the broadcasted signal termed as the peak-to-average power ratio (PAPR)."

"This paper proposes a MAPSO based MIMO-OFDM approach for optimal channel estimation."

Summary of Existing Surveys

"The biggest issue for the researchers was to achieve a minimum complexity energy-efficient PAPR minimization method for the massive MIMO antenna scheme [152]."

"Abdullah and others [153] proposed the selective codeword shift (SCS) and adapted selective mapping (SLM) methods in the MIMO-OFDM scheme for performance enhancement in PAPR and BER."

"Ramadan and others [154] proposed the joint low complexity regularized zero-forcing (JLRZF) equalizer depending on the successive interference cancellation for the OFDM system."

"Fan and others [155] proposed a couple of clipping interference cancellation detectors that were dependent on the soft information as well as the symbol feedback for the generalized space–time frequency index modulation (GSTFIM) methods."

"The symbol feedback based interference cancellation (SFIC) detector executes the iteration by the hard decision symbol."

"Venkateswara Rao and Venkateswarlu proposed a hybrid ABC-BAT algorithm for the interference cancellation in the MIMO-OFDM scheme."

System Design and Problem Formulation

"We formulate the problem that occurred during the transmission of the data."

"The channel between the base station and mobile station undergoes frequency selective fading and an adequate cyclic prefix is summoned to eliminate ISI."

"b. Self-interference channel scheme The uplink and the downlinks are worked under similar frequency so the MIMO-OFDM suffers from inter-symbol interference."

"When this limitation is dropped the below-constrained problem is achieved."

Proposed Methodology for the Optimal Channel Estimation

"The channel estimation process consists of the CDMA scheme utilized for the integral estimation and evaluates the minimum function to improve the performances of channel estimation for the short time block coded based on MIMO-OFDM methods in the fading atmosphere."

"The mutable weight-based adaptive particle swarm optimization (MAPSO) is utilized for hyperparameter tuning at every estimation level."

"The four phases involved in the optimal channel estimation of MIMO-OFDM systems are OFDM with MIMO, STBC based OFDM coding, signal recognition, and MC- CDMA."

"In order to obtain the improved precision value as well as to increase the speed of the particle movement, the ordinary weight function is replaced by the adaptive variable weight function."

"In order to manage the speed of the particle's movement as well as the precision factor, the adaptive variable weight is used in the modified particle swarm optimization, thus adaptive particle swarm optimization is formulated to obtain the value of optimal weight."

Experimental Results and Evaluations

"The experimental evaluations are depicted for evaluating the performances of the proposed approach in terms of the interference cancellation capability and compared with other traditional methods."

"Various simulation measures such as NMSE, MSE, BER, SNR, and INR are evaluated in this section and compared with various methods for the above-mentioned metrics."

"The metrics such as NMSE, MSE, SNR, SER, INR, and BER have been computed in the subsequent section. (i) Normalized mean square error (NMSE) NMSE represents the estimation of frequent deviations among the approximated and the computed values."

"If the BER value is low, then the noise in the proposed system is low which leads to better performance."

"The connection between the SER and the SNR value is evaluated and the value is compared with other existing approaches."

"The performance of the factor bit error rate (BER) is compared with other methods."

Conclusion

"The MAPSO based MIMO-OFDM approaches for the channel estimation technique are suggested in this paper to avoid the difficulty related to a bit error rate and system performances."

"The easy and very effective technique is proposed to provide excellent SER performance and reduced error rate when compared to other existing methods."

"From the evaluations, the results describe that the proposed MAPSO based MIMO-OFDM approach produces a better SNR and improved performance rate for the MIMO-OFDM frequency selective fading channels."

"Using our proposed method, the noise is reduced and it provides better performance compared to other existing systems."

Low-Complexity Near-Optimal Iterative Signal Detection Based on MSD-CG Method for Uplink Massive MIMO Systems [156]

This is a machine-generated summary of:

Albataineh, Zaid: Low-Complexity Near-Optimal Iterative Signal Detection Based on MSD-CG Method for Uplink Massive MIMO Systems [156]

Published in: Wireless Personal Communications (2020)

Link to original: https://doi.org/10.1007/s11277-020-07810-4

If you want to cite the papers, please refer to the original.

For technical reasons we could not place the page where the original quote is coming from.

Abstract-Summary

"Massive multiple-input multiple-output (MIMO) wireless system is increasingly becoming a vital factor in fifth-generation (5G) communication systems."

"In massive MIMO systems, the maximum likelihood detector achieve the optimum performance but it has exponential complexity for realistic antenna configurations systems, Moreover, Linear detectors commonly suffer from a matrix inversion which is not hardware-friendly."

"There is an increase in the computational complexity associated with the unique benefits of the massive MIMO systems."

"To reduce the data detection complexity, we investigate a linear detector based on the multiple search direction conjugate gradient (MSD-CG) in the massive MIMO uplink systems."

"Several theoretical iterative techniques that can be used to balance complexity and performance for massive MIMO detection have been proposed in the literature."

"Our results show that the proposed algorithm with realistic antenna configurations is superior to the existing methods in terms of computational complexity for large-scale MIMO systems."

Introduction

"The large-scale MIMO system has increased dimensions which help to lower the complexity detection signal algorithm in the uplink which in turn enables the system to achieve orders of magnitude increases in energy efficiency and spectrum [157, 158]."

"The complexity of the optimal detector i.e. the maximum likelihood detector increased up to a greater extent with the number of transmit antennas increases; eventually it no longer remain practical for the large scale MIMO systems."

"There are several linear detection approaches that necessitate a better tradeoff between the complexity and performance for massive MIMO uplink systems."

"The complexity of the maximum likelihood algorithm which is the optimal signal detection algorithm increases exponentially following an increase in the transmit antennas which hinders the working of the large-scale MIMO systems [38, 159]."

"The low-complexity linear detection approaches have been studied [160–162], but the high complexity remains to be impossible in massive MIMO system since the algorithms use an unfavorable matrix inversion."

System Model and Data Detection

"This section provides a summary of the efficient techniques for the detection of linear MMSE and introduces a model that is regarded as the OFDM-based uplink massive MIMO system."

"It should be noted that the frequency domain (FD) signals that reach the BS antenna are changed or transformed from the time signals after removing the cyclic prefixes."

"Solving the maximum likelihood (ML) problem helps in the minimization of the symbol error-rate for optimal data detection as per the model in equation (1) [163].It should be noted that even with the best-known decoding sphere algorithm, the solution of equation (2) quickly leads to prohibitive complexity regarding massive MU-MMO system [38, 159]."

"The linear MMSE data detection is the most abundant equalization-based data algorithm [164–166]."

"Besides, the equalization algorithm avoids the expansive computations; for instance, the problem resented by solving the regularized Gram matrix and its inverse."

The Proposed Algorithm Based on the MSD-CG Method
"Equation 2 is solved iteratively without the inversion of the matrix by exploiting MSD-CG method [167]."

"The MSD-CG is less complex than the traditional method of solving a matrix equation."

"The MMSE estimate has been iteratively realized without matrix inversion that has an algorithm with low complexity signal with the help of MSD-CG method."

"Computational complexity for the proposed method : The Computational complexity is analyzed as shown in Algorithm 1."

Simulation Results
"To confirm the working of the signal detection approach proposed in this paper, The simulation results of BER performance are carried out to compare the proposed method with the CG method which is the recently proposed algorithm [168]."

"One can observe that the proposed method performs more consistently and exhibits improvement in the performance as number of iterations increases."

"We observed that the proposed algorithm performs well with different number of antenna and users."

"It is also shown that the BER performance of the MMSE algorithm becomes closer to that of all conventional approaches when there is an increased number of iteration."

"When a similar number of iteration is employed, then the proposed method becomes more superior as compared to the other approaches."

"The performance of the all methods becomes better with an increase in the number of iteration although it still experiences a performance loss for the conventional ones."

Conclusions
"A low complexity signal detection method based on the MSD-CG algorithm is proposed."

"The proposed algorithm based on the MSD-CG method estimates the transmitted signal by iteratively solving the linear equation and avoiding the matrix inverse operation in the MMSE approach."

"The adapted MSD-CG algorithm iteratively estimates the transmitted signal due to eliminating the need of matrix inverse operation."

"With a small number of iterations, it was proved that the simulation results and the convergence of the MSD-CG method can attain the near-optimal achievement of the algorithm by the MMSE method."

Taylor-Based Least Square Estimation in MIMO-OFDM Systems for Multimedia Applications [169]

This is a machine-generated summary of:

Raut, Shital N.; Jalnekar, Rajesh M.: Taylor-Based Least Square Estimation in MIMO-OFDM Systems for Multimedia Applications [169]

Published in: Wireless Personal Communications (2021)

Link to original: https://doi.org/10.1007/s11277-021-08481-5

If you want to cite the papers, please refer to the original.

For technical reasons we could not place the page where the original quote is coming from.

Abstract-Summary

"An effective technique is introduced by proposing the Taylor-least square error algorithm (TLSE) to improve the performance of the MIMO-OFDM system in multimedia applications."

"The user admission control is done in multi user-MIMO (MU-MIMO) system using the priority-based scheduling based on Dolphin-rider optimization (DRO) algorithm that is integrated within the space–time block code (STBC) STBC-MIMO-OFDM system for efficient power allocation to ensure the energy efficiency."

"The DRO is the integration of rider optimization algorithm (ROA) and Dolphin Echolocation (DE)."

"Channel estimation is done using the novel optimization algorithm, termed TLSE, which is designed by modifying LSE with the Taylor series."

"The developed TLSE + DRO (QAM) outperformed other methods with minimal BER of 0.0001 based on channel-2 and maximal throughput of 0.9965 with respect to channel-1."

Introduction

"In the MIMO-OFDM system, the major concern in estimating the channel with high precision lies in maintaining the performance of the system."

"Multiple antennas in the MIMO-OFDM system made the accurate estimation of the channel a challenging task."

"The major requirement in the MIMO-OFDM system is the reduction of time and the accuracy during the channel estimation."

"The goal of the research is to design the channel estimation approach in the MIMO-OFDM system."

"Using the proposed Taylor- least square error algorithm (TLSE) in MIMO-OFDM for channel estimation, the performance can be improved because the LSE algorithm predicts the channel by minimizing the sum of squares of the residuals at every point, and Taylor series calculates the value of the entire function at every point."

"The main contribution of the research is: Proposed TLSE + DRO (QAM) for channel estimation in the MIMO-OFDM system: The TLSEis newly introduced for estimating the channels more effectively."

Literature Survey

"Huang and others [170] designed an adaptive greedy algorithm for estimating the channel in the MIMO-OFDM systems."

"This method reduced the complexity in computation and estimated the capacity of the channel effectively, but the running time of the iteration was higher in this method."

"Jeya and Amutha [20] modeled optimized semi-blind sparse (OSBS) algorithm for the channel estimation in the MIMO-OFDM systems."

"Singh and Bansal [171] developed a channel estimation method based on compressed sensing."

"This method had low computational complexity, but it failed to provide good performance in high mobility scenarios."

"The challenges faced during the channel estimation in the MIMO-OFDM system are as follows: The one-shot transmission regime is used in the sparsity-controlled multiple random-access methods for performing the channel estimation, user identification, and data decoding."

"The challenge in compressed sensing lies in improving the performance of the channel estimation for an increasing number of iterations."

System Model

"The users are scheduled using an optimization based scheduling approach."

"The antenna signals are initially encoded for assuring the diversity specified based on transmitted antennas that follow with the mapping of QAM."

"The mapped signals are transmitted and modulated over the channel, followed by receiving and decoding at receiving end."

"The channel estimation is carried out based on the proposed TLSE."

Proposed Taylor Least Square Algorithm for Channel Estimation in MIMO-OFDM Systems

"The user admission control is carried out based on the MU-MIMO system based on priority-based scheduling using the DRO algorithm."

"The user admission protocol is carried out using priority-based scheduling by DRO algorithm that is integrated within the STBC-MIMO-OFDM system for efficient power allocation to ensure energy efficiency."

"The user admission control is done using an optimization algorithm to the objective function, like priority, throughput, proportionally fair, and power."

"The user admission control is carried out based on DRO, which is designed by integrating ROA [172] and DE [173] for inheriting the advantages of both optimizations such that demerits of optimizations are balanced between each other."

"The standard equation of the LSE algorithm is given by, In order to obtain global optimal solutions in channel estimation, the Taylor series [174] is utilized in the algorithm."

"Once the channel estimation is carried out based on the proposed TLSE, the OFDM, QAM mapping, and decoder is performed."

Results and Discussion

"When SNR is 1 dB, the BER of methods computed by GA, JLSS, QoS guaranteed, TLSE + DRO (BPSK), TLSE + DRO (QPSK), and TLSE + DRO (QAM) are 0.00016, 0.00015, 0.00017, 7.97E–05, 0.00011, and 0.00012."

"When SNR is 1 dB, the BER of methods computed by GA, JLSS, QoS guaranteed, TLSE + DRO (BPSK), TLSE + DRO (QPSK), and TLSE + DRO (QAM) are 0.00018, 0.000173, 0.000196, 5.94E–05, 0.000103, and 0.000108."

"When SNR is 1 dB, the BER of methods computed by GA, JLSS, QoS guaranteed, TLSE + DRO (BPSK), TLSE + DRO (QPSK), and TLSE + DRO (QAM) are 0.000157, 0.000176, 0.000176, 6.88E–05, 0.0001, and 0.000107."

Conclusions

"This paper presents an approach for channel estimation in MIMO-OFDM systems using the proposed optimization algorithm."

"The user admission protocol is based on a priority-based scheduling algorithm incorporated within the STBC-MIMO-OFDM system for efficient power allocation to ensure energy efficiency."

"The channel estimation is carried out based on the proposed optimization algorithm, termed TLSEfor improving the system performance."

"The proposed TLSE estimates the channel very effectively in the MIMO-OFDM systems."

Channel Estimation of Massive MIMO-OFDM System Using Elman Recurrent Neural Network [175]

This is a machine-generated summary of:

Nandi, Shovon; Nandi, Arnab; Pathak, Narendra Nath: Channel Estimation of Massive MIMO-OFDM System Using Elman Recurrent Neural Network [175]

Published in: Arabian Journal for Science and Engineering (2021)

Link to original: https://doi.org/10.1007/s13369-021-06366-0

If you want to cite the papers, please refer to the original.

For technical reasons we could not place the page where the original quote is coming from.

Abstract-Summary

"The performance of MIMO networks dramatically depends upon the techniques exploited for channel estimations."

"The existing methods of channel estimation have failed to resolve the inter-symbol interference (ISI) effects efficiently."

"This paper presents a new channel estimation method based on machine learning."

"The presence of ISI in the MIMO-orthogonal frequency division multiplexing (MIMO-OFDM) network introduces errors in the decision device at the receiver."

"The Elman recurrent neural network (E-RNN) algorithm was employed herein to estimate the channel in MIMO-OFDM considering reliability and scalability."

"These channel estimation parameters exploiting recurrent neural network (RNN), artificial neural network (ANN), convolutional neural network (CNN), and deep neural network (DNN) methods are compared."

Introduction

"The OFDM has high bandwidth efficiency for high bit-rate transmission and can successfully be used over multipath fading channels [176]."

"This article employs the Elman recurrent neural network (E-RNN) algorithm to estimate the MIMO-OFDM channel, considering reliability and scalability."

"In an ideal situation, i.e., linear and stationary channel condition, the mean square error (MSE) estimation has produced the optimal performance."

"The optimal estimator's logical form is hard to be determined under actual complicated channel conditions."

"Deep learning-based channel estimation is exploited to improve the performance of the MIMO-OFDM system in a complicated channel environment."

"The contributions of the present work are as follows: A novel channel estimation scheme using machine learning is introduced."

Related Literature Review

"Parallel channels have been realized in the frequency domain on both the transmitter and receiver sides of various antennas, and similar timing and high information rates have been obtained without the need for additional bandwidth [177–179]."

"Minimum mean squared error (MSE)-based channel estimation was proposed by [180] and exploited the multidimensional nature of frequency domain analysis."

"A channel-estimation-algorithm-based MIMO-OFDM was proposed by [181], where channel state information (CSI) was generated by offline training of the learning network."

"In MIMO-OFDM systems, for high-mobility scenarios, the deep learning-based algorithm was more robust. [182] showed that the long-range fading forecast was necessary for an efficient adaptive OFDM wireless system that utilises the estimated CSI to predict the future CSI. [183] evaluated the MIMO optimisation system considering the Rayleigh fading channel. [184] used the joint swarm intelligence algorithm to minimise multiple access interference to the MIMO system's balance detection performance and complexity. [185] showed the actualisation of the Viterbi algorithm for the OFDM network by analysing the hard and soft decision Viterbi decodings. [186] proposed a technique for sparse channel estimations in massive MIMO systems."

"[187, 188] applied the Alamouti-based space-time block-coded OFDM with various adaptive modulation schemes to meet bandwidth capacity and ISI concealment requirements. [189] shown in their article that non-orthogonal multiple access (NOMA), in comparison with orthogonal multiple access (OMA) approaches, promotes ample connectivity, reduces latency, enhances user fairness and spectral efficiency, and increases reliability. [190] proposed a channel estimation network based on deep learning, termed a recurrent neural network (RNN)."

"A convolutional neural network (CNN) to implement a spatial-frequency-CNN-based channel estimation scheme exploits both frequency and spatial correlations; however, this channel estimation algorithm had high-performance losses."

Proposed Methodology

"This paper proposes minimization of the ISI in a MIMO-OFDM channel using the E-RNN."

"The E-RNN is based on robust feature and remembers information, which is helpful in time-series predictions; this following characteristic and the reliability and scalability of this approach are the reasons for choosing this machine learning algorithm in channel estimation."

"For the channel estimation, the E-RNN is applied on the receiver side."

"E-RNN estimates the channel, and the fast Fourier transform (FFT) is applied to these signals."

"We can achieve better outcomes of channel estimation by the E-RNN process."

"Each epoch is processed in our work is as follows- a. The input data sequence is introduced to the network, and its outputs are determined and compared to the target data to generate an error sequence."

"The E-RNN biases and weights are updated to train the channel [191–193]."

Results and Discussion

"The PAPR, throughput, BER, and capacity were compared to those obtained using existing techniques, such as ANN, CNN, DNN, and E-RNN approaches."

"Compared to the current algorithms, our proposed method achieved better performance."

"At 40 epochs, the obtained PAPR value for the proposed E-RNN is 0.1272, which is more stable than those of 1.127×10^4 for DNN, 2.1278×10^4 for CNN, and 2.627×10^4 for ANN algorithms."

"Our proposed method of E-RNN exhibits the lowest BER (At 5dB the measured BER is 10^{-3}) among different algorithms."

"Our proposed E-RNN exhibits high throughput (approx."

"At 5 bps/Hz, the CDF of our proposed method is lower than those of the existing algorithms."

"The achieved MSE for different neural network (NN) structures like ANN, CNN, DNN, and E-RNN are all non-negative values, and its performance progressively improves with increasing SNR."

Conclusion

"This paper proposed a novel channel estimation method based on machine learning algorithms."

"The PAPR, BER, capacity, throughput, and MSE of the proposed method are compared to existing neural networks such as an ANN, a CNN, a DNN, and an E-RNN."

"The proposed method achieved a low peak PAPR, low BER, high capacity, high throughput, and better MSE performance."

"Compared to the existing algorithms, our proposed method exhibited superior performance."

Multiuser Massive MIMO Channel Estimation and BER Analysis Using Convex Combination Based Algorithms [194]

This is a machine-generated summary of:

Sahoo, Madhusmita; Sahoo, Harish Kumar: Multiuser Massive MIMO Channel Estimation and BER Analysis Using Convex Combination Based Algorithms [194]

Published in: Wireless Personal Communications (2021)

Link to original: https://doi.org/10.1007/s11277-021-09275-5

If you want to cite the papers, please refer to the original.

For technical reasons we could not place the page where the original quote is coming from.

Abstract-Summary

"This paper proposes a computationally efficient adaptive algorithm based channel estimation model."

"These models are applicable to multiuser MIMO (MU-MIMO) systems, which are comprised of a large number of channel coefficients."

"Channel state information (CSI) is acquired from estimated channel coefficients of the UL channel, and CSI of DL is obtained through channel reciprocity."

"Convex combination-based modeling with the introduction of l0 norm penalty to Least mean squares (LMS) algorithm develops an efficient and stable sparse channel estimation model, which is the most significant contribution of the paper."

"The research work presented in the paper has simultaneously focused on channel estimation using the convex combination of momentum-fractional LMS approach and equalization for massive MIMO channels."

"The UL and DL performance of the system was investigated in terms of bit error rate (BER), mean square error (MSE), and channel capacity calculated from estimated CSI."

Introduction

"Massive MIMO systems encounter a major practical challenge, such as the estimation of CSI with a very high dimension."

"The estimation of CSI plays a vital role in modeling a massive MIMO system."

"To recover the transmitted data efficiently, the CSI should be known on the receiver side for UL and on the transmitter for DL."

"The paper proposes a new sparse estimation model using adaptive filtering techniques to obtain CSI for massive MIMO."

"The received data is equalized based on estimated CSI, which minimizes the BER."

Related Work and Motivation

"There are efficient approaches that can handle channel estimation either by designing training sequences or through blind and semi-blind techniques [58, 195]."

"Blind channel estimation is a promising solution to the pilot contamination problem, and it relies upon the higher-order statistics of the received symbols."

"The computational complexity of the blind estimation is much higher than the pilot-based method."

"Semi-blind estimation is a tradeoff of complexity between pilot-based and blind methods."

"Elina Nayebi and others developed tractable expectation-maximization (EM) algorithms that perform well for semi-blind channel estimation for unknown data symbols [195]."

"R. Jeya and others proposed the Optimized Semi-Blind Sparse channel estimation (OSBS-CE) technique."

"V. Singh and others presented an adaptive filtering approach used for the prediction of channel coefficient without any prior knowledge of statistical parameters and with a lower computational complexity as compared to Kalman filtering- based channel estimation [196]."

Main Contributions

"After an extensive study of recent research works published through articles on estimation and equalization for massive MIMO channels, precise contributions of our paper are summarized as follows: a) Convex combination-based adaptive filtering model is designed to estimate CSI for UL, and the same is used to calculate the channel capacity."

"b) The testing of the performance of the proposed model accessed by comparing BER performance and mean square error (MSE) under practical channel conditions."

"The previous existing literature proposed by researchers has not addressed the use of sparse adaptive filtering for estimation in massive MIMO systems, which is a significant contribution to the paper."

"The paper has addressed the TDD system, but the model can be extended to FDD massive MIMO systems"

System Model for MU MIMO OFDM

"For UL transmission, a base station with M antennas and K number of users with a single antenna is considered."

"In UL communication, the K number of users transmits information to the base station through a channel, and the received signal is passed through a detection block to recover the data from each user."

"In the DL, the data which have to be sent to each user are passed through a precoding block at the base station, and the signal of each base station antenna is transmitted through the channel."

"The resulted signal at the base station antennas is passed through a detection block to detect the data of each user."

"In DL, the data which have to be sent to each user are passed through a precoding block at the base station."

Proposed Convex Combination Based Sparse Adaptive Estimation

"The exploitation of sparseness in adaptive filters upgrades their performance in regards to convergence rate, robustness, steady-state performance, tracking ability, etc In this paper, we propose two forms of sparse convex algorithms for channel estimation: fractional LMS and convex m-fractional LMS."

"The convex combination- based modeling introduces sparsity through the norm penalty [197]."

"The combination of filters lies between moderately sparse to moderately non-sparse and thus efficiently tracks a sparse system like massive MIMO with time-varying sparse behavior [198]."

"The equivalent weight vector for the combined filter is"

Simulation Results and Discussion

"From an extensive analysis, it is clear that the channel capacity by using the estimated channel matrix obtained through convex m-Fractional LMS and m-Fractional LMS algorithm are higher than all other algorithms."

"Considering all other aspects such as MSE and BER, it can be observed that m-Fractional LMS is performing better."

"From these figures, it is clear that path loss affects performance quality, and it is evident that the BER performance of the m-fractional convex model is better as compared to other methods."

"To highlight the effectiveness of our suggested methods, we compared their performance to the Feature LMS algorithm [199] in a channel estimation scenario."

"It is fair to say that the proposed methods for MU-MIMO systems outperform the Feature-LMS algorithm in terms of BER , MSE and channel capacity performance."

"Proposed methods have better BER performance in comparison to Feature LMS."

Conclusion

"The implementation of massive MIMO systems in TDD mode is the key motivation behind this paper, and the convex combination approach for designing a sparse estimation model acts as a catalyst to get better performance."

"From the extensive comparison results and performance analysis, it has been observed that the convex sparse filtering model comes up as a robust estimator in comparison to existing traditional filters."

"Numerical results indicate the effectiveness of sparse channel estimation to generate CSI for UL and DL in massive MIMO systems."

Deep Learning Channel Estimation for OFDM 5G Systems with Different Channel Models [200]

This is a machine-generated summary of:

Mohammed, Aliaa Said Mousa; Taman, Abdelkarim Ibrahim Abdelkarim; Hassan, Ayman M.; Zekry, Abdelhalim: Deep Learning Channel Estimation for OFDM 5G Systems with Different Channel Models [200]

Published in: Wireless Personal Communications (2022)

Link to original: https://doi.org/10.1007/s11277-022-10077-6

If you want to cite the papers, please refer to the original.

For technical reasons we could not place the page where the original quote is coming from.

Abstract-Summary

"The most common methods are Decision-Directed Channel Estimation, Pilot-Assisted Channel Estimation (PACE) and blind channel estimation."

"The main objective of this paper is to assess the efficiency of DL-based CE compared to the conventional PACE techniques, such as least-square (LS) and minimum mean-square error (MMSE) estimators."

"A simulation environment to evaluate OFDM performance at different channel models has been used."

"A DL process that estimates the channel from training data is also employed to get the estimated impulse response of the channel."

"From the simulation results, the trained DL estimator provides better results in estimating the channel and detecting the transmitted symbols compared to LS and MMSE estimators although, the complexity of the proposed LSTM estimator exceeds the equivalent LS estimator."

"The DL estimator also demonstrates its effectiveness with various pilot densities and with different cyclic prefix periods."

Introduction

"The focus of the fifth generation (5G) mobile cellular system is to support higher data rates and provide seamless services across a multitude of wireless systems and networks as 5G is mostly a MIMO system that is composed of multiple SISO channels between the different antennas."

"High data rates are possible through adjusting the modulation order and/or transmitted power individually according to the channel response and noise background for each sub-band."

"We discuss DL channel estimation for OFDM 5G Systems with different channel models."

"The main contributions of this paper are to: Evaluate and compare MMSE, LS and DL methods in estimating the impulse response of different channels for next generation 5G cellular mobile communications based on OFDM modulation."

"The LSTM-based neural network is trained for a generated OFDM signal, where the Bit error rate (BER) is calculated and compared with the LS and MMSE estimations."

Literature Review

"In [181], the DL-based estimation algorithm for doubly selective channels is proposed by using the DNN and the numerical results show that the proposed DL-based algorithm perform better than the current estimator through both efficiency and robustness, essentially when the channel statistics are time-varying."

"In [201], the author introduced a DL-based CE algorithm in communication systems to find the full channel state based on pilot values, it regarded a fading channel time–frequency response as a 2D-image and applied image super-resolution and a denoising image restoration algorithm."

"By examining wireless fading channels time and frequency correlation, the authors of [202] offer a novel DL-based approach in OFDM systems for channel estimation and signal identification."

"In [203], the author uses PACE in OFDM systems using DL LSTM neural networks, referred as DLLSTM-based channel state estimator (CSE) and three different optimization algorithms for DL to evaluate the proposed estimator performance at each."

Background

"Channel estimator specification requirements are to minimize the Mean Squared Error (MSE) and Computational Complexity [204]."

"One design challenge for PACE is thus to jointly minimize the pilot numbers while precisely estimate the channel [204]."

"Current research focuses on reaching and evaluating CE techniques such as MMSE Channel Estimation, LS Channel Estimation, and DL channel estimation techniques for 5 G channel models using PACE [205]."

"Without using any channel statistics knowledge, the LS estimators are estimated with very low complexity but obviously it suffers from a high MSE."

"The LS method is used to obtain initial channel estimates at the pilot subcarriers, that are then further improved by various methods [206]."

"Minimum Mean Square Error (MMSE) Estimator: to minimize the MSE, the MMSE estimator uses the channel conditions second-order statistics."

"Using Gradient Descent, the weights between neurons are changed after every iteration through the data set to reduce the cost function [207]."

LSTM Based Channel Estimation

"Input layer interfaces with input data, while the LSTM cells perform the function of hidden layer."

"In the offline training first step, trained with the received OFDM samples which were generated with different data series and under various channel conditions with some statistical properties."

"OFDM modulation and the wireless channels as hidden data, the model is trained."

"The received signal and the original signal were collected as the training data."

"1 pilot block and 1 block of data are given as DL model input."

"In the training phase, those feature vectors are fed to the LSTM model with the corresponding target symbols as batches."

"In testing/prediction phases, the trained model predicts the corresponding symbol for the extracted input feature vector."

"The model is trained to reduce variance between neural network output and original message."

Simulation

"Channel estimation The channel coefficients are estimated in pilot subcarriers by using LS algorithm, MMSE algorithm and DL algorithm Three MATLAB script files are written to explain the DL methods performance for joint CE and symbol detection in OFDM wireless communication systems."

"The total number of multiplications required to estimate the channel in DL method is 17572 per OFDM symbol."

"N = 64, so the total number of multiplications required to estimate the channel in LS method is 4,160 per OFDM symbol [208]."

"N = 64, so the total number of multiplications required to estimate the channel in MMSE method is 528,384 per OFDM symbol [208]."

"The total number of multiplications required to estimate the channel in DL method is 17,572 per OFDM symbol."

Conclusion

"A DL channel estimation technique based on LSTM Training algorithm has been developed for OFDM systems over different channel models."

"BER analysis of the DL based estimator is obtained and compared with the LS, MMSE estimation techniques."

"Both DL and MMSE methods prove to be robust against shorter CP's compared to LS."

"The reduction of the required computational power required for estimation, together with the time required to train the model are two crucial subjects towards consideration of DL in real-time wireless communications."

A Low Complexity Near Optimal Signal Detection for Large Scale MIMO SC-FDMA Uplink System [209]

This is a machine-generated summary of:

Selvaraj, Kandasamy; Anandaraj, Mahalingam; Judson, David; Ganeshkumar, Perumal; Rajkumar, Krishnan: A Low Complexity Near Optimal Signal Detection for Large Scale MIMO SC-FDMA Uplink System [209]

Published in: Wireless Personal Communications (2022)

Link to original: https://doi.org/10.1007/s11277-022-09742-7

If you want to cite the papers, please refer to the original.

For technical reasons we could not place the page where the original quote is coming from.

Abstract-Summary

"This article proposes an improved Newton algorithm as a low complexity signal detection scheme for linear receiver in large scale multiple- input multiple- output (LS-MIMO) single carrier frequency division multiple access (SC-FDMA) uplink system, where a large number of antennas are set up at the base station and active users are with a single antenna system."

"For symbol detection process, LS-MIMO SC-FDMA system with linear detector requires to perform a large matrix inverse computation."

"Even though linear detectors such as zero forcing (ZF) and minimum mean square error (MMSE) can achieve near-optimal detection performance, they still introduce high computational complexity and obliviously involve in the computation of matrix inversion."

"A design of complexity reduction algorithm based near-optimal detector for LS-MIMO SC-FDMA system attains research interest."

"The improved Newton algorithm is employed to obtain linear detection solution which iteratively performs matrix free-inversion operation."

"Simulation results illustrate that the proposed scheme outperforms the conventional detection schemes and exhibits near-optimal performance with a small number of iterations."

Introduction

"This paper aims to design and implement a low complexity near-optimal symbol detection based on improved Newton iteration for LS-MIMO SC-FDMA uplink system."

"An improved Newton iteration technique [210] proposes to avoid matrix inversion involved in ZF/MMSE detector for uplink LS-MIMO SC-FDMA system."

"The performance trade-offs associated with Newton iteration scheme and exact linear data detection in the uplink SC-FDMA based LS-MIMO is analyzed."

"The proposed design works well on linear receiver for medium to large ratios between BS antennas and users, while maintaining near-optimal performance in uplink LS-MIMO SC-FDMA system."

"A low complexity linear detection based on improved Newton algorithm for LS-MIMO SC-FDMA is presented whereas [211] CG approximation is employed on linear MIMO SC-FDMA detection system."

"The proposed algorithm based ZF and MMSE detector performance of uplink LS-MIMO SC-FDMA system is assessed and compared with method [212] whereas regularized ZF based detection is investigated for uplink MIMO SC-FDMA system."

Large Scale MIMO SC-FDMA System Model

"The mth subcarrier SC-FDMA symbol for the kth (k = 1,..., K) user is described as [213] Following the DFT, the transmitter performs subcarrier assignment."

"At the M receiving antennas, it is assumed that the detector has precise time and frequency synchronization for the received MIMO SC-FDMA signal."

"The received signal is transformed into frequency domain by DFT."

"The linear receiver is designed to obtain the received signal via frequency domain operation after subcarrier demapping operation."

Conventional Signal Detection Techniques

"Interference cancellations based on ZF and MMSE detection carry out a $P \times P$ matrix inversion when P is larger in many practical systems."

"MMSE equalization needs an inversion of a non-diagonal matrix with large dimension which is related to total number of subcarriers."

"As in [212, 214], the banded system model is exploited to implement ZF and MMSE solutions with low-complexity as the inverse of a banded matrix is realized with few iterations than direct matrix inverse."

"In order to avoid exact matrix inversion of ZF/MMSE filtering, algorithm is implemented."

"CG based linear receiver is suggested for MIMO SC-FDMA systems [214], where authors achieve low complexity ZF and MMSE detection via the unique modification of interference matrix."

"CG method is proposed to evaluate the performance of ZF/MMSE filtering matrix iteratively instead of using exact matrix inversion."

"The ZF/MMSE filtering matrix is approximated to utilize NSE by a matrix polynomial."

Proposed Signal Detection Algorithm for LS-MIMO SC-FDMA

"Low-complexity algorithm is proposed for multiuser linear receiver in the uplink LS-MIMO SC-FDMA system."

"The new detection employs improved Newton method that achieves ZF/MMSE solution in an iterative way with lower complexity."

"Equation (20) converges quadratically and its number of order of iterations decides the complexity of Newton method."

"The proposed signal detection algorithm obtains ZF/MMSE solutions iteratively without direct matrix inverse operation."

"The number of complex multiplications required for estimation of data in various LS-MIMO SC-FDMA detection techniques are evaluated and compared with the proposed detection method."

"The proposed improved Newton algorithm is compared with direct ZF/MMSE, regularized detection [212], NSE [162], CG algorithm [211] and Newton method [215]."

"It can be proven that complexity of the proposed linear receiver approach is much superior to the conventional detection algorithms, especially for the interleaved SC-FDMA system."

Simulation Results

"The proposed improved Newton iterative algorithm performance is examined with the existing detection algorithms such as direct ZF/MMSE, regularized detection [212], NSE [162], CG algorithm [211] and Newton method [215], respectively."

"Simulations are performed for BER performance of the proposed signal detection scheme with SC-FDMA based LS-MIMO uplink configurations."

"The channel's SNR value is 20 dB. It is clearly seen that the simulation result of the proposed iterative Newton based MMSE detection improves when M increases, and it can obtain close match with performance of the direct MMSE detection with a small number of iterations (i.e., $l = 6$)and antennas of various size at BS."

"Further observations reveal that the proposed iterative Newton MMSE algorithm outperforms the existing detection methods in large-scale MIMO systems."

"It is verified that with the increased number of iterations, the proposed algorithm's BER performance is closer to that of direct MMSE detection."

Conclusions

"This article proposes a low-complexity near-optimal receiver symbol detection algorithm for LS-MIMO SC-FDMA uplink system with 16-QAM modulation."

"In order to achieve low complexity, a linear detector is employed based on improved Newton algorithm that reduces its complexity and estimates detection solution iteratively to avoid inversion operation."

"The numerical results reveal that the proposed detection algorithm converges to exact matrix inversion detection with a minimal number of iterations, while maintaining low complexity."

"The simulation shows effectiveness of the proposed scheme, compared to recently reported detection algorithm in terms of BER performance and computational complexity, and attains superior performance."

FPGA Design of an Efficient EEG Signal Transmission Through 5G Wireless Network Using Optimized Pilot Based Channel Estimation: A Telemedicine Application [216]

This is a machine-generated summary of:

Kumar, K. B. Santhosh; Sujatha, B. R.: FPGA Design of an Efficient EEG Signal Transmission Through 5G Wireless Network Using Optimized Pilot Based Channel Estimation: A Telemedicine Application [216]

Published in: Wireless Personal Communications (2021)

Link to original: https://doi.org/10.1007/s11277-021-09305-2

If you want to cite the papers, please refer to the original.

For technical reasons we could not place the page where the original quote is coming from.

Abstract-Summary

"Electroencephalogram (EEG) signifies a neurophysiologic measurement, which perceives the electrical activity of brain via making a record of EEG signal from the electrodes positioned on the scalp."

"Mixing of EEG signal with some other biological signal is referred as artifacts."

"Hybrid multi resolution discrete wavelet transform based delayed error normalized least mean square is proposed to eradicate motion artifact from the recorded EEG signal."

"Encryption is the process of scrambling the plain EEG signal in to Chipper format."

"Telemedicine system can be used to transmit medical data transmission and it requires an optimal channel estimation method (ESSA) to reduce BERs."

"The main aim of ESSA algorithm is to optimally place the pilot symbols and in-order to enable the automatic estimation of state of the channel."

"The proposed optimal based channel estimation is simulated under Xilinx platform with Verilog coding."

Introduction

"The data attained from the patient has evade due to the energy disputes such as life span of battery, power etc of the sensor nodes based on body sensor networks (BSNs) or wireless body-area networks (WBANs)."

"In order to empower remote health monitoring, WBAN should use the gateway node that forwards medical bio signal to wireless network and it transforms the composed data to the health centre."

"Turbo encoder takes place in the transmission side of wireless network in order to enhance the transmitted EEG data with minimum BER of the signal."

"In 5G wireless network, the data will be transmitted through some frequency-selective fading channels."

"To develop a pilot insertion based optimized channel estimation we are developing a ESSA optimization model for transmitting the de-noised EEG signal through 5G wireless network."

Related Work

"Satheeskumaran Sabrigiriraj and others [217] had introduced Least Mean Square (LMS) filtering method for elimination of noise and DWT based R-peak detection in EEG signals."

"Karthigaikumar and others [218] proposed NLMS filtering technique for efficient noise removal in biomedical signals."

"Venkatesan and others [219] had implemented FPGA based DENLMS adaptive filtering tactic for the deletion of noise from EEG signal."

"DENLMS filtering method was studied to eradicate white Gaussian noise present in EEG signal."

"Behzad Hejrati and others [220] had proposed a lossless EEG signal firmness based on channel clustering."

"Raghunatharao and others [221] introduced optimal pilot based channel estimation in Cognitive Radio in OFDM."

"The optimal pilot based channel estimation was optimized through Rider Grey Wolf Optimization (RGWO) and it was transmitted through OFDM modulation."

"This paper proposed an efficient noise reduction algorithm and an optimal transmission strategy for EEG signal transmission through wireless 5G network."

Proposed Methodology

"This work proposed a new filtering method namely hybrid multi resolution discrete wavelet transform based delayed error normalized least mean square (MRDWT-DENLMS) algorithm to reduce the motion artifacts from the recorded EEG signal."

"Optimal pilot insertion method is proposed to build the performance of channel estimation. Thus, the received EEG signal will be decoded without any error from the channel."

"This frequency domain channel estimation method uses pilot symbols at the optimal locations in the GFDM-IM symbol blocks."

"The received EEG signal is decoded by Turbo decoder based de-interleaver without any error from the channel."

"After constellation mapping the pilot are inserted in to the data blocks and the pilot selection process based on channel estimation is processed by Enhanced Squirrel Search (ESSA) Optimization Algorithm."

"An FPGA design for an enhanced squirrel search algorithm optimized pilot based channel estimation method is proposed to transmit the biomedical EEG signal through 5G wireless network."

Result and Discussion

"The proposed optimal pilot based channel estimation to transmit biomedical EEG signal has been simulated in the Xilinx tool using Verilog coding."

"An efficient motion artifacts removal method is required before transmitting the EEG signals through wireless 5G network."

"Performance analysis of Mean Square Error (MSE), Signal to Noise Ratio (SNR) and Root Mean Square Error (RMSE) have been analysed for different noise removal filtering methods."

"In our work GFDM-IM flexible multi carrier modulation technique have been implemented to transmit the de-noised EEG signals through 5G wireless network."

"The graph illustrates that the proposed method have minimum Bit Error Rate (BER) and we have noticed that the pilot insertion based on ESSA perform best when compared with some other existing techniques."

"This is due to the fact that, the proposed optimal estimation of channel and GFDM-IM based modulation scheme removes different interferences and decreases the bit error rate (BER) efficiently."

Conclusion

"De-noised EEG signal is transmitted through 5G wireless network in telemedicine application via GFDM-IM modulation technique."

"The main challenge of channel estimation is insertion of pilot symbols and it has been progressed through ESSA Algorithm."

"The ESSA optimization algorithm had been used to optimize the pilot symbol insertion with SVD-LMMSE channel estimation technique."

"The performance analysis of our proposed method have been analysed by comparing it with different filtering methods, modulation methods, interleaving techniques and channel estimation methods."

Robust Beamforming and Spatial Precoding for Quasi-OSTBC Massive MIMO Communications [222]

This is a machine-generated summary of:

Lee, Ju-Hong; Sun, Wei-En: Robust beamforming and spatial precoding for quasi-OSTBC massive MIMO communications [222]

Published in: EURASIP Journal on Wireless Communications and Networking (2019)

Link to original: https://doi.org/10.1186/s13638-019-1373-9

If you want to cite the papers, please refer to the original.

For technical reasons we could not place the page where the original quote is coming from.

Abstract-Summary

"This paper considers the problem of massive multiple-input-multiple-output (MIMO) wireless communication systems with quasi-orthogonal space-time block code (QOSTBC) transmission in the presence of spatial correlation effect (SCE) and mutual coupling effect (MCE)."

"It is difficult to make a quasi-orthogonal space time block coded massive MIMO system to achieve full transmission rate."

"The proposed method incorporates a beamforming scheme and spatial precoding to formulate an appropriate optimization process to effectively alleviate the considered problems."

"During the optimization process, the proposed method finds the optimal beamforming coefficients of the precoding matrix, the optimal normalized positions of transmit and receive antenna elements for the case of using linear antenna arrays, and the optimal angle differences of transmit and receive antenna elements for the case of using circular arrays."

"Based on the proposed method, we are able to cure the performance degradation of a quasi-orthogonal space time block coded massive MIMO system due to the SCE and MCE."

"The proposed method makes QOSTBC MIMO communications with full transmission rate for any number of transmit antennas achievable."

"Several simulation examples are presented to show the superior bit error rate (BER) performances of QOSTBC wireless MIMO scenarios with linear as well as circular antenna arrays by using the proposed method as compared to the existing methods."

Introduction

"The robust method proposed in this paper explores the joint design of beamforming and precoding for QOSTBC massive MIMO systems with any number of transmit antennas and receive antennas to deal with the SCE and MCE."

"Through the proposed beam selection spatial precoding scheme in conjunction with the appropriate optimization formulation, the considered problems can be effectively alleviated as shown by simulation examples. (II) The proposed joint beamforming and spatial precoding method can preserve the advantage of reducing the overheads for downlink training and feedback of channel state information between the base station (BS) and user equipments (UEs) for the frequency-division duplex (FDD) mode of downlink transmission in MIMO systems. (III) In the literature, it is well known that the QOSTBC can achieve 1/2 of the full transmission rate for any number of transmission antennas as well as the codes with 3/4 of the full transmission rate for the specific cases of three and four transmit antennas."

System Model and Fundamentals

"The space-time block encoder generates the simplest type of spatial temporal codes to exploit the transmission diversity provided by multiple transmit antennas in wireless communications."

"For downlink transmission, the STBC at BS with M antennas encodes the data to be transmitted into M transmitted code symbols."

"In order to search for similar schemes available for MIMO systems with more than two transmit antennas to achieve diversity level higher than two, orthogonal space-time block code (OSTBC) was then introduced by [223]."

"It is also shown [223] that a complex orthogonal design and the corresponding space-time block code which provides full diversity and full transmission rate is not possible for more than two transmit antennas."

"A typical example of a full rate quasi-orthogonal code that achieves full transmission rate for four transmit antennas is as follows: where s_i, i = 1,2,3,4, are the symbols selected based on the input bits."

The Robust Beamforming and Spatial Precoding Method

"The precoder W_{pk} is given by: Moreover, T_k represents the power constraint matrix associated with the selected N beams and is given by: where t_{ki} denotes the power distributed to the ith beam selected by the kth UE and the total power transmitted to the N beams is constrained by: To incorporate the advantages of full transmission rate and low ML decoding complexity over the conventional multiuser beamforming methods (e.g., zero-forcing beamforming and MMSE beamforming) by using a simple 4×4 QOSTBC with the proposed joint beamforming and spatial precoding method, we can easily set the number of beam groups of the proposed beamforming scheme to 4, i.e., B = 4."

"For the second-step optimization process, we develop an appropriate fitness function based on the bit error rate (BER) performance for implementing the CCPSO to search the optimal spatial precoding matrix and the power constraints for downlink communication in massive MIMO channels."

Simulation Results and Discussion

"The simulation results for Example 2 (a massive MIMO case) is presented for the case of $N_B = 32$ and $N_U = 2$."

"The number of array elements deployed at BS is set to $N_B = 32$ under the consideration of massive wireless MIMO scenarios."

"From the simulation results showing that using circular antennas provides better BER performance than using linear antennas under the same number of antenna elements, we reason that this is due to the following phenomena: (a) Because only small amount of power transmitted by utilizing circular antenna array would be reflected back, circular antenna array provides better return loss as compared to linear antenna array. (b) Because the percentage of bandwidth and the half-power beamwidth for circular antenna array are wider than those for linear antenna array, circular antenna array provides higher gain than linear antenna array. (c) To mitigate MCE, circular antenna array is a more suitable geometrical arrangement than linear antenna array."

Conclusion

"The proposed method can significantly mitigate the mutual coupling and spatial correlation problems for downlink transmission, particularly when massive antennas are deployed at the BS."

"Simulation results show that the proposed method can indeed achieve much better BER performance than the existing methods under the scenarios of massive antennas with antenna mutual coupling and spatial correlation."

"We reason that the proposed method offers a more appropriate beamforming and spatial precoding scheme for future QOSTBC massive wireless MIMO communications."

"To obtain the complete mathematical analysis regarding the proposed method, we are currently conducting the further research on the exact analytical results on the achievable performance gains, mathematical analysis of the proposed schemes/antenna placements in the presence of mutual coupling and spatial correlation effects."

Bibliography

1. Kaur, H., & Kansal, A. (2023). Closed-form analysis of RZF in multicell massive MIMO over correlated Rician channel. *Wireless Personal Communications*. https://doi.org/10.1007/s11277-023-10518-w
2. Kazemi, M., Göken, Ç., & Duman, T. M. (2023). Robust joint precoding/combining design for multiuser MIMO Systems with calibration errors. *IEEE Transactions on Wireless Communications*. https://doi.org/10.1109/TWC.2022.3232140
3. Zhang, Q., Jin, S., Wong, K. K., Zhu, H., & Matthaiou, M. (2014). Power scaling of uplink massive MIMO systems with arbitrary-rank channel means. *IEEE Journal of Selected Topics in Signal Processing, 8*(5), 966–981. https://doi.org/10.1109/JSTSP.2014.2324534
4. Sanguinetti, L., Kammoun, A., & Debbah, M. (2017, March). Asymptotic analysis of multicell massive MIMO over Rician fading channels. In *2017 IEEE International Conference on Acoustics, Speech and Signal Processing* (ICASSP). IEEE. (pp. 3539–3543). https://doi.org/10.1109/ICASSP.2017.7952815
5. Sanguinetti, L., Kammoun, A., & Debbah, M. (2018). Theoretical performance limits of massive MIMO with uncorrelated Rician fading channels. *IEEE Transactions on Communications, 67*(3), 1939–1955. https://doi.org/10.1109/TCOMM.2018.2884003
6. Kammoun, A., Sanguinetti, L., Debbah, M., & Alouini, M. S. (2019). Asymptotic analysis of RZF in large-scale MU-MIMO systems over Rician channels. *IEEE Transactions on Information Theory, 65*(11), 7268–7286. https://doi.org/10.1109/TIT.2019.2923401
7. Sanguinetti, L., Björnson, E., & Hoydis, J. (2019). Toward massive MIMO 2.0: Understanding spatial correlation, interference suppression, and pilot contamination. *IEEE Transactions on Communications, 68*(1), 232–257. https://doi.org/10.1109/TCOMM.2019.2945792

8. Özdogan, Ö., Björnson, E., & Larsson, E. G. (2019). Massive MIMO with spatially correlated Rician fading channels. *IEEE Transactions on Communications, 67*(5), 3234–3250. https://doi.org/10.1109/TCOMM.2019.2893221
9. Rajkumarsingh, B., Rajarai, R. R., & Hosany, M. A. (2023). Performance analysis of hybrid filtering technique for reduction of PAPR in Alamouti coded MIMO-OFDM systems. *Wireless Personal Communications*. https://doi.org/10.1007/s11277-023-10348-w
10. Zhang, S.-Y., & Shahrrava, B. (2021). A SLM scheme for PAPR reduction in polar coded OFDM-IM systems without using side information. *IEEE Transactions on Broadcasting, 67*(2), 463–472. https://doi.org/10.1109/TBC.2020.3039696
11. Sandeep Kumar, V. (2020). Joint Iterative filtering and companding parameter optimization for PAPR reduction of OFDM/OQAM signal. *AEU—International Journal of Electronics and Communications, 124*, 153365. https://doi.org/10.1016/j.aeue.2020.153365
12. Dash, L., & Thampy, A. S. (2023). Channel estimation using hybrid optimizer based recurrent neural network long short term memory for MIMO communications in 5G network. *SN Applied Sciences*. https://doi.org/10.1007/s42452-022-05253-z
13. Yang, B., Yu, Z., Lan, J., Zhang, R., Zhou, J., & Hong, W. (2018). Digital beamforming-based massive MIMO transceiver for 5G millimeter-wave communications. *IEEE Transactions on Microwave Theory and Techniques, 66*(7), 3403–3418. https://doi.org/10.1109/TMTT.2018.2829702
14. Kim, K-S, S-L Ju, H-R Choi (2016) Performance evaluation for 5G NR based uplink millimeter-wave MIMO systems under urban micro cell. In *2019 2nd International Conference on Communication Engineering and Technology (ICCET)* (pp. 48–51). IEEE, 2019. https://doi.org/10.1109/ICCET.2019.8726917
15. Liu, A., Huang, X., Liu, Y., Xiang, H., & Luo, W. (2009). Capacity bounds of MIMO channels with asymmetric channel state information at transmitter. *IEEE Communications Letters, 13*(8), 564–566. https://doi.org/10.1109/LCOMM.2009.090991
16. Sharma, M (2013) Novel adaptive channel state information feedback for multiuser MIMO in wireless broadband communications. In *2013 IEEE 14th International Symposium on A World of Wireless, Mobile and Multimedia Networks (WoWMoM)* (pp. 1–2). IEEE, 2013. https://doi.org/10.1109/WoWMoM.2013.6583428.
17. Dash, L, Anand ST (2022) Performance evaluation of coding schemes for 5G Communication under different channel setting. In *2022 International conference on innovative computing, intelligent communication and smart electrical systems (ICSES)* (pp. 1–5). IEEE, 2022, https://doi.org/10.1109/ICSES55317.2022.9914100
18. Mehrabi, M., Mohammadkarimi, M., Ardakani, M., & Jing, Y. (2019). Decision directed channel estimation based on deep neural network k -step predictor for MIMO communications in 5G. *IEEE Journal on Selected Areas in Communications, 37*(11), 2443–2456. https://doi.org/10.1109/JSAC.2019.2934004
19. Dai, R., Liu, Y., Wang, Q., Yu, Y., & Guo, X. (2021). Channel estimation by reduced dimension decomposition for millimeter wave massive MIMO system. *Physical Communication, 44*, 101241. https://doi.org/10.1016/j.phycom.2020.101241
20. Jeya, R., & Amutha, B. (2019). Optimized semiblind sparse channel estimation algorithm for MU-MIMO OFDM system. *Computer Communications, 146*, 103–109. https://doi.org/10.1016/j.comcom.2019.07.018
21. Koudouridis, G. P., He, Q., & Dán, G. (2022). An architecture and performance evaluation framework for artificial intelligence solutions in beyond 5G radio access networks. *EURASIP Journal on Wireless Communications and Networking*. https://doi.org/10.1186/s13638-022-02164-w
22. ITU-T FG-ML5G-ARC5G (2019). Unified architecture for machine learning in 5G and future networks, in Telecommunication Standardization Section of ITU.
23. O-RAN Working Group 2. AI/ML workflow description and requirements
24. 5G PPP Architecture Working Group. (2021, October). *View on 5G Architecture, version 4.0.*

25. 3GPP Technical Specification Group Core Network and Terminals. (2022, June) TS 29.520 5G System; Network Data Analytics Services; Stage 3 (Release 17). ver.V17.7.0
26. 3GPP Technical Specification Group Radio Access Network. (2022, June) TS 38.420 NG-RAN; Xn general aspects and principles (Release 17), ver.17.1.0.
27. 3GPP Technical Specification Group Radio Access Network. (2022, June). TS 38.410 NG-RAN; NG general aspects and principles (Release 17), ver.17.1.0.
28. Wang, C., Di Renzo, M., Stanczak, S., Wang, S., & Larsson, E. G. (2020). Artificial intelligence enabled wireless networking for 5G and beyond: Recent advances and future challenges. *IEEE Wireless Communications, 27*(1), 16–23.
29. Calabrese, F. D., Wang, L., Ghadimi, E., Peters, G., Hanzo, L., & Soldati, P. (2018). Learning radio resource management in RANs: Framework, opportunities, and challenges. *IEEE Communications Magazine, 56*(9), 138–145.
30. Hakeem, A., Shimaa, A., Hady, A. A., & Kim, H. W. (2020). 5G-V2X: Standardization, architecture, use cases, network-slicing, and edge-computing. *Wireless Networks*. https://doi.org/10.1007/s11276-020-02419-8
31. Xu, Q., Mak, T., Ko, J., & Sengupta, R. (2004). Vehicle-to-vehicle safety messaging in DSRC. In *Proceedings of the 1st ACM international workshop on vehicular ad hoc networks* (pp. 19–28). ACM.
32. Al-Kahtani, M. S. (2012). Survey on security attacks in vehicular ad hoc networks (VANETs). In *2012 6th international conference on signal processing and communication systems* (pp. 1–9). IEEE.
33. Dai, L., Wang, B., Yuan, Y., Han, S., Chih-Lin, I., & Wang, Z. (2015). Non-orthogonal multiple access for 5G: Solutions, challenges, opportunities, and future research trends. *IEEE Communications Magazine, 53*(9), 74–81.
34. Fifth Generation Communication Automotive Research and innovation. (2020). *Deliverable D1.3 5GCAR final project report version: V1.0 2019-07-31*. Retrieved May 9, 2020, from https://5gcar.eu/wp-content/uploads/2019/08/5GCAR_D1.3_v1.0.pdf.
35. Condoluci, M., Gallo, L., Mussot, L., Kousaridas, A., Spapis, P., Mahlouji, M., et al. (2019). 5G V2X system-level architecture of 5GCAR project. *Future Internet, 11*(10), 217.
36. Wang, Y., & Fortier, P. (2022). Polynomial expansion-based MMSE channel estimation and precoding for massive MIMO-GFDM systems. *Wireless Personal Communications*. https://doi.org/10.1007/s11277-022-09943-0
37. Shariati, N., Björnson, E., Bengtsson, M., & Debbah, M. (2014). Low-complexity polynomial channel estimation in large-scale MIMO with arbitrary statistics. *IEEE Journal of Selected Topics in Signal Processing, 8*(5), 815–830.
38. Rusek, F., Persson, D., Lau, B. K., Larsson, E. G., Marzetta, T. L., Edfors, O., & Tufvesson, F. (2012). Scaling up MIMO: Opportunities and challenges with very large arrays. *IEEE Signal Processing Magazine, 30*(1), 40–60. https://doi.org/10.1109/MSP.2011.2178495
39. Ehsanfar, S., Matthé, M., Zhang, D., & Fettweis, G. (2017). In *2017 IEEE Wireless Communications and Networking Conference (WCNC)*. San Francisco: CA, USA.
40. Ehsanfar, S., Matthé, M., Zhang, D., & Fettweis, G. (2016). *In 20th International ITG Workshop on Smart Antennas (WSA 2016)*. Germany, Munich.
41. Ehsanfar, S., Matthé, M., Zhang, D., & Fettweis, G. (2016). *In 2016 IEEE Global Communications Conference (GLOBECOM)*. Washington: DC, USA.
42. Panda, S. (2020). Spectral efficiency optimization of massive MIMO system under channel varying conditions. *Wireless Personal Communications*. https://doi.org/10.1007/s11277-020-07924-9
43. Panda, S. (2020a). Joint user patterning and power control optimization of MIMO–NOMA systems. *Wireless Personal Communications, 112*, 2557–2573.
44. Lu, L., Li, G. Y., Swindlehurst, A. L., Ashikhmin, A., & Zhang, R. (2014). An overview of massive MIMO: Benefits and challenges. *IEEE Journal of Selected Topics in Signal Processing, 8*(5), 742–758.

45. Narasimhan, T., et al. (2014). Channel hardening-exploiting message passing (CHEMP) receiver in large-scale MIMO systems. *IEEE Journal of Selected Topics in Signal Processing, 8*(5), 847–860.
46. Ravi Babu, T., Dharma Raj, C., Adinarayana, V., & Murali Krishna, K. (2021). Estimation of sparse channel using Bayesian Gaussian mixture and CS-aided techniques for pilot contaminated massive MIMO system. *Wireless Personal Communications*. https://doi.org/10.1007/s11277-020-07927-6
47. Larsson, E. G., Edfors, O., Tufvesson, F., & Marzetta, T. L. (2014). Massive MIMO for next generation wireless systems. *IEEE Communications Magazine, 52*(2), 186–195. https://doi.org/10.1109/MCOM.2014.6736761
48. Marzetta, T. L. (2010). Noncooperative cellular wireless with unlimited numbers of base station antennas. *IEEE Transactions on Wireless Communications, 9*(11), 3590–3600. https://doi.org/10.1109/TWC.2010.092810.091092
49. Bajwa, W. U., Haupt, J., Sayeed, A. M., et al. (2010). Compressed channel sensing: A new approach to estimating sparse multipath channels. *Proceedings of the IEEE, 98*(6), 1058–1076.
50. Rao, X., & Lau, V. K. N. (2015). Compressive sensing with prior support quality information and application to massive MIMO channel estimation with temporal correlation. *IEEE Transactions on Signal Processing, 63*(18), 4914–4924.
51. Bjornson, E., et al. (2018). Massive MIMO networks: Spectral, energy, and hardware efficiency. *Foundationsand Trends in Signal Processing, 11*(3–4), 154–655.
52. Noh, S., & Zoltowski, M. D. (2014). Pilot beam pattern design for channel estimation in massive MIMO systems. *IEEE Journal on Selected Topics in Signal Processing, 8*(5), 787–801.
53. Hoydis, J., Hoek, C., et al. (2012). Channel measurements for large antenna arrays. In *Proceedings, IEEE international symposium on wireless communication systems (ISWCS'12), Paris, France* (pp. 811–815).
54. Tamilarasan, N., Lenin, S. B., & Nithyanandan, L. (2022). Performance of MIMO MC-CDMA system with relay using adaptive pilot channel estimation. *Journal of The Institution of Engineers (India): Series B*. https://doi.org/10.1007/s40031-022-00773-z
55. Wang, R., Tao, M., Mehrpouyan, H., & Hua, Y. (2015). Channel estimation and optimal training design for correlated MIMO two-way relay systems in colored environment. *IEEE Transactions on Wireless Communications, 14*(5), 2684–2699.
56. Guo, X., Chen, S., Zhang, J., Xiaomin, M., & Hanzo, L. (2016). Optimal pilot design for pilot contamination elimination/reduction in large-scale multiple-antenna aided OFDM systems. *IEEE Transactions on Wireless Communications, 15*(11), 7229–7243.
57. Chen, H., Zhang, R., Zhai, W., Liang, X., & Song, G. (2018). Interference-free pilot design and channel estimation using ZCZ sequences for MIMO-OFDM-based C-V2X communications. *China Communications, 15*(7), 47–54.
58. Shin, C., Heath, R. W., & Powers, E. J. (2007). Blind channel estimation for MIMO OFDM systems. *IEEE Transactions on Vehicular Technology, 56*(2), 670–685.
59. Ghavami, K., & Naraghi-Pour, M. (2018). Blind channel estimation and symbol detection for multi-cell massive MIMO systems by expectation propagation. *IEEE Transactions on Wireless Communications, 17*(2), 943–954.
60. Zhang, J., Yuan, X., & Zhang, Y.-J. A. (2018). Blind signal detection in massive MIMO: Exploiting the channel sparsity. *IEEE Transactions on Wireless Communications, 66*(2), 700–712.
61. Chen, L., & Yuan, X. (2019). Blind multiuser detection in massive MIMO channels with clustered sparsity. *IEEE Transactions on Wireless Communications Letters, 8*(4), 1052–1055.
62. Afshar, A., & Vakili, V. T. (2021). Spatial based pilot allocation (SBPA) in crowded massive MIMO systems. *Wireless Personal Communications*. https://doi.org/10.1007/s11277-021-08205-9

63. Sørensen, J. H., De Carvalho, E., Stefanovic, Č., & Popovski, P. (2018). Coded pilot random access for massive MIMO systems. *IEEE Transactions on Wireless Communications, 17*(12), 8035–8046.
64. Björnson, E., De Carvalho, E., Sørensen, J. H., Larsson, E. G., & Popovski, P. (2017). A random access protocol for pilot allocation in crowded massive MIMO systems. *IEEE Transactions on Wireless Communications, 16*(4), 2220–2234.
65. Miranda, R. K., da Costa, J. P. C., Guo, B., de Almeida, A. L., Del Galdo, G., & de Sousa, R. T. (2019). Low-complexity and high-accuracy semi-blind joint channel and symbol estimation for massive MIMO-OFDM. *Circuits, Systems, and Signal Processing, 38*(3), 1114–1136.
66. Bakulin, M. G., Kreindelin, V. B., & Pankratov, D. Y. (2021). Algorithms for automatic selection of antennas in a MIMO system without enumeration of combinations. *Journal of Communications Technology and Electronics*. https://doi.org/10.1134/s1064226921120032
67. Bakulin, M. G., Varukina, L. A., & Kreindelin, V. B. (2014). *MIMO technology: Principles and algorithms*. Goryachaya Liniya–Telekom. [in Russian].
68. Jerry, R. (2014). *Hampton, introduction to MIMO communications*. University Press.
69. Kreindelin, V. B., & Yu Pankratov, D. (2012). Sharing of technologies of the autochoice of antennas on transfer and space–time coding. *Elektrosvyaz, 2*, 22–24.
70. Molisch, A., & Win, M. (2004). MIMO systems with antenna selection. *IEEE Microwave Magazine, 5*, 46–56.
71. Boccuzzi, J. (2012). *Processing of signals for wireless communication*. McGraw Hill. Tekhnosfera, Moscow.
72. Bakulin, M. G., Kreindelin, V. B., & Khazov, M. L. (2018). Criteria of the autochoice of antennas in the MIMO systems. *Elektrosvyaz, No., 10*, 78–81.
73. Kreindelin, V. B., & Khazov, M. L. (2017). Algorithms of switching of antennas in the MIMO systems. *Elektrosvyaz', No., 10*, 59–64.
74. Bakulin, M. G., Kreindelin, V. B., & Pankratov, D. Y. (2018). *Technologies in the systems of a radio communication on the way to 5G*. Goryachaya Liniya–Telekom. [in Russian].
75. Khurshid, K., Khan, A. A., Siddiqui, M. H., Hadi, M. U., Rashid, I., & Imran, M. (2021). Optimality of linear MIMO detection for 5G systems via 1-opt local search. *Journal of Electrical Engineering & Technology*. https://doi.org/10.1007/s42835-020-00625-0
76. Verdu, S. (1998). *Multiuser detection*. Cambridge University Press.
77. Xie, X., Yue, D., & Park, J. H. (2019). Observer-based fault estimation for discrete-time nonlinear systems and its application: A weighted switching approach. *IEEE Transactions on Circuits and Systems I: Regular Papers, 66*(11), 4377–4387.
78. Mohades, Z., & Vakili, T. (2021). Vahid deep neural network for compressive sensing and application to massive MIMO channel estimation. *Circuits, Systems, and Signal Processing*. https://doi.org/10.1007/s00034-021-01675-z
79. Palangi, H., Ward, R. K., & Deng, L. (2017). Convolutional deep stacking networks for distributed compressive sensing. *Signal Processing, 131*, 181–189.
80. A. Mousavi, G. Dasarathy, R.G. Baraniuk (2017). Deepcodec: Adaptive sensing and recovery via deep convolutional neural networks. In *55th Annual Allerton Conference on Communication, Control, and Computing* (pp. 3–6)
81. Lu, H., & Bo, L. (2019). WDLReconNet: Compressive sensing reconstruction with deep learning over wireless fading channels. *IEEE Access, 7*, 24440–24451.
82. Ma, X., Gao, Z., & Bruckstein, A. (2020). Data-driven deep learning to design pilot and channel estimator for massive MIMO. *IEEE Transactions on Vehicular Technology, 69*(5), 5677–5682.
83. Eldar, Y. C., Kuppinger, P., & Bolcskei, H. (2010). Block-sparse signals: Uncertainty relations and efficient recovery. *IEEE Transactions on Signal Processing, 58*(6), 3042–3054.
84. Dai, W., & Olgica, M. (2009). Subspace pursuit for compressive sensing signal reconstruction. *IEEE Transactions on Signal Processing, 55*(5), 2230–2249.

85. Zhou, Y., Herdin, M., & Bonek, E. (2007). Experimental study of MIMO channel statistics and capacity via the virtual channel representation. *University of Wisconsin-Madison, Madison, WI, USA, Technical Report, 5*, 10–15.
86. Kansal, L., Sharma, V., & Singh, J. M. (2019). Massive MIMO-OFDM system incorporated with diverse transformation for 5G applications. *Wireless Personal Communications*. https://doi.org/10.1007/s11277-019-06707-1
87. Li, Y., Winters, J. H., & Sollenberger, N. R. (2009). MIMO-OFDM for wireless communications: Signal detection with enhanced channel estimation. *IEEE Transactions on Communications, 50*(9), 1471–1477.
88. Lakshmanan, M. K., & Nikookar, H. (2006). A review of wavelets for digital wireless communication. *Wireless Personal Communications, 37*, 387–420.
89. Xiaodong, Z., Pingping, X., Guoan, Z., & Guangguo, B. (2002). Study on complex wavelet packet based OFDM modulation (CWPOFDM). *Acta Electronica Sinica, 30*(4), 477–479.
90. Asif, R., Abd-Alhameed, R. A., Oanoh, O., Dama, Y., Migdadi, H. S., Noars, J.M. et al. (2012). Performance comparison between DWT-OFDM and FFT-OFDM using time domain zero forcing equalization. In *2012 international conference on telecommunication and multimedia* (pp. 175–179).
91. Kumbasar, V., & Kucur, O. (2012). Performance comparison of wavelet based and conventional OFDM systems in multipath Rayleigh fading channels. *Digital Signal Processing, 22*, 841–846.
92. Kansal, A., Singh, K., & Saxena, R. (2014). FrFT based OFDM system for wireless communications. *An International Journal of Engineering Sciences, 10*(6), 43–48.
93. Kansal, A., Singh, K., & Saxena, R. (2015). Bit error rate analysis of FrFT appended OFDM system. *Optik, 126*, 715–718.
94. Vieira, J., Malkowsky, S., Nieman, K., Miers, Z., Kundargi, N., Liu, L., et al. (2014). A exible 100-antenna testbed for massive MIMO. In *Proceedings of IEEE global communications conference (GLOBECOM) workshop on massive MIMO: From theory to practice, December 2014*.
95. Kansal, L., Sharma, V., & Singh, J. (2016). BER assessment of FFT-OFDM against WHT-OFDM over different fading channels. *Wireless Networks*. https://doi.org/10.1007/s11276-016-1283-2
96. Kansal, L., Sharma, V., & Singh, J. (2016). Performance evaluation of FFT-WiMAX against WHT-WiMAX over Rayleigh fading channel. *Optik, 127*, 4514–4519.
97. Kansal, A., Singh, K., & Saxena, R. (2014). Performance analysis of FrFT based OFDM system with 1024-PSK and 1024-QAM modulation under various wireless fading channels. *International Journal of Systems Assurance Engineering and Management*. https://doi.org/10.1007/s13198-014-0297-3
98. Singh, S. K., & Gandhi, A. S. (2019). Preamble based timing offset estimation and correction in OFDM assisted massive MIMO systems in the presence of inter-user-interference. *Wireless Personal Communications*. https://doi.org/10.1007/s11277-019-06216-1
99. Singh, S. K., Rathkanthiwar, A. P., & Gandhi, A. S. (2017). New algorithm for time and frequency synchronization in MIMO-OFDM systems. *Wireless Personal Communications, 96*(3), 3283–3295.
100. Xu, K., & Shen, Y. (2010). Timing synchronization using cross ambiguity function for MIMO OFDM systems with distributed antennas. *AEU - International Journal of Electronics and Communications, 64*(1), 82–86.
101. Zhang, J., Tian, L., Wang, Y., & Liu, M. (2014). Selection transmitting/maximum ratio combining for timing synchronization of MIMO-OFDM systems. *IEEE Transactions on Broadcasting, 60*(4), 626–636.
102. Luo, R., Yang, J., Li, R., Hu, S., & Duan, J. (2015). A new timing and frequency synchronization algorithm for distributed MIMO-OFDM systems. *Wireless Personal Communications, 82*(3), 1685–1696.

103. Liu, G., Ge, J. H., & G, Y. (2010). Time and frequency offset estimation for distributed multiple-input multiple-output orthogonal frequency division multiplexing systems. *IET Communications, 4*(6), 708–715.
104. Guo, F., Li, D., Yang, H., & Cai, L. (2006). A novel timing synchronization method for distributed MIMO-OFDM system. In *IEEE 63rd vehicular technology conference* (pp. 1933–1936).
105. Bogale, T. E. & Le, L. B. (2014). Beamforming for multiuser massive MIMO systems: Digital versus hybrid analog-digital. In *IEEE global communications conference* (pp. 4066–4071).
106. Yoshioka, S., Suyama, S., Okuyama, T., Mashino, J., & Okumura, Y. (2017). 5G massive MIMO with digital beamforming and two-stage channel estimation for low SHF band. In *Wireless days* (pp. 107–112).
107. Gao, Y., Khaliel, M., Zheng, F., & Kaiser, T. (2017). Rotman lens based hybrid analog digital beamforming in massive MIMO systems: Array architectures, beam selection algorithms and experiments. *IEEE Transactions on Vehicular Technology, 66*(10), 9134–9148.
108. Singya, P. K. & Gandhi, A. (2015). A data-aided timing offset estimator for OFDM synchronization. In *IEEE international conference on signal processing, informatics, communication and energy systems (SPICES)* (pp. 1–5).
109. Schmidl, T. M., & Cox, D. C. (1997). Robust frequency and timing synchronization for OFDM. *IEEE Transactions on Communications, 45*(12), 1613–1621.
110. Li, Y. G., & Stuber, G. L. (2006). *Orthogonal frequency division multiplexing for wireless communications*. Springer.
111. Akbarpour-Kasgari, A., & Ardebilipour, M. (2019). Improving MIMO relay compressed sensing-based channel estimation and pilot allocation. *EURASIP Journal on Wireless Communications and Networking*. https://doi.org/10.1186/s13638-019-1384-6
112. Akbarpour-Kasgari, A., & Ardebilipour, M. (2018). Probability-based pilot allocation for MIMO relay distributed compressed sensing based channel estimation. *EURASIP Journal on Advances in Signal Processing, 2018*(1), 18.
113. He, X., Song, R., & Zhu, W. P. (2015). Pilot allocation for distributed compressed sensing based sparse channel estimation in MIMO-OFDM systems. *IEEE Transactions on Vehicular Technology, 65*(5), 2990–3004.
114. Duarte-Carvajalino, J. M., & Sapiro, G. (2009). Learning to sense sparse signals: Simultaneous sensing matrix and sparsifying dictionary optimization. *IEEE Transactions on Image Processing, 18*(7), 1395–1408.
115. L. Xu, K. Niu, Z. He, W. Xu, Z. Zheng (2013). In *2013 9th International conference on information, communications & signal processing*. MIMO channel estimation based on distributed compressed sensing for LTE-advanced (IEEE, 2013) (pp. 1–5).
116. Younas, T., Li, J., Tulu, M. M., Munir, H. M., & Liaqat, M. (2019). Study of single-cell massive MIMO systems with channel aging and prediction. *Wireless Networks*. https://doi.org/10.1007/s11276-019-02030-6
117. Appaiah, K., Ashikhmin, A., & Marzetta, T. L. (2010). Pilot contamination reduction in multi-user TDD systems. In *Proceedings of the IEEE ICC* (pp. 1–5).
118. Ashikhmin, A., & Marzetta, T. L. (2012). Pilot contamination and precoding in multi-cell large scale antenna systems. In *Proceedings of the IEEE international symposium on information theory, Cambridge, MA* (pp. 1142–1146).
119. Fernandes, F., Ashikhmin, A., & Marzetta, T. (2012). Interference reduction on cellular networks with large antenna arrays. In *Proceedings of the IEEE ICC, Ottawa, Canada* (pp. 5773–5777).
120. Gopalakrishnan, B., & Jindal, N. (2011). An analysis of pilot contamination on multi-user MIMO cellular systems with many antennas. In *Proceedings of the international workshop on signal processing advances in wireless communications* (pp. 381–385).
121. Jose, J., Ashikhmin, A., Marzetta, T. L., & Vishwanath, S. (2011). Pilot contamination and precoding in multi-cell TDD systems. *IEEE Transactions on Wireless Communications, 10*(8), 2640–2651.

122. Ngo, H. Q., Marzetta, T. L., & Larsson, E. G. (2011). Analysis of the pilot contamination effect in very large multi-cell multiuser MIMO systems for physical channel models. In *Proceedings of the IEEE international conference acoustics speech signal processing* (pp. 3464–3467).
123. Baddour, K. E., & Beaulieu, N. C. (2005). Autoregressive modeling for fading channel simulation. *IEEE Transactions on Wireless Communications, 4*(4), 1650–1662.
124. Shen, Z., Andrews, J. G., & Evans, B. L. (2003). Short range wireless channel prediction using local information. In *Proceedings of the Asilomar conference on signals, systems & computers* (pp. 1147–1151).
125. Boche, H., & Schubert, M. (2002). A general duality theory for uplink and downlink beamforming. In *Proceedings of the IEEE 56th vehicular technology conference (VTC-Fall)* (pp. 87–91).
126. Björnson, E., Larsson, E. G., & Debbah, M. (2015). Massive MIMO for maximal spectral efficiency: How many users and pilots should be allocated? *IEEE Transactions on Wireless Communications, 15*(2), 1293–1308.
127. Du, H., & Xu, G. (2019). Optical wireless multiple-input multiple-output system based on avalanche photodiode receiver. *Annals of Telecommunications*. https://doi.org/10.1007/s12243-019-00736-4
128. Chan, V. W. S. (2006). Free-space optical communications [J]. *IEEE Journal of Lightwave Technology, 24*(12), 4750–4762.
129. Koonen, A. M. J., & Tangdiongga, E. (2014). Photonic home area networks [J]. *Journal of Lightwave Technology, 32*(4), 591–604.
130. Ge, X., Tu, S., Mao, G., Wang, C. X., & Han, T. (2016). 5G ultra-dense cellular networks [J]. *IEEE Wireless Communications, 23*(1), 72–79.
131. Agiwal, M., Roy, A., & Saxena, N. (2016). Next generation 5G wireless networks: A comprehensive survey [J]. *IEEE Communications Surveys & Tutorials, 18*(3), 1617–1655.
132. Grubor, J., Randel, S., Dieter Langer, K., & Walewski, J. W. (2008). Broadband information broadcasting using LED-based interior lighting [J]. *IEEE Journal of Lightwave Technology, 26*(24), 3883–3892.
133. Khalighi, M. A., & Uysal, M. (2014). Survey on free space optical communication: A communication theory perspective [J]. *IEEE Communications Surveys & Tutorials, 16*(4), 2231–2258.
134. Fath, T., & Haas, H. (2013). Performance Comparison of mimo techniques for optical wireless communications in indoor environments [J]. *IEEE Transactions on Communications, 61*(2), 733–742.
135. Cossu, G., Corsini, R., Khalid, A. M., & Ciaramella, E. (2014). Bi-directional 400 Mbps LED-based optical wireless communication for non-directed line-of-sight transmission combined deterministic and modified Monte Carlo method for calculating impulse responses of indoor optical wireless channels [J]. *Journal of Lightwave Technology, 32*(18).
136. Sugiura, S., & Iizuka, H. (2016). Element-by-element full-rank optical wireless MIMO systems using narrow-window angular filter designed based on one-dimensional photonic crystal [J]. *Journal of Lightwave Technology, 34*(24), 5601–5609.
137. Koonen, T. (2018). Indoor optical wireless systems—Technology, trends, and applications [J]. *Journal of Lightwave Technology, 36*(8), 1459–1467.
138. Wilson, S. G., Brandt-Pearce, M., Cao, Q., & Leveque, J. (2005). Free-space optical MIMO transmission with Q-ary PPM [J]. *IEEE Transactions on Communications, 53*(8), 1402–1412.
139. Hranilovic, S. (2005). *Wireless optical communication systems [M]*. Springer Press.
140. McIntyre, R. J. (1972). The distribution of gains in uniformly multiplying avalanche photodiodes: Theory [J]. *IEEE Transactions on Electron Devices, 19*(6), 703–713.
141. Conradi, J. J. (1972). The distribution of gains in uniformly multiplying avalanche photodiodes: Experimental [J]. *IEEE Transactions on Electron Devices, 19*(6), 714–718.
142. Webb, P. P., McIntyre, R. J., & Conradi, J. (1974). Properties of avalanche photodiodes. *RCA Review, 35*, 234–278.

143. Glover, I. A., & Grant, P. M. (2004). *Digital communications [M]*. Prentice-Hall Press.
144. Li, S., Qiu, H., Zheng, L., & Yang, C. (2022). Performance analysis of spatial multiplexing MIMO-MFSK based on energy detection for fast-fading environments. *EURASIP Journal on Wireless Communications and Networking*. https://doi.org/10.1186/s13638-022-02116-4
145. Chowdhury, M., Manolakos, A., & Goldsmith, A. 2016 Scaling laws for noncoherent energy-based communications in the SIMO MAC. *IEEE Transactions on Information Theory, 62*(4), 1980–1992.
146. Manolakos, A., Chowdhury, M., & Goldsmith, A. (2016). Energy-based modulation for non-coherent massive SIMO systems. *IEEE Transactions on Wireless Communications, 15*(11), 7831–7846.
147. Gómez-Cuba, F., Chowdhury, M., Manolakos, A., Erkip, E., & Goldsmith, A. J. (2019). Capacity scaling in a non-coherent wideband massive SIMO block fading channel. *IEEE Transactions on Wireless Communications, 18*(12), 5691–5704.
148. Zhang, Y.-Y., Zhang, J.-K., & Yu, H.-Y. (2018). Physically securing energy-based massive MIMO MAC via joint alignment of multi-user constellations and artificial noise. *IEEE Journal on Selected Areas in Communications, 36 4*, 829 844.
149. Venkateswarlu, C., & Rao, N. V. (2022). An efficient MAPSO model for interference cancellation and optimal channel estimation in MIMO-OFDM system. *Wireless Personal Communications*. https://doi.org/10.1007/s11277-022-09955-w
150. Kang, X. F., Liu, Z. H., & Yao, M. (2022). Deep learning for joint pilot design and channel estimation in MIMO-OFDM systems. *Sensors, 22*(11), 4188.
151. Mashhadi, M. B., Yang, Q., & Gündüz, D. (2020). CNN-based analog CSI feedback in FDD MIMO-OFDM systems. In *ICASSP 2020–2020 IEEE International Conference on Acoustics, Speech and Signal Processing (ICASSP)* (pp. 8579–8583). IEEE.
152. Rakshit, M., Bhattacharjee, S., Garai, G., & Chakrabarti, A. (2020). A novel distributive population-based differential evolution algorithm for SLM scheme to reduce PAPR in massive MIMO-OFDM systems. *SN Computer Science, 1*(5), 1–7.
153. Abdullah, E., Idris, A., & Saparon, A. (2017). Papr reduction using SCS-SLM technique in stfbc mimo-ofdm. *ARPN Journal of Engineering and Application Science, 12*(10), 3218–3221.
154. Ramadan, K., Dessouky, M. I., & Abd El-Samie, F. E. (2020). Performance enhancement of OFDM systems with lower-complexity using DST based on successive interference cancellation. *Digital Signal Processing, 102*, 102739.
155. Fan, S., Xiao, Y., Fang, S., Zhao, Y., & Zhou, X. (2020). Clipping noise cancellation for signal detection of GSTFIM systems. *IEEE Access, 8*, 33830–33837.
156. Albataineh, Z. Low-complexity near-optimal iterative signal detection based on MSD-CG method for uplink massive MIMO systems. *Wireless Personal Communications*. https://doi.org/10.1007/s11277-020-07810-4
157. Ngo, H., Larsson, E., & Marzetta, T. (2012). Energy and spectral efficiency of very large multiuser MIMO systems. *IEEE Transactions on Communications, 61*(4), 1436–1449.
158. Andrews, J. G., Buzzi, S., Choi, W., Hanly, S. V., Lozano, A., Soong, A. C. K., & Zhang, J. C. (2014). What will 5G be? *IEEE Journal on Selected Areas in Communications, 32*(6), 1065–1082.
159. Dai, L., Wang, Z., & Yang, Z. (2013). Spectrally efficient time-frequency training OFDM for mobile large-scale MIMO systems. *IEEE Journal on Selected Areas in Communications, 31*(2), 251–263.
160. Golub, G. H., & Van Loan, C. F. (2012). *Matrix computations*. JHU Press.
161. Barbero, L. G., & Thompson, J. S. (2008). Fixing the complexity of the spheredecoder for MIMO detection. *IEEE Transactions on Wireless Communications, 7*(6), 2131–2142.
162. Wu, M., Yin, B., Wang, G., Dick, C., Cavallaro, J. R., & Studer, C. (2014). Large-scale MIMO detection for 3GPP LTE: Algorithms and FPGA implementations. *IEEE Journal on Selected Topics in Signal Processing, 8*(5), 916–929.

163. Paulraj, A., Nabar, R., & Gore, D. (2008). *Introduction to space-time wireless communications*. Cambridge Univ. Press.
164. Wuet, M., et al. (2016). High-throughput data detection for MassiveMU-MIMO-OFDM using coordinate descent. *IEEE Transactions on Circuit and System I, 63*(12).
165. Gesbert, D., Shafi, M., Shiu, D., Smith, P. J., & Naguib, A. (2003). From theory to practice: An overview of MIMO space-time coded wireless systems. *IEEE Journal on Selected Areas in Communications, 21*(3), 281–302.
166. Seethaler, D., Matz, G., & Hlawatsch, F. (2004). An efficient MMSE-based demodulator for MIMO bit-interleaved coded modulation. In *Proceedings of Global Telecommunications conference (GLOBECOM)* (Vol. 4. pp. 2455–2459).
167. Gu, T., Liu, X., Mo, Z., & Chi, X. (2004). Multiple search direction conjugate gradient method I: Methods and their propositions. *International Journal of Computer Mathematics, 81*(9), 1133–1143.
168. Yin, B., Wu, M., Cavallaro, J.R., & Studer, C. (2014). *Conjugate gradient-based soft-output detection and precoding in massive MIMO systems*. arXivpreprint:1404.0424v1.
169. Raut, S. N., & Jalnekar, R. M. (2021). Taylor-Based Least Square Estimation in MIMO-OFDM Systems for Multimedia Applications. *Wireless Personal Communications*. https://doi.org/10.1007/s11277-021-08481-5
170. Huang, Y., He, Y., Luo, Q., Shi, L., & Wu, Y. (2018). Channel estimation in MIMO-OFDM Systems based on a new adaptive greedy algorithm. *IEEE Wireless Communications Letters, 8*, 29–32.
171. Singh, H., & Bansal, S. (2017). Channel estimation with ISFLA based pilot pattern optimization for MIMO-OFDM system. *AEU International Journal of Electronics and Communications, 81*, 143–149.
172. Binu, D., & Kariyappa, B. S. (2018). RideNN: A new rider optimization algorithm-based neural network for fault diagnosis in analog circuits. *IEEE Transactions on Instrumentation and Measurement, 99*, 1–25.
173. Borkar, G. M., & Mahajan, A. R. (2017). A secure and trust based on-demand multi-path routing scheme for self-organized mobile ad-hoc networks. *Wireless Networks, 23*(8), 2455–2472.
174. Mangai, S. A., Ravi Sankar, B., & Alagarsamy, K. (2014). Taylor series prediction of time series data with error propagated by artificial neural network. *International Journal of Computer Applications, 89*(1), 0975–8887.
175. Nandi, S., Nandi, A., & Pathak, N. N. (2021). Channel estimation of massive MIMO-OFDM system using Elman recurrent neural network. *Arabian Journal for Science and Engineering*. https://doi.org/10.1007/s13369-021-06366-0
176. Wu, S., Kuang, L., Ni, Z., Huang, D., Guo, Q., & Lu, J. (2016). Message-passing receiver for joint channel estimation and decoding in 3D massive MIMO-OFDM systems. *IEEE Transactions on Wireless Communications, 15*(12), 8122–8138. https://doi.org/10.1109/TWC.2016.2612629
177. Chen, L., & Yuan, X. (2018). Massive MIMO-OFDM channel estimation via structured turbo compressed sensing. *IEEE International Conference on Communications, 2018*, 1–6. https://doi.org/10.1109/ICC.2018.8422538
178. Gao, Z., Hu, C., Dai, L., & Wang, Z. (2016). Channel estimation for millimeter-wave massive MIMO with hybrid precoding over frequency-selective fading channels. *IEEE Communications Letters, 20*(6), 1259–1262. https://doi.org/10.1109/LCOMM.2016.2555299
179. Lin, X., Wu, S., Jiang, C., et al. (2018). Estimation of broadband multiuser millimeter-wave massive MIMO-OFDM channels by exploiting their sparse structure. *IEEE Transactions on Wireless Communications, 17*(6), 3959–3973.
180. Araújo, D. C., De Almeida, A. L., Da Costa, J. P., & de Sousa, R. T. (2019). Tensor-based channel estimation for massive MIMO-OFDM systems. *IEEE Access, 7*, 42133–42147. https://doi.org/10.1109/ACCESS.2019.2908207

181. Liao, Y., Hua, Y., & Cai, Y. (2019). Deep learning based channel estimation algorithm for fast time-varying MIMO-OFDM systems. *IEEE Communications Letters, 24*(3), 572–576. https://doi.org/10.1109/LCOMM.2019.2960242
182. Kapoor, D. S., & Kohli, A. K. (2018). Channel estimation and long range prediction of fast fading channels for adaptive OFDM system. *International Journal of Electronics, 105*(9), 1451–1466. https://doi.org/10.1080/00207217.2018.1460871
183. Palanimuthu, S. J., & Muthial, C. (2016). An enhanced multi-channel bacterial foraging optimization algorithm for MIMO communication system. *International Journal of Electronics, 104*(4), 608–623. https://doi.org/10.1080/00207217.2016.1242027
184. Hu, F., Du, D., Zhang, P., & Wang, Z. (2014). A joint swarm intelligence algorithm for multiuser detection in MIMO-OFDM system. *International Journal of Electronics, 101*(11), 1478–1494. https://doi.org/10.1080/00207217.2013.869769
185. Nandi, S., Pathak, N. N., & Nandi, A. (2020). Analysis of hard decision and soft decision decoding mechanism using Viterbi decoder in presence of different adaptive modulations. *International Journal of Future Generation Communication and Networking, 13*(3), 3002–3012.
186. Xu, W., & Zhao, Y. (2019). Reducing MMV-based OMP channel estimation for massive MIMO OFDM systems. In *2019 International Conference on Communications, Information System and Computer Engineering* (pp. 289–293). https://doi.org/10.1109/CISCE.2019.00072
187. Nandi, S., Nandi, A., & Pathak, N. N. (2017). Performance analysis of Alamouti STBC MIMO OFDM for different transceiver system. *IEEE International Conference on Intelligent Sustainable Systems, 2017*, 883–887. https://doi.org/10.1109/ISS1.2017.8389305
188. Nandi, S., Pathak, N. N., & Nandi, A. (2020). A novel adaptive optimized fast blind channel estimation for cyclic prefix assisted space-time block coded MIMO OFDM systems. *Wireless Personal Communications*. https://doi.org/10.1007/s11277-020-07629-z
189. Shankar, R., Nandi, S., & Rupani, A. (2021). Channel capacity analysis of non-orthogonal multiple access and massive multiple-input multiple-output wireless communication networks considering perfect and imperfect channel state information. *Journal of Defense Modeling and Simulation*. https://doi.org/10.1177/15485129211000139
190. Liao, Y., Hua, Y., Dai, X., Yao, H., & Yang, X. (2019). Chanestnet: A deep learning based channel estimation for high-speed scenarios. *IEEE International Conference on Communications, 2019*, 1–6. https://doi.org/10.1109/ICC.2019.8761312
191. Şeker, S., Ayaz, E., & Türkcan, E. (2003). Elman's recurrent neural network applications to condition monitoring in nuclear power plant and rotating machinery. *Engineering Applications of Artificial Intelligence, 16*(7–8), 647–656. https://doi.org/10.1016/j.engappai.2003.10.004
192. Madhumita, S., & Shila, G. (2020). Development of a secured optical code-division multiple access system by implementing hybrid 2D-modified Walsh code. *Optical Engineering, 59*(10), 106107. https://doi.org/10.1117/1.OE.59.10.106107
193. Elman, J. L. (1990). Finding structure in time. *Cognitive Science, 14*, 179–211.
194. Sahoo, M., & Sahoo, H. K. M. (2021). Massive MIMO channel estimation and BER analysis using convex combination based algorithms. *Wireless Personal Communications*. https://doi.org/10.1007/s11277-021-09275-5
195. Nayebi, E., & Rao, B. D. (2018). Semi-blind channel estimation for multiuser massive MIMO systems. *IEEE Transactions on Signal Processing, 66*(2), 540–553.
196. Singh, V., Srivastava, A., & Jagannatham, A. K. (2020). Superimposed pilots based adaptive time-selective channel estimation in MU-MIMO systems. *IEEE National Conference on Communication, 23*, 1–6.
197. Gu, Y., Jin, J., & Mei, S. (2009). l_0 Norm constraint LMS algorithm for sparse system identification. *IEEE Signal Processing Letters, 16*, 774–777.
198. Das, B. K., & Chakraborty, M. (2014). Sparse adaptive filtering by an adaptive convex combination of the LMS and the ZA-LMS algorithms. *IEEE Transactions on Circuits and Systems-I, 61*(5), 1499–1507.

199. Yazdanpanah, H., Diniz, P. S. R., & Lima, M. V. S. (2020). Feature adaptive filtering: Exploiting hidden sparsity. *IEEE Transactions on Circuits and Systems-I, 67*(7), 2358–2371.
200. Mohammed, A. S., Taman, A. I., Hassan, A. M., & Zekry, A. (2022). Deep learning channel estimation for OFDM 5G systems with different channel models. *Wireless Personal Communications*. https://doi.org/10.1007/s11277-022-10077-6
201. Wang, S., Yao, R., Tsiftsis, T. A., Miridakis, N. I., & Qi, N. (2020). Signal detection in uplink time-varying OFDM systems using RNN with bidirectional LSTM. *IEEE Wireless Communications Letters, 9*(11), 1947–1951.
202. Honkala, M., Korpi, D., & Huttunen, J. M. (2021). DeepRx: Fully convolutional deep learning receiver. *IEEE Transactions on Wireless Communications, 20*(6), 3925–3940.
203. Wang, Z., Pu, F., Yang, X., Chen, N., Shuai, Y., & Yang, R. (2020). Online LSTM-Based channel estimation for HF MIMO SC-FDE system. *IEEE Access, 8*, 131005–131020.
204. Kim, H. (2015). *Wireless communications systems design* (1st ed.). Wiley.
205. Mehmood, A., & Cheema, W. A. (2009). *Channel estimation for lte downlink*. (M.S. thesis). Dept. Electrical Eng., Blekinge Institute of Technology.
206. Wang, F. (2011). *Pilot-based channel estimation in OFDM system*. (Doctoral dissertation). University of Toledo.
207. https://www.freecodecamp.org/news/want-to-know-how-deep-learning-works-heres-a-quick-guide-for-everyone-1aedeca88076/ [Online].
208. Alam, J., & Shaha Mohammed, G. A. S. (2010). *Low complexity channel estimation of OFDM systems based on LS and MMSE estimators*. (M.Sc Thesis). Electrical Engineering, Blekinge Institute of Technology.
209. Selvaraj, K., Anandaraj, M., Judson, D., Ganeshkumar, P., & Rajkumar, K. A low complexity near optimal signal detection for large scale MIMO SC-FDMA uplink system. *Wireless Personal Communications*. https://doi.org/10.1007/s11277-022-09742-7
210. Jin, F., Liu, Q., Liu, H., & Wu, P. (2019). A low complexity signal detection scheme based on improved Newton iteration for massive MIMO systems. *IEEE Communications Letters, 23*(4), 748–751.
211. Selvaraj, K., Ganeshkumar, P., & Anandaraj, M. (2016). Iterative MMSE equalization and CFO compensation for the uplink SC-FDMA transmission. *International Journal of Communication Systems, 29*(7), 1323–1337.
212. Al-kamali, F. S., Dessouky, M. I., Sallam, B. M., & Abd El-Samie, F. E. (2012). Regularized MIMO equalization for SC-FDMA systems. *Circuits, Systems, and Signal Processing, 31*(4), 1423–1441.
213. Myung, H. G., Lim, J., & Goodman, D. J. (2006). Single carrier FDMA for uplink wireless transmission. *IEEE Vehicular Technology Magazine, 1*(3), 30–38.
214. Farhang, A., Marchetti, N., Doyle, L. E., & Boroujeny, B. (2015). Low complexity CFO compensation in uplink OFDMA systems with receiver windowing. *IEEE Transactions on Signal Processing, 63*(10), 2546–2558.
215. Tang, C., Liu, C., Yuan, L., & Xing, Z. (2016). High precision low complexity matrix inversion based on Newton iteration for data detection in the massive MIMO. *IEEE Communications Letters, 20*(3), 490–493.
216. Kumar, K. S., & Sujatha, B. R. (2021). FPGA design of an efficient EEG signal transmission through 5G wireless network using optimized pilot based channel estimation: A telemedicine application. *Wireless Personal Communications*. https://doi.org/10.1007/s11277-021-09305-2
217. Satheeskumaran, S., & Sabrigiriraj, M. (2014). A new LMS based noise removal and DWT based R-peak detection in ECG signal for biotelemetry applications. *National Academy Science Letters, 37*(4), 341–349.
218. Venkatesan, C., & Karthigaikumar, P. (2018). An efficient noise removal technique using modified error normalized LMS algorithm. *National Academy Science Letters, 41*(3), 155–159.

219. Venkatesan, C., Karthigaikumar, P., & Varatharajan, R. (2019). FPGA implementation of modified error normalized LMS adaptive filter for ECG noise removal. *Cluster Computing, 22*(5), 12233–12241.
220. Hejrati, B., Fathi, A., & Abdali-Mohammadi, F. (2017). Efficient lossless multi-channel EEG compression based on channel clustering. *Biomedical Signal Processing and Control, 31*, 295–300.
221. Raghunatharao, D., Prasad, T. J., & Prasad, M. N. G. (2020). Optimal pilot-based channel estimation in cognitive radio. *Wireless Personal Communications, 114*(4), 2801–2819.
222. Lee, J.-H., & Sun, W.-E. (2019). Robust beamforming and spatial precoding for quasi-OSTBC massive MIMO communications. *EURASIP Journal on Wireless Communications and Networking*. https://doi.org/10.1186/s13638-019-1373-9
223. Tarokh, V., Jafarkhani, H., & Calderbank, A. R. (1999). Space-time block codes from orthogonal designs. *IEEE Transactions on Information Theory, 45*, 1456–1467.

Chapter 4
Planning and Dimensioning of WiMAX Networks

Milind Pande, Anand J. Kulkarni, and Apoorva S. Shastri

Introduction by the Editor

The chapter discusses the complexities involved in the Planning and Dimensioning of Worldwide Interoperability for Microwave Access (WiMAX) Networks. Before choosing the supported features of WiMAX that best fit the operator's business model dimensioning and planning are essential. Several difficulties and challenges exist when comparing WiMAX network dimensioning and planning to traditional 2G/3G cellular network planning. This chapter discusses topics such as optimization of energy and delay of wireless sensor networks, V2X side link, link-level, and system-level simulation, Analysis of macro-diversity, Cross-Layer Resource Allocation Scheme for IEEE 802.16 Broadband Wireless Networks, optimization of multichannel allocation for hybrid overlay/underlay cognitive radio, controller placement problem for wireless SDN, interference mitigation scheme with D2D, Hybrid DSRC/WiMAX Technology, adaptive WiMAX network incorporating diverse transforms. This chapter also addresses the application of algorithms such as the Throughput Based Discovery Algorithm (TDA) that is used for D2D Discovery in PSN for Post-Disaster scenarios and the Genetic Algorithm for resource allocation problems in 5G UDN.

Cellular vehicle-to-everything (C-V2X) communication supports data exchange in autonomous driving and vehicle networking. It uses a band on the 5.9 GHz spectrum and may thus suffer from both co-channel and adjacent channel interference,

M. Pande
Dr. Vishwanath Karad MIT World Peace University, Pune, Maharashtra, India
e-mail: milind.pande@mitwpu.edu.in

A. J. Kulkarni · A. S. Shastri (✉)
Institute of Artificial Intelligence, Dr. Vishwanath Karad MIT World Peace University, Pune, Maharashtra, India
e-mail: anand.j.kulkarni@mitwpu.edu.in

M. Pande et al. (eds.), *Optimization Methods in Mobile Communication Systems*, https://doi.org/10.1007/978-981-95-1810-4_4

which cannot be negligible, especially in urban scenarios. The vehicular network was initially implemented by using 4G LTE and later upgraded to 5GNR. C-V2X includes side link radio technology. It is a stage relay transmission model that uses a power loss angle technique. The energy and delay optimizing polynomial time algorithm, named as EDO algorithm, constructs an energy and delay optimized routing topology on heterogeneous bi-directional connected WSNs where nodes are allowed to vary their transmission range within their assigned power interval. The radio access network (RAN) of LPWAN supports macro reception diversity: a packet transmitted by a given device can be received by several base stations (BS). A series of maximum ratio combining (MRC) function blocks are installed within the central unit so that several transmitted packets in the radio network can be simultaneously decoded. A bit is assumed to be correctly received if the Signal to Noise and Interference ratio (SINR) is not less than a threshold value. The objective of macro reception diversity is to find the maximum capacity of the system and we thus consider it in high load conditions: interferences are much larger than the background noise. This configuration is called Pa-Aloha (pure Aloha with average interference).

In a WiMAX network, the Medium Access Control (MAC) protocol deals with resource allocation to different types of traffic. The key components that ensure Quality of Service (QoS) guarantees in a WiMAX network include Call Admission Control (CAC), Bandwidth and Burst allocation. CAC and Bandwidth allocation are dealt in the MAC layer, while Burst allocation in the Physical layer. The History based Bandwidth Allocation (HBA) scheme deals with allocating bandwidth based on Consumption and Equity measures. The proposed tightly coupled Delay Tolerance based Scheduler (DTS) and Bucket based Burst Allocator (BBA) allocate resources by prioritizing flows with least delay tolerance.

Prioritization of Subscriber terminals (ST) during opportunistic allocation of channels is critical as properties like energy level. Multi-channel (m-channel) allocation performance is a maximization algorithm works for low-power mobiles and improves channel distribution performance. The controller placement problem is threefold in nature: the number of controllers to be placed in a network, the locations of these controllers and the assignment function of controllers to switches. algorithm solves the WCPP for a specific and constant number of controllers but obviously by changing the number of the controllers we may achieve better solution, with lower objective value. Algorithm is extended to find the number of controllers and their placement that minimize the objective value. Device-to-Device (D2D) communication paradigm is being recognized as one of the promising techniques which could help with matching the data demands. Interference Mitigation in Multicast D2D networks in this D2D transmitter in each group multicasts the same data to the D2D receivers in the group. Interference Mitigation in D2D 5G Network. The Machine Learning Model could be employed in 5G D2D networks to reduce the threat of interference.

Cluster-Based Hybrid DSRC-Mobile Network Communication System optimizes communication efficiency of vehicles. Dedicated short-range communication (DSRC) manages low latency communication between clusters and IEEE 801.11

handles high data rate communication, combining both to optimize network efficiency. The adaptive algorithm realizes a spectral-efficient WiMAX system that enables the system to select the suitable transform strategy (DWT/FrFT) with minimum SNR requirements to achieve the target BER of 10^{-4}. A Throughput-based Discovery Algorithm (TDA) is proposed to find D2D peers among devices on a Public Security Network (PSN) in a post-disaster situation. Device-to-device (D2D) discovery and communication provides communication services in such difficult situations without the requirement of network infrastructure and by reducing call traffic in cellular communication networks. D2D pairs are determined when they exceed the minimum threshold of throughput and then the best pair with maximum throughput values is found. D2D pair with the minimum distance is selected and if this pair meets the minimum (signal-to-interference and noise ratio) SINR threshold, it is determined as the best pair.

Machine Generated Summaries

Disclaimer: The summaries in this chapter were generated from Springer Nature publications using extractive AI auto-summarization: An extraction-based summarizer aims to identify the most important sentences of a text using an algorithm and uses those original sentences to create the auto-summary (unlike generative AI). As the constituted sentences are machine selected, they may not fully reflect the body of the work, so we strongly advise that the original content is read and cited. The auto generated summaries were curated by the editor to meet Springer Nature publication standards. To cite this content, please refer to the original papers.

Machine generated keywords: allocation, wimax, resource, resource allocation, vehicle, spectrum, assign, metric, discovery, cellular, hybrid, bss, radio resource, controller, cell

On Potential Challenges of V2X Sidelink Relaying Under Interference: Link-Level and System-Level Simulation with Neural Network Assisted [1]

This is a machine-generated summary of:

Zhao, Chonghao; Wu, Gang: On potential challenges of V2X sidelink relaying under interference: link-level and system-level simulation with neural network assisted [1]

Published in: EURASIP Journal on Wireless Communications and Networking (2023)

Link to original: https://doi.org/10.1186/s13638-023-02244-5

Copyright of the summarized publication:

If you want to cite the papers, please refer to the original.

For technical reasons we could not place the page where the original quote is coming from.

Abstract-Summary

"C-V2X users dedicated band on the 5.9 GHz spectrum may thus suffer from both co-channel and adjacent channel interference, which cannot be negligible, especially in urban scenarios."

"3GPP has standardized relay technology in New Radio (NR) V2X sidelink to extend the transmission range under interference."

"Through a link-level and system-level simulation study, we evaluate the sidelink performance in relaying scenarios under different interference."

"The simulation also reveals that the interference incurred by NR on unlicensed spectrum (NR-U) signals and other sidelink signals is intractable to be suppressed, which may bring potential challenges in future works."

Introduction

"Due to the rapid growth of autonomous driving and networked vehicle holdings, vehicle-to-everything (V2X) communication has been envisioned as one of the compelling technology for providing wireless connectivity to support the data exchange among vehicles and other network entities."

"In that sense, safety and non-safety applications can be supported well in C-V2X. As aforementioned, 3GPP initiated the first standard in the Rel-14 [2], known as 4G Long Term Evolution (LTE) C-V2X. This standard was further upgraded to the Release-17 (Rel-17) [3], named 5G New Radio (NR) V2X. NR C-V2X also introduces sidelink radio technology, known as the PC5 interface, into vehicle communication."

"It is necessary to investigate how significant signals of different protocols impact sidelink performance and then design the corresponding mitigating methods, notwithstanding there exist some proactive regulations such as listen-before-talk (LBT), maximum channel occupation time (MCOT), etc In this paper, through a link-level and system-level simulation study, we evaluate sidelink relaying performance in an urban scenario and provide direct insight into the challenges of optimizing overall communication performance."

Related Work

"In [4], the authors design an LTE V2X link simulator to analyze physical layer transmission performance on highways and obtain the performance curves of signal-noise ratio (SNR) and BLER at different vehicle speeds."

"Sattiraju and others [5] investigate the BLER performance using IEEE 802.11p and LTE V2X. This work evaluates BLER performance with MCSs in the additive white Gaussian noise (AWGN) channel and with different SNRs in 5 kinds of V2V links."

"It is noted that the V2V link simulation in LTE V2X of work [4, 5] cannot accurately describe the link performance of NR V2X in most scenarios. [6] cannot directly reflect the impact of physical layer parameters on the BLER, notwithstanding it reduces simulation time and algorithm complexity."

"Different from the above existing works, in this paper, we evaluate the sidelink relaying transmission scheme in a non-line of sight (NLoS) scenario at a crossroad with the consideration of both large-scale and small-scale fading."

Methods and System Model

"Vehicle R, in the second stage, transfer the received signal to vehicle S. Please note that all the receiver in the relay transmission scheme will be affected by interference."

"The mth subcarrier vehicle R receives can be defined asSimilarly, as for the vehicle R, the SINR of the received signal can be defined asIn the second stage, vehicle R transfers the received signal from vehicle S and it is easily concluded that the mth subcarrier vehicle D receives can be defined asLikewise, the SINR of vehicle D isTo simplify the discussion, here we use the PLA technique to provide an application interface between link-level and system-level simulation."

"Every time the user, i.e., vehicle S, transmits data, the wireless channel is updated and the receiver at vehicle R or vehicle D will use the large-scale fading model to obtain the power of the target signal and interference signal received."

Neural Network Architecture

"The signal of different protocols may have the same modulation schemes and similar structures."

"It is significantly difficult to design a model-based method to analyze the impact of different interference signals and the difficulty of mitigation."

"Since the same neural network has the same expression ability for different signals, its classification results can provide a benchmark for interference evaluation."

"We propose a novel neural network to classify the signal based on their protocol, rather than modulation."

"The majority of the existing work about signal identification is based on CNNs or more complicated CNN-LSTM combinations."

"To make the signal consistent with the input of the neural network, short-time Fourier transform (STFT) has been adopted to transform the signals into a time-frequency distribution matrix, which contains the features of frequency varying with time."

Simulation Results and Discussion
"The network has the worst performance on NR-U since both the precision and recall are close to 50%, which means it nearly cannot detect any of the interference incurred by NR-U. We conclude that this is because NR sidelink is designed based on the basic NR signal and has the most identical protocol."

"The above analysis indicates that different interference has a notably different impact on sidelink signal, as Bluetooth and Wi-Fi are easier to be detected while NR-U is extremely intractable when mixed with NR sidelink."

"Please note that the neural network proposed is absolutely not state-of-the-art, but it provides a uniform standard for us to evaluate the impact of diverse interference on sidelink signals in different channel environments."

"3 to compare direct transmission and relaying schemes in both interference scenarios and interference-free scenarios."

Conclusion
"We have implemented link-level and system-level simulations to investigate parameter optimization and performance evaluation in the C-V2X sidelink relaying scenario."

"The simulation results provide direct insight into the system performance in relaying scenarios and reveal some challenges."

"Further, transmission scheme selection, relay selection and resource allocation are also challenges when there exist multiple relays and UEs."

Optimization of Energy and Delay on Interval Data Based Graph Model of Wireless Sensor Networks [7]

This is a machine-generated summary of:

Kavra, Radhika; Gupta, Anjana; Kansal, Sangita: Optimization of energy and delay on interval data based graph model of wireless sensor networks [7]

Published in: Wireless Networks (2023)

Link to original: https://doi.org/10.1007/s11276-023-03292-x

If you want to cite the papers, please refer to the original.

For technical reasons we could not place the page where the original quote is coming from.

Abstract-Summary

"Despite a plethora research on energy and delay optimization in this field, work on developing delay minimum and energy efficient connected routing topology on heterogenous bi-directional WSNs where each node is assigned with a power interval, remain untouched."

"Considering this problem, we have introduced 'Interval data based graph model (IDGM)' and 'Sorted-interval data based graph model (S-IDGM)' of WSNs to explicitly deal with nodes' power interval and proposed 'Energy and Delay Optimization (EDO)' algorithm to optimize S-IDGM such that the maximum topology delay, total topology delay and maximum node's power interval become minimum in polynomial time complexity."

"t-test analysis shows that the proposed EDO algorithm achieves optimal energy saving of nodes at 5% level of significance along with optimal minimization of max and total topology delay at 2% level of significance on S-IDGMs."

Introduction

"These existing gaps motivate us to bring out a methodology to obtain delay minimum energy efficient connected routing topology considering interference factor which simultaneously work to minimize nodes' power consumption and topology delay on bidirectional connected heterogenous WSNs where each sensor node has been allowed to vary its transmission range within a power interval."

"The manifold contributions of the proposed work are as follows: (1) Introduction of 'Interval data based graph model (IDGM)' and 'Sorted-interval data based graph model (S-IDGM)' for type of bi-directional connected heterogenous WSNs where each node is weighted with a power interval so that nodes can vary their transmission range in the power interval according to communication requirement. (2) A new weighted 'Delay estimation (DE) function' based on distance and interference factors has been formulated after showing dependency analysis between link distance (hop length) and link interference towards optimality. (3) A polynomial time 'Energy and Delay Optimization (EDO) algorithm' has been proposed to obtain an energy and delay optimized connected routing topology on S-IDGMs using DE function."

Related Work

"Authors worked with different strategies and introduced many methodologies to either optimize transmission energy or routing delay or both with the involvement of different estimation metrics."

“In [8], an energy-aware multi-hop centralized routing scheme can be seen where static as well as mobile gateway nodes have been engaged to handle packet delivery tasks to minimize energy consumption and end-to-end transmission delay.”

“A concentric circular bands (CCBs) based data forwarding scheme from source to sink node optimizing transmission energy or routing delay or both has been proposed in [9].”

“Authors in [10] proposed DCEMRA algorithm to minimize energy consumption in cluster-based routing without increasing end-to-end delay.”

“Two factors that are distance between the nodes [10] and link interference model, introduced in [11], have been considered to estimate topology delay to obtain delay minimized and energy efficient reliable routing topology on an interval data based graph model of WSNs.”

Proposed Work

“The type of WSNs whose underlying graph models show the interval weight on each node, do not require to satisfy interval graph properties, are called ‘interval data based graph model (IDGM)’ of WSNs.”

“Let G(V, E) be a S-IDGM, then Now introducing Link-level interference model which has been proposed by Sun and others in 2015 [11] to compute link interference in bidirectional WSNs.”

“To minimize both the factors simultaneously, a weighted function of link distance and link interference needs to formulate. (Discussion of weighted function is carried forward to next section.) Link-level interference is more depending on network density than link distance.”

“After analysing the results obtained in previous section, a conclusion is made to construct a new function which optimally balances the minimization of link interference with hop length (link distance) when any algorithm is applied after it.”

Problem Definition and Proposed Algorithm

“Energy and Delay Optimization (EDO) algorithm To the defined optimization problem, a polynomial time algorithm has been proposed, named as ‘Energy and Delay Optimization (EDO)’ algorithm.”

“Step 7 Initialization of while loop brings execution of steps from 8 to 16 until UncoverV becomes empty.”

“Step 29 PC becomes empty for the given illustration, therefore end of while loop.”

“The proof carries forward by assuming the contradicting situation in the same way as we did in case(i) which finally results that T is the energy and delay optimal topology, obtained by EDO algorithm.”

“Steps 15 and 16 take constant time and here’s the end of first while loop.”

Performance Evaluation

“Let us now evaluate the performance of EDO algorithm in term of topology delay minimization and energy efficiency by computing max node’s power interval and percentage of energy saving and show the comparison with predefined algorithms, given in [12, 13] on interval data based graph models of WSNs.”

"Using t-test, we show that topology formed by EDO algorithm has no significant difference in optimizing power consumption to save network energy, comparing with the energy efficient topology obtained by algorithm [13]."

"This is because algorithm, proposed in [13] has been an energy optimal in providing connected routing topology on interval graphs without considering delay factors and EDO algorithm which works to obtain an energy and delay minimized connected routing topology, considers DE function depending on link interference and distance factor which helps in minimizing topology delay and power consumption simultaneously."

Conclusion

"We have proposed an energy and delay optimizing polynomial time algorithm, named as EDO algorithm, to construct an energy and delay optimized connected routing topology on heterogenous bi-directional connected WSNs where nodes are allowed to vary their transmission range within their assigned power interval."

"A sorted interval data based graph model (S-IDGM) is introduced to show nodes power interval explicitly and proposed EDO algorithm runs on S-IDGM to result an optimized topology which minimizes max topology delay, total topology delay and max nodes' power interval."

"Extensive simulation work, graphical and t-test analysis have been carried out to compare the performance of EDO algorithm with delay minimum spanning tree (DE MST) and algorithms [12, 13] on S-IDGMs of WSNs."

"Work on energy and delay optimization can be extended to dynamic WSNs where topologies are required to be more adaptable and stable with respect to node's movement."

Dimensionality Reduction to Solve Resource Allocation Problem in 5G UDN Using Genetic Algorithm [14]

This is a machine-generated summary of:

Reyna-Orta, Adolfo; Andrade, Ángel G.: Dimensionality reduction to solve resource allocation problem in 5G UDN using genetic algorithm [14]

Published in: Soft Computing (2021)

Link to original: https://doi.org/10.1007/s00500-020-05473-8

If you want to cite the papers, please refer to the original.

For technical reasons we could not place the page where the original quote is coming from.

Abstract-Summary

"5G ultra-dense network (UDN) systems consist of massive deployment of small cells."

"The resource allocation (RA) algorithms distribute the available spectrum resources with the least interference."

"The strategy is based on allocating a single block of channels to either users or cells instead of disjoint channels across the available spectrum."

"We call them user block allocation and cell block allocation, respectively."

"The scenario evaluation consists of an unplanned UDN and a uniform small cell deployment, where at least one active user is present for each cell."

"The results obtained through the genetic algorithm solution on the network's spectral efficiency, cell's average capacity, and subchannel allocation rate show that the proposed arrangements alleviate the high complexity of the channel allocation problem and find feasible solutions for UDN scenarios."

Introduction

"Appropriate interference mitigation methods and resource allocation (RA) strategies should be utilized to maximize the benefits of 5G network densification [15]."

"RA is usually modeled as an optimization problem whose objective is to maximize the aggregate throughput of the network [16]."

"Allocating the different spectral resources at once results exceedingly complex due to the RA problem belonging to non-convex and NP-hard problems[17]."

"Some works reported in the literature describe solutions for the RA problem applying metaheuristic algorithms; however, they are solved for low-cell-density scenarios [16–21]."

"Due to the high dimensionality, extensive search space, and aggressive frequency reuse in UDN, the RA optimization problem quickly becomes too complicated to be solved efficiently with approximated algorithms [22]."

Related Work

"Consider that the bandwidth is distributed in a low number of channels, leading to inefficient use of the resources by allocating excessive bandwidth to users and increasing interference."

"Distributing the available spectrum in a lower number of channels reduces the CA problem's search space and dimensionality but at the cost of increasing interference, as reported in [23]."

"In that case, if an interference threshold is set, it is possible to determine the number of channels required to reach users' demands, to bound the bandwidth allocated to them [20]."

"Our proposal consists of reducing the RA problem's dimensionality by allocating the entire required bandwidth of each user in a single channel."

"Unlike reported in [23], this approach avoids the aggregated interference by allocating channels with specific bandwidths regarding users' demands."

System Model

"Users are served by one cell at a time, which is chosen based on the highest signal strength they receive, known as path loss association [24]."

"The RA algorithm is executed in each time slot, considering that the surrounding conditions, such as user locations and channel propagation characteristics, are known by the network and remain static during the time it takes to perform this task [25]."

"The data rate that each cell provides to its users is affected when other cells share the same communication channel."

"This interference, which aggravates with cell densification, decreases the perceived signal-to-noise-plus-interference ratio (SNIR) and the UE data rate."

Proposed Channel Allocation Strategy in UDN

"Solutions that allocate few channels to users make it impossible to fulfill their requirements, while over-allocations result in unnecessary interference in the network, thereby degrading the network capacity."

"One known and effective method to alleviate the CA optimization problem complexity is to control the number of channels to allocate to each user before the allocation."

"The high complexity on the resource allocation is one of the main issues that limits the benefits of the network densification in UDN, our proposal considers three aspects to reduce the computational complexity of the CA optimization problem: 1."

"Given the channel invariance property, a vector solution representation can be designed to focus the CA problem on selecting users who could share a channel efficiently, regardless of its channel."

"The proposed UBA and CBA strategies reduce the dimensionality and search space of the CA optimization problem."

Solving the CA Problem Applying Genetic Algorithm

"At each iteration, called generation, the chromosome population is modified by applying evolutionary operators: parents selection, crossover, mutation, and survival selection."

"The pair of solutions interchange their genetic information by the crossover operator, producing two new (offspring) solutions."

"When the new population of M solutions is generated, the fitness function is applied to these to calculate their fitness value."

"The survival selection step chooses between the old and new generations to maintain a fixed number of solutions in the population, based on the analogy of limited resources."

"Parent solutions are selected using the well-known tournament selection operator."

"Two random solutions of the population are chosen for each required parent, and the one with the highest fitness is considered."

"As suggested in the literature [26], we considered a high probability of occurrence for the crossover process (0.9), while the mutation process occurs on 5% of the generated solutions."

Performance Results

"The convergence results show that, for dense cell deployments and high network demands, the GA with the proposed CBA scheme offers significantly better performance than ICA and UBA in terms of network capacity and users QoS (which is considered in the fitness calculation)."

"Experiment, the dimensionality of the three channel allocation schemes is: (i) higher than 1,000 for ICA, (ii) 500 dimensions for UBA as 500 users where considered, and (iii) a range from 50 to 200 dimensions for CBA with regard to cell density."

"This result is remarkable as ICA with 100 SC cannot to reach the network capacity demands at high user densities, which justify increasing the cell densification."

"When the user data rate demand is 10 Mbps, the CBA scheme assigns a smaller number of channels to each SC, reducing the SAR values."

Conclusion

"As the network grows, the allocation becomes complicated due to the high number of cells and users deployed in the system, and the ever-increasing interference."

"A block channel allocation is proposed to reduce network interference and the complexity of solving the channel allocation with the GA for high densification levels."

"The first experimental results show that GA with the proposed UBA and CBA makes more efficient use of the network's channels, reaching higher network capacities and using lower computational resources."

"As pointed out in these experimental results, GA/ICA seems unable to attain the network's aggregated demand at 500 users, which implies that further densification is advisable."

"GA/UBA and GA/CBA network capacity shows that using this demand is achievable without increasing the densification."

"It can be observed that even if the cell densification grows linearly with the network demands, the user's capacity remains monotonically descendent."

Analysis of Macro Diversity Based on Maximum Ratio Combining in Long Range ALOHA Networks [27]

This is a machine-generated summary of:

Song, Qipeng; Nuaymi, Loutfi; Lagrange, Xavier: Analysis of macro diversity based on maximum ratio combining in long range ALOHA networks [27]

Published in: Telecommunication Systems (2019)

Link to original: https://doi.org/10.1007/s11235-019-00579-3

If you want to cite the papers, please refer to the original.
For technical reasons we could not place the page where the original quote is coming from.

Abstract-Summary
"In low power wide area networks, packets transmitted by a device are not addressed to a specific base station: any surrounding station can receive them."

"As opposed to independent decoding of packets received by each base station, we propose using maximum ratio combining (MRC) to improve system performance."

"We use a curve-fitting approach based on the simulation results to get a closed-form formula of the packet loss probability with 2 receivers."

"This curve-fitting approach is also applicable to other cases in which more receivers are leveraged by MRC."

Introduction
"The radio access network (RAN) of LPWAN supports macro reception diversity: a packet transmitted by a given device can be received by several base stations (BS)."

"A series of MRC function blocks are installed within the central unit so that several transmitted packets in the radio network can be simultaneously decoded."

"The authors of [28] confirmed the interest of applying pure Aloha with receiver diversity in wide range sensor networks and proposed an analytic model for packet interference calculation in asynchronous random access protocols using diversity."

"The performance of SC macro diversity and MRC macro diversity was studied for GSM networks with a remote antenna system by using hexagonal topology [29], but analytical formulas were not available."

"We evaluate the performance of Aloha with MRC macro diversity in terms of packet loss probability in relation to the system load."

System Model
"A bit is assumed to be correctly received if the Signal to Noise and Interference ratio (SINR) is not less than a threshold value."

"Our objective is to find the maximum capacity of the system and we thus consider it in high load conditions: interferences are much larger than the background noise."

"This configuration is called Pa-Aloha (pure Aloha with average interference)."

"LoRa uses forward error correction (FEC) with different coding rates and can be considered as a Pa-Aloha system."

"We denote this case as Pm-Aloha (pure Aloha with maximum interference)."

"Sigfox only includes an error detection code and can be regarded to use Pm-Aloha."

Preliminary Study Based on Simulations

"We study the impact of SIR independence hypothesis by simulations."

"We first considered a realistic simulation, denoted by R-SIM, in which the interference on each RU is computed without any independence assumption."

"In M-SIM, we consider independent interference for different RUs, which is compliant with our numerical framework."

"Path loss exponent also has impact for interference-independence assumption."

"The larger the path loss exponent, the more important the effect of the interference correlation."

"We would like to study the network packet loss probability of LPWAN network supporting MRC under different number of involved RUs."

"We consider the following scenario according to the number of RUs that are leveraged by MRC: (1) only the best RU; (2) the two best RUs; (3) the six best RUs; (4) the twenty best RUs; (5) all the RUs on the infinite plane."

Curve Fitting with Respect to Simulation Results

"It depends on the normalized load, path loss exponent and Aloha type."

"To have an accurate analysis, the intuitive solution is to analyze with dependent interference."

"The related parameters can be determined by curve fitting with respect to simulation results."

"We take the case in which only the 2 RUs that receive the signal with the best SNRs are involved by MRC mechanism."

"This methodology is applicable for other cases, with 3, 6, 20 RUs, or even all the RUs in the infinite plane (i.e. the ideal case)."

Performance Analysis with the 2 Best Receivers

"The effectiveness of proposed analytical model, based on curve fitting technique, is studied."

"In all the following figures, for two types of Aloha, we observe that the disparity between R-SIM simulation results and analytical model can be neglected, especially with the increase of the normalized load."

"To better illustrate the accuracy of the curve-fitting-based analytical model, we take Pa-Aloha as an example."

"We study the system capacity gain brought by MRC compared with SC."

Conclusion

"We proposed an analytical framework to evaluate the network packet loss rate of LPWAN systems applying MRC macro diversity."

"We showed that MRC macro diversity can multiply the capacity by a factor greater than 1.26 if MRC leverages 2 nearest RUs."

"The capacity gain will be much higher if more RUs are involved."

Fair Adaptive Cross-Layer Resource Allocation Scheme for IEEE 802.16 Broadband Wireless Networks [30]

This is a machine-generated summary of:

Deva Priya, M.; Sangeetha, M.; Christy Jeba Malar, A.; Dhivyaprabha, E.; Kiruthiga, N.; Rajarajeswari, PL.: Fair Adaptive Cross-Layer Resource Allocation Scheme for IEEE 802.16 Broadband Wireless Networks [30]

Published in: Wireless Personal Communications (2019)

Link to original: https://doi.org/10.1007/s11277-019-06929-3

If you want to cite the papers, please refer to the original.

For technical reasons we could not place the page where the original quote is coming from.

Abstract-Summary

"In a WiMAX network, the Medium Access Control (MAC) protocol deals with resource allocation to different types of traffic."

"The key components that ensure Quality of Service (QoS) guarantees in a WiMAX network include Call Admission Control (CAC), Bandwidth and Burst allocation."

"CAC and Bandwidth allocation are dealt in the MAC layer, while Burst allocation in the PHYsical layer."

"The History based Bandwidth Allocation (HBA) scheme deals with allocating bandwidth based on Consumption and Equity measures."

"The proposed tightly coupled Delay Tolerance based Scheduler (DTS) and Bucket based Burst Allocator (BBA) allocate resources by prioritizing flows with least delay tolerance."

Introduction

"The BS evaluates a MSS' request and grants resources based on the service level agreement."

"The BS keeps track of the number of MSSs that contend for the medium and uses this information to predict the number of slots in the next frame for sending Bandwidth REQuests (BW-REQs)."

"The BS grants resources based on the QoS, and the amount of requested and available resources."

"If the requested bandwidth is not granted within timeout, the MSS starts a resolution process deciding that the request is corrupted."

"WiMAX offers 2 modes of resource grants namely, Grant Per Subscriber Station (GPSS) and Grant Per Connection (GPC) [31]."

"Grant Per Subscriber Station (GPSS): In GPSS, the BS apportions bandwidth to each subscriber rather than a connection."

"As each MSS requests bandwidth for each connection, this mode is suitable when the number of users per MSS is less."

Related Work

"A dynamic CAC scheme that allocates channel bandwidth based on recent requests is proposed by Chaudhry and Guha [32]."

"Afzali and others [33] have proposed a greedy based resource allocation scheme, wherein Learning Automata based Greedy Channel Assignment (LAGCA) and Learning Automata based Greedy Centralized Scheduling (LAGCS) algorithms are proposed to deal with interference related issues and increase spatial reuse."

"A Channel-aware Subchannel Renumbering (CSR) scheme and a Channel-aware Downlink Burst Allocation (CDBA) scheme are proposed by Shih and others [34]."

"A subchannel-aware Burst Fragmentation, Packing and Scheduling (BFPS) algorithm that involves burst allocation, fragmentation, packing and swapping of bursts is proposed by Chen and others [35] to position bursts, reduce unused slots and adjust the positions of bursts."

"Hailan and others [36] have proposed a scheme wherein, bandwidth for each subscriber is allocated based on the access rate."

Cross-Layer Framework

"A Cross-layer framework that couples a Priority based Scheduler (PS) and a Bucket based Burst Allocator (BBA) is designed by Liang and others [37] for organizing resources and reducing the number of Information Elements (IEs)."

"DTS schedules the traffic for the Bucket based Burst Allocator (BBA)."

"BBA organizes bursts to traffic flows based on scheduler defined priority."

"Further, it guides the scheduler to effectively allocate bursts to traffic flows."

History Based Bandwidth Allocation (HBA)

"History based Bandwidth Allocation (HBA) dynamically assigns bandwidth to traffic flows based on various factors from the previous round."

"The Equity measure of a flow is based on the amount of the requested and allocated bandwidth in the previous round."

"The total consumption measure shows the utilization of the assigned bandwidth."

"In the current round, bandwidth is allocated to each flow based on Equity and Consumption measures."

"If the resource assigned to a flow exceeds the demand, then the values of Consumption and Equity measures are high."

"If the resource assigned to a flow is less than the demand, then both the measures show lesser values."

History Based Call Admission Control (HCAC)

"Calls are accepted based on the factor that, accepting them will not contribute to Contention Delay (CD)."

"When the medium is busy, CD increases and hence the channel is analyzed for collision."

"After performing bandwidth and fair utilization tests, the MSSs are accepted based on the amount of CD."

"New calls are accepted based on the BW-REQ and its availability, behavior of traffic, network conditions and CD."

"Variance of Contention Delay (VCD) is the deviation of CD at an MSS."

"This increases the number of requests contending for the medium, thus increasing CD."

Delay Tolerance Based Scheduler (DTS) for Burst Allocation

"Scheduling is done based on the priority assigned to the incoming traffic [37]."

"DTS schedules flows based on the amount of delay that can be tolerated by a traffic flow."

"The IEs and bursts are arranged by the BBA and the information regarding burst allocation is sent to the DTS."

"For UGS flows, priority is assigned based on delay tolerance and data rate."

"Similar to UGS, the traffic priority is based on delay tolerance."

"The scheduler defines the priority based on which bursts are allocated to the traffic by the BBA."

"As the bursts are arranged in the form of buckets of size 'Δbkt', the assignments for traffic are taken as a multiple of 'Δbkt'."

"The Bucket based Burst Allocator (BBA) is wholly based on the delay tolerance of flows at the MSS and involves less computation complexity."

"BBA assigns bursts and IEs based on the information from the scheduler."

Implementation and Performance Analysis

"The proposed Delay Tolerance based Scheduler (DTS) offers better performance in contrast to the benchmarked Priority Scheduler (PS)."

"The Packet Loss Ratio (PLR) of PS is 1.78 times more than that of DTS."

"The Average Delay of DTS is 2.34 times less than that of PS."

"It is seen that HRA yields better results when compared to HIRA and DTS performs well in contrast to PS."

Conclusion

"The History based Resource Allocation (HRA) scheme proposed in this Paper accepts calls only when the incoming call does not affect the QoS of the network."

"This prevents degradation of QoS to a greater extent in contrast to the existing schemes."

"Bursts are assigned based on the priority and delay tolerance of requests."

QoE-Oriented Resource Allocation for Dense Cloud NOMA Smallcell Networks [38]

This is a machine-generated summary of:

Shao, Hongxiang; Sun, Youming; Du, Zhiyong; Cai, Jihao; Duan, Zhentao: QoE-oriented resource allocation for dense cloud NOMA smallcell networks [38]

Published in: Wireless Networks (2022)

Link to original: https://doi.org/10.1007/s11276-022-02929-7

If you want to cite the papers, please refer to the original.

For technical reasons we could not place the page where the original quote is coming from.

Abstract-Summary

“We investigate the resource allocation for dense cloud non-orthogonal multiple access smallcell networks (NOMA SCN), aimed at maximizing the users’ quality of service (QoE).”

“We formulate the QoE-oriented channel allocation and user pairing problem in NOMA SCN as a local cooperation game.”

“The optimal pure strategy Nash equilibrium (PNE) in the proposed game can maximize the network QoE level.”

“To achieve the optimal PNE in the proposed game, we redesign a directed-hypergraph-based multi-agent learning algorithm, which allows multiple non-coupled agents in directed hypergraph to simultaneously update their actions.”

Introduction

“Optimizing network resources in ultra-dense networks (UDN) is extremely complicated as resource allocation and user association are coupled, especially to deal with complex interference relations among multiple entities.”

“We investigate the resource allocation for dense cloud NOMA SCN based on the directed-hypergraph interference model, aiming at maximize users’ QoE level in networks.”

“Cui and others proposed a three-dimensional (3D) matching based resource allocation scheme for NOMA networks to maximize the sum mean opinion scores (MOSs) of users, where MOS is adopted to evaluate the QoE level of users [39].”

“Different from the existing literature, this paper focuses on the resource allocation problem of the NOMA based ultra dense network, and three factors, i.e., NOMA, QoE and hypergraphcmodel, are considered at the same time.”

System Model and Problem Formulation

“Each UE is equipped with SIC receiver and can share the same channel with other users in NOMA mode.”

“We apply the directed hypergraph to simultaneously capture the cumulative interference effect and asymmetric interference for NOMA SCN, while the traditional binary conflict graph is based on pair-wise strong interference links.”

"To avoid the head set to be empty and meaningless as shown in [40], we introduce a common virtual node 0 (CVN 0) in head set of the specific hyperarcs for NOMA SCN, in which each node is interfered by their co-channel transmission."

"We show how to construct a directed hypergraph in detail for a given NOMA SCN."

"We can obtain the directed hypergraph for a given multi-cell NOMA network by constructing the directed binary edges and hyperedges for each vertex."

"Since SBS can multiplex multiple UEs in one channel and certain UEs may suffer intra-cell interference in NOMA SCN."

Game Model and Equilibrium Analysis

"According to definition 4, the proposed game G is an EPG."

"The EPG has two significant properties as follows [41]: Every EPG possesses at least one PNE."

"Either global or local maxima of the potential function can constitute a PNE."

Multi-Agent Learning Algorithm

"The spatial adaptive play (SAP) based learning algorithm has attracted significant attention for its good property, which can achieve the optimal PNE with an arbitrarily high probability."

"Given a directed interference hypergraph for NOMA SCN, we redesign an multi-agent SAP based learning algorithm (MALL)."

"To fully utilize the spatial structure of the directed hypergraph to accelerate the learning process, the MALL admits multiple non-coupled agents simultaneously updating their strategies at each learning step."

"In [41], it is proved that SAP based learning algorithm can achieve the global maxima of the potential function with an arbitrarily high probability."

"MALL can be viewed as the concurrent version of SAP learning algorithm for the directed hypergraph."

"The information of each SBS's coupled neighbors and concurrent SBSs can be pre-stored in the cloud and these information is periodically updated with the change of directed interference hypergraph, which can significantly reduce computation burden and accelerate implement of MALL."

Numerical Results

"Based on the hypergraph interference model of NOMA SCN, we can see that the proposed MALL (marked as MALL-NOMA-Hypergraph) can obtain higher network MOS and converge faster than SAP (marked as SAP-NOMA-Hypergraph) and BRD (marked as BRD-NOMA-Hypergraph)."

"Note that the performance of all NOMA schemes are significantly superior to OMA scheme in current network configuration, where BRD and hypergraph interference model are adopted for OMA system (marked as BRD-OMA-Hypergraph)."

"Due to the interference effect caused by multiple weak interfering sources in dense networks is ignored, the binary interference graph based scheme (marked as SAP-NOMA-Binary graph) seems to be unstable in the given network configuration and the performance of SAP-NOMA-Binary graph is inferior to other hypergraph-based NOMA schemes."

"The first one is with the network MOS maximization goal, marked as MALL-QoE, and we have applied it in previous simulations; the second one is with the network throughput maximization goal, marked as MALL-Capacity, which can be viewed as the extension of the proposed algorithm [42] in hypergraph."

Conclusion

"We aim at maximizing the users' quality of service (QoE) for dense cloud non-orthogonal multiple access smallcell networks (NOMA SCN) and investigate the QoE-oriented resource allocation problem."

"Since the optimization goal is to maximize the sum of users' MOS in the networks."

"Compared with other algorithms that target capacity maximization, it can achieve significant improvements from the perspective of users realistic QoE."

Retracted Article: Throughput Maximization of Multichannel Allocation Mechanism Under Interference Constraint for Hybrid Overlay/Underlay Cognitive Radio Networks with Energy Harvesting [43]

This is a machine-generated summary of:

Karaca, Hakan Murat: RETRACTED ARTICLE: Throughput maximization of multichannel allocation mechanism under interference constraint for hybrid overlay/underlay cognitive radio networks with energy harvesting [43]

Published in: Wireless Networks (2020)

Link to original: https://doi.org/10.1007/s11276-020-02305-3

If you want to cite the papers, please refer to the original.

For technical reasons we could not place the page where the original quote is coming from.

Abstract-Summary

"Motivated by this and improved spectrum reuse by combined use of overlay/underlay modes of cognitive radio networks (CRNs), this paper proposes a novel multi-channel (m-channel) allocation performance maximization algorithm for low-power mobiles."

"Prioritization of STs during opportunistic allocation of channels is critical as properties like energy level and harvesting capability improve channel distribution performance."

"A novel metric is proposed that prioritizes STs based on initial energy levels, harvesting capability, and number of channels through which they can transmit."

"For comparison, three algorithms were considered: a greedy mechanism for m-channel allocation of hybrid CRNs without harvesting, the proposed m-channel allocation schemes based on maximum independent sets (MIS), and the proposed metric of hybrid CRNs with harvesting capability."

"The simulations show that the proposed m-channel allocation method based on MIS outperforms the greedy algorithm."

"The proposed m-channel allocation using the proposed metric on hybrid CRNs with energy harvesting ability produced the best performance of the three methods, proving the superiority of the proposed algorithm."

Introduction

"Lee et al. [44–48] showed that energy from ambient radio frequency (RF) from primary transmitters can be harvested by SUs, and the transmission probability and final throughput under the outage constraint were found."

"Lee et al. [48] represented the mobile equipment in a secondary network opportunistically by either harvesting RF energy from transmissions of nearby devices in a primary network, or transmitting data if outside of the interference range of any other primary network."

"In this journal, a hybrid m-channel allocation mechanism is proposed that maximizes the m-channel allocation efficiency in EH-based CRNs, where a ST is capable of harvesting energy from the RF signals transmitted by a PT."

"For the CRN with mobile energy harvesting ST, this paper is the first attempt to maximize the efficiency of m-channel allocation by dividing the coverage area into overlay mode area, underlay mode area, and harvesting and guard zones for the throughput analysis given the hybrid underlay-overlay scheme under the interference constraint."

System Model

"In the overlay mode area, energy cannot be harvested from the RF signals, and ST transmits in overlay mode when the licensed spectrum is sensed as free, but it is not permitted to transmit in underlay mode when the licensed spectrum is found busy because of the high level secondary interference of underlay mode transmission to the PR unless transmission power level is decreased."

"After this point, if the residual energy is sufficient and the licensed spectrum is tested as busy, ST may transmit without the need to decrease power level, which implies overlay mode transmission, that is why the region is called here overlay mode area."

"If the ST cannot make sensing and is located in any HZ of an active PT, the ST harvests energy from RF signals or stays idle if it is located in the underlay, overlay mode area, or in the guard zone."

Greedy and Proposed Algorithms

"A greedy mechanism is taken as a basis for channel allocation to compare the proposed mechanisms and highlight its performance."

"The aim of all these mechanisms is to improve multi-channel allocation performance when utilizing energy harvesting."

"This algorithm's allocation mechanism is based on the method proposed previously [49]."

Simulation Results

"Parameters used in this experiment can be summarized as: Average percentages of STs in the HZ is 21%, and the number of PTs and STs are 2 and 20, respectively, when the SINR (outage) threshold fixed to 1."

"Parameters used in this experiment can be summarized as: Average percentage of STs in HZ is fixed to 21% and number of PTs and STs are 2 and 10, respectively, when the SINR (outage) threshold fixed to 1."

"Experiment 10 In the tenth experiment, ChDistFairness results are compared for greedy, proposed maximum independent set (MIS)-based, and proposed metric based allocation mechanisms against time slots under the same average percentage of STs in HZ."

"Average acquired channels of each ST per joule and ChDistFairness per joule are compared for greedy, proposed maximum independent set (MIS)-based, and proposed metric based allocation mechanisms against signal to interference and noise ratio (SINR) under the same average percentage of STs in HZ."

Conclusions

"A novel algorithm was proposed to allocate multiple channels in a hybrid cognitive radio network where STs can harvest from RF signals."

"This is the first work to investigate the assignment of multi-channels to SUs with multi-channel transmission capability and harvesting ability in a hybrid cognitive radio network to maximize channel allocation performance."

"The simulation results indicate that addition of harvesting capability to allocation scheme markedly improves m-channel allocation performance when comparing MIS-based and the proposed metric-based allocation mechanisms with the greedy one."

"The results prove the superiority of the proposed mechanism with the proposed metric because, if an ST hosted: With more initial energy, Having energy harvesting capability, and Equipped with multi-channel transmission ability is prioritized for channel assignment, metric ChDistFairness will be minimized, thereby maximizing allocation performance."

Graph-Based Radio Resource Sharing Schemes for MTC in D2D-Based 5G Networks [50]

This is a machine-generated summary of:

Hamdoun, Safa; Rachedi, Abderrezak; Ghamri-Doudane, Yacine: Graph-Based Radio Resource Sharing Schemes for MTC in D2D-based 5G Networks [50]

Published in: Mobile Networks and Applications (2020)
Link to original: https://doi.org/10.1007/s11036-020-01527-1

If you want to cite the papers, please refer to the original.

For technical reasons we could not place the page where the original quote is coming from.

Abstract-Summary

"In order to offer the ability to connect billion of devices to propel the society into a new era of connectivity in our homes, officies and smart cities, we design novel radio resource sharing algorithms in a H2H/M2M coexistence case to accommodate M2M communications while not severely degrading existing H2H services."

"We propose group-based M2M communications that share the same spectrum with H2H communications through device-to-device (D2D) communication, as one of the technology components of 5G architecture."

"To overcome the computational complexity of the optimal solution, we model the resource sharing problem as a bipartite graph, then propose a novel interference-aware graph-based resource sharing scheme using a fixed M2M transmit power."

"To further enhance the protection of H2H services, we introduce an adaptive power control mechanism into the interference-aware graph-based resource sharing scheme."

"The latter is proposed within the aim to assure the desired quality-of-service (QoS) of H2H users and increase the efficiency of M2M spectrum usage."

"Simulation results show that adaptive M2M radio resource sharing scheme using fuzzy logic is the one that achieves the best compromise."

"Simulation results also show that in spite of its quite good performance, semi-distributed M2M resource sharing instantiation achieves them with a decline of up to 10% in terms of H2H throughput compared to the centralized instantiation."

Introduction

"Hamdoun et al. [51], we investigated from a MAC layer perspective the radio resource sharing problem in a H2H/M2M coexistence scenario where group-based MTC underlying cellular network transmit traffic through D2D technology."

"The novelties and main contributions of this paper are summarized as follows: First, we consider that M2M communications are underlying H2H communications, and formulate the radio resource sharing issue as a sum-rate maximisation, problem for which the optimal solution is non-deterministic polynomial-time hard (NP-hard).We model the radio resource sharing problem for M2M underlying H2H communications as a bipartite graph to overcome the high computational complexity of the optimal solution, and consider a two-stage radio resource allocation mechanism."

"We develop an interference-aware graph-based resource sharing scheme for M2M communications underlying cellular networks."

"We introduce a novel adaptive power control mechanism that we integrate into our interference-aware graph-based resource sharing approach to further guarantee H2H applications and maximize the efficiency of M2M services."

Background

"We introduce briefly two major controllers namely PID and fuzzy logic that we integrate into our proposed adaptive power control-based radio resource sharing scheme to adjust M2M transmit power."

"PID and fuzzy logic are considered the two major controllers in the control field that are widely used to address many problems in different application areas such as leveling security services or managing the radio resources in wireless networks [52–55]."

"In our work, we integrate either PID or fuzzy logic aiming at controlling efficiently the M2M transmit power in order to mitigate the co-channel interference in a H2H/M2M coexistence scenario [56]."

"We are motivated in this paper to integrate one among PID and fuzzy logic controllers into our adaptive interference-graph based radio resource sharing scheme with the aim to guarantee H2H services and maximize M2M spectrum usage."

Related Works

"In [57], authors have proposed an interference limited area control scheme to prevent cellular user equipments located in a predefined area from sharing radio resources assigned to D2D users."

"Previous works in [58–60] have considered the joint optimization of power allocation and resource scheduling for D2D communications underlying a cellular infrastructure."

"Convinced by the strength of the graph theory, we propose a novel interference-aware-graph-based radio resource sharing scheme for M2M. We enable M2M to transmit traffic through D2D technology, making MTC an underlay to traditional cellular communications."

"We introduce a low computational complexity interference-aware graph-based resource sharing scheme for M2M that efficiently re-assign MTDs one of the available radio resources already allocated to H2H users."

"None of the mentioned works, proposing radio resource sharing algorithms for MTC underlying cellular networks, jointly and successfully considered the impact on H2H services as well as spectrum usage efficiency."

System Model and Problem Formulation

"To increase spectrum usage efficiency, we assume that MTDs are sharing radio resources with H2H users (UEs) through D2D technology."

"MTDs of the same cluster cannot share the radio resource RB."

"Radio resource sharing among UEs and MTDs as well as among MTDs of different clusters is not forbidden owing to the MTD's low transmit power."

"This latter represents the interference generated from MTDs of different clusters to the MTH when sharing the radio resource, RB_k."

"Let us as a next step, after describing our system model, investigate the uplink resource sharing problem for group-based MTDs underlying cellular network through D2D technology."

"The bipartite graph is a powerful tool to model radio resource sharing problem in a H2H/M2M coexistence scenario."

"We first assume that radio resource, RBs, are assigned to traditional H2H users."

Fixed Radio Resource Sharing Algorithms (F-RRSA)

"Based on the knowledge of the path loss and fading for the potential interference links on each RB_k, the eNB calculates the sum-throughput that will be associated to the weight of each edge."

"Different from the centralized feature, presented in Sect. 5.1, that requires to gather in a single point, eNB, all accurate CSI of communication and interference links leading to a significant communication overhead, only the positions of MTDs belonging to its cluster are required for each MTH (cluster head), in semi-Distributed feature, which is not greedy in terms of resources."

"The interference calculation remains the key parameter to determine the scheduling operator, eNB for the centralized feature and MTHs for the the semi-distributed feature of radio resource sharing algorithm."

"The computational complexity of the proposed two-stage resource assignment approach consists of the sum of the complexity of both algorithms performed in each stage."

Adaptive Power Control-Based Radio Resource Sharing Algorithms (A-RRSA)

"Two power control options are designed based respectively on (i) PID controller and (ii) fuzzy logic controller."

"Through adjusting appropriately MTD transmit power, not only the QoS of H2H users is assured, but also the efficiency of M2M spectrum usage is maximized."

"In each scheduling period (TTI), the new MTD transmit power is determined by multiplying actual MTD transmit power with power control ratio R_{pc} derived from the output of the PID system."

"Unlike MTD transmit power control process using a PID controller where a specific and a precise output is given through adopting a mathematical model, fuzzy controllers approximate the mathematical solution and hence require less computational complexity, making the prototyping process and implementation easier."

"We design a fuzzy power controller to dynamically adjust MTD transmit power in order to assure the QoS of H2H users as well as to maximize the efficiency of the MTD spectrum usage."

Performance Evaluation

"The network sum-throughput is decreased nearly about 40% and 30% compared to the optimal resource allocation scheme when using the proposed adaptive radio resource sharing algorithm with the centralized instantiation along with the RR and PF scheduling algorithms, respectively."

"The worst performance in a H2H/M2M is obtained when using the random M2M radio resource allocation algorithm where the total H2H throughput for 80 UEs is decreased about 50% and 44% compared to the throughput obtained in an exclusive H2H scenario when using RR and PF schedulers, respectively."

"Even though, F-C-RRSA achieves better performance compared to the random M2M radio resource allocation in a H2H/M2M coexistence scenario, the H2H throughput remains significantly degraded."

"What is quite interesting to note here, however, is that the proposed adaptive M2M radio resource sharing algorithms in a H2H/M2M coexistence scenario with both instantiations, centralized and semi-distributed, maintain the level of fairness of the existing RR and PF schedulers in an exclusive H2H scenario."

Conclusion

"We first proposed a novel fixed interference-aware graph-based resource sharing algorithm where the M2M transmit power is set to a fixed value."

"To further guarantee H2H services while maximizing the efficiency of M2M communications, we introduced a novel adaptive power control feature using one among the two following alternative mechanisms, namely, either proportional integral derivative (PID) or fuzzy logic controllers and integrate it into our interference-aware graph-based resource sharing algorithm."

"The fuzzy logic-based adaptive M2M radio resource sharing algorithm yields the best compromise between guaranteeing H2H performance and satisfying M2M services."

"For future work, we propose to combine fuzzy logic with other intelligent methods such as minimum variance and predictive strategies in order to further improve the performance of our adaptive radio resource sharing scheme."

The Controller Placement Problem for Wireless SDN [61]

This is a machine-generated summary of:

Dvir, Amit; Haddad, Yoram; Zilberman, Aviram: The controller placement problem for wireless SDN [61]

Published in: Wireless Networks (2019)

Link to original: https://doi.org/10.1007/s11276-019-02077-5

If you want to cite the papers, please refer to the original.

For technical reasons we could not place the page where the original quote is coming from.

Abstract-Summary

"The controller placement problem is threefold in nature: the number of controllers to be placed in a network, the locations of these controllers and the assignment function of controllers to switches, with all of them important for the design of an efficient control plane."

"Most of the existing literature focuses on the placement problem assuming the medium between the controllers and the switches is wired."

"We model the problem of determining the placement of wireless controllers in software defined networking."

Introduction

"Some research has already proposed a wireless control plane architecture in which inter control plane communication is wireless and the communication on the SBi is also wireless [62–65] but none of them investigated the influence of the SBi wireless metrics on the solution of the CPP."

"We investigate the wireless controller placement problem(WCPP), taking into consideration the complexities of the wireless medium between the control plane and the data plane."

"In order to investigate the efficiency of the wireless control plane, we propose an objective function which is a weighted sum of three objectives: latency, link failure probability and transparency."

"We propose a new metric for the WCPP, namely the transparency defined as the marginal data plane average latency caused by the additional interference that the controllers add to the data plane."

Related Work

"Tanha and others [66] proposed a solution for the resilient CPP with the objective of maximizing the control plane resiliency."

"Based on these models, they introduced the WCPP and proposed an objective function that minimizes the number of SDN wireless controllers."

"Their proposed model is based on a TDMA mechanism where both the control plane and the SBi are wireless."

"Johnston and Modiano [67] considered the impact of delayed state information on the performance of centralized wireless scheduling algorithms and proposed a dynamic controller placement framework, in which the controller is relocated using delayed queue length information at each node, and where transmissions are scheduled based on channel and queue length information."

"Liu and others [65] proposed a joint placement of controllers and gateways in SDN-enabled 5G-satellite integrated networks and showed the advantage of a wireless control plane."

System Model

"In the control plane, there are numbers of SDN controllers which communicate with other controllers and with the data plane APs through a Wi-Fi based medium."

"Wireless links between each AP and its assigned controller."

"The average latency on the SBi is: The transparency metric is the marginal average latency in the data plane caused by the interference from the controllers, meaning, how the conversation between the controllers and the APs interfering in the data plane due to the fact that all the transmitting nodes use the same wireless medium and technology."

"We claim that the aforementioned metrics behave differently, for example, adding controllers shortens the distance between the AP and the controller and therefore, the link failure probability decreases but then the interference grows and the latency and the transparency increase."

Wireless Controller Placement Algorithms

"The previously proposed K-Median algorithm uses as an input a set A of APs located randomly in the plane where A is divided into k = 3 clusters such that all the APs in the same cluster are assigned with a controller which is co-located in one of the AP locations in the cluster."

"The K-Median algorithm chooses the clustering and the locations of the controllers such that the sum of weights of all links is minimized."

"The proposed algorithm solves the WCPP for a specific and constant number of controllers but obviously by changing the number of the controllers we may achieve better solution, with lower objective value."

"We extend the algorithm such that we find the number of controllers and their placement that minimize the objective value."

Experimental Results

"That in a wired network the latency depends mainly on the Euclidean distance, therefore, each cluster is organized such that all the switches and their assigned controller are geographically close."

"The simulations show that the clustering of the data plane APs and their assignment to controllers have no prominent relation to the geographical distance and the APs may be assigned to more distant controllers since the link to a close controller may have low SINR."

"In a wired network, proximity to the controller lowers the latency and improves the performance, while in a wireless network this adds more interference on the data plane and may increase the transparency."

"Installing controllers imposes interference on the data plane and changes the SINR, but when the network grows this overhead is less significant and the transparency of the control plane improves."

Conclusions

"SDN with a wireless control plane is a challenging new concept that has not yet been thoroughly investigated."

"WSDN requires different way of thinking than SDN for wired networks."

"We introduced the WCPP and showed the complexity of implementing SDN for wireless networks."

"We modeled the different metrics that affect the solution for the WCPP and proposed an objective function and constraints for the problem."

Performance Evaluation of Interference Mitigation Scheme with D2D Users in Wireless Networks [68]

This is a machine-generated summary of:

Dhilipkumar, S.; Arunachalaperumal, C.: Performance Evaluation of Interference Mitigation Scheme with D2D Users in Wireless Networks [68]

Published in: Wireless Personal Communications (2021)

Link to original: https://doi.org/10.1007/s11277-021-08413-3

If you want to cite the papers, please refer to the original.

For technical reasons we could not place the page where the original quote is coming from.

Abstract-Summary

"The existing communication techniques could not cope with the proliferating demands."

"The major hazard for both the existing and upcoming communication system is interference."

"Device-to-Device (D2D) communication paradigm is being recognised as one of the promising techniques which could help with matching the data demands."

"This paper analyses and reviews the number of interference mitigation techniques."

Introduction

"Adaptive resource allocation in OFDMA can be optimised through subcarrier assignment and power loading."

"The resource allocation was made adaptive based on knowing the Channel State Information (CSI)."

"The main idea of resource allocation to interference mitigation will eventually increase data rate or decrease power."

"By this adaptive resource allocation, we can model a system which adapts data rate and power accordingly."

Spectrum Sensing in D2D Networks

"The spectrum sensing for interference free communication was handled by modelling the position of base stations (BSs) and cellular users (CU) as Poisson Point Process (PPP)."

"By sensing the users in close proximity, communication is established between D2D users."

"Unlike spectrum sensing in cognitive radio network, the spectrum holes are detected in the spatial domain and the protection regions are detected without the help of BSs in the D2D network."

"A Fractional Frequency reuse with Almost blank subframe (ABS) schemes were experimented for interference mitigation in D2D communication by Jeehyeong and others, in [69]."

"The eNB controls both CUEs and D2D users which requires the spectrum usage."

"The D2D receivers get affected by interference from cellular users and the co-channel D2D users."

"ABS scheme for interference mitigation performs well in closer distances though at a larger distance they perform almost equivalently."

Cellular System with Interference Calculation

"On considering multi user detection, the interference cancellation is categorized into two modes (i) parallel and (ii) successive."

"In [70] successive interference cancellation the receivers detect one user at once."

"On utilising traditional power control algorithms without considering the errors in channel state information, the system capacity is better off without successive interference cancellation."

"If the channel estimation error increases to 50 percent, modified power control algorithm that considers the error statistics is only able to provide doubled capacity with successive interference cancellation."

Graph Based Dynamic Frequency Reuse in Femtocell Networks

"The author proposes a graph-based dynamic frequency reuse method, in which the base station continuously monitors and discovers the neighbouring base stations."

"This can be done by a graph based dynamic frequency reuse algorithm."

"BS utilise one subband out of the six subbands based on a graph colouring method."

"Frequency reuse method shows better performance, only 50 percent of the users experience SINR exceeding 0 dB. All the user equipments are able to achieve better SINR exceeding 5 dB in graph-based dynamic frequency reuse."

Game Theoretical Model in Femtocell

"Femtocell access is confined to primary users in closed access mode and not allowed to the remaining unauthorized users called cognitive users."

"It is found that by providing open access to cognitive users, the interference between microcell and femtocell can be mitigated."

"The user equipments which are in the vicinity of both the microcell base station and femtocell base station are considered as the players."

"The subscriber utility function increases as the number of players in the femtocell increases."

Low Duty Operation Interference Mitigation

"In the IEEE mobile WiMax standard, the interference mitigation is handled by switching the operation modes between (AI) available interval and (UAI) unavailable interval."

"In [71] Helena Widiarti and others proposed a method to group neighbourhood femtocells and reduce the unnecessary available interval thereby to reduce the interference significantly."

"The interference reduction time is also reduced in LDO method resulting in reduced energy consumption."

"On comparing the conventional scheme with LDO scheme, the interference time is small for all the values of available interval."

"90% of reduction in interference time is also achieved as the unavailable interval increases, the total interference time is also decreased."

Interference Avoidance in OFDMA Based Femtocell Network

"The signal from the femtocell base station should be in synchronous with the signal from microcell base stations to ensure there is no inter carrier interference."

"To achieve this synchronisation, the authors Su Huan and others in [72] proposed orthogonal subcarrier allocation to the femtocell base stations."

"It can be seen that the CIR is greatly affected if the femtocells and macrocell BS transmit signals at the same time."

Interference Mitigation in Multicast D2D Networks

"The Frequency reuse method can also be employed to mitigate interference in multicast D2D networks [73]."

"Each multicast D2D cluster can reuse only one uplink cellular link at a time."

"D2D transmitter in each group multicasts the same data to the D2D receivers in the group."

"These pair of D2D user links forms the multicast group."

Interference Mitigation in D2D 5G Network

"Game theory models and Machine learning models [74] are suitable for interference mitigation in 5G networks."

"Price based modelling [75] sets a price for each users interference based on the Nash game and strategies are formulated for required QoS and data requirements, In Mean field game model [76–78] the D2D transmitter considers the interference dynamics of all other players and thus the desired power control is achieved, Non-cooperative Game model [79] approaches the interference problem by making each user in the network to maximise the energy efficiency."

"The machine Learning Model could possibly be employed in 5G D2D networks to reduce the threat of interference."

"Cooperative online learning model [80] (Q-learning) is utilised to find the best action to be taken at a given state."

Conclusion

"Interference mitigation is enabled by different approaches by the researchers."

"The novel methodology for interference mitigation includes decreasing the power and frequency reuse."

"This paper analyses the next generation techniques which moves ahead from novelty such as frequency reuse to adaptive frequency reuse; Successive Interference cancellation to Multimode Successive Interference Cancellation; Graph colouring algorithm to Graph based dynamic frequency reuse."

"All these methodologies have been proved to mitigate interference significantly."

A Novel Approach to D2D Discovery in PSN for Post-Disaster: Throughput Based Discovery Algorithm (TDA) [81]

This is a machine-generated summary of:

Marttin, Vedat; Yüzgeç, Uğur; Bayılmış, Cüneyt; Küçük, Kerem: A Novel Approach to D2D Discovery in PSN for Post-Disaster: Throughput Based Discovery Algorithm (TDA) [81]

Published in: Wireless Personal Communications (2021)

Link to original: https://doi.org/10.1007/s11277-021-08407-1

If you want to cite the papers, please refer to the original.

For technical reasons we could not place the page where the original quote is coming from.

Abstract-Summary

"Device to device (D2D) discovery and communication provides communication services in such difficult situations without the requirement of network infrastructure and by reducing call traffic in cellular communication networks."

"Considering the fact that almost everyone uses a smart mobile device (UE-User Equipment), the D2D communication method seems the most suitable solution for a post-disaster situation."

"A Throughput-based Discovery Algorithm (TDA) is proposed to find D2D peers among devices on Public Security Network (PSN) in a post-disaster situation."

"In terms of the outage SINR and average pair SINR metrics, the proposed TDA showed better performance than the other discovery (MSNA, SDA, and MSLA) algorithms for D2D discovery in the disaster area."

Introduction

"Of Osman [82], some D2D discovery algorithms are proposed according to the SINR metrics and distances between UEs."

"MSNA is a discovery algorithm that finds D2D pairs according to the maximum SINR without limit distance between UEs."

"MSLA is a discovery algorithm that searches D2D pairs according to high SINR values under limit distance."

"The communication probability of UEs increases if there are many matching D2D pairs in the post-disaster area."

"We proposed a Throughput-based Discovery Algorithm (TDA) which finds D2D pairs according to the throughput metric between UEs."

"The results show that TDA provides more D2D pairs for the disaster area compared to the discovery algorithms that use the distance and SINR value between UEs."

"The focus of our study and motivation is to find the most matching UEs with the lowest outage SINR among D2D peers on the PSN in the out-of-disaster area."

Preliminary

"Of Doppler and his colleagues, D2D communication, which forms the basis of the 3GPP LTE-Advanced (LTE-A) cellular network, was considered as a facilitator of regional services with restricted interference effect on the primary cellular network [83]."

"Of Tang and her colleagues, the problem of neighbour discovery was addressed to enable D2D communication in standard LTE cellular networks."

"Of Vasudevan and his colleagues, the algorithm was proposed and analyzed for the discovery of neighbouring nodes in wireless networks."

"Of Yang and his colleagues, a device discovery scheme was proposed similar to the FlashLinqQ structure using a marker in conjunction with the conventional packet-based scheme suitable for LTE cellular network architecture [84]."

"Yuan and his colleagues presented dynamically select multi-path routes together in a fully loaded cellular network for D2D communication."

D2D Discovery System Model

"D2D pairs are determined when they exceed the minimum threshold of throughput and then the best pair with maximum throughput values is found."

"D2D pair with the minimum distance is selected and if this pair meets the minimum SINR threshold, it is determined as the best pair [82]."

"MSNA discovers the D2D pairs according to the SINR values among the UEs, and the best pair with maximum SINR is determined."

"UEs with SINR values greater than the threshold level are detected in the network, and then, the best D2D pair with maximum SINR values is found."

"MSLA finds the best D2D pair with maximum SINR value among the UEs in the network based on a distance constraint between UEs."

"In MSLA, the devices are detected according to the distance values greater than the threshold value among UEs."

Results and Discussion

"According to the SINR result, the average SINR value of active pairs was calculated as 10.959 dB. In 2nd subfigure, the number of D2D pairs obtained by TDA among 100 UEs varies within the range of 31–39."

"Table's results show that TDA has better service quality than other compared algorithms by the obtained best (24%), worst (40%), and mean (32.58%) outage SINR values."

“The throughput values of device pairs that are close to each other are naturally high in post-disaster area, so this situation causes high energy efficiency and high capacity values in short distance.”

“In the case where the number of devices is more than 12, it seems that the results of TDA and the other discovery algorithms are close to each other.”

“From the capacity CDF results for TDA and the other discovery algorithms, the proposed TDA has a capacity between 25 and 150 Mbps for 50 devices.”

Conclusion and Future Works

“UE's which are distributed randomly on simulation are examined in terms of the SINR outage percentage, average energy, capacity, throughput values, the number of the matching D2D pairs for the disaster area.”

“According to the statistical results, SDA and MSLA (50 m) give better results in terms of throughput, energy efficiency, and capacity metrics; but SDA discovers only two pairs at the closest distance and MSLA (50 m) finds D2D pairs within 50 m according to the SINR values of UE pairs, in this way, throughput, energy efficiency and capacity results were obtained as high.”

“These values fall behind the TDA's outage SINR values for the disaster area scenario in terms of D2D discovery and service quality.”

“The TDA produces 32.58% for mean outage SINR, and 33.71 dB for mean SINR of the discovered D2D pairs, so these are better results than the other discovery algorithms.”

Extended Communication Range for Autonomous Vehicles Using Hybrid DSRC/WiMAX Technology [85]

This is a machine-generated summary of:

Dhilip Kumar, V.; Kanagachidambaresan, G. R.; Chyne, Phidahunlang; Kandar, D.: Extended Communication Range for Autonomous Vehicles using Hybrid DSRC/WiMAX Technology [85]

Published in: Wireless Personal Communications (2021)

Link to original: https://doi.org/10.1007/s11277-021-09242-0

If you want to cite the papers, please refer to the original.

For technical reasons we could not place the page where the original quote is coming from.

Abstract-Summary

"In order to enhance the performance of vehicular communication, cluster-based hybrid DSRC/WiMAX technology plays a crucial role in creating communications between vehicle-to-infrastructure (V2I) and vehicle-to-vehicle (V2V) (V2V)."

"The authors proposed a variety of communication scenarios, and the effectiveness of a hybrid system was analysed by comparing possible physical layer features."

"When compared to various communication scenarios, the NS2 simulation result indicates the capability of the cluster-based hybrid DSRC-IEEE802.16d system."

Introduction

"Clustering Mechanism is important for making communication decisions between vehicles depending on characteristics including vehicle speed, distance, and direction."

"WiMAX, also known as WMAN (Wireless Metropolitan Area Networks), is often used as a Road Side Unit to provide a wide range of vehicular communications, up to 50 kilometers in a fixed framework and 5 to 15 kilometers in a dynamic framework."

"Due to the availability of the cluster head (CH) within each cluster, which can commune with the cell framework and hand-off the data to the cluster mechanisms, clustering in the vehicular system situation provides scalability, longer-range communication, and lower overhead at the BSs."

"clustering algorithms are classified into the following categories in this paper: Direction, speed, beacon, neighborhood follow, force, location, density, direction, certificate based, and routing are examples of mobility and non-mobility-based metrics."

"Different algorithms fall into this category, where the vehicle's motion is the most important factor in determining the cluster head."

Related Works

"In [86], the author proposed multiple interface OBUs for automotive networks by integrating DSRC and LTE technology with protocol management and cluster-head selection."

"Sang-Eon Kim and others [87] proposed using mobile WiMAX technology to develop vehicle safety applications."

"With the use of integrated mobile WiMAX and WLAN technology, Thomas Mangel and others [88] provided a review on the comparative analysis of LTE and UMTS for inter vehicle communication."

"In terms of road safety applications, M. Hayashi and others [89] proposed a WAVE design to broaden the range of wireless communication from IEEE 802.11p to WiMAX and 3G for future DSRC applications."

"William J. Song and others [90] suggested an IP-based handover solution for hybrid communication networks between mobile WiMAX and 3GPP networks."

"In [91], the authors suggested an n-casting handover technique for smooth vehicular communication using WiMAX and WLAN technology."

Proposed Cluster-Based Hybrid DSRC-Mobile Network Communication System

"In Hybrid DSRC and IEEE 802.16d Mobile technology, there are a variety of communication possibilities to consider."

"Since the IEEE 802.16d fixed-based station has a range of 10 km, DSRC hubs in the area are capable of obtaining and transmitting frames to Mobile Base Stations (BS) through clustering head by considering various situations like distance, speed and position of vehicles from [92, 93]."

"A fixed WiMAX with 256 FFT transmitter transmits a signal to a 64 FFT DRSC recipient across a DSRC and IEEE 802.16d channel, and vice versa."

"The suggested structure will be used to communicate between 64 and 256 FFT subcarriers using IQ signals in the OFDM symbol: For the IEEE 802.11P: For IEEE 802.16d: The number of bits per symbol is calculated using a known estimation of the modulation type used in the current OFDM symbol."

Performance Analysis for Hybrid Cluster-Based Vehicular Communication

"The beacon message is transmitted and gathered at each beacon interval, which includes the vehicle's identifier, current position and speed, moving course, vehicle's current status, and CH's identify if it is in a particular vehicular clusters."

"Route failure and packet drop in increased payload and high speed mobility are caused by rising vehicle movement and packet density [94, 95]."

"Cluster-based Hybrid vehicular communication achieves a greater packet delivery ratio than existing systems."

"As the speed of vehicular nodes increases, a greater number of routes will be unable to reach, resulting in an increase in packet drops and a rise in control overhead."

"The results showed that the hybrid system achieves the lowest overhead with lower overhead in cluster-based conventions than the present non-clustering system."

"The graphs above can be used to estimate the efficiency of Cluster-based Hybrid Vehicular Communication in terms of packet delivery ration, average end-to-end delay, and control [96] overhead [97–109]."

Conclusion

"The goal of the proposed cluster-based hybrid vehicular communication system is to increase efficiency while increasing data capacity and coverage range."

"Existing DSRC-DSRC communication and Cluster-based hybrid DSRC and IEEE802.16d Mobile Communication situations are different."

"Outage likelihood varies over time in the cluster-based hybrid vehicular communication scenario due to network congestion and routing in high-speed mobility situations."

"To improve the effectiveness of ITS applications, the proposed Combined cluster-based hybrid vehicular communication system is being developed."

Model of Adaptive WiMAX Network Incorporating Diverse Transforms [110]

This is a machine-generated summary of:

Kansal, Lavish; Sharma, Vishal; Singh, Jagjit: Model of adaptive WiMAX network incorporating diverse transforms [110]

Published in: Wireless Networks (2019)

Link to original: https://doi.org/10.1007/s11276-019-01945-4

If you want to cite the papers, please refer to the original.

For technical reasons we could not place the page where the original quote is coming from.

Abstract-Summary

"The physical layer of a WiMAX network is composed of orthogonal frequency-division multiplexing (OFDM) technology, which allows it to provide elevated data rates with minimum distortion in fading environments."

"Discrete wavelet transforms (DWTs) and fractional Fourier transforms (FrFTs) are used to augment the reliability and efficiency of an adaptive WiMAX system."

"The results show considerable improvements in both bit error rate (BER) reduction and spectral efficiency enhancement at a given signal-to-noise ratio (SNR)."

"The proposed adaptive algorithm realizes a spectral-efficient WiMAX system that enables the system to select the suitable transform strategy (DWT/FrFT) with minimum SNR requirements to achieve the target BER of 10^{-4}."

Introduction

"Wide area access technologies will soon be replaced by broadband wireless access (BWA) networks because of their simpler equalisation mechanisms, higher spectral efficiencies and resilience in multipath fading environments."

"The standardised version of BWA is WiMAX, which is also known as the IEEE 802.16 standard, and it has the capability to provide higher data rates to many users, even during high traffic scenarios."

"The various well-known versions of IEEE 802.16 are as follows [111]: (a) IEEE 802.16: Designed for use in line-of-sight (LOS) environments and operating in the 10–66 GHz frequency range."

"(b) IEEE 802.16a: Designed for use in non-line-of-sight (NLOS) environments and operating in the 2–11 GHz frequency range."

"(c) IEEE 802.16-2004: Described as the air interface for fixed broadband wireless access, and is also recommended by the IEEE for use in local and metropolitan networks."

Related Work

"Because of the application of the time-domain equalisation method, wavelet-based OFDM systems provide better BER performance over both additive white Gaussian noise (AWGN) channels and multipath Rayleigh fading channels [112]."

"Sharma and others [113] reported the utilization of wavelet transformation to analyse BER performance of OFDM based wireless system over AWGN channels only."

"Kansal and others [114] reported FFT and FrFT to analyse the BER performance of an OFDM system using M-PSK and M-QAM modulation techniques and claims reduction of SNR requirements to achieve a maximum BER of 10^{-3}."

"The proposed system established the performance of WiMAX system using DWT and FrFT over the rayleigh fading channel and exhibits a significant improvement of spectral efficiency and PAPR reduction."

"Kaur and others [115] established DWT based OFDM over fading environments for mobile WiMAX in terms of BER performance under the joint effect of path-loss, fading and noise without computing the spectral efficiency and PAPR reduction."

Model Description

"The transmitter side of the WiMAX physical layer involves the following processes: (i) Randomization: This is required to scramble the data to ensure that the effectiveness of the coding scheme can be optimised."

"The FEC process will add some redundancy to the data and this is directly responsible for the error reduction that occurs at the receiver end."

"Convolutional codes are mostly used in the WiMAX physical layer as FEC codes."

"In the WiMAX physical layer, block interleaving is used with a block size that is equal to the encoded block size in bits."

"The interleaving procedure dispenses the bits in the frequency, time, and time-frequency domains to achieve a minimum burst error rate at the receiver end."

"After this stage, the bits are rearranged by a block interleaver, and this is followed by FEC decoding to mitigate the redundancies that are present in the data."

Simulation Parameters and Channel Model

"The channel model that was used to evaluate the performances of the FFT-WiMAX, DWT-WiMAX, and FrFT-WiMAX systems was the Rayleigh fading channel."

"This is a statistical model that is used to approximate the effects of multipath fading on the radio signal, which assumes that no LOS path is available between the transmitter and the receiver."

"The Rayleigh fading model also states that the signal strength varies continuously based on a Rayleigh distribution."

"The received signal at the receiver side can be written as follows: Here, s(t) is the signal from the transmitter side, n(t) is the AWGN, h(t) represents the Rayleigh fading channel's impulse response, and '$\otimes$' is the convolution operator."

Proposed Methodology

"The proposed system simulations, as defined in Algorithm 1, operate on the principle of finding the optimum value of 'a', at which the BER performance is better than that for all other values of 'a'."

"The use of the FrFT in place of the FFT results in a significant improvement in the BER performance of the WiMAX system."

"Algorithm 2 represents the principle of using the optimum wavelet from the wavelet family to ensure the best possible BER performance."

"The use of wavelets has led to good enhancements in the BER performances of WiMAX systems."

Results and Discussion

"The analysis clearly shows that the FrFT based-WiMAX system performs better than the FFT based-WiMAX system."

"The DWT based-WiMAX system outperforms the system performance achieved when using both the FFT and the FrFT."

"The DWT-WiMAX system can perform better than the FFT-WiMAX and FrFT-WiMAX systems because the orthogonality between the subcarriers in the DWT-WiMAX system is maintained at a better level when compared with the FrFT-WiMAX and FFT-WiMAX systems [113, 116–118]."

"It is very much apparent from the preceding discussion that because the BER performance of the DWT-WiMAX system is better than that of the FFT-WiMAX and FrFT-WiMAX systems, the spectral efficiency of the DWT-WiMAX system will also be better than that of the FFT-WiMAX and FrFT-WiMAX systems."

"The reduction in the PAPR in the FrFT-WiMAX system case is greater when compared with that of the FFT-WiMAX system, but it is less than that of the DWT-WiMAX system."

Conclusions

"The overall performance outcome illustrates that the incorporation of diverse transforms, for instance, DWTs and FrFTs despite of conventional FFTs in WiMAX systems proffer significant improvement in terms of BER reduction, spectral-efficiency enhancement, and PAPR reduction."

"Further, the reported work realizes an adaptive WiMAX system equipped with both the potential transformation strategies."

"It has been observed that the use of FrFT with the optimum value of transform kernel 'a' improves the performance of WiMAX system."

Bibliography

1. Zhao, C., & Wu, G. (2023). On potential challenges of V2X sidelink relaying under interference: Link-level and system-level simulation with neural network assisted. *EURASIP Journal on Wireless Communications and Networking*. https://doi.org/10.1186/s13638-023-02244-5

2. 3GPP: Evolved universal terrestrial radio access (E-UTRA) and evolved universal terrestrial radio access network (E-UTRAN); overall description; stage 2. Technical Specification (TS) 36.300, 3rd Generation Partnership Project (3GPP) (December 2016). Version 14.1.0. Retrieved from https://portal.3gpp.org/desktopmodules/Specifications/SpecificationDetails.aspx?specificationId=2430
3. 3GPP: Study on NR sidelink relay. Technical Report (TR) 38.836, 3rd Generation Partnership Project (3GPP) (March 2021). Version 17.0.0. Retrieved from https://portal.3gpp.org/desktopmodules/Specifications/SpecificationDetails.aspx?specificationId=3725
4. Wang, D., Sattiraju, R. R., Qiu, A., Partani, S., & Schotten, H. D. (2019). Methodologies of link-level simulator and system-level simulator for C-V2X communication. In *2019 IEEE 2nd International Conference on Electronics and Communication Engineering (ICECE)* (pp. 178–184). IEEE.
5. Sattiraju, R., Wang, D., Weinand, A., & Schotten, H. D. (2020). Link level performance comparison of C-V2X and ITS-G5 for vehicular channel models. In *2020 IEEE 91st Vehicular Technology Conference (VTC2020-Spring)* (pp. 1–7). IEEE.
6. Anwar, W., Dev, S., Kumar, A., Franchi, N., & Fettweis, G. (2021). PHY abstraction techniques for V2X enabling technologies: Modeling and analysis. *IEEE Transactions on Vehicular Technology, 70*(2), 1501–1517.
7. Kavra, R., Gupta, A., & Kansal, S. (2023). Optimization of energy and delay on interval data based graph model of wireless sensor networks. *Wireless Networks*. https://doi.org/10.1007/s11276-023-03292-x
8. Akkaya, K., & Younis, M. (2004). Energy-aware delay constrained routing in wireless sensor networks. *International Journal of Communication Systems, 17*(6), 663–687.
9. Ammari, H. M., & Das, S. K. (2008). A trade-off between energy and delay in data dissemination for wireless sensor networks using transmission range slicing. *Computer Communications, 31*(9), 1687–1704.
10. Selvi, M., Velvizhy, P., Ganapathy, S., Nehemiah, H. K., & Kannan, A. (2019). A rule based delay constrained energy efficient routing technique for wireless sensor networks. *Cluster Computing, 22*(5), 10839–10848.
11. Sun, G., Zhao, L., Chen, Z., & Qiao, G. (2015). Effective link interference model in topology control of wireless ad hoc and sensor networks. *Journal of Network and Computer Applications, 52*, 69–78.
12. Li, X., Feng, H., Jiang, H., & Zhu, B. (2016). A polynomial time algorithm for finding a spanning tree with maximum number of internal vertices on interval graphs. *International Workshop on Frontiers in Algorithmics*, 92–101.
13. Kavra, R., Gupta, A., & Kansal, S. (2021). Interval graph based energy efficient routing scheme for a connected topology in wireless sensor networks. *Wireless Networks, 27*(8), 5085–5104.
14. Reyna-Orta, A., & Andrade, Á. G. (2021). Dimensionality reduction to solve resource allocation problem in 5G UDN using genetic algorithm. *Soft Computing*. https://doi.org/10.1007/s00500-020-05473-8
15. Chuang MC, Chen MC, S Y (2015) Resource management issues in 5g ultra dense small cell networks. In: 2015 International Conference on Information Networking (ICOIN), IEEE, https://doi.org/10.1109/icoin.2015.7057875
16. Martínez-Vargas, A., Domínguez-Guerrero, J., Andrade, Á. G., Sepúlveda, R., & Montiel-Ross, O. (2016). Application of NSGA-II algorithm to the spectrum assignment problem in spectrum sharing networks. *Applied Soft Computing, 39*, 188–198. https://doi.org/10.1016/j.asoc.2015.11.010
17. Bhardwaj, P., Panwar, A., Ozdemir, O., Masazade, E., Kasperovich, I., Drozd, A. L., Mohan, C. K., & Varshney, P. K. (2016). Enhanced dynamic spectrum access in multiband cognitive radio networks via optimized resource allocation. *IEEE Transactions on Wireless Communications, 15*(12), 8093–8106. https://doi.org/10.1109/twc.2016.2612627

18. Zhao, Z., Peng, Z., Zheng, S., & Shang, J. (2009). Cognitive radio spectrum allocation using evolutionary algorithms. *IEEE Transactions on Wireless Communications, 8*(9), 4421–4425. https://doi.org/10.1109/twc.2009.080939
19. Yuan Gao, H., & Long Cao, J. (2013). Non-dominated sorting quantum particle swarm optimization and its application in cognitive radio spectrum allocation. *Journal of Central South University, 20*(7), 1878–1888. https://doi.org/10.1007/s11771-013-1686-5
20. Marshoud, H., Otrok, H., Barada, H., Estrada, R., Jarray, A., & Dziong, Z. (2015). Realistic framework for resource allocation in macro–femtocell networks based on genetic algorithm. *Telecommunication Systems, 63*(1), 99–110. https://doi.org/10.1007/s11235-015-9976-x
21. Estrada, R., Otrok, H., & Dziong, Z. (2013). Resource allocation model based on particle swarm optimization for OFDMA macro-femtocell networks. In *2013 IEEE International Conference on Advanced Networks and Telecommunications Systems (ANTS)*. IEEE. https://doi.org/10.1109/ants.2013.6802850
22. Xu L, Mao Y, Leng S, Qiao G, Zhao Q (2017) Energy-efficient resource allocation strategy in ultra dense small-cell networks: A stackelberg game approach. In: 2017 IEEE International Conference on Communications (ICC), IEEE, https://doi.org/10.1109/icc.2017.7997289
23. Cao, J., Peng, T., Qi, Z., Duan, R., Yuan, Y., & Wang, W. (2018). Interference management in ultradense networks: A user-centric coalition formation game approach. *IEEE Transactions on Vehicular Technology, 67*(6), 5188–5202. https://doi.org/10.1109/tvt.2018.2799568
24. Fooladivanda, D., Daoud, A. A., & Rosenberg, C. (2011). Joint channel allocation and user association for heterogeneous wireless cellular networks. In *2011 IEEE 22nd international symposium on personal, indoor and mobile radio communications*. IEEE. https://doi.org/10.1109/pimrc.2011.6139988
25. Peng, C., Zheng, H., & Zhao, B. Y. (2006). Utilization and fairness in spectrum assignment for opportunistic spectrum access. *Mobile Networks and Applications, 11*(4), 555–576. https://doi.org/10.1007/s11036-006-7322-y
26. Yang, X. (2010). Engineering optimization: An introduction with metaheuristic applications. In X.-S. Yang (Ed.), *Chapter 11: Genetic algorithms* (pp. 171–180., ISBN: 9780470582466). https://doi.org/10.1002/9780470640425.ch11
27. Song, Q., Nuaymi, L., & Lagrange, X. (2019). Analysis of macro diversity based on maximum ratio combining in long range ALOHA networks. *Telecommunication Systems*. https://doi.org/10.1007/s11235-019-00579-3
28. Meloni, A., & Murroni, M. (2016). Interference calculation in asynchronous random access protocols using diversity. *Telecommunication Systems, 63*(1), 45–53.
29. Jung, P., Steiner, B., & Stilling, B. (1996). Exploitation of intracell macrodiversity in mobile radio systems by deployment of remote antennas. In *Proceedings of IEEE 4th international symposium on spread spectrum techniques and applications* (Vol. 1, pp. 302–307). IEEE.
30. Deva Priya, M., Sangeetha, M., Christy Jeba Malar, A., Dhivyaprabha, E., Kiruthiga, N., & Rajarajeswari, P. L. (2019). Fair adaptive cross-layer resource allocation scheme for IEEE 802.16 broadband wireless networks. *Wireless Personal Communications*. https://doi.org/10.1007/s11277-019-06929-3
31. Munoz, R. J., & Guice, R. J. (2004). *IEEE 802.16 commercial off the shelf (cots) technologies as a compliment to ship to objective maneuver (stom) communications*. (Doctoral dissertation). Naval Postgraduate School.
32. Chaudhry, S. B., & Guha, R. K. (2007). Adaptive connection admission control and packet scheduling for QoS provisioning in mobile WiMAX. In *Proceedings of IEEE international conference on signal processing and communications* (pp. 1355–1358).
33. Afzali, M., AbuBakar, K., & Lloret, J. (2019). Adaptive resource allocation for WiMAX mesh network. *Wireless Personal Communications, 107*(2), 849–867.
34. Shih, K. P., Chen, H. C., Chiang, C. T., & Hsieh, T. H. (2010). Channel-aware subchannel renumbering and downlink burst allocation for IEEE 802.16 OFDMA systems. In *Proceedings of IEEE wireless communication and networking conference* (pp. 1–6).

35. Chen, H. C., Shih, K. P., Chiang, C. T., & Chen, C. L. (2011). A subchannel-aware burst fragmentation, packing and scheduling (BFPS) algorithm for downlink traffic in IEEE 802.16 OFDMA systems. In *Proceedings of 7th IEEE international wireless communications and mobile computing conference* (pp. 1141–1146).
36. Hailan, A., Al-Janabi, H. D. K., & Hojayev, O. (2015). Research of factors influencing the subchannel allocation to subscriber stations in WiMAX.
37. Liang, J. M., Chen, J. J., Wang, Y. C., & Tseng, Y. C. (2011). A cross-layer framework for overhead reduction, traffic scheduling, and burst allocation in IEEE 802.16 OFDMA networks. *IEEE Transactions on Vehicular Technology, 60*(4), 1740–1755.
38. Shao, H., Sun, Y., Du, Z., Cai, J., & Duan, Z. (2022). QoE-oriented resource allocation for dense cloud NOMA smallcell networks. *Wireless Networks*. https://doi.org/10.1007/s11276-022-02929-7
39. Cui, J., et al. (2018). QoE-based resource allocation for multi-cell NOMA networks. *IEEE Transactions on Wireless Communications, 17*(9), 6160–6176.
40. Sun, Y., Du, Z., Xu, Y., et al. (2018). Directed-hypergraph-based channel allocation for ultra-dense cloud D2D communications with asymmetric interference. *IEEE Transactions on Vehicular Technology, 67*(8), 7712–7718.
41. Xu, Y., Wu, Q., et al. (2013). Opportunistic spectrum access with spatial reuse: Graphical game and uncoupled learning solutions. *IEEE Transactions on Wireless Communications, 12*(10), 4814–4826.
42. Chen, C., Wang, B., Zhang, R., et al. (2019). Interference hypergraph-based resource allocation (IHG-RA) for NOMA-integrated V2X networks. *IEEE Internet of Things Journal, 6*(1), 161–170.
43. Karaca, H. M. (2020). Retracted article: Throughput maximization of multichannel allocation mechanism under interference constraint for hybrid overlay/underlay cognitive radio networks with energy harvesting. *Wireless Networks*. https://doi.org/10.1007/s11276-020-02305-3
44. Lee, S., Huang, K., & Zhang, R. (2012). Cognitive energy harvesting and transmission from a network perspective. In *Proceedings of IEEE international conference on communication systems (ICCS)* (pp. 225–229).
45. Plata, D. M. M., & Reátiga, A. G. A. (2012). Evaluation of energy detection for spectrum sensing based on the dynamic selection of detection-threshold. *Procedia Engineering, 35*, 135–143.
46. Stoopman, M., Keyrouz, S., Visser, H. J., Philips, K., & Serdijn, W. A. (2014). Co-design of a CMOS rectifier and small loop antenna for highly sensitive RF energy harvesters. *IEEE Journal Solid-State Circuits, 49*(3), 622–634.
47. Zhang, Y., Han, W., Li, D., Zhang, P., & Cui, S. (2015). Power versus spectrum 2-D sensing in energy harvesting cognitive radio networks. *IEEE Transactions Signal Processing, 63*(23), 6200–6212.
48. Lee, S., Zhang, R., & Huang, K. (2013). Opportunistic wireless energy harvesting in cognitive radio networks. *IEEE Transactions on Wireless Communications, 12*(9), 4788–4799.
49. Karaca, H. M., Kurt, T., Dicle, S. Z., & Anarim, E. (2013). Auction-based throughput maximization in cognitive radio networks under interference constraint. *Wireless Personal Communications, 72*, 1259–1275.
50. Hamdoun, S., Rachedi, A., & Ghamri-Doudane, Y. (2020). Graph-based radio resource sharing schemes for MTC in D2D-based 5G networks. *Mobile Networks and Applications*. https://doi.org/10.1007/s11036-020-01527-1
51. Hamdoun, S., Rachedi, A., & Ghamri-Doudane, Y. (2015). Radio resource sharing for MTC in LTE-a: An interference-aware bipartite graph approach. In *IEEE Global Communication Conference (IEEE Globecom)*.
52. Rachedi, A., & Hasnaoui, A. (2015). Advanced quality of services with security integration in wireless sensor networks. *Wireless Communications and Mobile Computing, 15*(6), 1106–1116.

53. Ku, M.-L. (2012). *Cognitive radio and interference management: Technology and strategy: Technology and strategy*. IGI Global.
54. Ksentini, A., Hadjadj-Aoul, Y., & Taleb, T. (2012). Cellular-based machine-to-machine: Overload control. *Network, IEEE, 26*(6), 54–60.
55. Cho, J.-w., & Chong, S. (2005). Stabilized max-m in flow control using PID and PII 2 controllers. *IEICE Transactions on Communications, 88*(8), 3353–3364.
56. Hamdoun, S., Rachedi, A., Tembine, H., & Ghamri-Doudane, Y. (2016). Efficient transmission strategy selection algorithm for M2M communications: An evolutionary game approach. In *2016 IEEE 15th International Symposium on Network Computing and Applications (NCA)* (pp. 286–293). IEEE.
57. Min, H., Lee, J., Park, S., & Hong, D. (2011). Capacity enhancement using an interference limited area for device-to-device uplink underlaying cellular networks. *IEEE Transactions on Wireless Communications, 10*(12), 3995–4000.
58. Belleschi, M., Fodor, G., & Abrardo, A. (2011). Performance analysis of a distributed resource allocation scheme for D2D communications. In *GLOBECOM workshops (GC Wkshps)* (pp. 358–362). IEEE.
59. Qianxi, L., Qingyu, M., Fodor, G., & Brahmi, N. (2013). Radio resource management for network assisted D2D in cellular uplink. In *2013 IEEE/CIC International Conference on Communications in China-Workshops (CIC/ICCC)* (pp. 159–164). IEEE.
60. Yu, G., Xu, L., Feng, D., Yin, R., Li, G. Y., & Jiang, Y. (2014). Joint mode selection and resource allocation for device-to-device communications. *IEEE Transactions on Communications, 62*(11), 3814–3824.
61. Dvir, A., Haddad, Y., & Zilberman, A. (2019). The controller placement problem for wireless SDN. *Wireless Networks*. https://doi.org/10.1007/s11276-019-02077-5
62. Abdel-Rahman, M. J., Mazied, E. A., MacKenzie, A., Midkiff, S., Rizk, M. R., & El-Nainay, M. (2017). On stochastic controller placement in software-defined wireless networks. In *2017 IEEE Wireless Communications and Networking Conference (WCNC)* (pp. 1–6). https://doi.org/10.1109/WCNC.2017.7925942
63. Abdel-Rahman, M. J., Mazied, E. A., Teague, K., MacKenzie, A. B., & Midkiff, S. F. (2017). Robust controller placement and assignment in software-defined cellular networks. In *In 2017 26th International Conference on Computer Communication and Networks (ICCCN)* (pp. 1–9).
64. Akyildiz, I. F., Lin, S. C., & Wang, P. (2015). Wireless software-defined networks (W-SDNs) and network function virtualization (NFV) for 5G cellular systems: An overview and qualitative evaluation. *Computer Networks, 93*, 66–79.
65. Liu, J., Shi, Y., Zhao, L., Cao, Y., Sun, W., & Kato, N. (2018). Joint placement of controllers and gateways in SDN-enabled 5G-satellite integrated network. *IEEE Journal on Selected Areas in Communications, 36*, 221–232.
66. Tanha, M., Sajjadi, D., Ruby, R., & Pan, J. (2018). Capacity-aware and delay-guaranteed resilient controller placement for software-defined WANs. *IEEE Transactions on Network and Service Management, 15*, 991–1005.
67. Johnston, M., & Modiano, E. (2017). Controller placement in wireless networks with delayed CSI. *IEEE/ACM Transactions on Networking, 25*(3), 1775–1788. https://doi.org/10.1109/TNET.2017.2651808
68. Dhilipkumar, S., & Arunachalaperumal, C. (2021). Performance Evaluation of Interference Mitigation Scheme with D2D Users in Wireless Networks. *Wireless Personal Communications*. https://doi.org/10.1007/s11277-021-08413-3
69. Kim, J., Karim, N. A., & Cho, S. (2017). An interference mitigation scheme of device-to-device communications for sensor networks underlying LTE-A. *Journal on Sensors, 17*(5), 1088–1105.
70. Andrews, J. G. (2005). Interference cancellation for cellular systems: A contemporary overview. *IEEE Wireless commuications, 12*, 19–29.

71. Helena, W., Sung-Yeop, P., & Dong-Ho, C. (2010). *Interference mitigation based on femtocells grouping in low duty operation*. 978–1–4244–3574–6/10/$25.00. IEEE.
72. Huan, S., Linling, K., & Jianhua, L. (2009). Interference avoidance in OFDMA-based femtocell network. In *IEEE Youth conference on information, computing and telecommunication*. 978–1–4244–5076–3/09/$26.00. IEEE.
73. Ningombam, D. D., Lee, C. G., & Shin, S. (2019). *Interference mitigation for multicast D2D communications underlay cellular networks*. 978–1–5386–7822–0/19/$31.00. IEEE.
74. Subra, S. S., & Hazra, R. (2020). Interference mitigation methods for D2D communciations in 5G network.Cognitive informatics and soft computing. In *Advances in intelligent systems and computing* (Vol. 1040). Springer.
75. Yin, R., Yu, G., Zhang, H., Zhang, Z., & Li, G. Y. (2015). Pricing-based interference coordination for D2D communications in cellular networks. *IEEE Transactions on Wireless communications., 14*(3), 1519–1532.
76. Yang, C., Li, J., Sheng, M., Anpalagan, A., & Xiao, J. (2017). Mean field game-theoretic framework for interference and energy-aware control in 5G ultra-dense networks. *IEEE Wireless Communications, 25*(1), 114–121.
77. Yang, C., Li, J., Semasinghe, P., Hossain, E., Perlaza, S. M., & Han, Z. (2016). Distributed interference and energy-aware power control for ultra-dense D2D networks: A mean field game. *IEEE Transactions on Wireless Communication, 16*(2), 1205–1217.
78. Mkiramweni, M. E., Yang, C., Li, J., & Han, Z. (2018). Game-theoretic approaches for wireless communications with unmanned aerial vehicles. *IEEE Wireless Communications, 25*(6), 104–112.
79. Zhou, Z., Dong, M., Ota, K., Shi, R., Liu, Z., & Sato, T. (2015). Game-theoretic approach to energy- efficient resource allocation in device-to-device underlay communications. *IET Communications, 9*(3), 375–385.
80. AlQerm, I., & Shihada, B. (2016). A cooperative online learning scheme for resource allocation in 5G systems. *IEEE International Conference on Communications (ICC)*, 1–7.
81. Marttin, V., Yüzgeç, U., Bayılmış, C., & Küçük, K. (2021). A novel approach to D2D Discovery in PSN for post-disaster: Throughput based Discovery Algorithm (TDA). *Wireless Personal Communications*. https://doi.org/10.1007/s11277-021-08407-1
82. Osman, E. A. M. (2018). *Device discovery methods in D2D communications for 5G communications system*. Tallinn University of Technology.
83. Doppler, K., Rinne, M., Wijting, C., Ribeiro, C. B., & Hugl, K. (2009). Device-to-device communication as an underlay to LTE-advanced networks. *IEEE Communications Magazine, 47*(12), 42–49.
84. Yang, K., Wang, M., Zou, K., Hua, M., Hu, J., Zhang, J., Sheng, W., & You, X. (2014). Device discovery for multihop cellular networks with its application in LTE. *IEEE Wireless Communications, 21*(5), 24–34. https://doi.org/10.1109/MWC.2014.6940430
85. Dhilip Kumar, V., Kanagachidambaresan, G. R., Chyne, P., & Kandar, D. (2021). Extended communication range for autonomous vehicles using hybrid DSRC/WiMAX technology. *Wireless Personal Communications*. https://doi.org/10.1007/s11277-021-09242-0
86. Remy, G., Senouci, S. M., Jan, F., & Gourhant, Y. (2011). LTE4V2X: LTE for a centralized VANET organization. In *Proceedings of the IEEE global telecommunications conference (GLOBE-COM 2011)* (pp. 1–6). IEEE Computer Society. https://doi.org/10.1109/GLOCOM.2011.6133884
87. Kim, S. E., Kim, H. L., Jin, J. S., & Lee, S. C. (2009). A new approach for Vehicle Accident Prevention on the highway using Mobile WiMAX. In *IEEE Transactions on Vehicular Technology, Wireless technology Department, Central R&D Laboratory, KT* (p. 17).
88. Mangel, T., Kosch, T., & Hartenstein, H. (2010, December), A comparison of UMTS and LTE for vehicular safety communication at intersections. In *Vehicular Networking Conference (VNC)* (pp. 293–300) IEEE.
89. Hayashi, M., Fukuzawa, S., Ichikawa, H., Kawato, T., Yamada, J., Tsuboi, T., Matsui, S., & Maruyama, T. (2007). Development of vehicular communication (WAVE) system for safety

applications. In *Telecommunications. ITST'07. 7th International conference on ITS, June, 2007* (pp. 1–5). IEEE.

90. Song, W. J., Chung, J. M., Lee, D., Lim, C., Choi, S., & Yeoum, T. (2009). Improvements to seamless vertical handover between mobile WiMAX and 3GPP UTRAN through the evolved packet core. *IEEE Communications Magazine, 47*(4).
91. Yap, K. K., Kobayashi, M., Sherwood, R., Huang, T. Y., Chan, M., Handigol, N., & McKeown, N. (2010). OpenRoads: Empowering research in mobile networks. *ACM SIGCOMM Computer Communication Review, 40*(1), 125–126.
92. Khan, A., Sadhu, S., & Yeleswarapu, M. (2009). *A comparative analysis of DSRC and 802.11 over vehicular Adhoc networks* (pp. 1–8). Project Report, Department of Computer Science, University of California.
93. Andrews, Jerey G., Arunabha Ghosh, & Rias Muhamed (2007). *Fundamentals of WIMAX: Understanding Broadband wireless networking no.57.*
94. Mehta, V., & Gupta, D. N. (2012). Performance analysis of qos parameters for wimax networks. *International Journal of Engineering and Innovative Technology (IJEIT), 1*, 105–110.
95. Anwar, F., Azad, M. S., Rahman, M. A., & Uddin, M. M. (2008). Performance analysis of Ad hoc routing protocols in mobile WiMAX environment. *IAENG International Journal of Computer Science, 35*(3).
96. Al-Sherbaz, A., Jassim, S., & Adams, C. (2009). *WiMAX- WiFi convergence with OFDM bridge*. Society of Photo Optical Instrumentation Engineers (SPIE).
97. Papadimitratos, P., De La Fortelle, A., Evenssen, K., Brignolo, R., & Cosenza, S. (2009). Vehicular communication systems: Enabling technologies, applications, and future outlook on intelligent transportation. *IEEE Communications Magazine, 47*(11), 84.
98. Lee, W.-H., Tseng, S.-S., & Shieh, W.-Y. (2010). Collaborative real-time traffic information generation and sharing framework for the intelligent transportation system. *Information Sciences, 180*(1), 62–70.
99. Li, Y. J. (2010). An overview of the DSRC/WAVE technology. In *International conference on heterogeneous networking for quality, reliability, security and robustness* (pp. 544–558). Springer.
100. Dedicated Short Range Communications Report and Order, U.S. Federal Communications Commission, R&O FCC 03–324, Dec. 17
101. IEEE Standard for Information Technology—Telecommunications and Information Exchange between Systems-Local and Metropolitan Area Networks-Specific Requirements (2010). Part 11: Wireless LAN Medium Access Control (MAC) and Physical Layer (PHY) Specifications; Amendment 6: Wireless Access in Vehicular Environments, IEEE Std. 802.11p, Jul. 2010.
102. Asif, Saad Z (2010). *OFDM and OFDMA. Next generation mobile communications ecosystem: Technology management for mobile communications* (pp. 7–17). Wiley Online Library.
103. So-In, C., Jain, R., & Tamimi, A.-K. (2009). Scheduling in IEEE 802.16 e mobile WiMAX networks: Key issues and a survey. *IEEE Journal on Selected Areas in Communications, 27*(2), 156–171.
104. Kandar, D., Kumar, V. D., & Chyne, P. (2016). OFDM simulation in combined WiMAX/DSRC mechanism for intelligent transportation system. In *Emerging research in computing communication and applications* (pp. 495–502). Singapore: Springer.
105. Wang, F., Ghosh, A., Sankaran, C., Fleming, P. J., Hsieh, F., & Benes, S. J. (2008). Mobile WiMAX systems: Performance and evolution. *IEEE Communications Magazine, 46*(10), 41.
106. Hsieh, F., Wang, F., & Ghosh, A. (2008). Link performance of WiMAX PUSC. In *Wireless communications and networking conference, 2008. WCNC 2008* (pp. 1143–1148). IEEE.
107. Mach, P., & Bestak, R. (2007). WiMAX throughput evaluation of conventional relaying. In *Personal wireless communications* (pp. 75–86). Springer.
108. Rajeshkumar, V., & Sivakumar, P. (2013). Comparative study of AODV, DSDV and DSR routing protocols in MANET using network simulator-2. *International Journal of Advanced Research in Computer and Communication Engineering, 2, no. 12.*

109. Hasan, M. S., Harding, C., Yu, H., & Grifiths, A. (2005). Modeling delay and packet drop in networked control systems using network simulator NS2. *International Journal of Automation and Computing, 2*(2), 187–194.
110. Kansal, L., Sharma, V., & Singh, J. (2019). Model of adaptive WiMAX network incorporating diverse transforms. *Wireless Networks*. https://doi.org/10.1007/s11276-019-01945-4
111. Part 11: Wireless LAN medium access control (MAC) and physical layer (PHY) specifications: High-speed physical layer in the 5 GHz Band, IEEE Standard 802.11a 1999. Retrieved from https://ieeexplore.ieee.org/document/817038/
112. Asif, R., Abd-Alhameed, R. A., Oanoh, O., Dama, Y., Migdadi, H. S., Noars, J. M., Hussaini, A. S., & Rodriquez, J. (2012). Performance comparison between DWT-OFDM and FFT-OFDM using time domain zero forcing equalization. In *2012 international conference on telecommunication and multimedia* (pp. 175–179).
113. Sharma, V., & Singh, G. (2014). On BER assessment of conventional- and wavelet-OFDM over AWGN channel. *Optik, 125*, 6071–6073.
114. Kansal, A., Singh, K., & Saxena, R. (2015). Bit error rate analysis of FrFT appended OFDM system. *Optik, 126*, 715–718.
115. Kaur, H., Kumar, M., Sharma, A. K., & Singh, H. P. (2016). Performance analysis of DWT based OFDM over fading environments for mobile WiMAX. *Optik, 127*(2), 544–547.
116. Kumbasar, V., & Kucur, O. (2012). Performance comparison of wavelet based and conventional OFDM systems in multipath Rayleigh fading channels. *Digital Signal Processing, 22*, 841–846.
117. Kansal, L., Sharma, V., & Singh, J. (2016). BER assessment of FFT-OFDM against WHT-OFDM over different fading channels. *Wireless Networks*. https://doi.org/10.1007/s11276-016-1283-2
118. Kansal, L., Sharma, V., & Singh, J. (2016). Performance evaluation of FFT-WiMAX against WHT-WiMAX over Rayleigh fading channel. *Optik, 127*, 4514–4519.

Chapter 5
Error Correction Coding

Milind Pande, Anand J. Kulkarni, and Apoorva S. Shastri

Introduction by the Editor

One of the challenges in the case of data storage is data corruption. To avoid data corruption while transmitting the message Error Correcting Codes (ECC) are used. ECC is an encoding technique that helps to protect data while transmitting the message as a binary number. The message can be recovered even if some bits are mistakenly flipped. This chapter discusses several contributions associated with turbo codes in short block lengths, binary turbo coded BICM scheme for JPWL image transmission, Modified Viterbi Decoder with Convolution Codes, Flip Left-to-Right Approach Based Inverse Tree, Combined Double Binary Turbo Coding and Colour Shift Keying Technique, Improvement of CDMA using Raptor Codes. Several low-cost devices cannot afford to run high-power forward error control coding algorithms, such as Turbo codes. To overcome this Patching Based Extra Short Packet Forward Error Control Coding is also discussed. Additionally, Parallel Concatenation of LDPC Coded SISO-GFDM System, iterative detection and decoding, and Iterative Reception Algorithms for Product Codes are also explored.

The interleaver generates decorrelation in the external input information to parallel recursive convolutional encoders, with sufficient data propagation to reduce the Bit Error Rate (BER). A basic turbo encoder consists of two Recursive Systematic Convolutional (RSC) encoders in parallel are used, and the second encoder is preceded by an interleaver. Each constituent encoder is terminated independently by

M. Pande
Dr. Vishwanath Karad MIT World Peace University, Pune, Maharashtra, India
e-mail: milind.pande@mitwpu.edu.in

A. J. Kulkarni · A. S. Shastri (✉)
Institute of Artificial Intelligence, Dr. Vishwanath Karad MIT World Peace University, Pune, Maharashtra, India
e-mail: anand.j.kulkarni@mitwpu.edu.in

M. Pande et al. (eds.), *Optimization Methods in Mobile Communication Systems*, https://doi.org/10.1007/978-981-95-1810-4_5

tail bits. For a rate one-half constituent encoder, the block interlaces the three streams—the systematic and parity bits stream of the first encoder and only the parity bit streams of the second encoder—and multiplexes the tail bits to the end of the encoded data streams. The first set of six bits that is three systematic bits, and three parity bits are the output tail bits of the first constituent encoder. The interleaver helps to optimize distance of the turbo codes and their performance. A good propagation property of the S-random interleaver provides significantly improves BER performance compared to the random interleaver. To reduce the correlation between the extrinsic information of the first encoder and the data stream of information in the input of the second decoder, an S-random circular displacement is incorporated into the output vector of the interleaver. Performance of the turbo code is dependent on interleaver and the typical S-Random interleaver. A recently developed novel interleaver i.e. 'flip left–right approach-based inverse-tree interleaver' (FLRITI) or simply 'inverse tree interleaver', has been explored for two unconventional integrated interleave division multiple access techniques i.e. single carrier frequency division multiple access cum interleave division multiple access (SCFDMA-IDMA) and orthogonal frequency division multiplexing based interleave division multiple access (OFDM-IDMA). This unconventional integration of CIDMA with SCFDMA and OFDM techniques in presence of FLRITI improves the overall system performance in terms of bit-error rate, memory footprint and computation complexity. According to the channel state information (CSI) knowledge at the transmitter side, to maximize the visual quality of the received JPWL images. It is required to determines the number of sub-channels that can be used and, for each sub-channel, optimal values of the modulation order, the rate of the channel coder and the source bit-rate. The optimization algorithm defines the power needed to achieve the target BER permitting to use a given BICM configuration, the configuration with the highest channel coding rate, the highest modulation order and the smallest number of iterations is favoured in the current sub-channel. To maximize the number of bits allocated for the source coding and to minimize the computational time in the turbo decoding, this ranking is carried out such as the configuration with a higher modulation order, higher turbo code rate and less iteration number is favoured.

The existing real-time routing system involves multiple hops with time varying mobility channels for the fastest data propagation is greatly degraded with power utilization factor through congestion traffic queue. Viterbi decoder with convolution codes maximizes throughput with minimum delay the QoS in real-time data communication. With IEEE 802.11 WLAN physical layers to enhance performance, it is tricky to select desired transmission rate by engaging various modulations and channel coding schemes. Each node needs to be capable of choosing the optimized transmission rate so that the system performance can be improved with very minimum power with high packet transmission ratio with minimum traffic rate to improve QoS.

Visible Light Communication (VLC) with light emitting diodes (LEDs) used as wireless access points and modulated at high speeds invisible to the human eye, is based on the dual use of LEDs for lighting and communication purposes. The spectral distribution of the flickering effect is induced by a light stimulator that fluctuates over time and causes the impression of imbalance in the visual sense, and it is mostly caused by the modulation technique used in the physical layer (PHY) during data transmission in VLC systems. Double binary Turbo coding (DBTC) technique with color shift keying (CSK) modulation, is a solution to the flicker problem of VLC systems. DBTC-CSK method is tested on optical OFDM systems from multi-carrier wireless communication systems and has better BER performances.

Spread spectrum-based code division multiple access (CDMA) has played an important role in cellular and personal communication systems. Direct-sequence code-division multiple-access (DS-CDMA), the most common type, In DS-CDMA technology, each user is given a specific code, which is crucial for modulating the signal as it affects how well the system works. A critical issue in traditional DS-CDMA systems is how to reduce multiple access interference (MAI) from other users. To achieve good transmission performance in digital communication systems, effective channel coding is required. Raptor coding improves the performance of the DS-CDMA multi-user system. It is a robust method in which accurate data gets retrieved though packet loss occurs.

Parallel concatenation of low-density parity-check (PC-LDPC) codes in a generalized frequency division multiplexing (GFDM) system. PC-LDPC codes are obtained by dividing the long and high complexity single LDPC codes into small two lower complexity codes, and these designed codes are applied to the 5G-GFDM waveform. Since the GFDM signal transmits data in both the time and frequency domain, these PC-LDPC codes can deal with two-dimensional errors. PC-LDPC code outperforms LDPC code with a coding gain of 2 dB.

Product codes are low-density codes based on finite Euclidean geometry and finite projective geometry. By using mathematical properties of geometry these codes offer efficient error correction and reduce complexity. Iterative reception algorithm uses massage passing repeatedly to refine estimates of transmitted data to improve accuracy in data communication. In this way, it helps to improve the reliability of data transmission on noisy channels and also improve data integrity in storage devices.

Machine Generated Summaries

Disclaimer: The summaries in this chapter were generated from Springer Nature publications using extractive AI auto-summarization: An extraction-based summarizer aims to identify the most important sentences of a text using an algorithm and uses those original sentences to create the auto-summary (unlike generative AI). As the constituted sentences are machine selected, they may not fully reflect the body of the work, so we strongly advise that the original content is read and cited. The auto generated summaries were curated by the editor to meet Springer Nature publication standards. To cite this content, please refer to the original papers.

Machine generated keywords: code, turbo, turbo code, block, reception, convolutional, convolutional code, class, channel code, code rate, window, construction, modulation, decoder, ofdm signal.

Design of an Interleaver with Criteria to Improve the Performance of Turbo Codes in Short Block Lengths [1]

This is a machine-generated summary of:

Urrea, Claudio; Kern, John; López-Escobar, Ricardo: Design of an interleaver with criteria to improve the performance of turbo codes in short block lengths [1]

Published in: Wireless Networks (2022)

Link to original: https://doi.org/10.1007/s11276-022-02912-2

If you want to cite the papers, please refer to the original.

For technical reasons we could not place the page where the original quote is coming from.

Abstract-Summary

"This paper presents the design of an interleaver for short Parallel Turbo Codes (PTC) with short block lengths in the order of 64, 128, and 256 bits."

"The objective is to design an interleaver that generates decorrelation in the external input information to parallel recursive convolutional encoders, with sufficient data propagation to reduce the Bit Error Rate (BER)."

"An excellent reduction in BER is achieved for the three short block lengths, according to the quality indicator defined by the International Telecommunication Union (ITU)."

Introduction

"The Code-matched design in [2] is validated with a block length of 1024 bits, against comparative testing with the S-random and X-random interleavers."

"The bit error rate and the frame error rate performance of turbo codes are improved, and the error floor is reduced considerably."

"The construction of interleavers in [3] considers a Parallel Turbo Code (PTC) design with block lengths in the order of 64 and 128 bits."

"A design of an S-random interleaver is presented with the purpose of applying criteria to obtain an improvement in turbo code performance in data lengths of 64, 128, and 256 bits."

"Based on the interleaver, the performance of the turbo code is compared, assessing the results of the proposed interleaver and the typical S-Random interleaver, while maintaining the characteristics of an iterative encoder and decoder."

Turbo Codes

"A basic turbo encoder is shown in Figure, in which two Recursive Systematic Convolutional (RSC) encoders in parallel are used, and the second encoder is preceded by an interleaver."

"Each constituent encoder is terminated independently by tail bits."

"For a rate one-half constituent encoder, the block interlaces the three streams—the systematic and parity bits stream of the first encoder and only the parity bit streams of the second encoder—and multiplexes the tail bits to the end of the encoded data streams."

"The first set of six bits (three systematic bits and three parity bits) are the output tail bits of the first constituent encoder."

Criteria Considered in the Development of the Interleaver

"The interleaver is key for the minimum distance of the turbo codes and thus for their performance [4, 5]."

"Interleaver parameters are selected so that the minimum distance of the code is maximized, and the S-random interleaver that separates adjacent input symbols after interleaving is used."

"A good propagation property of the S-random interleaver provides significantly improves BER performance compared to the random interleaver [6]."

"The concept of propagation to design an S-random interleaver was proposed in [7]."

"To reduce the correlation between the extrinsic information of the first encoder and the data stream of information in the input of the second decoder, an S-random circular displacement is incorporated into the output vector of the interleaver."

Experimental Testing of the Interleaver

"It is configured so that its parameters correspond to the values of the Eb/No ratio, interleaver indexes, the length of the code block, and the number of decoding iterations."

"The sampling time used in this block is the same as the sampling time in its input signal."

"Error Rate Calculation: This block enables the comparison of the input data of a transmitter with the input data of a receiver."

"Model Parameters: This block incorporates the three parameters used in the simulation of the communication system, i.e., the Eb/No ratio, the length of the code block, and the number of decoding iterations."

"Turbo Encoder: This block enables encoding a binary input signal using a parallel concatenated encoding scheme that uses two identical convolutional encoders and one internal interleaver provided as a parameter."

"Turbo Decoder: This block enables decoding the input signal using a parallel concatenated decoding scheme."

Conclusions and Future Work

"We have presented the design and construction of an S-random interleaver with circular propagation constraints and circular displacement."

"The experimental results obtained showed that, with the use of a classic S-random interleaver in short block frames, the performance of the turbo codes degrades, and this degradation increases as the block size decreases."

"Both applied criteria allowed for reducing the BER of the turbo code performance with the classical S-random and the improved S-random interleaver."

"Performance improvement is only compared to the performance of the same S-random interleaver."

Enhanced Adaptive Code Modulation for Rainfall Fade Mitigation in Ethiopia [8]

This is a machine-generated summary of:

Woldamanuel, Eyob Mersha; Diba, Feyisa Debo: Enhanced adaptive code modulation for rainfall fade mitigation in Ethiopia [8]

Published in: EURASIP Journal on Wireless Communications and Networking (2022)

Link to original: https://doi.org/10.1186/s13638-021-02085-0

If you want to cite the papers, please refer to the original.

For technical reasons we could not place the page where the original quote is coming from.

Abstract-Summary

"Since the average raindrop size is bigger and the rainfall rate is high in magnitude in tropical regions than that of non-tropical areas, this prediction model is not suitable for the measured rain data."

"A rain fade mitigation technique based on local rain data has not been adequately studied in tropical regions."

"Work, locally collected 1-min rain rate data is used to determine the rain attenuation."

"A comparison of the performance of this proposed scheme is with the non-adaptive technique, and fuzzy-based adaptive modulation and coding technique is carried out."

"MATLAB simulation result showed that lower-order quadrature amplitude modulation (QAM) scheme with a lower convolutional coding rate is better in maintaining link availability in bad weather conditions."

"Spectral efficiency is improved by utilizing a larger constellation size of quadrature amplitude modulation (QAM) scheme with a higher convolutional coding rate when the channel is not affected by rain."

Introduction

"ITU-R recommends to consider rain attenuation in radio link design at a frequency of operation above 5-GHz [9], rain fades is quite noticed in tropical equatorial and temperate climates for a frequency spectrum higher than 7 and 10-GHz, respectively [10, 11]."

"Recent researches also depict that for radio links operating at a frequency higher than 10-GHz rainfall is the dominant factor that results in deep fading and network outage [12–16]."

"A rain fade mitigation technique based on local rain data has not been adequately studied in equatorial and tropical countries."

"A rain attenuation countermeasure based on the local rain data model must be investigated for frequency spectrum above 10-GHz radio links."

"Investigation of the mitigation technique of rain fade on microwave and millimeter-wave LOS terrestrial communication in Ethiopia based on local data is the main motivation for this proposed work."

"The adapting capability of soft computing-based ACM technique to mitigate the effects of rain fade in single-frequency carrier communication link has not been adequately studied."

Methodology

"A method used to calculate rain attenuation and neuro-fuzzy-based ACM is presented."

"A rainfall attenuation based on locally recorded rain data is determined and used as input to mitigate the rain fade."

"According to ITU-R rain attenuation prediction procedures [9], the efficacious way of determining cumulative rainfall distribution is done by making a direct measurement at the site of interest."

"Work, the ITU-R rain attenuation model is used to determine rain fade."

"ITU-R P.530-17 Model The ITU-R P.530-17 [9] gives a simple procedure that may be used for predicting the long-term data of rain attenuation."

"The computation is based on the values given in the link profile and the result from rain attenuation determination using ITU-R model."

"The received signal is additionally degraded by the total rain attenuation amount."

Implementation of Adaptive Coding and Modulation

"Using this hybrid soft computing method, an initial fuzzy logic model with its input parameters is first obtained from the input-output data of the BER performance simulation of different modulation-code pairs."

"The steps involved in the ANFIS training process are: (1) Loading the I/O training data; (2) Generate an initial fuzzy inference system model; (3) View the FIS model structure; (4) Select the FIS model optimization method (hybrid method); (5) Choose the training epochs and training error tolerances; (6) Train ANFIS and view adjusted membership functions and the output surface."

"In other word, the multiplication of the number of membership function of the four input parameters (SNR, BER, coding rate, and modulation order) result in 225 fuzzy linguistic rules."

"The proposed fuzzy sets have three membership function for the input variables SNR and BER designated as low, medium and high and five membership function for the input variable modulation order (MOD) and code rate represented as very low, low, medium, high, and very high."

Results and Discussion

"Even if there is a degradation of the received signal when there is rain in between this link, it can achieve a high data rate in a rain rate below 20-mm/h. The spectral efficiency with various SNR ranges for different modulation and coding techniques over the AWGN channel is presented in this section."

"In other word, 512 QAM and 1024 QAM with coding rate 2/3 and 3/4 is used when the rain rate is less (i.e. 0–25 mm/h) to achieve a high spectral efficiency and high QoS. By using a medium constellation size modulation order 64, 128, and 256 QAM, with coding rate 1/2 and 2/3, the link availability is sustained and achieve a moderate spectral efficiency in the channel condition when the rain rate is between 30 and 80 mm/h. However, while the rain rate is increased above 80 mm/h the adaptive coding and modulation system is forced to use a lower order modulation scheme (i.e. 32, 16, and 8 QAM) to keep the wireless link alive by reducing the data rate."

Conclusion

"The performance of neuro-fuzzy-based ACM systems in terms of spectral efficiency and BER using various coding rates and QAM modulation schemes under the AWGN channel was analyzed and compared to fixed and adaptive techniques."

"Whereas, in rainy time (bad channel condition), the more robust modulation schemes like 4, 8, and 16 QAM (lower modulation order QAM) are better in sustaining link availability."

"In the rainy condition, it is better to use a lower order modulation with a lower coding rate."

"At a rain rate above 105 mm/h, the signal to noise ratio level is reduced to around 6 dB. This implies that using high order modulation scheme, which demands high SNR, such as 128, 256, 512, and 1024 QAM along with high code rate like 2/3 and 3/4 result in link reliability reduction."

Shaping Duo Binary Turbo-Coded BICM Scheme for JPWL Image Transmission Using a Link Adaptation Strategy Over Wireless Channels [17]

This is a machine-generated summary of:
Hadji, Ahmed; Bouazza, Boubakar S.; Boeglen, Hervé; Perrine, Clency; Nouri, Keltouma; Chatellier, Christian; Pousset, Yannis: Shaping duo binary turbo-coded BICM scheme for JPWL image transmission using a link adaptation strategy over wireless channels [17]
Published in: EURASIP Journal on Wireless Communications and Networking (2021)
Link to original: https://doi.org/10.1186/s13638-021-01949-9

If you want to cite the papers, please refer to the original.

For technical reasons we could not place the page where the original quote is coming from.

Abstract-Summary

"According to the CSI (channel state information) knowledge at the transmitter side, an algorithm based on unequal error protection (UEP) and unequal power allocation (UPA) is used to select the transmitter key parameters (source/channel encoder rate, modulation order, power, number of quality layers and number of iterations of the Turbo decoder) to achieve the target Quality of Service (QoS)."

"The proposed DBTC-shaping BICM scheme reaches a shaping gain of 1.2 dB for a 256 QAM modulation over a SISO Gaussian channel, whereas only 0.7 dB of shaping gain can be achieved in a scheme that uses the LDPC shaping BICM scheme for the same modulation order."

"Based on a DBTC shaping BICM scheme and an adaptive algorithm, the proposed MIMO-OFDM strategy achieves better performance compared to a strategy using an iterative process between an RS (Reed-Solomon) and arithmetic decoders."

"The adaptive number of iterations in the proposed strategy can minimize the computational complexity of the turbo decoding compared to a scheme using four iterations whatever the channel conditions."

Introduction

"This scheme takes into account the content of the JPWL images, and all the transmission parameters (transmitted power, source/channel coder rates and modulation order) are considered in the optimization process."

"The link adaptation scheme introduced in [18] and [19] and the shaping BICM strategy proposed in [20] are investigated in order to achieve high visual image quality for the transmission of JPWL images over realistic wireless channels."

"The duo binary shaping BICM technique will allow to achieve low BERs at low SNRs even if high-order modulations are used in poor channel conditions, thus improving spectral efficiency and minimizing power consumption."

"The optimization block determines the optimal parameters (source and channel code rates, modulation order, number of layers and number of iterations of the DBTC decoding) before the transmission of the JPWL layers."

The Proposed Strategy

"According to the CSI knowledge at the transmitter side, the optimization block determines the number of sub-channels that can be used and, for each sub-channel, determines the optimal values of the modulation order, the rate of the channel coder and the source bit-rate to maximize the visual quality of the received JPWL images."

"The optimization algorithm starts by defining the SNR (the power) needed to achieve the target BER permitting to use a given BICM configuration (the configuration with the highest channel coding rate, the highest modulation order and the smallest number of iterations is favored) in the current sub-channel."

"In order to maximize the number of bits allocated for the source coding and to minimize the computational time in the turbo decoding, this ranking is carried out such as the configuration with a higher modulation order, higher turbo code rate and less iteration number is favored."

Simulation Results

"The proposed shaping BICM scheme presents a better performance compared to the BICM scheme without shaping code whatever the channel conditions."

"To compare the performance of the proposed adaptive strategy with the work presented in [18] and [19], the realistic MIMO channel is decomposed into 4 SISO sub-channels, thus allowing the transmission of 1–4 JPWL quality layers."

"This allows achieving the target BER at a lower SNR compared to the scheme without shaping code and consequently transmitting more quality layers, thus improving the quality of the received JPWL image."

"The proposed scheme uses only 21%, 18% and 13% of the power budget for, respectively, poor, average and good channel conditions."

"For poor, average and good channel conditions, the degradation is, respectively, of about 2.09 dB, 2.15 dB and 2.95 dB. The unreliable CSI does not allow to allocate the optimal power in order to reach the target BER, which degrades the performance."

Conclusion
"The proposed scheme based on an optimization process between all the transmitter parameters (source and channel encoder rates, modulation order, quality layers and number of iterations) improves the quality of the received JPWL images, and a good QoS is obtained even for poor channel conditions."

"Higher-order modulations can be used for the transmission of JPWL images, in which less power is used to achieve the target QoS. This gives a power consumption gain of 78% if the same target QoS is fixed as in the strategy that uses iterative process between RS and arithmetic decoders."

"The gain provided by the proposed strategy can be either in terms of QoS, or in terms of power consumption for a comparable QoS. In addition to this, the adaptive iterations number in the proposed strategy can minimize the decoding complexity compared to strategy that uses RS and arithmetic decoders with 4 iterations whatever the channel conditions."

Low Complexity Modified Viterbi Decoder with Convolution Codes for Power Efficient Wireless Communication [21]

This is a machine-generated summary of:

Devi, T. Kalavathi; Priyanka, E. B.; Sakthivel, P.; Sagayaraj, A. Stephen: Low Complexity Modified Viterbi Decoder with Convolution Codes for Power Efficient Wireless Communication [21]

Published in: Wireless Personal Communications (2021)

Link to original: https://doi.org/10.1007/s11277-021-08919-w

If you want to cite the papers, please refer to the original.

For technical reasons we could not place the page where the original quote is coming from.

Abstract-Summary
"To attain high quality of service (QoS) with efficient power consumption with minimum delay through Wireless Local Area Network (WLAN) through mesh network is an important research area."

"The existing real-time routing system involves multiple hops with time varying mobility channels for fastest data propagation is greatly degraded with power utilization factor through congestion traffic queue."

"In order to achieve high throughput with minimum delay the QoS in real-time data communication have to be concentrated by using Viterbi decoder with convolution codes."

"By undertaking IEEE 802.11 WLAN physical layers afford multiple transmission rates by engaging various modulations and channel coding schemes, major point arises to pinpoint the desired transmission rate to enhance the performance."

"In order to validate the real-time traffic with power consumption and average delay communication, an improved Viterbi decoder is designed with convolution codes to determine accurate channel estimation based on learning the utilization ration of the needed to execute the current wireless channel optimization."

"Each node is capable to choose the optimized transmission rate, so that the system performance can be improved with very minimum power with high packet transmission ratio with minimum traffic rate to improve QoS. The proposed scheme also offers an appealing combination of the allocation of transmission rate and the current link condition."

Introduction

"Different convolution encoders of various code rate and limitation lengths are utilized in different kinds of data communication channels."

"Convolution code is commonly utilized in a few computerized interchanges frameworks in applications as various as Physical Downlink Control Channel (PDCCH), CDMA (Code Division Multiple Access) and GSM (Global System for Mobile Communication) advanced cell, dial-up modems, satellite, profound space correspondences, and 802.11 remote LANs (Local Area Networks) [22]."

"Convolutional coding has been utilized in correspondence frameworks including profound space interchanges and remote correspondences."

"IS-95, a remote advanced cell standard for Code Division Multiple Access (CDMA), utilizes convolutional coding [23]."

"The proposed work explains the execution of Viterbi decoder with convolution code to advance the force dispersion the cross-section organization of WLAN with reproduction results and approve the mathematical examination of wanted correspondence boundaries."

Encoding and Decoding of Viterbi Algorithm

"The trellis diagram for the decoder 16 message bits was taken."

"The 16-bit message used is Y = 1,101,110,010,001,101."

"The decoding of the first message bit for Y = 11 is as follows."

"One branch is at next hub A1 addressing decoded signal as 00 and the other branch is at B1 addressing decoded signal as 11."

"The branch from A0B1 addresses decoded yield as 11 which is same as gotten signal at that hub."

"The encoded number almost a hub shows the way metric (PM) arriving at the hub [24]."

"Disentangling of the subsequent message bit for Y = 01 is clarified as follows."

"The four ways with lower measurements are put away in the decoder and the interpreting proceeds to next got bits."

"Starting with A0 the 16-bit lattice graph is handled bringing about the base way metric being chosen and any remaining measurements ended [25]."

Enhanced Vetribo Decoder Design with Modified Sleepy Keeper Design
"The calculation depends on the past state (branch) metric of each hub (2 k-1 hubs) and every conceivable way."

"An encoder having a code pace of 1/2, there are two ways from each past state portrayed by the branch metric [26]."

"Every hub in the new state is raised from two likely metric ways."

"In the event that (PM(i) t − 1 + BM(i,p)t) is more noteworthy than (PM(j) t − 1 + BM(j,p)t), at that point the way from j to p is the survivor way for p. The pointer j is moved to one side, and the spot, which causes the survivor way progress from j to p (for this situation 0), is attached to the LSB."

"The pointer which had conveyed the worth j presently conveys the worth p. At that point, the decoded bit is attached to the substance of the register whose pointer esteem is changed from j to p [27]."

Simulation and Analysis of Viterbi Decoder Using Enhanced CMOS Circuits
"The VD utilizing CMOS circuits is planned by coordinating every one of the units like BMU, VDU and SMU."

"Despite the fact that the basic way season of the third stage of proposed enhanced framework is same as the second phase of the current system, current strategy accomplishes better defer decrease as it has lesser number of registers and look-into tables (LUT's)."

"Rate 2/3 convolution coder has just two branch metric qualities, so it requires just one phase of two info comparator."

"To work with the determination between rate 3/4 and rate 2/3 mode, input select unit is embedded between the comparator stages."

"Contingent on the info mode, the information select unit will acknowledge the contributions of the required mode and it will take care of the acknowledged qualities to the remainder of the calculation [28]."

Experimental Results and Comparisons
"The proposed reconfigurable VD for the requirement length 7 with coding rate 1/4, 2/4 and 3/4 has huge equipment overhead contrasted with the proposed nonreconfigurable VD."

"The subsequent reconfigurable VD design is discovered to be effective as far as region and has practically a similar planning and force results as the non-reconfigurable VD architecture."

"We analyze the exhibition of the proposed calculation with two diverse rate control plans."

"In the interim the normal pace of the proposed conspire is about 650 kbps, which is higher than those of fixed rate and modified VD scheme in WLAN mesh topology plans about 231 and 53.26 kbps, individually."

"It shows that the proposed plan can adaptively modify rate and exceptionally improve the goodput of the framework on schedule."

Conclusions and Future Scope

"This paper presents a novel reconfigurable Viterbi decoder design, which meets the rapid and low force necessities of most famous remote norms, without utilizing any equal handling and pipelining."

"The proposed engineering is a proficient convolution decoder for limitation lengths 3–7 with code-rate 1/2 and 1/3 explicitly for 802.11a remote neighborhood, 3G cell code division various access conditions, UMTS and EDGE."

"Tried FPGA results show that the proposed design burns-through less force and accomplishes more speed than the detailed models with less power utilization."

"As future work, the proposed design will be validated in real-time data traffic zone by implementing mesh network of any public WLAN network connection to enhance the multi-node interconnect wireless data transmission with high speed with minimum packet loss."

Flip Left-to-Right Approach Based Inverse Tree Interleavers for Unconventional Integrated OFDM-IDMA and SCFDMA-IDMA Systems [29]

This is a machine-generated summary of:

Yadav, Manish; Shokeen, Vinod; Singhal, Pramod Kumar: Flip Left-to-Right Approach Based Inverse Tree Interleavers for Unconventional Integrated OFDM-IDMA and SCFDMA-IDMA Systems [29]

Published in: Wireless Personal Communications (2019)

Link to original: https://doi.org/10.1007/s11277-019-06133-3

If you want to cite the papers, please refer to the original.

For technical reasons we could not place the page where the original quote is coming from.

Abstract-Summary

"Several interleavers have been proposed for conventional interleave division multiple access (CIDMA) systems which provides a mean to control burst errors and to reduce multi user and multiple access interferences."

"A recently developed novel interleaver i.e. 'flip left–right approach based inverse-tree interleaver' (FLRITI) or simply 'inverse tree interleaver', has been explored for two unconventional integrated interleave division multiple access techniques i.e. single carrier frequency division multiple access cum interleave division multiple access (SCFDMA-IDMA) and orthogonal frequency division multiplexing based interleave division multiple access (OFDM-IDMA)."

"The results and analysis reveal that this unconventional integration of CIDMA with SCFDMA and OFDM techniques in presence of FLRITI improves the overall system performance in terms of bit-error rate, memory footprint and computation complexity."

Introduction

"Interleaving pattern used here remains same for all the users and is an optional field."

"In contrary to this, CIDMA achieves the spreading of bandwidth with the use of either a repetition code or through a PN sequence which remains the same for all users."

"The only mean of user separation now is the use of user specific interleaving patterns."

"In CIDMA, users are recognized with the help of their respective interleaving patterns only [30]."

"Tree based interleavers proposed in [31] demonstrates significant reduction in computation complexity and memory requirement without any degradation in CIDMA system performance."

"Flip left–right approach based inverse tree interleaver (FLRITI), explored here for IIDMA, is a category of derived interleaver."

"FLRITIs are obtained from some design changes made in tree interleavers (TI) [30, 31]."

"These can be viewed as a modified version of tree based interleavers."

Block Models of Unconventional Integrated IDMA Systems

"An unconventional integrated IDMA refers to the purposeful integration of CIDMA with one or more special multiplexing/multiple access schemes such as single carrier frequency division multiple access (SCFDMA) and/or orthogonal frequency division multiplexing (OFDM) etc. [30, 32–38]."

"OFDM has the provision to split the entire available transmission bandwidth into multiple orthogonal subcarriers and then each transmitted data symbol is carried by an individual subcarrier only."

"Due to the combine effect of OFDM-IDMA layers, specific user's data symbol can be easily identified following that user's interleaving pattern [34, 35]."

"The remaining portion of OFDM-IDMA receiver consists of an elementary signal estimator (ESE), user specific interleaver/de-interleaver ($\prod_k$, $\prod$) pairs and a bank of a posteriori probability decoders (APP-DECs)."

"In SCFDMA-IDMA approach, interleaved data symbols from different users are spreaded over the entire available transmission bandwidth of MAC with the help of number of group of subcarriers."

FLRITI: Structure and Its Functionality

"In recently proposed FLRITI design in [39], users are allocated based on specific sequence numbers."

"By this way, those users who have been allocated a sequence number are arranged in two different branches of a tree."

"The second user permitted to access the MAC is allocated even sequence number i.e. '2'."

"As the number of users increases, this technique also reduces the computation complexity because very less number of computation cycles is required to generate an interleaver for a higher sequence number user."

"The rules to be followed in generating the different combinations of $\{\prod_1, \prod_2\}$ for different sequence number of users have been described in [40] and rules for placement of users (based on their odd/even sequence number) in a tree is available in [31, 41] in detail."

"It is simply exemplifying the generation mechanism of interleaved chip-vectors for first four users accessing the MAC with the help of their respective chip-sequence and allocated FLRITIs."

Performance Analysis

"The case of K = 16 users, for lower values of E_b/N_0 (i.e. up to 4 dB) both SCFDMA-IDMA and OFDM-IDMA provide almost same value of BER."

"As the value of E_b/N_0 is increased above 4 dB corresponding value of BER for SCFDMA-IDMA starts decaying and after $E_b/N_0 = 6$ dB no bit error is observed."

"The value of BER is still constant for OFDM-IDMA systems and it starts decaying only after $E_b/N_0 = 12$ dB. No bit error is observed after $E_b/N_0 = 14$ dB in this system."

"Further, performance of OFDM-IDMA is better than CIDMA as it offers lower BER than CIDMA for the same value of E_b/N_0."

"In SCFDMA-IDMA, single group of subcarrier can only be utilized by the limited number of users; thus, reduced POC of ICI and ISI and better BER performance."

Conclusion

"Performance analysis of both the unconventional IIDMA systems i.e. SCFDMA-IDMA and OFDM-IDMA systems in presence of FLRITI proves the worthiness of this interleaver as a competitive alternative for cellular communication technologies particularly 4G and beyond i.e. B4G. These IIDMA systems with FLRITI demonstrate almost the similar BER versus E_b/N_0 performance as they demonstrate with RI and TI."

"SCFDMA-IDMA with FLRITI performs significantly better than OFDM-IDMA and CIDMA considering equal number of users accessing the MAC."

"Use of FLRITI with unconventional integration of two or more technologies is highly recommended."

"This research work can be extended for more such unconventional integrations with/without FLRITI."

A Novel Combined Double Binary Turbo Coding and Color Shift Keying Technique for Flicker Mitigation in Multi Carrier Visible Light Communications [42]

This is a machine-generated summary of:

Savaşcıhabeş, Asuman: A Novel Combined Double Binary Turbo Coding and Color Shift Keying Technique for Flicker Mitigation in Multi Carrier Visible Light Communications [42]

Published in: Wireless Personal Communications (2022)

Link to original: https://doi.org/10.1007/s11277-022-10053-0

If you want to cite the papers, please refer to the original.

For technical reasons we could not place the page where the original quote is coming from.

Abstract-Summary

"It is proposed to combine the double binary Turbo coding (DBTC) technique with color shift keying (CSK) modulation, which does not have a flickering problem, as a solution to the flicker problem of VLC systems."

"The proposed combined DBTC-CSK method is tested on optical OFDM systems from multi-carrier wireless communication systems."

"Monte-Carlo simulations are performed to verify the performance of the proposed DBTC-CSK-OFDM system in terms of bit-error-rate (BER) over AWGN channel, single-path optical LOS channel and multi-path optical channel."

"According to the simulation results it is observed that, DBTC-CSK-OFDM system has better BER performance than that of RS–CC coded CSK-OFDM and uncoded CSK-OFDM systems, providing 5.5 dB and 14 dB SNR gain at multi-path optical channel environments, respectively."

Introduction

"VLC with the light emitting diodes (LEDs) used as wireless access points and modulated at high speeds invisible to the human eye, is based on the dual use of LEDs for lighting and communication purposes [43]."

"The spectral distribution of the flickering effect is induced by a light stimulator that fluctuates over time and causes the impression of imbalance in the visual sense, and it is mostly caused by the modulation technique used in the physical layer (PHY) during data transmission in VLC systems."

"The flickering effect of the light is due to the modulation techniques used in VLC systems causing variations on the average transmitted power and in this case, the fluctuation effect on the brightness of the LEDs, as stated above."

"In VLC systems, CSK is often used as the multiple light source modulation scheme which uses red, green and blue LEDs and it modulates the information signal with the intensity of RGB rays [44]."

M-CSK-OFDM System Model

"After matching the subcarrier for each color with the modulated data obtained at the output of the CSK modulator, time domain samples are obtained by subjecting them to inverse fast Fourier transform in the IFFT block."

"OFDM signals obtained at the CP insertion block output are parsed into their real and imaginary parts in the Real/Imaginary separator block."

"Since the obtained OFDM signals are real but bipolar signals, it is sufficient for the IM/DD method to make the information signals sent to each component using DC level with positive values as seen from the block diagram."

"The obtained real positive OFDM signals are transmitted over the multipath optical channel through RGB LEDs and collected by photo diodes at the receiver side and are distorted with additive white Gaussian noise (AWGN)."

Channel Coding

"The normal data string enters into the encoder when the switch is in position 1."

"In the second, when the switch is in position 2, the interleaved data string is entered into the encoder."

"To perform the coding process of the data string, two cycle states consisting of one conversion state for each component encoder are determined and the data string is encoded 4 times."

"Parity 1 and Parity 2 are the channel values of the encoder parity output bits."

"In the Turbo decoder, the sequential input bits are split into information and parity bits along the trellis multiplexer."

"Soft input Soft Output (SISO) information generated by the decoder, parity bits and a priori information are used in decoding."

"For the use of double binary Turbo code in QPSK, the odd indices of the coded data string are sent from the I channel and the even indices from the Q channel."

Computer Simulation Studies

"Un-coded M-CSK-OFDM, 1/3 classical TC coded, RS–CC coded and DBTC coded M-CSK-OFDM system performances are compared in all simulation results for 48 sub-carrier and 64-point FFT process."

"Of the study, BER-SNR performances of un-coded OFDM using 4-CSK, 8-CSK and 16-CSK modulation, 1/3 RS–CC, classical TC and DBTC coded OFDM systems over AWGN channel are compared."

"It is seen that the classical TC encoder at 1E-3 BER provides approximately 5 dB coding gain from the un-coded BER performance in the first iteration."

"In the given simulations, BER-SNR performances of un-coded OFDM using 4-CSK, 8-CSK and 16-CSK modulation, 1/3 RS–CC, classical TC and DBTC coded OFDM systems over single-tap optical LOS channel are compared."

"In simulation studies, BER-SNR performances of un-coded OFDM, 1/3 RS–CC, classical TC and DBTC encoded OFDM systems employing 4-CSK, 8-CSK, and 16-CSK modulations and the results are compared over the multipath optical channel."

Conclusions

"It is verified that the proposed DBTC coded M-CSK-OFDM transmission systems for VLC channels in this paper."

"The proposed DBTC coded M-CSK-OFDM, TC coded M-CSK-OFDM, RS–CC coded M-CSK-OFDM and un-coded M-CSK-OFDM communication systems were compared employing BER-SNR achievement benchmark over AWGN channel, single-path optical LOS channel and multi-path optical channel simulations."

"From the computer simulation results, it is observed that the acquired performance in case of using DBTC in M-CSK OFDM systems is better than the obtained performance with other encoders."

"The obtained performances with this proposed method show that DBTC encoders can be used in new generation wireless communication systems."

Performance Improvement of CDMA Wireless Sensor Networks in Low SNR Channels Based on Raptor Codes [45]

This is a machine-generated summary of:

Farhan, Ikhlas M.; Zaghar, Dhafer R.; Abdullah, Hadeel N.: Performance Improvement of CDMA Wireless Sensor Networks in Low SNR Channels Based on Raptor Codes [45]

Published in: Wireless Personal Communications (2023)

Link to original: https://doi.org/10.1007/s11277-023-10387-3

If you want to cite the papers, please refer to the original.

For technical reasons we could not place the page where the original quote is coming from.

Abstract-Summary

"Low signal to noise ratio (SNR) channels require careful design of communication systems in order to operate reliably in such extreme channels."

"Channel coding is a powerful tool to aid against adverse channel conditions including low SNR."

"We studied the performance of a system that combines Raptor codes with direct sequence code division multiple access (DS-CDMA) in low SNR scenarios."

"Combining Raptor codes with DS-CDMA showed an improvement of approximately 6 dB when compared with other schemes using DS-CDMA in low SNR conditions."

Introduction

"Spread spectrum-based code division multiple access (CDMA) has played an important role in cellular and personal communication systems [46]."

"Direct-sequence code-division multiple-access (DS-CDMA), the most common type, will be discussed."

"In DS-CDMA technology, each user is given a specific code, which is crucial for modulating the signal as it affects how well the system works."

"A critical issue in traditional DS-CDMA systems is how to reduce multiple access interference (MAI) from other users."

"To achieve good transmission performance in digital communication systems, effective channel coding is required."

"Various alternative coding techniques are available, including CDMA-specific "trellis codes," very low-rate convolutional codes, and conventional convolutional codes [47]."

"Convolutional error-correcting codes have extensively been used to provide reliable communication over channels with a low signal-to-noise ratio (LSNR) [48]."

"This study suggested Raptor coding as a performance improvement method for the DS-CDMA multi-user system."

Related Works

"In [49], the examination of CDMA under the AWGN channel is highlighted after the performance and evaluation of turbo code are properly presented."

"For use over a Rayleigh fading channel, a convolutional code for the IS-95 standard CDMA system and a turbo-code for the CDMA 2000 system were also proposed in [50]."

"This method has demonstrated that, based on the simulation results, the BER of the polar channel-coded with DS-CDMA system performs significantly better than the convolutional DS-CDMA system (about 4 dB gain at E_b/N_o of 4 dB)."

"The performance of the LDPC-Code DS-CDMA system is examined in the presence of LDPC/Turbo coding and decoding methods when an ideal multi-user detector is used in the recovery of the information data, and the performance is enhanced with a BER = 10^{-3} for SNR = 4 dB in an AWGN channel."

"Comparing Raptor codes to Turbo/convolutional when coupled with codes direct-sequence CDMA systems on BIAWGN channel has been conducted in [51]."

Raptor Code

"Using error-correction block codes, a collection of input symbols is divided into (n, k) blocks to produce intermediate symbols."

"The pre-code protects the source packets and corrects erasure packets that the LT code cannot recover."

"A Tanner graph with intermediate variable nodes (VNs) is decoded using the sum-product algorithm (SPA), and the output symbol represents the check nodes (CNs) to an inner decoder."

"Since the SPA algorithm is used for both the encoding and decoding steps of the Raptor code, the overall code rate (R) and the code rates of the inner (R_i) and outer (R_0) codes are related [52]: where k in the number of the bits per packet entered into the encoder, k′ is the number of bits per block at the output of the inner stage of the coder, and n is the number of the bits per block at the output of the channel encoder."

Proposed Model of Hybrid Channel Coding with CDMA System

"In order to evaluate the impact of incorporating Raptor coding into the CDMA transmission system, the proposed scheme needs to be implemented and tested while creating a general model for the communication system in a fair environment."

"The performance of this code is partially influenced by the pre-code parity-check matrix, H, of the LDPC codes acting as the first layer of the Raptor structure."

"Two recursive systematic convolutional (RSC) encoders are concatenated in parallel to create the turbo code's encoding stage, as described in [53]."

"When using a turbo code, the parity bits are punctured alternately to change the standard 1/3 rate to a rate of 1/2."

"These are the two encoders used by the Turbo/convolutional coded DS-CDMA system."

"One of encoders has the constraint length of 7, the code rate of 1/2 and generators 133, 171, expressed in octal notation [54]."

Performance Analysis of Proposed System Based on Hybrid Raptor Code with DS-CDMA

"The performance characteristics of the proposed scheme, which consists of a Raptor channel code in combination with a DS-CDMA multiple access technique, are examined and compared with other channel coding algorithms used with the same DS-CDMA system in this work."

“We implemented a DS-CDMA system with an 8-bit spreading code and evaluated its performance before adding the channel coding stages for comparison in order to study the performance of the Raptor coded DS-CDMA system.”

“The Raptor-coded CDMA scheme receiver, which uses a spreading factor of length 4, performs better than the traditional DS-CDMA receiver, which uses a higher spreading factor of length 8.”

“The Raptor-coded DS-CDMA scheme offers superior performance in low Eb/No channels, and it is therefore strongly advised to be used in these channels.”

“This graph shows how the code length affects channel codes in general as well as the Raptor code’s performance with DS-CDMA.”

Performance Metrics of the Proposed Raptor Code DS-CDMA System

“This section will describe the main performance metrics that were used in this study to assess the suggested system (Raptor code DS-CDMA for WSN).”

“This metric is used to assess the error performance of the provided scenarios; it calculates the number of lost packets percentage (NLPP) to estimate the amount of data loss.”

“Throughput or capacity is the number of bits that can be processed by a decoder, or the amount of data that the decoder can process in a given time [55].”

“Computational complexity represents the number of computations performed by an algorithm, e.g., encoder or decoder, to process a data array.”

“The required operations for the encoding or decoding processes are referred to as the computational complexity of coding scheme.”

“The combination between the DS-CDMA transmission technique and Raptor channel coding will create a reliable communication system for sending data in LSNR channels.”

Conclusions

“The decoding BER performance was tested and compared with that of a design combining Turbo/Convolutional codes with DS/CDMA in low SNR conditions over the BIAWGN channel.”

“The obtained results show that the DS-CDMA with Walsh code design offers better performance compared to other schemes when operating in low SNR ranges.”

“The simulation results concluded that Raptor codes combined with DS-CDMA using Walsh code length of 4 can achieve better performance than conventional DS-CDMA using Walsh code length of 8.”

“Simulation results showed that the Raptor codes with DS-CDMA system is better than the conventional CDMA with Turbo/convolutional coding transmission system under the same channel conditions.”

Error Analysis of Grouped Multilevel Space-Time Trellis Coding with the Combined Application of Massive MIMO and Cognitive Radio [56]

This is a machine-generated summary of:

Chopra, Shakti Raj; Gupta, Akhil: Error Analysis of Grouped Multilevel Space-Time Trellis Coding with the Combined Application of Massive MIMO and Cognitive Radio [56]

Published in: Wireless Personal Communications (2021)

Link to original: https://doi.org/10.1007/s11277-020-07878-y

If you want to cite the papers, please refer to the original.

For technical reasons we could not place the page where the original quote is coming from.

Abstract-Summary

"Multilevel coding scheme with trellis codes as component codes provides flexible data transmission rates, coding gain, diversity gain with improved spectral efficiency and low decoding complexity."

"This paper discussed space-time coding with beamforming and antenna grouping according to the channel state information."

"Multilevel Space-time coding is based on multi-level Quadrature Amplitude Modulation signaling and beamforming to mitigate the effect of primary users for the enactment of secondary users in Cognitive Radio."

"Our transmission use Quadrature Amplitude Modulation based signals, with an adaptive grouping of antenna which weight according to the optimization, which inherently depends upon the resource allocation of the secondary user."

Introduction

"The error execution of un-coded MIMO, distinctive diversity brushing strategies, space-time block, and trellis codes are dissected utilizing various parameters like several receiving antenna, M-array modulations, trace criteria, rank and determinant criteria, to determine the performance of STTC."

"In the current scenario, the number of users increases drastically, so in the limited spectrum, data speed and link quality play a major role to improve the quality of service, this paper tries to investigate to get the solution by implementing the space-time coding in massive MIMO, further cognitive radio techniques will allow the system to enhance the spectrum utilization of limited available spectrum."

"GMLSTTC sends error-free data with better link quality, Cognitive radio provides the better utilization of scarce spectrum and finally, the use of Massive-MIMO technology in channel coding is to extend broadband access."

System Model and Problem Description

"At the time 't' 'N' no of 4-QAM STTC codes are represented as C(1), C(2), …C(I) and written as: Symbols in C(1) and C(I) helps to get the most significant bits and least significant bits, respectively."

"It also satisfied with the power normalization constraints: the code communicated from the jth transmission antenna at the time 'h,' will be represented as: Where J = 1, 2 … N K_j is the weight of the antenna."

"The received signal at receive antenna at times 't' is written as The conditional (PDF), of rt of all I encoder, the outputs maybe then written as: Where The multi-stage detector, where x_t (I) is decoded first and is passed to the next stage, where x_t (I − 1) is decoded, and so on."

"Let X_f be the length of the frame, X_s is the time of detection, and X_t is the time of CR transmission."

Realization and Representation of Objectives

"Algorithms are divided into two sections, Algorithm 1 involves GMLSTTC planning and realization, while Algorithm 2 involves GMLSTTC implementation in combined application of Massive MIMO and cognitive radio."

"Algorithm 2 requires the initialization and realization of a case."

"The secondary phase of Algorithm 2 involves the application of the cognitive radio is applied and the results are compared."

"Algorithm 1 includes computing GMLSTTC, which is used to describe the polynomial generator and then add 64 QAM constellation and produce the transmitted signal using a convolutionary encoder."

Simulation and Results

"For a system that processes input information at four stages, the output is evaluated."

"The encoders used at various levels are coevolutionary encoders having a rate of 2/3."

"The sub information flows are interpreted at their respective level by a coevolutionary encoder."

"The receiver demodulates the signal used a parallel Viterbi decoder, independent decoding is used to detect the information at each point."

Conclusion

"This paper discusses an MLC system with trellis codes to mitigate the detrimental consequences of multipath fading."

"This scheme delivers enhancement in the enactment of wireless communication systems by providing enhanced spectral efficiency, data transmission rate, diversity, and coding gain, with less decoding complexity."

"The efficiency of the system, further increases by the application of GMLSTTCs with combined application of cognitive radio and Massive MIMO."

"Future Research Direction: The performance of the system can also be further be analyzed by applying the GMLSTTC in the only Multi-user Massive MIMO to prepare the system for the next generation."

Patching Based Extra Short Packet Forward Error Control Coding for Ultra-Reliable Low Latency Communications (URLLC) in 5G [57]

This is a machine-generated summary of:

Woo, Tai-Kuo: Patching Based Extra Short Packet Forward Error Control Coding for Ultra-Reliable Low Latency Communications (URLLC) in 5G [57]

Published in: Wireless Personal Communications (2021)

Link to original: https://doi.org/10.1007/s11277-021-08815-3

If you want to cite the papers, please refer to the original.

For technical reasons we could not place the page where the original quote is coming from.

Abstract-Summary

"Extra short packets are common in the 5G Internet which provides low latency, ultra-reliable services for applications such as autonomous driving, remote surgery, industry automation, to name just a few."

"The multitude number of low cost devices cannot afford to run high power forward error control coding algorithms, such as Turbo codes, LDPC codes, Polar codes, and so on, to meet the stringent requirements of ultra-reliability and extremely low latency."

"We propose using patching as the basis for forward error control coding of extra short packets, where replicas of a packet along with its checksum are transmitted adaptively."

"Despite the fact that the proposed coding scheme provides lower coding rate for a fixed packet error rate in comparison to the Polar codes and LDPC codes, the main advantages lie in its flexibility and simplicity of processing ideal for asymmetric configuration with low power receiving devices providing ultra-reliable low latency communications."

Introduction

"Within the context of the second approach to achieve the goals of URLLC services, the length of packets must be extra short in terms of tens of bits for the following three reasons."

"In regard to the latency, the transmission time for extra short packets is considered trivial in relative to the processing time Thirdly, from the reliability point of view, extra short packets are naturally low in packet error rate."

"The probability of retransmission is related to the packet error rate, extra short packet reduces the PER and raises the efficiency and quality of the error control coding."

"To meet the stringent requirements of URLLC, the error control coding with extra short packets is essential."

"It is a reasonable to assume that minimizing the length of the packets and reducing the BER with flexible FEC coding, improve the reliability and cut the latency in URLLC."

The Patching Based FEC Coding for Extra Short Packets, Papa's Codes

"If the checksum is null, a non-faulty data packet has been recovered."

"Suppose a full packet of thirteen symbols (data + checksum), is transmitted twice, i.e., two replicas."

"If the checksum is null, we then output the full packet as non-faulty packet."

"To meet the requirement of ultra-low latency, bit level granularity is assumed, and the packet length is kept to the minimum, where the data portion of the packet is enough for the URLLC applications and the checksum portion of the packet is short enough to capture error bits in the packet."

"Using CRC for detecting non-faulty packets, we need to know that two types of errors may incur."

"The second type is the checksum error, where the data portion is non-faulty, but it cannot be detected due to the checksum error."

"If the CRC checksum is all zero, we have identified the non-faulty packet."

[Section 3]

"To obtain a valid code word of a data word (packet), we simply compute, and append the checksum to the data word."

"The appended packet is replicated to obtain the corresponding code word."

"If a "dirty bit" exists, one cannot form a non-faulty packet for the following reason."

"If the common point is a "dirty bit," then the OR operations cannot yield a non-faulty bit, therefore the combined packet cannot be non-faulty."

Foundation of Using FPP

"The key point is that the more symbols (bits) you replace for each formation of a packet, the more likely that a CRC checking with the result of non-faulty replica can happen."

"Since the error bits of these two replicas are in different locations, a non-faulty packet can be established."

"One may notice that the subpackets corresponding to the sets with a common point 7, (3,4,7), (1,6,7), and (2,5,7) can form a non-faulty packet."

"The non-faulty subpackets of the first replica include the subpackets corresponding to the sets (4,5,9,13), (4,6,10,11), and (4,7,8,12)."

"The non-faulty subpackets include the subpackets corresponding to the sets (1,2,3,4), (1,5,6,7), and (1,11,12,13) in replica 2."

"The subpacket of the first replica corresponding to (1,2,3,4) can be combined with the subpackets, corresponding to sets (4,5,9,13), (4,6,10,11), and (4,7,8,12) of replica 2, to form a non-faulty packet since these four subpackets share a common point 4."

Performance Evaluation

"A packet may become discarded if it encounters hostile channel conditions."

"The approach of forming a packet on the fly based on the FPP can significantly reduce the cost of error control coding and is well suitable to extra short packets."

"In case of a ruined packet, each retransmission also provides additional error correction probability."

"This feature is important in HARQ, where packets are transmitted multiple times, and cyclic redundancy checks are required to verify if the packet is free of errors."

"The first type is the bit error patterns in the packet that are not detected."

"Using a longer checksum with a careful selection of the CRC polynomial can minimize the type I error. (undetected faulty packet)."

"The second type of error is the non-faulty data that cannot be detected due to error bits in the checksum."

Conclusions

"Flexibility refers to the fact that both the length of a packet and the number of replicas can be modified on the fly, whereas adaptability indicates that the changes can be made according to the changing channel condition and the user requirements."

"The criteria of selecting a CRC polynomial is to meet the requirement that a faulty packet will not slip through without being detected."

"The length of CRC checksum must be short enough such that detection of non-faulty subpackets remains high."

"One could use extra short packets to reduce latency and increase reliability."

Performance Analysis of Parallel Concatenation of LDPC Coded SISO-GFDM System for Distinctive Pulse Shaping Filters Using USRP 2901 Device and Its Application to WiMAX [58]

This is a machine-generated summary of:

Telagam, Nagarjuna; Lakshmi, S.; Kandasamy, Nehru: Performance Analysis of Parallel Concatenation of LDPC Coded SISO-GFDM System for Distinctive Pulse Shaping Filters using USRP 2901 Device and its Application to WiMAX [58]

Published in: Wireless Personal Communications (2021)

Link to original: https://doi.org/10.1007/s11277-021-08865-7

If you want to cite the papers, please refer to the original.

For technical reasons we could not place the page where the original quote is coming from.

Abstract-Summary

"This article presents an investigation of the parallel concatenation of low-density parity-check codes (PC-LDPC) in the fifth generation proposed waveform candidate called generalized frequency division multiplexing (GFDM)."

"PC-LDPC codes are obtained by dividing the long and high complexity single LDPC codes into small two lower complexity codes, and these designed codes are applied to the 5G-GFDM waveform."

"Since the GFDM signal transmits data in both the time and frequency domain, these PC-LDPC codes can deal with two-dimensional errors."

"The PC-LDPC coded GFDM outperforms LDPC by 6.5 dB in the RRC filter for roll-off factor rate 0.5 under the Rayleigh fading channel."

"PC-LDPC code outperforms LDPC code with a coding gain of 2 dB was observed in IEEE 802.16 Transceiver."

Introduction

"In order to overcome that, concatenation of LDPC codes are proposed with the same iterative decoding algorithm, which is used in turbo codes as are the right choice for better performance [59]."

"A 71% improvement in the complexity of receiver design was observed in computer-based simulation for LDPC code concatenated with staircase codes [60]."

"A significant reduction in BER was observed in 5G-GFDM when the concatenation of LDPC code with Turbo codes are used as channel coding schemes [61]."

"The investigation of PC-LDPC codes with a 5G GFDM system over Gaussian noise and flat fading Rayleigh channel for distinctive prototype filters is analyzed in this paper."

"To propose the novel GFDM transceiver design with high coding gain and less complexity decoder in receiver."

"Passage 3 explains the mathematical analysis of GFDM and parallel concatenation of LDPC codes using diagrams."

Channel Coded GFDM Model

"The corresponding bits are converted to complex data symbols with QAM or BPSK modulation schemes."

"The received data is a combination of AWGN channel or Rayleigh fading channel along with environmental noise."

"These data are processed through the FPGA receiver, and these IQ data samples are transferred through the GFDM receiver section."

"In the receiver section, removing of cyclic prefix symbol, if zero padding is included in the transmitter, the zero-padding is also removed in the receiver."

"The sequence of steps followed by the transmitter, receiver its sub-VI's are mentioned in the above Figure."

List of VI's in GFDM Transmitter

"The source encoder instance VI maps the input bits to complex symbols."

"Due to the Intersymbolic interference problem in the channel, the zero-forcing equalizer is performed to get the desired signal."

"The desired signal can be obtained from the noisy channel using zero forcing equalizer is explained mathematically using equations from (14), (15), (16), (17) and (18)."

"The encoding process of PC-LDPC codes is explained with the parity check matrix H matrix of size 6 × 12."

"The two channels are employed to GFDM complex waveform, and the two channels are time-varying channels."

"These channel VI's are already inbuilt in the LabVIEW programming models; the environmental noise is also added to the GFDM waveform along with two channels."

"The data type of signal is pink color, a complex IQ GFDM signal applied to MT Apply IQ impairments."

"The GFDM complex signal is applied to selective fading."

USRP 2901 Radio Transceiver Working Sequence for PC-LDPC Codes with GFDM

"This section explains the two USRP 2901 devices used, which transmits the IQ data from the verto antenna and receives the combination of IQ data with AWGN channel noise along with environmental noise."

"This USRP radio transceiver is tested for different channel coding techniques for determining the performance analysis of the GFDM system under the AWGN channel."

"The exact process is again repeated for the Rayleigh fading channel."

Transceiver Blocks in USRP Device

"The Low noise amplifier amplifies the signal, mixer down-converts the IQ samples and Analog to digital converter to convert the signal into user-specified rate [62]."

"We proposed to convert the GFDM system into the OFDM model and then apply PC-LDPC codes are forward error-correcting codes before the interleaving section in the scenario."

"To convert the GFDM system into the OFDM model, the number of subsymbols should be equal to 1."

"i.e. (M = 1), then K increases to N, Number of subcarriers(K) goes to symbols(N), (without altering bandwidth), then the sufficient spacing between the subcarriers is present, then GFDM is converted to OFDM."

"VI, the IQ rate, carrier frequency, device names, a gain of the antenna in decibels, and active antennas numbers are user-defined."

"The number of subsymbols is equal to one, making the GFDM system model converted to the OFDM model."

Result Analysis

"The best value of SNR is 16 dB for less BER value in PC-LDPC."

"The best value of SNR is 12 dB for less BER value in PC-LDPC."

"The best value of SNR is 18 dB for less BER value in PC-LDPC."

"The best value of SNR is 12 dB for less BER value in PC-LDPC."

"The BCH coded GFDM, Convolutional, Golay, single LDPC code system is not efficient, i.e., BER = 10^{-3} were 21 dB, respectively, the BCH coded GFDM has worst BER performance for SNR below 18 dB values."

"The remaining prototype filters with roll-off factor value 0.5 provides BER = 10^{-4} were SNR = 18 dB, the coding gain of 6.5 dB is observed in PC-LDPC codes with Xia 4th order corresponding to PC-LDPC codes for other prototype filters."

"The best value of SNR is 8 dB for less BER value in PC-LDPC."

Conclusions

"The RRC provides out of band power as −37 dB, and Xia's filter generates out of band power as −40 dB, the improvement of 7 dB OOB is observed in RRC, Xia prototype filters corresponding to RC filter."

"The coding gain of 6.5 dB was observed in the RRC prototype filter based GFDM system under Rayleigh fading channel in PC-LDPC codes corresponding to LDPC codes here BER = 10^{-4} were 12 dB was observed respectively."

"PC-LDPC-GFDM system in the combination of Xia prototype filter with roll-off factor value 0.5 outperforms other PC-LDPC-GFDM of RC, RRC, Gaussian filters with a coding gain of 3 dB maximum iterations kept as 100."

"The coding gain of 12 dB was observed in PC-LDPC codes with Xia 4th order filter corresponding to other prototype filters."

Continuous Phase Modulation with 1-Bit Quantization and Oversampling Using Iterative Detection and Decoding [63]

This is a machine-generated summary of:

de Alencar, Rodrigo R. M.; Landau, Lukas T. N.; de Lamare, Rodrigo C.: Continuous phase modulation with 1-bit quantization and oversampling using iterative detection and decoding [63]

Published in: EURASIP Journal on Wireless Communications and Networking (2020)

Link to original: https://doi.org/10.1186/s13638-020-01856-5

If you want to cite the papers, please refer to the original.

For technical reasons we could not place the page where the original quote is coming from.

Abstract-Summary

"A channel with continuous phase modulation and 1-bit ADC with oversampling is considered."

"Due to oversampling, higher-order modulations yield a higher achievable rate and this study presents methods to approach this with sophisticated channel coding."

"Besides the strategy based on conventional channel coding techniques, a tailored bit mapping is proposed where two bits can be detected reliably at high SNR and only the third bit is affected by the uncertainties from the coarse quantization."

"It is shown that the proposed method based on the tailored bit mapping corresponds to a lower bit error rate in comparison with the case of conventional channel coding."

Introduction

"In [64], where a bandlimited channel is considered, a marginal benefit of 1-bit quantization and oversampling at the receiver in terms of achievable rate has been reported."

"The achievable rate for channels with 1-bit quantization and oversampling at the receiver has been studied in [65, 66], where bandlimited and not strictly bandlimited signals are considered, respectively."

"We consider the design and analysis of CPM schemes with 1-bit quantization and oversampling at the receiver, employing channel coding in scenarios with higher modulation order."

"The main contributions of this paper are the following: Iterative detection and decoding scheme for CPM systems with 1-bit quantization and oversampling at the receiver; Novel phase-state-dependent bit mapping, which is designed for the 1-bit ADC problem; Channel coding scheme suited for the bit stream separation into sub-channels."

System Model Overview

"On the transmit path, the channel encoder receives information bits and generates an encoded message adding redundant information."

"The encoded message is interleaved to protect the coded information against burst errors."

"On the receive path, the signal is filtered and quantized by a 1-bit ADC."

"The soft information is given to the channel decoder, which returns extrinsic soft information to the detection algorithm via an interleaver and a soft mapper."

CPM with 1-Bit Quantization

"The phase response function shapes the phase transition between the phase states."

"In order to reduce the complexity at the receiver, a time invariant trellis is constructed by tilting the trellis according to the decomposition approach in [67]."

"A discrete-time system model description is considered which implies that the CPM phase is represented in a vector notation."

"A higher sampling grid in the waveform signal, in the noise generation and in the filtering is adopted to adequately model the aliasing effect."

Soft Detection and Decoding

"This work uses an iterative detection and decoding procedure that consists of the exchange of soft information between the detector and the channel decoder."

"The iterative detection and decoding process relies on the feedback of soft information from the decoder to adjust the transition probabilities of the soft detector, which becomes aware of the underlying code."

"This message passing algorithm is done iteratively with soft information exchange between the detection and decoding steps."

"In the final step of the channel decoder, i.e., after the execution of the required number of iterations, the soft information on the information bits is computed where R is the used code rate."

Modified Sub-channel Coding

"The Gray mapping shuffles the uncertainties brought by the coarse quantization for every bit sub-channel, whereas the advanced mapping aims to concentrate the uncertainty on the third bit, by doing a circular shift on the Gray mapping when phase transition happens from a phase state of "odd" order."

"With the proposed advanced mapping, it is known apriori on the transmitter side which bits in terms of binary sub-channels require protection by forward error correction."

"It is proposed a modified coding scheme that has the same structure of the conventional one, but applies different code rates for the bit sub-channels separately, i.e., in transmission, bit sub-channels streams that are more sensitive to the coarse quantization, noise or channel impairments would be protected by a stronger channel code."

Results and Discussion

"Channel codes with larger constraint length and longer blocks have better performance in general, but it requires more computation resources on the decoder."

"While the performance in terms of achievable rate is comparable, the system that uses 1-bit ADCs does not require automatic gain control."

"The BER performance indicate that a higher resolution in time is beneficial at low and medium SNR and that a higher resolution in amplitude is beneficial in the high SNR regime."

"In this last scenario, the use of the proposed coding scheme brings a consistent performance gain from low to high SNR levels."

"While the system with 1-bit quantization requires a higher SNR for achieving a low BER, it has practical benefits such as the relaxed requirements on the automatic gain control and the low noise amplifier."

Conclusions
"Iterative detection and decoding applied to a CPM system with 1-bit quantization and oversampling at the receiver has been studied."

"Different code rates have been considered, and it turns out that channel coding is beneficial in all the cases."

"Additional performance gain can be achieved by using the proposed tailored bit mapping strategy in combination with a coding scheme that considers different binary sub-channels separately."

Analysis of Noise Immunity of Optimal Symbol-by-Symbol Reception of Phase-Keyed Signals with Correcting Codes in Non-Binary Galois Fields [68]

This is a machine-generated summary of:

Nazarov, L. E.; Batanov, V. V.: Analysis of Noise Immunity of Optimal Symbol-by-Symbol Reception of Phase-Keyed Signals with Correcting Codes in Non-Binary Galois Fields [68]

Published in: Journal of Communications Technology and Electronics (2022)

Link to original: https://doi.org/10.1134/s1064226922080137

If you want to cite the papers, please refer to the original.

For technical reasons we could not place the page where the original quote is coming from.

Abstract-Summary
"The paper describes an algorithm for optimal symbol-by-symbol reception of signal structures based on signals with multiphase keying and block correcting codes in non-binary Galois fields."

Introduction
"The main requirements for these codes are dictated by the need to make the probabilistic characteristics of the reception of signal structures close to the limiting probabilistic characteristics of the Shannon throughput of noisy channels, as well as the acceptable complexity of the reception algorithms being implemented [69]."

"There are two general classes of algorithms for the optimal reception of signal structures known in the literature: algorithms that implement symbol-by-symbol reception with minimized error probability per code symbol and algorithms that implement the maximum likelihood rule with minimized probability of error in code words [69, 70]."

"Algorithms for symbol-by-symbol reception underlie the computational procedures that implement iterative processing when receiving signal structures based on correction codes, most efficient in terms of the above-formulated criteria (based on block and convolutional turbo codes, low-density codes) [69, 71]."

"This approach makes it possible to extend the class of efficient signal and code structures and is consistent with the direction of development of the theory of the class of error-correcting codes in non-binary fields [69, 72]."

4. Simulation Results

"The probabilistic characteristics are obtained by simulating the above algorithm for optimal symbol-by-symbol reception when transmitting code words with an information size of 60 bits."

"This also shows the efficiency of the optimal symbol-by-symbol reception algorithm considered."

Conclusions

"Symbol-by-symbol reception minimizes the error probability per code symbol, in contrast to the well-known maximum likelihood rule, which minimizes the error probability per code word."

"The resulting complexity of the symbol-by-symbol reception algorithm is determined by the dimension of the dual code, which makes it promising for error-correcting codes with a high code rate (with low redundancy), in particular, for parity-check codes with the addition of only one parity symbol."

"It has been shown that the use of the symbol-by-symbol reception algorithm provides an energy gain of up to 1.5–2.0 dB over transmission and reception of phase-shift keyed signals without coding."

"The research and development of computational procedures for iterative reception based on the considered symbol-by-symbol reception algorithm for efficient code structures, e.g., for block turbo codes, formed on the basis of block codes in non-binary fields, is a promising direction of research."

Study of Iterative Reception Algorithms for Product Codes Based on Low-Density Codes of Finite Geometries [73]

This is a machine-generated summary of:

Nazarov, L. E.: Study of Iterative Reception Algorithms for Product Codes Based on Low-Density Codes of Finite Geometries [73]

Published in: Journal of Communications Technology and Electronics (2020)
Link to original: https://doi.org/10.1134/s1064226920060212

If you want to cite the papers, please refer to the original.

For technical reasons we could not place the page where the original quote is coming from.

Abstract-Summary

"We report the results of studies of iterative algorithms for symbol-by-symbol reception of product codes (block turbo codes) formed using the constituent low-density codes based on finite geometries (Euclidean geometry and projective geometry)."

Introduction

"At an increase in the information volumes of product codes and the application of iterative reception algorithms, probabilistic characteristics at receiving that are close to the limit theoretical characteristics are achieved [69, 74, 75]."

"The complexity of the developed algorithms for iterative reception of product codes is determined by the complexity of the algorithms for receiving constituent block codes."

"Series of derived product codes, iterative reception algorithms, which are modifications of the developed iterative reception algorithms that implement the principle of turbo decoding, were developed."

"Another direction of expanding the class of product codes is based on the use of constituent block low-density codes [69, 74], in particular, codes of finite geometries (Euclidean and projective geometries) [76, 77]."

"An actual problem is the development of reception algorithms for this class of product codes, as well as the study and comparative analysis of their probabilistic characteristics and complexity of implementation."

1. Problem Statement

"For product codes based on the constituent low-density codes, another series of iterative reception algorithms developed for low-density codes can be applied [69, 74]."

"The rationale for this statement is based on the fact that these product codes are also included in the class of low-density codes [76]."

"The essence of the problem is the development and study of the probabilistic characteristics of iterative reception algorithms for signal structures corresponding to product codes based on the constituent low-density codes of finite Euclidean and projective geometries."

"The results of modeling iterative reception algorithms in the presence of a channel interference model in the form of AWGN for a series of studied product codes are presented."

2. Low-Density Codes of Finite Geometries and Product Codes on Their Basis

"Are descriptions of low-density codes based on finite Euclidean geometry and finite projective geometry."

"We consider the most effective algorithms for iterative symbol-by-symbol reception of low-density codes: BP (belief propagation) algorithm and APP (a posteriori probability) algorithm and their modifications [69, 74, 78]."

4. App Iterative Reception Algorithm

"If the condition for stopping the algorithm operation is not fulfilled, then the next iteration is performed, starting from step 1."

"This simplifies the implementation complexity of the m-APP algorithm with respect to the APP algorithm."

"Regarding the required amount of memory, the m-APP algorithm is the most efficient."

5. Modeling Results

"The application of the m-BP algorithm determines the energy gain of up to 0.1 dB in relation to the BP reception algorithm."

"As was noted above, the m-APP algorithm is twice as efficient as the m-BP algorithm in the required number of computational operations."

"Curves 4 and 5 correspond to the application of the m-BP algorithm, and these curves practically coincide with the corresponding curves 2 and 3 for m-APP."

"Curves 4 and 5 correspond to the application of the m-BP algorithm, and these curves practically coincide with the corresponding curves 2 and 3 for m-APP."

Conclusions

"Iterative reception algorithms developed for the general class of low-density codes, in particular, BP, APP algorithms and their modifications m-BP and m-APP can be applied."

"For a series of considered product codes, iterative reception algorithms were modeled."

"For product codes from the considered series with a code rate of R > 0.3, the iterative reception algorithms m-BP and m-APP are also shown to be almost equivalent."

"The probability curves for the m-BP and m-APP algorithm (20 iterations) are close to the probability curves of the upper boundary of random coding."

Efficiency Study of Kravchenko Weight Windows at Receiving Signal Structures Based on OFDM Signals in the Presence of Spectrum-Concentrated Interference [79]

This is a machine-generated summary of:

Kravchenko, V. F.; Nazarov, L. E.; Pustovoit, V. I.: Efficiency Study of Kravchenko Weight Windows at Receiving Signal Structures Based on OFDM Signals in the Presence of Spectrum-Concentrated Interference [79]

Published in: Journal of Communications Technology and Electronics (2019)

Link to original: https://doi.org/10.1134/s1064226919100073

If you want to cite the papers, please refer to the original.

For technical reasons we could not place the page where the original quote is coming from.

Abstract-Summary

"We consider signal constructions based on OFDM signals and error-correcting codes that are resistant to the effects of spectrum-concentrated interference."

"An algorithm for receiving these signal constructions using weight windows and a criterion for optimal weight windows are given."

Introduction

"Orthogonal frequency-division multiplexing (OFDM) signals are basic signals for a number of accepted standards for broadband access (IEEE 802.12.11 is package transmission over local network channels (WLAN), and IEEE 802.16 is package information transmission over urban WMAN channels), as well as for a number of wireless data transmission standards (DAB and DVB are digital broadcasting and television, DVB-SH is digital satellite television, and 3GPP LTE is 4G cellular communication) [80, 81]."

"The prospects for using these signals in applications are determined by high spectral efficiency, as well as the possibility of organizing reliable information transmission over channels with multipath frequency-selective fading of signals, intersymbol interference, and nonstationarity of transmission channels [81]."

"For applications, it is important to increase the noise resistance of information transmission using OFDM signals in the presence of a class of spectrum-concentrated interference (SCI) [82, 83]."

"Specifying the optimality criteria for weight windows and their practical choice, determining the maximal noise resistance of information transmission using the signal structures under study, as well as the analysis of noise resistance during reception for the interference considered, is a topical problem."

1. Problem Statement

"The algorithms developed for receiving these signal constructions are based on the use of weight processing of input implementations [82, 84, 85]."

"Hamming and Kaiser windows were shown to be most effective for weight processing of the signal structures in the presence of SCIs [84, 85]."

"We analyze properties of the class of Kravchenko weight windows formed based on atomic functions [86, 87], which showed promise for use in the reception algorithms developed."

"The effect of this class of windows on the increase in interference resistance of the signal structures in the presence of SCIs was also studied by modeling the reception algorithm."

2. Description of Signal Constructions of OFDM Signals

"When applying weight processing, the signal/interference values at the demodulator output are reduced with respect to matched filtering using a rectangular window, and the orthogonality of the partial signals in (1) is disrupted, which causes the occurrence of interchannel interference (ICI) [80, 81]."

"When using a rectangular window (curve 1), a significant excess of side lobes of interference with respect to the frequency counts of the OFDM signal is seen."

"When using the Kaiser window [84, 88] (curve 2), a significant decrease in side lobe values (almost to −70 dB relative to the main lobe) causing an increase in interference resistance with respect to a rectangular window is observed."

"The optimality of the weight windows when solving the problem under consideration is based on the following criteria:—Ensuring the minimal values of window side lobes in the frequency domain.—".

4. Research Results of Kravchenko Weight Windows

"This code is included in the error-correcting codes recommended for satellite communication systems [89]."

"A convolutional coder is represented by a trellis diagram while optimal reception of signals corresponding to convolutional codes is performed based on the Viterbi algorithm, i.e., a dynamic search of possible paths is performed in a trellis diagram with the choice of the most plausible path [90]."

"Using the Kravchenko window in this case provides an energy gain of up to 0.75 dB relative to the Kaiser window."

Conclusions

"Signal constructions based on OFDM-signals and error-correcting codes that are resistant to the SCI effect are described."

"The algorithms developed for their reception in the presence of channel SCIs are based on an increase in the base of these signals relative to the base of the initial OFDM signals and based on the compensation of interference in the frequency domain."

"The reception algorithms include: an FFT with a weight window, a threshold limit for the values of spectral components, and calculation of solutions based on them that are used when receiving error-correcting codes."

"The modeling was performed in the presence of an AWGN channel and up to five SCIs with a signal-to-noise ratio of −20 dB. This model adequately describes a number of satellite transmission channels."

Signal Constructions Based on OFDM Signals Resistant to the Influence of Spectrum-Concentrated Interference [91]

This is a machine-generated summary of:

Nazarov, L. E.: Signal Constructions Based on OFDM Signals Resistant to the Influence of Spectrum-Concentrated Interference [91]

Published in: Journal of Communications Technology and Electronics (2019)
Link to original: https://doi.org/10.1134/s106422691907012x

If you want to cite the papers, please refer to the original.

For technical reasons we could not place the page where the original quote is coming from.

Abstract-Summary

"A description is given of a class of signal structures that are resistant to the effects of narrowband interference based on OFDM signals and error-correcting coding."

"Algorithms for receiving these signal structures are given using the weighted processing of input implementations in the presence of channel interference under consideration and the criterion of weight windows is given."

Formulation of the Problem

"The known methods of reducing the efficiency of spectrum-concentrated interference are based on an increase in the base of the signals used or the use of methods to compensate for this interference [92]."

"We can increase the reliability of the information transfer in the presence in the channel of the considered interference concentrated on the spectrum of interference using signal constructions based on OFDM signals."

"In the presence of a group of a lumped interference spectrum, the considered signal constructions and their processing algorithms are more effective."

"The purpose of this work is to analyze the noise immunity of the considered signal constructions, formed based on the OFDM signals and error-correcting coding in combination with the algorithms for the compensation of the concentrated spectrum interference, which is the core of the problem."

Signal Constructions Based on OFDM Signals

"The considered noise compensation efficiency and noise immunity of the signal structures in combination with the reception algorithms is determined by the choice of the weight windows [84, 88]."

"The use of weight windows in the processing of input implementations leads to a decrease in the side lobe values relative to the main lobe value in the frequency domain."

"The main characteristics of these windows (the width of the main lobe and the values of the amplitudes of the side lobes) are set in terms of the parameters, varying the values of which makes it possible to construct the considered weight windows with the properties of the petals being close to the optimal properties."

"Consider the Kaiser and Dolph–Chebyshev windows used in the reception algorithms of the signal structures under study."

Results of the Studies of the Error-Performance Characteristics of Signal Structures

"The error-performance characteristics were studied when receiving signal constructions based on OFDM signals and error-correcting coding for a convolutional code with the code rate 1/2."

"A trellis diagram is a convolutional code coder and is given by generator polynomials."

"The optimal reception of the signals corresponding to the convolutional code is performed using the Viterbi algorithm: a dynamic search is performed for possible paths along a code lattice with a choice of the most plausible path."

"The convolutional code used in the study is specified by the generating polynomials in the octal representation (133, 171)."

"P_{ICI} was also evaluated based on the correlation of the partial signal and OFDM signals without this partial signal, taking into account the influence of the weight window."

Conclusions

"The reception algorithms include an FFT with a weighted window, a threshold limit on the values of spectral components, and the calculation of solutions based on them that are used when receiving error-correcting codes."

"The noise immunity analysis for the signal structure generated using a 1/2 by convolutional code has been made by simulating the developed reception algorithms using the Kaiser and Dolph–Chebyshev weight windows."

"When using simpler processing with a hard limitation from the output of the signal demodulator, the additional energy loss is 0.4–2.0 dB. The study of the error-performance characteristics of the considered signal structures formed using effective error-correcting codes (codes with iterative decoding algorithms and block codes using soft reception solutions), the search and analysis of a wider class of weight windows with optimality properties is the subject of future studies."

[Section 5]

"INTRODUCTION Using signals with orthogonal frequency multiplexing, known in the literature as OFDM signals (orthogonal frequency division multiplexing), it is possible to organize communication with multipath propagation, which causes frequency-selective fading of signals, intersymbol interference, and nonstationarity of transmission channels [2–4]."

"This determines the effectiveness of the influence of interference of this type on the noise immunity of information transmission using OFDM signals."

"The problem of increasing the noise immunity of information transmission systems using OFDM signals in the presence of a class of interference data [93] is relevant."

"The studies [88, 93, 94] suggested signal constructions based on OFDM signals, as well as error-correcting coding and interleaving, which are received using frequency-domain interference compensation algorithms."

Bibliography

1. Urrea, C., Kern, J., & López-Escobar, R. (2022). Design of an interleaver with criteria to improve the performance of turbo codes in short block lengths. *Wireless Networks*. https://doi.org/10.1007/s11276-022-02912-2
2. Feng, W., Yuan, J., & Vucetic, B. S. (2002). A code-matched interleaver design for turbo codes. *IEEE Transactions on Communications, 50*, 926–937. https://doi.org/10.1109/TCOMM.2002.1010612
3. Jerkovits, T., & Matuz, B. (2016). Turbo code design for short blocks. In *2016 8th Advanced Satellite Multimedia Systems Conference and the 14th Signal Processing for Space Communications Workshop (ASMS/SPSC)* (pp. 1–6). IEEE. https://doi.org/10.1109/ASMS-SPSC.2016.7601543
4. Trifina, L., Tarniceriu, D., Ryu, J., & Rotopanescu, A. M. (2020). Further upper bounds on the minimum distance for turbo codes using CPP interleavers. *Physics Communication, 43*, 101212. https://doi.org/10.1016/j.phycom.2020.101212
5. Trifina, L., Tarniceriu, D., Ryu, J., & Rotopanescu, A. M. (2020). Some lengths for which CPP interleavers have weaker minimum distances than QPP interleavers. *Journal of the Franklin Institute, 357*, 3097–3112. https://doi.org/10.1016/J.JFRANKLIN.2020.02.031
6. Sharma, S., Sau, P. C., & Shukla, A. (2014). Performance analysis of S-Random interleaver for IDMA system using MRRC diversity scheme. In *2014 Seventh International Conference on Contemporary Computing (IC3)* (pp. 248–253). IEEE. https://doi.org/10.1109/IC3.2014.6897181
7. Hu, X. Y., Eleftheriou, E., & Arnold, D. M. (2005). Regular and irregular progressive edge-growth tanner graphs. *IEEE Transactions on Information Theory, 51*, 386–398. https://doi.org/10.1109/TIT.2004.839541
8. Woldamanuel, E. M., & Diba, F. D. (2022). Enhanced adaptive code modulation for rainfall fade mitigation in Ethiopia. *EURASIP Journal on Wireless Communications and Networking*. https://doi.org/10.1186/s13638-021-02085-0
9. ITU-R. (2017). *Propagation data and prediction methods required for the design of terrestrial line-of-sight systems*. Recommendations. ITU-R P.530-17.
10. Freeman, R. L. (2006). *Radio system design for telecommunications* (3rd ed.). Wiley.
11. Crane, R. K. (2003). *Propagation handbook for wireless communication system design*. CRC Press.
12. Diba, F. D., Samad, M. A., & Choi, D.-Y. (2021). The effects of rain on terrestrial links at K, Ka and E-bands in South Korea: Based on supervised learning. *IEEE Access, 9*, 9345–9355.
13. Luini, L., Roveda, G., Zaffaroni, M., Costa, M., & Riva, C. G. (2019). The impact of rain on short E-band radio links for 5G mobile systems: Experimental results and prediction models. *IEEE Transactions on Antennas and Propagation, 68*(4), 3124–3134.
14. Shrestha, S., & Choi, D.-Y. (2017). Rain attenuation statistics over millimeter wave bands in South Korea. *Journal of Atmospheric and Solar-Terrestrial Physics, 152*, 1–10.
15. Lam, H. Y., Luini, L., Din, J., Capsoni, C., & Panagopoulos, A. D. (2012). Investigation of rain attenuation in equatorial Kuala Lumpur. *IEEE Antennas and Wireless Propagation Letters, 11*, 1002–1005.
16. Lutz, E., Werner, M., & Jahn, A. (2012). *Satellite systems for personal and broadband communications*. Springer.
17. Hadji, A., Bouazza, B. S., Boeglen, H., Perrine, C., Nouri, K., Chatellier, C., & Pousset, Y. (2021). Shaping duo binary turbo-coded BICM scheme for JPWL image transmission using a link adaptation strategy over wireless channels. *EURASIP Journal on Wireless Communications and Networking*. https://doi.org/10.1186/s13638-021-01949-9
18. Abot, J., Olivier, C., Perrine, C., & Pousset, Y. (2000). A link adaptation scheme optimized for wireless JPEG, transmission over realistic MIMO systems. *Signal Processing: Image Communication, 27*, 2012.

19. Mhamdia, M., Perrine, C., Zribia, A., Pousset, Y., Olivier, C., & Bouallegue, A. (2017). Soft decoding algorithms for optimized JPEG 2000 wireless transmission over realistic MIMO-OFDM systems. *Signal Processing: Image Communication, 52*, 41–53.
20. Hadji, A., Bouazza, B. S., Boeglen, H., Perrine, C., Nouri, K., Chatellier, C., & Pousset, Y. (2018). Duo binary turbo coded 256 QAM with constellation shaping. *IEEE International Conference on Telecommunications*. https://doi.org/10.1109/ICT.2018.8464875
21. Devi, T. K., Priyanka, E. B., Sakthivel, P., & Sagayaraj, A. S. (2021). Low complexity modified Viterbi decoder with convolution codes for power efficient wireless communication. *Wireless Personal Communications*. https://doi.org/10.1007/s11277-021-08919-w
22. Raviv, T., Park, S., Shlezinger, N., Simeone, O., Eldar, Y. C., & Kang, J. (2021). Meta-ViterbiNet: Online meta-learned Viterbi equalization for non-stationary channels. arXiv preprint arXiv:2103.13483.
23. Cavallaro, J., & Vaya, M. (2003). VITURBO: A reconfigurable architecture for Viterbi and Turbo decoding. In *IEEE International Conference on Acoustics Speech and, Signal Processing* (pp. 497–500).
24. H. Singh, R. K. Vyas and Deepak Raghuvanshi 2012 Comparative analysis of modified register exchange method and trace back method of Viterbi decoder for wireless communication". International Journal of Advances in Engineering & Technology, 4(2):148–155
25. Santhi, M., Lakshminarayanan, G., Sundaram, R., & Balachandar, N. (2019). Synchronous pipelined two stage Radix-4 200Mbps MB-OFDM UWB Viterbi decoder on FPGA. *IEEE ISOCC, 89*(12), 468–471.
26. Guo, M., Ahmad, M., Swamy, M., & Wang, C. (2005). FPGA design and implementation of a low-power systolic array-based adaptive Viterbi decoder. *IEEE Transactions on Circuits and Systems, I*(52), 350–365.
27. McDonald, E. (2008). Runtime FPGA partial reconfiguration. In *IEEE Aerospace Conference* (pp. 1–7).
28. Dadoria, A. K., Khare, K., Gupta, T. K., & Khare, N. (2017). Integrating sleep and pass transistor logic for leakage power reduction in FinFET circuits. *Journal of Computational Electronics, 16*, 867–874.
29. Yadav, M., Shokeen, V., & Singhal, P. K. (2019). Flip left-to-right approach based inverse tree interleavers for unconventional integrated OFDM-IDMA and SCFDMA-IDMA systems. *Wireless Personal Communications*. https://doi.org/10.1007/s11277-019-06133-3
30. Yadav, M., Shokeen, V., & Singhal, P. K. (2016). BER versus BSNR analysis of conventional IDMA and OFDM-IDMA based systems with tree interleaving. In *IEEE 2nd International Conference on Advances in Computing, Communication and Automation (ICACCA)-Fall'16.*
31. Shukla, M., Srivastava, V. K., & Tiwari, S. (2008). Analysis and design of tree based interleaver for multiuser receivers in IDMA scheme. In *Proceedings of the IEEE 16th International Conference on Networks ICON'08, New Delhi, India* (pp. 978–981).
32. Yadav, M., Gautam, P. R., Shokeen, V., & Singhal, P. K. (2017). Modern Fisher-Yates shuffling based random interleaver design for SCFDMA-IDMA system. *Wireless Personal Communications, 97*(1), 63–73.
33. Yadav, M., Shokeen, V., & Singhal, P. K. (2017). Uncoded integrated interleave division multiple access systems in presence of power interleavers. *Radioelectronics and Communications Systems, 60*(11), 503–511.
34. Mahafeno, I., Langlais, C., & Jego, C. (2006). OFDM-IDMA versus IDMA with ISI cancellation for quasi-static Rayleigh fading multipath channels. In *Proceedings of the 4th International Symposium Turbo Codes* (pp. 140–144).
35. Ping, L., Guo, Q., & Tong, J. (2007). The OFDM-IDMA approach to wireless communication systems. *IEEE Wireless Communications, 14*, 18–24.
36. Wang, Z. H., & Yu, G. F. (2013). Study on simulation performance of OFDM-IDMA system. *Applied Mechanics and Materials, 380–384*, 4120–4123.
37. Bie, H., & Bie, Z. (2006). A hybrid multiple access scheme: OFDMA-IDMA. In *IEEE 1st International Conference on Communication and Networking in Beijing, China* (pp. 1–3).

38. Xiong, X., & Luo, Z. (2011). SC-FDMA-IDMA: A hybrid multiple access scheme for LTE uplink. In *Proceedings of the 7th IEEE International Conference on Wireless Communication Networking and Mobile Computing (WiCOM)'11, Wuhan, China* (pp. 1–5).
39. Yadav, M., Shokeen, V., & Singhal, P. K. (2017). Flip left right approach based inverse tree interleaver design for IDMA scheme. *AEÜ: International Journal of Electronics and Communications, 81*, 182–191.
40. Wu, H., Ping, L., & Perotti, A. (2006). User-specific chip-level interleaver design for IDMA systems. *IEEE Electronics Letters, 42*(4), 233–234.
41. Shukla, M., Srivastava, V. K., & Tiwari, S. (2009). Analysis and design of optimum interleaver for iterative receivers in IDMA scheme. *Wireless Communication and Mobile Computing, 9*, 1312–1317.
42. Savaşcıhabeş, A. A. (2022). Novel combined double binary turbo coding and color shift keying technique for flicker mitigation in multi carrier visible light communications. *Wireless Personal Communications*. https://doi.org/10.1007/s11277-022-10053-0
43. Qiu, Y., Chen, H.-H., & Meng, W.-X. (2016). Channel modeling for visible light communications—A survey. *Wireless Communications Mobile Computing, 16*, 2016–2034.
44. IEEE Standard Association. (2011). IEEE standard for local and metropolitan area networks–part 15.7: Short-range wireless optical communication using visible light. IEEE Std 802.15.7-2011 (pp. 1–309).
45. Farhan, I. M., Zaghar, D. R., & Abdullah, H. N. (2023). Performance improvement of CDMA wireless sensor networks in low SNR channels based on raptor codes. *Wireless Personal Communications*. https://doi.org/10.1007/s11277-023-10387-3
46. Marcu, I., & Halunga, S. V. (2016). Implication of LDPC technique in non-ideal multiuser communication system. *Wireless Personal Communications, 87*, 797–814. https://doi.org/10.1007/s11277-015-2627-4
47. Rasadurai, K., & Kumaratharan, N. (2012). Performance enhancement of MC-CDMA system through Turbo multi-user detection K. Rasadurai. In *2012 International Conference on Computer Communication and Informatics (ICCCI-2012).*
48. Shirvanimoghaddam, M., & Johnson, S. (2016). Raptor codes in the low SNR regime. *IEEE Transactions on Communications, 64*(11), 4449–4460. https://doi.org/10.1109/TCOMM.2016.2606410
49. Babiker, M., Khalifa, O. O., Hashim, A. H. A., Salami, M. J. E., & Ahmed, M. Z. (2017). Performance of turbo code in CDMA under AWGN channel. *International Journal of Future Generation Communication and Networking, 10*(5), 19–28. https://doi.org/10.14257/ijfgcn.2017.10.5.02
50. Shah, D. J., Patel, V. K., & Patel, H. A. (2010). Performance analysis of turbo code for CDMA 2000 with convolutional coded IS-95 system in wireless communication system. In *ICECT 2010—Proceedings of the 2010 2nd International Conference on Electronic Computer Technology* (pp. 42–45). https://doi.org/10.1109/ICECTECH.2010.5479994
51. Farhan, I. M., Zagha, D. R., & Abdullah, H. N. (2022). Enhancement of code division multiple access system performance using raptor codes. *Indonesian Journal of Electrical Engineering and Computer Science, 26*(3), 1460–1468. https://doi.org/10.11591/ijeecs.v26.i3.pp1460-1468
52. Etesami, O., & Shokrollahi, A. (2006). Raptor codes on binary memoryless symmetric channels. *IEEE Transactions on Information Theory, 52*(5), 2033–2051. https://doi.org/10.1109/TIT.2006.872855
53. Chatzigeorgiou, I. A., Rodrigues, M. R. D., Wassell, I. J., & Carrasco, R. A. (2007). Comparison of convolutional and turbo coding for broadband FWA systems. *IEEE Transactions on Broadcasting, 53*(2), 494–503. https://doi.org/10.1109/TBC.2007.893387
54. Berrou, C., & Glavieux, A. (1996). Near optimum error correcting coding and decoding: Turbo-codes. *IEEE Transactions on Communications, 44*(10), 1261–1271.
55. Lee, Y. L., Loo, J., & Chuah, T. C. (2015). *Modeling and performance evaluation of resource allocation for LTE femtocell networks* (pp. 683–716). Elsevier Inc..

56. Chopra, S. R., & Gupta, A. (2021). Error analysis of grouped multilevel space-time Trellis coding with the combined application of massive MIMO and cognitive radio. *Wireless Personal Communications*. https://doi.org/10.1007/s11277-020-07878-y
57. Woo, T.-K. (2021). Patching based extra short packet forward error control coding for ultra-reliable low latency communications (URLLC) in 5G. *Wireless Personal Communications*. https://doi.org/10.1007/s11277-021-08815-3
58. Telagam, N., Lakshmi, S., & Kandasamy, N. (2021). Performance analysis of parallel concatenation of LDPC coded SISO-GFDM system for distinctive pulse shaping filters using USRP 2901 device and its application to WiMAX. *Wireless Personal Communications*. https://doi.org/10.1007/s11277-021-08865-7
59. Mostari, L., & Taleb-Ahmed, A. (2018). High performance short-block binary regular LDPC codes. *Alexandria Engineering Journal, 57*(4), 2633–2639.
60. Barakatain, M., & Kschischang, F. R. (2018). Low-complexity concatenated LDPC-staircase codes. *Journal of Lightwave Technology, 36*(12), 2443–2449.
61. Merah, H., Mesri, M., & Tahkoubit, K. (2018). Significant reduction of bit error rate in 5G-Gfdm system using concatenated turbo code with low density parity check code in fading channels. *Journal of Electrical Systems, 14*(1), 118–129.
62. National Instruments. *USRP-2901 block diagram*. http://www.ni.com/documentation/en/usrpsoftware-defined-radio-device/latest/usrp2901/block-diagram/
63. de Alencar, R. R. M., Landau, L. T. N., & de Lamare, R. C. (2020). Continuous phase modulation with 1-bit quantization and oversampling using iterative detection and decoding. *EURASIP Journal on Wireless Communications and Networking*. https://doi.org/10.1186/s13638-020-01856-5
64. Gilbert, E. N. (1993). Increased information rate by oversampling. *IEEE Transactions on Information Theory, 39*(6), 1973–1976.
65. Landau, L., Dörpinghaus, M., & Fettweis, G. P. (2017). 1-Bit quantization and oversampling at the receiver: Communication over bandlimited channels with noise. *IEEE Communications Letters, 21*(5), 1007–1010.
66. Landau, L. T. N., Dörpinghaus, M., & Fettweis, G. P. (2018). 1-Bit quantization and oversampling at the receiver: Sequence-based communication. *EURASIP Journal on Wireless Communications and Networking, 17*, 83.
67. Rimoldi, B. E. (1988). A decomposition approach to CPM. *IEEE Transactions on Information Theory, 34*(2), 260–270.
68. Nazarov, L. E., & Batanov, V. V. (2022). Analysis of noise immunity of optimal symbol-by-symbol reception of phase-keyed signals with correcting codes in non-binary Galois fields. *Journal of Communications Technology and Electronics*. https://doi.org/10.1134/s1064226922080137
69. Johnson, S. J. (2010). *Iterative error correction: Turbo, low-density parity-check and repeat-accumulate codes*. Cambridge University Press.
70. Ping, L., Chan, S., & Yeng, K. L. (1997). *Electronics Letters, 33*(19), 1614.
71. Nazarov, L. E. (2002). *Journal of Communications Technology and Electronics, 47*, 1346.
72. Steiner, F., Bocherer, G., & Liva, G. (2018). *IEEE Communications Letters, 22*, 2210.
73. Nazarov, L. E. (2020). Study of iterative reception algorithms for product codes based on low-density codes of finite geometries. *Journal of Communications Technology and Electronics*. https://doi.org/10.1134/s1064226920060212
74. Li, J., Lin, S., Abdel-Chaffar, K., et al. (2017). *LDPC code designs, constructions and unification*. Cambridge University Press.
75. Nazarov, L. E., Batanov, V. V., & Kuznetsov, O. O. *Zhurnal Radioelektroniki, 9*. Retrieved from http://jre.cplire.ru/jre/sep14/1/text.pdf
76. Nazarov, L. E., & Shishkin, P. V. (2018). *Inform Tekhnol, 24*, 427.
77. Nazarov, L. E., & Shishkin, P. V. *Zhurnal Radioelektroniki, 5*. Retrieved from http://jre.cplire.ru/jre/may18/1/text.pdf
78. Sunil, K., Jayaraj, P., Soman, K. P. (2012). *Message Passing Algorithm: A Tutorial Review IOSR Journal of Computer Engineering, 2*(3), PP 12–24.

79. Kravchenko, V. F., Nazarov, L. E., & Pustovoit, V. I. (2019). Efficiency study of Kravchenko weight windows at receiving signal structures based on OFDM signals in the presence of spectrum-concentrated interference. *Journal of Communications Technology and Electronics*. https://doi.org/10.1134/s1064226919100073
80. Bakulin, M. G., Kreindelin, M. G., Shloma, A. M., & Shumov, A. P. (2016). *Technology OFDM*. GoryachayaLiniya-Telekom. [in Russian].
81. Vishnevskii, V. M., Lyakhov, A. I., Portnoi, S. L., & Shakhnovich, I. V. (2005). *Broadband networks of transfer*. Tekhnosfera. [in Russian].
82. Darsena, D., & Verde, F. (2008). *IEEE Signal Processing Letters, 15*(873).
83. Al-Tous, H., Barhumi, I., & Al-Dhahir, N. (2016). *Proc. 12th IEEE Int. Conf. on Wireless and Mobile Computing, Networking and Communications (WiMob 2016), New York, USA, Oct. 17–19*. IEEE.
84. Zudilin, A. A., & Nazarov, L. E. (2017). *Journal of Radio Electronics, 11*. Retrieved from http://jre.cplire.ru/jre/nov17/4/text.pdf
85. Nazarov, L. E., & Zudilin, A. S. (2018). *Fiz Osnovy Priborostroen, 7*(3), 26.
86. Kravchenko, V. F., & Kravchenko, O. V. (2018). *Constructive methods of algebra of logic, atomic functions, wavelets, fractals in problems of physics and the equipment*. Tekhnosfera. [in Russian].
87. Kravchenko, V. F., Konovalov, Y. Y., & Pustovoit, V. I. (2015). *Journal of Communications Technology and Electronics, 60*, 986.
88. Nazarov, L. E., & Zudilin, A. S. (2018). *Journal of Radio Electronics, 3*. Retrieved form http://jre.cplire.ru/jre/mar18/4/text.pdf
89. *TM synchronization and channel coding—Summary of concept and rationale* (2006). Inf. Rep. CCSDS 130.1-G-1, Green Book.
90. Zyuko, A. G., Fal'ko, A. I., Panfilov, I. P., et al. (1985). *Noise stability and efficiency of systems of information transfer*. Radio iSvyaz'. [in Russian].
91. Nazarov, L. E. (2019). Signal constructions based on OFDM signals resistant to the influence of spectrum-concentrated interference. *Journal of Communications Technology and Electronics*. https://doi.org/10.1134/s106422691907012x
92. Borisov, V. I., Zinchuk, V. M., Limarev, A. E., & Shestopalov, V. I. (2011). *Interference protection of radio communication systems by means of signal spectrum spreading provided by modulation of the carrier frequency with a pseudorandom sequence* (2nd ed.). Radiosoft. [in Russian].
93. Liu, H., & Li, G. (2005). *OFDM-based broadband wireless networks*. Wiley.
94. Nazarov, L. E., & Zudilin, A. S. (2013). *Izvestiya Vysshikh Uchebnykh Zavedenii, Elektronika, 6*, 45.

Chapter 6
Optimization of Network latency

Milind Pande, Anand J. Kulkarni, and Apoorva S. Shastri

Introduction by the Editor

For any communication system network latency, also called network lag, is the time it takes for a request to travel from the sender to the receiver and for the receiver to process that request. Networks with smaller communication delays are called low-latency networks. In the case of high latency networks delays are longer affecting the performance of the system. So, to have the best user experience, optimization of the network latency is very much needed. The chapter discusses the Modified dropping-random early detection (MD-RED) algorithm, Minimum transmission delay algorithm for multi-hop nodes, Hybrid Fuzzy-Genetic Algorithm, Congestion avoidance with source location privacy, multi-objective-based deployment of throw boxes, Taxonomy of controller placement problem (CPP) optimization, JOSP: Joint Optimization of Flow Path Scheduling and Virtual Network Function Placement, etc. The chapter also reviews the optimization methods for unstructured, scalable, and 6G networks. In addition, the chapter also analyses some protocols related to WSN viz., energy efficient routing protocols with optimization, Asynchronous Anycast Cross-layer Protocol, Cross-Layer Protocol, and Data Fusion and Cluster-Based Multi-hop Relay-Assisted Routing Protocol.

Network congestion often results in increase in end-to-end delay. To address this issue active queue management (ACM) algorithm and random early detection is used at the router, however, scope remains to reduce average queue size and

M. Pande
Dr. Vishwanath Karad MIT World Peace University, Pune, Maharashtra, India
e-mail: milind.pande@mitwpu.edu.in

A. J. Kulkarni · A. S. Shastri (✉)
Institute of Artificial Intelligence, Dr. Vishwanath Karad MIT World Peace University, Pune, Maharashtra, India
e-mail: anand.j.kulkarni@mitwpu.edu.in

M. Pande et al. (eds.), *Optimization Methods in Mobile Communication Systems*, https://doi.org/10.1007/978-981-95-1810-4_6

network delay improvement. New version of modified dropping-RED uses a different approach based on traffic type for better control over network congestion by reducing network delay. If the traffic type is light and moderate it applies linear drop function and if traffic type is highly congested, then it applies exponential drop function. This algorithm helps to manage traffic effectively and enhance the performance of the network.

The requirement of high QoS performance with the constraint of limited node power demands an energy-efficient data routing protocol for WBAN systems. Fuzzy-Genetic Algorithm is use for cluster head selection in a group of clusters. Each cluster node is examined for cluster head selection cost and residual energy. CH selection cost of a node is the forecasted value of network energy consumption if the node acts as a cluster head. This hybrid approach utilizes a fuzzy-TOPSIS function which makes a quantitative comparison of cluster nodes and selects the cluster head node possessing the attributes closest to their ideally desired values. This algorithm has advantage of equal sharing of routing load and increase network lifetime.

Networks containing IoT nodes requires efficient utilization of networks in such a way that packet loss rate and transmission delay is less. In emergency routing, emergency data should get more priority than normal data to reduce delay for emergency data. Minimum transmission delay for multi hop nodes contains prioritization mechanism which lead to more resource allocation to emergency packets and thus reduction in transmission delay of these date. According to this algorithm priority is set at header of emergency packets and spatial roots gets reserved for these packets resulting in reduced end-to-end delay. Energy efficient routing protocols with optimization helps to boost the network life, increase power efficiency and resolve the problems of fault tolerance in wireless sensor networks, we reviewed numerous literature surveys concerning energy optimize routing protocols.

The optimal path for the data transmission process is obtained using the proposed octopus structure-based routing (OSBR) scheme. This routing protocol considered the distance between the CH and sink as the main objective of finding the optimal path for data transmission. This protocol is based on two techniques. First is energy consumption in the network occurs whenever there is a data transmission in a node and second is When there is a higher arrival rate of data packets to the sink node exceeding the service rate, the arriving packets must wait for some time until the sink can process them.

IoT applications in a smart city paradigm collect air quality data and temperature measurements through the sensor nodes that can be buffered until they find an opportunity (throwbox) to forward. Dedicated stationary nodes, called throwboxes are employed to equip the IoT networks for better communication Anticipating movements of the nodes is a significant challenge in IoT networks. The entire IoT network lacks in movements of nodes globally and finally the network becomes disconnected with a number of partitions. Multi-Objective Differential Evolution and a popular Non-Dominated Sorting Genetic Algorithm-II helps in finding the optimal deployment location of throwboxes by maximizing the coverage of all the throwboxes, minimizing the average delay and maximizing the delivery ratio among all the nodes.

Software Defined Network is a network architecture which is programmable controllable and flexible though the scalability is the most important challenge. Multiple controllers are a necessity of current SDN, so the optimum number of controllers and their placement is a problem called controller placement problem (CPP). It is NP hard problem

As the service requests are highly dynamic and different services require distinct bandwidth occupation amount and time, Virtual network function need to place to schedule the paths of flows which guarantee the performances of network applications and maximize the utilization of the underlying network is a challenging problem. Joint Optimization of Flow Path Scheduling (JOSP) explores the best utilization of bandwidth from two different aspects to reduce the network delay, first is delay scaling strategy that adds the penalty to the link bandwidth occupation that may cause congestion in accordance with the network placement locations. Second is reinforcement learning framework and use both the flow path delay and the network function-related delay to calculate the reward of placing VNFs adaptively.

Latency is a problem for asynchronous MAC due to sleep-delay, since each relay must wait for the receiver to wake-up. This problem is addressed by anycast cross-layer protocol that takes advantage from geographic routing and preamble sampling for low latency communication in duty cycled networks.

Wireless sensor network is more prone to energy consumption and highly susceptible to loss of connectivity. Implementation of Power efficient compressive data fusion and cluster based multihop relay assisted protocol can strengthen the network's communication reliability and efficiency. K means algorithm is used to nominate cluster head based on multi weight functions. Then relay is chosen from k cells which for broadcasting.

Machine Generated Summaries

Disclaimer: The summaries in this chapter were generated from Springer Nature publications using extractive AI auto-summarization: An extraction-based summarizer aims to identify the most important sentences of a text using an algorithm and uses those original sentences to create the auto-summary (unlike generative AI). As the constituted sentences are machine selected, they may not fully reflect the body of the work, so we strongly advise that the original content is read and cited. The auto generated summaries were curated by the editor to meet Springer Nature publication standards. To cite this content, please refer to the original papers.

Machine generated keywords: node, sensor, iot, sensor network, protocol, route, wsn, wsns, wireless sensor, route protocol, cluster, packet, sink, sensor node, congestion

Modified Dropping-Random Early Detection (MD-RED): A Modified Algorithm for Controlling Network Congestion [1]

This is a machine-generated summary of:

Hassan, Samuel; Rufai, Adewole: Modified dropping-random early detection (MD-RED): a modified algorithm for controlling network congestion [1]

Published in: International Journal of Information Technology (2023)

Link to original: https://doi.org/10.1007/s41870-023-01201-1

If you want to cite the papers, please refer to the original.

For technical reasons we could not place the page where the original quote is coming from.

Abstract-Summary

"This problem of high end-to-end delay is due in part to the presence of a congested network."

"One algorithm with a high hype in the research area is the random early detection (RED) proposed well over 2 decades ago."

"Previous enhancements to RED are unable to convincingly reduce the average queue size and improve network delay."

"Reports from experiments conducted in ns-3 open-source simulation tool confirms that MD-RED suitably controls network congestion and offers an appreciable improvement in network delay."

Introduction

"Queue management algorithms are implemented at the routers and addresses congestion by means of dropping or accommodating packets whenever its necessary at the buffer [2, 3]."

"AQM algorithm offers a vast improvement over passive queue management or PQM which drop packets only when the buffer is fully saturated."

"It is worth noting in passing that the choice of a suitable drop function used in an AQM algorithm is of prime importance, as it to a large extent, impacts its effectiveness in handling network congestion [4, 5]."

"Even though, these existing dropping function-based AQM algorithms achieved a better performance than RED algorithm, there is still a need for a modified RED-dependent algorithm with an improved dropping function that could provide a substantial shortening of the average queue size."

Related Works

"The algorithm is considered an enhanced RED version as it operates a pair of packet dropping functions (that is, a linear with one more linear)."

"To achieve an improved network throughput, researchers in Ref. [6] proposes an improved adaptive RED algorithm in which the linear drop function of RED was replaced with a cubic drop function."

"The drop probability is computed as follows:As could be noted from prior described algorithms that a significant and visible feature of RED is the use of a linear dropping function which has made it appealing to several scientists, thereby making RED a research focus."

"Sequel to the aforementioned premise, in this paper, wherefore, the RED algorithm is again researched with a view to developing a new dropping function-based AQM algorithm that yields a substantial shortening of average queue size and provides an effective control of network congestion."

Modified Dropping-Random Early Detection (MD-RED) Algorithm

"A new AQM algorithm which enhances RED is described, called modified dropping-random early detection or MD-RED."

"RED uses a linear drop function which causes the problem of large network delay."

"It therefore turns out that four situations are feasible: 1."

Simulation and Performance Evaluation

"An evaluation and benchmarking the performance of MD-RED against two improved RED proposals, namely TRED and NLRED algorithms were made under low, moderate, and high network traffic scenarios."

"In this simulation experiment, the number of TCP flows considered is 20 that represents a moderate congestion scenario."

"MD-RED model exhibited a good control of the average queue size when compared with NLRED and TRED algorithms."

"It is again observed that MD-RED provides an improved performance than both NLRED and TRED algorithms."

"In this simulation experiment, the number of TCP flows considered is 50 that represents a high congestion scenario."

Conclusion

"A new RED-based algorithm, named MD-RED has been developed."

"MD-RED operates two-dropping function compositions and further divides the queue region between the two thresholds of RED into two partitions depending on network traffic loads (i.e. low and moderate region, high region)."

"Simulation results from various performance evaluations carried out in ns-3 evidently shows that MD-RED is superior to two prior better-known amended versions of RED algorithm (namely NLRED and TRED) and effectively addresses network congestion."

"Work, effort will be geared regarding implementing MD-RED algorithm for queue management in wireless ad-hoc networks."

Congestion Avoidance with Source Location Privacy Using Octopus-Based Dynamic Routing Protocol in WSN [7]

This is a machine-generated summary of:

Singh, Mintu; Singh, Maheshwari Prasad: Congestion avoidance with source location privacy using octopus-based dynamic routing protocol in WSN [7]

Published in: Wireless Networks (2022)

Link to original: https://doi.org/10.1007/s11276-022-03165-9

If you want to cite the papers, please refer to the original.

For technical reasons we could not place the page where the original quote is coming from.

Abstract-Summary

"Wireless sensor network (WSN) is a collection of geographically distributed sensor nodes (SNs) responsible for data collection and transmission over the Internet."

"This article proposes an efficient dynamic routing protocol that addresses this issue with privacy solutions and congestion avoidance."

"An optimal node is elected as the cluster head for each cluster using the proposed adaptive jellyfish search optimizer algorithm."

"The optimal path for the data transmission process is obtained using the proposed octopus structure-based routing scheme."

Introduction

"Providing efficient routing with privacy-aware source location is identified as the unsettled and significant issue in WSN privacy [8]."

"Further, the mobile nodes in WSN make the network even worse by allowing the sensor nodes (SNs) to expose to different attackers from various geographical locations [9, 10]."

"The contextual information corresponding to each communication, like source location or message source, encrypted data can be easily identified by adversaries."

"To ensure privacy, the absence of location privacy may reveal network congestion, whether the data is encrypted or not."

"Many routing protocol based techniques for securing the location of the source node have been developed, but none have attained a promising result."

"Another important factor to consider in the field of WSN is the privacy of the source location."

"While transferring private information via the network, it is important to maintain the privacy of the source location."

Related Work

"The TN then collects the data from the routing path and sends it to the base station, assured that the attackers will not identify the location's position, and the data will be transmitted with less delay."

"The packets are delivered based on multiple sink nodes by generating many dynamic routings using a packet-slicing transmission strategy."

"One of the main causes of energy consumption in WSN is the dropping of data packets due to congestion in the network."

"Technological improvement may raise a question about security during the data transmission from the source to the sink node because the third party (enemy) will forge the data using a traffic analysis attack or eavesdropper."

"Some other issues, like congestion avoidance, spectral efficiency, power efficiency, etc , are also faced by various existing techniques while transmitting the data from the source to the sink node."

System Model of the Proposed Method

"A novel attacker model, known as target area-based assaults is introduced as a complementary inside model, allowing networking assaults, such as sink attacks, source attacks, and neighbour node assaults."

"Identifying target data packets can determine the sink node path by calculating sink movement or monitoring network flow data."

"Data packets can be detected, and attackers can capture the path the target node takes as it passes via the destination."

"Each cluster has at least one CH, which collects data from all SNs and transfers it to the receiver node or the BS."

"We tend to make the following assumptions: (1) Clusters of SNs are formed (2) Each cluster has a head node responsible for sending data from nearby head nodes and client nodes to the sink. (3) The sensible ability of all SNs is the same, and the initial battery power cannot be varied."

Proposed Methodology

"The proposed routing protocol considered the distance between the CH and sink as the main objective of finding the optimal path for data transmission."

"The mathematical formulations and fitness function of the proposed OSBR routing technique are detailed as follows: Energy consumption in the network occurs whenever there is a data transmission in a node."

"The mathematical formulation for energy consumption in the network is as follows: When there is a higher arrival rate of data packets to the sink node exceeding the service rate, the arriving packets must wait for some time until the sink can process them."

"In OSBR, information is gathered from the target region of the member nodes and sent to the sink via the CH to ensure that the transmitted data of other TNs confuse the attacker and prevent him from reaching the path positions."

"The octopus technique uses static network data to route through alternate paths less likely to be stopped."

Result and Discussion

"The effectiveness of the proposed method is compared with existing techniques such as SEHR, SecTrust-RPL, Heuristic Based energy-efficient routing, and PSO-based routing in terms of throughput, packet drop ratio, end-to-end delay, energy consumption, network overhead, and network lifetime for the number of nodes and data generation rate."

"The performance of the proposed method is evaluated in terms of throughput, packet drop ratio, end-to-end delay, energy consumption and network lifetime."

"Although the computational cost of cryptography-based techniques increases network overheads, simulation findings demonstrate that the OSBR protocol minimizes network overhead compared to other methods for various sites and data creation rates."

"In our proposed approach, the congestion performance is evaluated in terms of energy consumption, packet drop ratio and packet interarrival time."

"The graph shows that the network's energy consumption is reduced when the interarrival times are increased."

Conclusion

"Some of the limitations noted include model flexibility for different SNs."

"When the number of SNs is increased, the model faces certain difficulties because the number of clusters is increased and the time it takes to complete the entire implementation is delayed."

"The security of the data transmitted in the network is also not realizable if a massive data generation takes place between different SNs at the same time."

"The complexity of the model comes from the implementation of the routing protocol and the timing mechanism to find the optimal communication path."

"Medical data is highly vulnerable to multiple attacks, and security is becoming the most important requirement for this data in IoT networks."

"It aims to implement a new and effective security protocol that can offer higher security to medical data."

Efficient Data Transmission Using Trusted Third Party in Smart Home Environments [11]

This is a machine-generated summary of:

Panda, Niharika; Supriya, M.: Efficient data transmission using trusted third party in smart home environments [11]

Published in: EURASIP Journal on Wireless Communications and Networking (2022)

Link to original: https://doi.org/10.1186/s13638-022-02200-9

If you want to cite the papers, please refer to the original.

For technical reasons we could not place the page where the original quote is coming from.

Abstract-Summary

"A novel computationally tractable optimization technique that can accept cross-layer resource configurations and focus on network enhancement with longevity should be created for the smart home, one of the heterogeneous IoT applications."

"This study shows different smart home architectures in static and mobile environments, taking into account some of the challenges like orchestration, mobility, and range in IoT. For network communication, routing protocol over 6LoWPAN (RPL) is used."

"The proposed smart home architectures are analysed and compared based on different parameters such as packet reception ratio, network overhead, throughput, average latency, and total energy consumption."

"Even when some of the devices in the smart home are mobile, the modified smart home-optimized path (MSHOP) is found to achieve a packet reception ratio of 99.93%, minimum latency of 0.9 s, and the highest total energy usage in the network of 3373 millijoules."

"Proposed MSHOP outperforms all existing smart home architectures when considering network efficiency, time, and usability."

Introduction

"The nature of the sensor that gathers and disseminates data between source and sinks nodes which involve multi-hop communications affects the protocol stack design of wireless sensor networks (WSN) [12]."

"Applications of IoT include smart cities, vehicular networks, smart grids, smart agriculture, and cyber-physical systems, among which the smart home is the focus of this paper [13]."

"The gateway connects the smart nodes to send data from the wireless sensor network to the Internet [14]."

"Researchers can simulate realistic scenarios using standard application-layer protocols like message queue telemetry transport (MQTT) and constrained application protocol (CoAP) over 802.15.4 and low-power wireless personal area networks using IPv6 (6LoWPAN)."

"The contribution of this paper is as follows Proposed different smart home architectures with the help of IoT technologies."

"This work aims to propose different smart home architectures with the help of IoT technologies."

Related Work

"Kamgue and others [15] show the way to create and implement a routing protocol for low-power lossy networks using the node's remaining energy as the primary routing measure."

"RPL performance is found to be optimal for both OFs when the network density is between 30 and 40 nodes for Rx 60% using random or grid topology in a light density network [16]."

"The RPL routing combines node link quality and residual energy ETX-ENR to minimize network latency and grow the sensor node's life."

"RPL-MRHOF deemed to be a better solution for PDR and latency in networks with fewer nodes."

"While employing MRHOF with ETX + Energy, it was discovered that the network's performance improved in terms of PDR without influencing the network's power usage."

"An energy-efficient and path reliability-aware objective function (ERAOF) is designed for IoT applications that achieve a PDR of 93% for network size 20, 40, and 60 [17]."

Proposed Methodology

"The architectural layers of smart home as an IoT application include a trusted third party in the proposed work."

"This article aims to create a smart home environment where TTP acts as a middleman framework for communication and data transmission in wireless sensor networks, resulting in increased energy efficiency, longer lifetime, and faster data delivery."

"As smart homes develop, a number of challenges emerge including security and privacy, heterogeneity, reliability, large scale, legal and social aspects, sensor networks, multiple data fusion, and energy management."

"6LoWPAN standards enable IPv6 to be effectively used over low-power, low-rate wireless networks on basic embedded devices by providing an adaption layer and optimizing related protocols."

"It supports data path validation and loop detection routing optimization and groups nodes into a destination-oriented acyclic graph (DODAG), with each router identifying a set of parents."

Simulation Setup

"A typical smart home might be a network of smart gadgets all attempting to communicate data to a server."

"The TTP functions as a go-between for the client nodes and the server, forwarding packets from the devices to the sink."

"The simulation starts with a simple smart home with 10 smart devices and a sink node/home router."

"The simulation then moves on to a typical smart home with 20 devices and an advanced smart home with 30 devices, each with its own server node considering different topologies and ranges."

"This work considers the most responsive and necessary parameters, such as packet reception ratio, throughput, average latency, network overhead, convergence time, and total energy consumption, to evaluate network performance. (1) Packet reception ratio (PRR): The ratio of the total number of packets delivered to the sink by all client nodes to the total number of packets received by the server [18]."

Results and Discussion

"Packet reception ratio, throughput, average latency, network overhead, convergence time, and energy consumption are some of the parameters that have been studied in this work."

"The influence of mobile nodes is shown using PRR when considering the mobile environment in various smart homes."

"It is obvious that, despite having a higher number of nodes than earlier smart homes, both MSH and MSHOP are successful in lowering network overhead."

"Energy consumption has increased in both static and mobile environments as transmission range, network density, and PRR have all increased."

"The MSH consumes more energy than the smart homes with 11, 21, and 31 nodes."

"In a mobile context, the OF0 outperformed MRHOF in terms of PRR, minimizing latency, and network overhead."

"Energy usage rises as density and PRR rise; however, MSHOP in a mobile environment succeeded to lower it compared to a static setting."

Conclusion

"MSHOP has TTP working as a middleman framework and an event timer for efficient delivery of packets."

"This design has been put to the test in a variety of scenarios and compared to a number of LNN classified as basic, average, and advanced smart homes."

"The MSHOP routing protocol RPL is evaluated under heavy load by focusing on packet reception ratio and network efficiency."

"With the help of TTP and event timer, the PRR is almost 100%, network overhead is reduced, average latency of packets has a significant reduction, and throughput has an improvement."

"The IoT network is enormous, and there are also increasing dangers of loss."

Taxonomy of Controller Placement Problem (CPP) Optimization in Software Defined Network (SDN): A Survey [19]

This is a machine-generated summary of:

Shirmarz, Alireza; Ghaffari, Ali: Taxonomy of controller placement problem (CPP) optimization in Software Defined Network (SDN): a survey [19]

Published in: Journal of Ambient Intelligence and Humanized Computing (2021)

Link to original: https://doi.org/10.1007/s12652-020-02754-w

If you want to cite the papers, please refer to the original.

For technical reasons we could not place the page where the original quote is coming from.

Abstract-Summary

"There are defined different functions to optimize scalability, reliability, and others in SDN which consider various metrics in recent research papers."

"We survey the CPP optimization in recent well-known papers to extract optimization strategies, objective functions, and solutions."

"We finally reveal a new taxonomy of cutting-edge studies about solutions of CPP optimization in the SDN from different dimensions."

"CPP solutions based on objective functions, metrics and traits are classified in this article."

"We put forward the new challenges and metrics which can be reckoned in CPP to be optimized."

"This paper can be useful for those working on CPP optimization in SDN to expose the objective functions, variables and metrics."

Introduction

"These papers have introduced the number and placement of the controllers in SDN-based networks as major challenges for future work."

"Article, after reviewing the controller placements problem in SDN-based networks, we review the recent solutions presented to this issue and state the optimization strategies, and metrics in different SDN-based networks."

"The contributions of this paper are as follows: This paper investigates the controller placement problem from different points of view The solutions which have been proposed for the controller placement problem are inquired and categorized based on different required awareness like reliability, scalability, performance, load, etc The CPP optimization problems that have been suggested in recent research are examined to extract optimization strategy, approaches, network scale, etc. The metrics which have been used for CPP optimization research are summarized in this paper."

Related Works

"Lu and others investigate the controller placement problem and survey this topic from the perspective of optimization objectives."

"They classify literature of controller placement problem into four aspects which include latency, reliability, cost, and multi-objective depending on their objective and analyze specific algorithms in different application scenarios [20]."

"Wang, Guodong and others study controller placement problem objectives including minimizing the latency between controllers and their associated switches, enhancing reliability and resilience of the network, and minimizing deployment cost and energy consumption [21])."

"Kumari and others [20] delve the controller placement problem and state the CPP solution techniques, various performance metrics and characteristics of the available solutions."

Software Defined Network Architecture

"The main purpose of the project is to design a new architecture for enterprise networks that provides simple management and robust security [22]."

"The Software-defined Networking Research Group (SDNRG), in IETF, was designed to examine SDN to investigate this architecture from a variety of effects."

"Many research projects around the world offer different architectures for future networks [23]."

"SDN architecture is proposed as the main structure of the future networks [24]."

"SDN separates the control and data plane to change the traditional closed network architecture."

"This architecture provides a centralized controller and makes the network programmable [25, 26]."

"SDN has changed network design and management [26, 27]."

Controller Placement Problem

"If the controller does not work well, the performance of the network can be influenced by this defect."

"There are two methods for controllers' deployment to improve the scalability of these networks, which include centralized and distributed schemes [28–30]."

"In this architecture, the root controller has a global view of the entire network and performs routing operations."

"According to the paper [31], one controller is sufficient for network management in most SDN-based networks."

"Various metrics are used to select relevant controllers by network devices."

"To increase the performance of the SDN-based networks, it should be able to answer the following questions as the controller placement problem [31, 32]:1."

"The set V presents FEs and S represent the controllers in the networks."

"this metric presents the expected percentage of control path loss which is optimized in papers [33, 34]."

Controller Placement Solutions Taxonomy

"According to recent researches, in general, we cluster the CPP optimization into three general ways which are: o Methods of FEs or Switches assignment to the controllers o Optimization strategy o Placement Objective in CPP Optimization We describe the CPP optimization methods in the following subsections (5–1), (5–2), and (5–3)."

"The switch should be allocated to a controller, so this allocation can be done statically, or dynamically."

"The switch-to-controller allocation can be divided into static and dynamic which are defining in the bellow."

"Static controller allocation A switch should change its controller when the controller fails."

"Dynamic allocation In the dynamic method, switch assignment to the appropriate controller can be done dynamically, a switch can change its controller and allocate a new controller dynamically based on performance metrics like load balancing and so on."

"Clustering solutions. This approach can be used by the heuristic and ILP methods to optimize the network."

Survey on Existing Controller Placement Solutions

"The paper [35] proposes the algorithm to optimize the CPP based on the load and propagation delay in the multi-controller network."

"Liu and others [36] propose the Network Clustering Particle Swarm Optimization Algorithm (NCPSO) to optimize the CPP based on the controller load-balancing and latency."

"Zhang et al. [37] propose a multi-objective optimization model to maximize the network reliability, controller load balance ability, and minimize the latency between controllers and switches."

"Perrot and Reynaud [38] in, inquire CPP to provide "good" placement based on QoS. This approach proposes a formulation of resilient CPP to meet the optimal number of the controllers and their location with load balancing and QoS constraints."

"The authors in [39] propose a dynamic matching algorithm to optimize the utilization with controller placement problem solving based on load balancing."

Open Issues and Future Research Directions

"The controller placement problem is considered to solve reliability and scalability while it has its effect on the non-functional states of the networks."

"We have reviewed the papers which had stated the controller placement problem optimization to organize the solutions, metrics, and optimization strategy."

"There are the other non-functional objectives which are influenced by controller placement problem optimization."

"These objects can be security and QoS. The security in SDN should be modelled and the CPP optimization's effects should be investigated for future work."

"For the future work, the security functions and metrics for optimization can be extracted and proposed like this paper in CPP."

"For the future work, extracting network security and controller security parameters, attacks and solutions are the other background research for SDN optimization."

"It can be a future work to determine QoS optimization functions, methods, and metrics."

Conclusion

"This paper has studied the CPP optimization and sought to pave the way for the researchers to achieve knowledge about the controller placement problem optimization in SDN from various aspects."

"The papers which had optimized the CPP have been surveyed and have extracted the non-functional objectives that are under the effects of CPP directly or indirectly."

"The non-functional objects which have been extracted from these papers are performance and load-balancing."

"Some papers have been made of the combination of these objective functions to optimize in CPP."

"The metrics which have been used to model these non-functional objectives have been taken out from the recent related papers."

"The papers have made optimization with CPP approaches which are controller placement and the number of the controllers."

Multi-Objective Based Deployment of Throwboxes in Delay Tolerant Networks for the Internet of Things Environment [40]

This is a machine-generated summary of:

Chinnasamy, Poongodi; Udgata, Siba Kumar; K, Lalitha; A, Jeevanantham: Multi-objective based deployment of throwboxes in Delay Tolerant Networks for the Internet of Things environment [40]

Published in: Evolutionary Intelligence (2020)

Link to original: https://doi.org/10.1007/s12065-020-00474-w

If you want to cite the papers, please refer to the original.

For technical reasons we could not place the page where the original quote is coming from.

Abstract-Summary

"Increase in the connection opportunities in a throwbox assisted DTN-IoT environment depends mainly on the optimal deployment of throwboxes."

"The objective of this paper is to identify optimal deployment locations for placement of throwboxes in a throwbox assisted DTN-IoT environment by (1)

maximizing the coverage of all the throwboxes (2) minimizing the average delay and (3) maximizing the delivery ratio among all the nodes."

"We use an efficient Multi-Objective Differential Evolution and a popular Non-Dominated Sorting Genetic Algorithm-II for finding the optimal deployment location of throwboxes."

"The simulation results are compared to find a preferable strategy in throwbox deployment and enhance the performance of throwbox assisted DTN-IoT environment."

Introduction

"IoT applications in a smart city paradigm collect air quality data and temperature measurements through the sensor nodes that can be buffered until they find an opportunity (throwbox) to forward."

"DTNs make use of node's mobility to deliver data around the disconnected network partitions using multi-hop opportunistic routing with store-carry-and-forward paradigm."

"The routing protocols in a DTN based IoT application are dependent critically on the node movements that decides the contact opportunities."

"Anticipating movements of the nodes is a significant challenge in DTN-IoT networks."

"The entire IoT network lacks in movements of nodes globally and finally the network becomes disconnected with a number of partitions."

"These features bring new issues in DTN-IoTs and make relay node placement necessary for them."

"Dedicated stationary nodes, called throwboxes are employed to equip the IoT networks for better communication [41]."

Related Work

"Zhao and others [42] proposed the MF approach in DTN-IoT which possess enough message ferrying capability with their own mobility pattern, without modifying existing node mobility."

"Due to the difficulty of route design of MFs, it either requires long transmission range or affects the regular nodes' movement patterns which is a non-communication task [43]."

"Optimized Way-points is a MF route design algorithm without any need for online collaboration between the regular nodes and MFs and generates a route design which assures delivery."

"Fan and others [44] proposed a combined throwbox deployment and optimized routing problem for efficient delivery."

"A multi-objective problem along with diversity in the design-variable space is of much interest as the solution to find the better location for throwboxes involves different QoS mechanisms to achieve better performance."

"Javvaji and Udgata [45] and Chai and Liang [46] have discussed how multi-objective optimization methods can be applied for real time network related problems effectively."

Contributions

"Through extensive survey on the Throwbox deployment for DTN-IoT applications, we derive the following insights for the discussed problem and also try to contribute in addressing the challenges."

"Through throwbox deployment methods, we try to address this challenge."

"Evolutionary Algorithms (EAs) are not applied for deployment of throwboxes in existing methods."

"NSGA-II and MODE algorithms are simulated for a Throwbox deployment in a DTN-IoT network and the results are compared considering two and three objective cases at a time."

Network Model

"We assume a DTN-IoT composed of an L network of nodes and K throwboxes."

"Sensor nodes and throwboxes can buffer the data and can communicate via wireless technologies."

"This paper discusses identifying the optimal locations for placing throwbox nodes in the network to arrive at any destination based on the available link probability."

"Every node forwards messages in the interest of another source and destination."

"Communications among the nodes are contact opportunities, and these opportunities are assumed to be independent and follow STEPS mobility model recommended for DTN environment."

"Sensor nodes can store the messages in their buffer for infinite time."

"Mobility model attached to each node is called Spatio TEmporal Parametric Stepping (STEPS)."

"Local areas are modeled in which sensor nodes can move easily according to a random mobility model which is called a partition or zone."

Multi-Objective Optimization for Throwbox Deployment

"A single objective optimization problem for minimization of a function can be given as below:where f is a scalar function, and A is the set of constraints for equality and inequality and it can be defined as Multi-objective optimization for minimization of multiple objectives can also be described in mathematical terms as given belowwhere n > 1 and A is the same as the set of constraints given above."

"The multi-objective optimization can be successfully applied by locating the throwboxes in a feasible location of a given network region."

"In DTN-IoT, there are many objectives like node degree, delivery delay, delivery ratio etc It is needed to find where to deploy them by optimizing among multiple objectives."

"We now formally define the M-throwbox deployment problem along with two multi-objective optimization techniques NSGA-II and MODE."

"An individual's delivery delay is the average of queuing delay in all intermediate nodes, as given in Eq. (2)."

Throwbox Deployment Using NSGA-II Algorithm

"Its main properties are: All the individual solutions found are sorted according to the level of non-domination All non-dominated solutions are stored as elitism property and the convergence properties are enhanced It adapts a suitable method for guaranteed diversity and spread of solutions which is based on the crowding distance Constraints are implemented without penalty functions and use a modified definition of dominance First version of the NSGA method faced problems because of a few weaknesses."

"It uses the crowding distance method within the same Pareto Front for sorting the solutions to preserve the diversity."

"It uses the crowded comparison operator for the selection process of a uniformly spread Pareto-front set and crowding distance which estimates the density of solutions in the objective space [47]."

"NSGA-II sort and rank the solutions based on the current crowding distance and Pareto Front."

Throwbox Deployment Using MODE Algorithm

"Choosing an individual solution with the highest fitness value yields an optimal solution for a single-objective problem."

"The fitness functions are evaluated at the beginning for a random population of size N. The populations are ranked based on dominance concepts for each generation of the evolutionary search."

"DE functions are carried with the individuals of the population and then the fitness functions for the defined objectives of the trial vectors are evaluated."

"Based on the non-dominance sorting, the best N individual solutions are ranked and selected."

"In order to mimic the reproduction operator for reproduction of offspring for each individual in the population, two kinds of vectors are defined: the perturbation vectors and the differential vector."

"The set of non-dominated solutions can be identified with the highest ranks of the population at each generation of the evolutionary process."

Analysis of NSGA-II and MODE for Throwbox Deployment

"The trace file is taken for evaluating cost functions of the NSGA II and MODE algorithms."

"NSGA-II and MODE are able to find different non-dominated solutions by considering the following three objectives, 1."

"Both algorithms are tested with different number of nodes and throwboxes for getting the Pareto-optimal solutions."

"For both NSGA-II and MODE algorithms, computational complexity increases with the number of generations."

"NSGA-II and MODE algorithms are executed for a scenario of 20 throwboxes and 100 nodes in a grid size of 1000×1000."

"In case of NSGA II and MODE algorithms, the results are consistent and close to optimal."

"Compared to NSGA-II, MODE algorithm gives better optimum solutions as it additionally employs diversity in reproduction operations."

"The NSGA-II algorithm though do not perform better than MODE, but still it is not far behind."

Research Findings

"To model the problem as a multi-objective optimization problem considering three conflicting but important relevant objectives."

"To suitably modify the NSGA-II and MODE algorithms to make them suitable for throwbox deployment problem in a DTN-IoT environment."

"To find out and recommend a good multi-objective optimization algorithm for the throwbox deployment algorithm in a relative new DTN-IoT paradigm."

"Many such combinations can be obtained to help better decision making."

Conclusions and Future Scope

"Self-adaptive optimization techniques like NSGA-II and MODE can find better extended solutions and convergence near the true Pareto-optimal front."

"The Pareto Fronts generated by MODE and NSGA-II are better and cost functions are evaluated for their effectiveness."

"The numerical simulation results, related with all data obtained with multiple objectives, point to the dominance of the MODE over NSGA-II in terms of the converging point and better results of the final Pareto front."

"It can be summarized that MODE could be applied to the deployment of throwboxes in real time scenarios."

Minimum Transmission Delay Algorithm for Multi-hop Nodes in Internet of Things Based on Symmetric Algorithm [48]

This is a machine-generated summary of:

Hu, Chengbo; Liu, Yang; Tao, Fengbo; Lu, Yongling; Wang, Zhen: Minimum transmission delay algorithm for multi-hop nodes in Internet of Things based on symmetric algorithm [48]

Published in: Journal of Ambient Intelligence and Humanized Computing (2020)

Link to original: https://doi.org/10.1007/s12652-020-02603-w

If you want to cite the papers, please refer to the original.

For technical reasons we could not place the page where the original quote is coming from.

Abstract-Summary
"In order to improve the quality and performance of information transmission, a minimum transmission delay algorithm of multi-hop nodes in the Internet of Things is designed based on symmetric algorithm."

"In the perception layer of the Internet of Things, based on the revocable symmetric encryption algorithm, the multi-hop nodes in the perception layer are encrypted; The link information transmission delay under the direct transmission, cooperative transmission and multi-hop transmission modes is compared, and the gradient projection algorithm is used to solve the problem of minimizing the delay of multi-hop nodes' information transmission in the Internet of Things."

"The simulation test shows that the algorithm can guarantee the security of data transmission of multi-hop nodes in the Internet of Things, and the transmission delay is the minimum."

Introduction
"Due to the constraints of cost and hardware resources, how to optimize the network protocol of sensor nodes in these application scenarios has become a critical research issue (Krishnan and Chaporkar 44).The hop transmission delay between nodes is a quantitative index to improve the operation efficiency of the Internet of Things."

"With the increase of the amount of information transmitted in the Internet of Things, the transmission delay of sensor's multi-hop nodes in the Internet of Things does not increase, which greatly reduces the efficiency of data transmission in the Internet of Things."

"To reducing the transmission delay of multi-hop nodes in the Internet of Things, enhancing the confidentiality of node transmission of wireless sensor network in the Internet of Things is another key issue to be improved [49]."

"Information leakage in the process of data acquisition and transmission in wireless sensor networks of the Internet of Things will bring immeasurable losses."

Materials and Methods
"The content of this paper is the minimum transmission delay algorithm of multi-hop nodes in the Internet of Things."

"The application of symmetric encryption algorithm in multi-hop node transmission communication in the sensing layer of the Internet of Things is described in detail."

"When symmetric encryption algorithm is used to process the ciphertext data access in the Internet of Things, the same key k is used for encryption and decryption."

"According to the fingerprints entered by users in the ciphertext data access of Internet of Things, the revocable key generation algorithm consists of three main processes: data feature extraction process, template generation process and key generation process. (1) Data feature extraction."

"The information transmission delay of multi-hop nodes in the Internet of Things is related to the link information flow and link capacity."

"By using this algorithm, multi-hop nodes can properly allocate the relevant link information flow and effectively reduce the transmission delay."

Results

"The simulation experiment platform ONE (opporenvironmen) is used to complete the transmission delay experiment of multi-hop nodes in the Internet of Things."

"For the simulation system, the main objects of simulation are various IoT multi-hop nodes."

"There should be an IoT multi-hop node entity in the simulation environment, which will occupy computing resources during operation and network resources when sending messages."

"The functional efficiency of the proposed algorithm, RSA encryption algorithm and AES encryption algorithm in different stages is analyzed."

Discussions

"It can be seen that according to the experimental results of transmission success rate, the minimum delay algorithm based on the stability probability has little change in the transmission success rate under different node densities, because the minimum delay algorithm based on the stability probability consumes a lot of energy; the minimum delay algorithm based on MTDA increases the transmission success rate greatly under low and high density conditions; the algorithm in this paper increases significantly, when the number of nodes reaches more than 250, and the transmission success rate can be achieved."

"The transmission delay remains stable under the conditions of high and low node density, and is 2/3 of the transmission delay of MTDA minimum delay algorithm."

Conclusions

"In the data transmission part of the Internet of Things, a revocable symmetric encryption algorithm is proposed to ensure the security of multi-hop nodes in the perception layer."

"Based on the analysis and comparison of the link delay relationship among direct transmission, cooperative transmission and multi-hop transmission, a gradient projection algorithm is proposed to solve the problem of minimizing information transmission delay under multi-routing paths."

"The simulation results show that compared with the same type of minimum delay algorithm, the proposed algorithm not only minimizes the transmission delay of multi-hop nodes in the Internet of Things, but also creates a secure data transmission environment."

An Empirical Investigation Based Quality of Service Aware Transmission Power Prediction in Low Power Networks [50]

This is a machine-generated summary of:

Vidhya, S S; Senthilkumar, Mathi; Anantha Narayanan, V: An empirical investigation based quality of service aware transmission power prediction in low power networks [50]

Published in: Sādhanā (2022)

Link to original: https://doi.org/10.1007/s12046-022-01982-4

If you want to cite the papers, please refer to the original.

For technical reasons we could not place the page where the original quote is coming from.

Abstract-Summary

"The commonly used power control schemes maintain the transmission power based on the received signal strength indicator (RSSI) that depends on the interference in the environment."

"It is necessary to consider interference for retaining the lowest transmission power since low-energy network signals are vulnerable to interference changes."

"The earlier investigations suggested only linear models for power prediction in low-power networks."

"This paper investigates a classification-based transmission power prediction approach with the presence of interference."

"The approach works for linear and non-linear models based on RSSI, link quality indicator, neighbour node distance, and receiver power to maintain reliable communication with low energy consumption."

"The investigation results demonstrate that it is possible to build a classification-based power prediction for linear and non-linear models by considering different spatial effects with 99% accuracy."

Introduction

"The efficient utilization of available energy will increase life and reduce the power consumption of IoT networks."

"The second energy-conserving approach reduces power consumption through optimized energy management in different network activities."

"The environments with the deployment of IoT networks vary based on the applications such as military, medical care, home automation, etc Due to the node power constraints and the changing environmental conditions, reliable communication in a low-power network becomes difficult."

"This feature can reduce the transmission energy by the intelligent selection of transmit power in each node during the communication."

"TPC in the low power network is a cross-layer approach since the power level details are physical layer parameters."

"The commonly used link quality estimators in low power network communication are RSSI and link quality indicator (LQI)."

"The present work proposes a classification-based predictive model that envisages the power level needed for reliable communication with the neighbouring node."

Energy Management in Low Power Networks—An Overview

"The power level can be managed so that it is enough to guarantee the communication quality so that the transmission energy can be saved."

"In most of the research works, a feedback-based mechanism is adopted to manage power levels."

"Another fuzzy-based approach stated in [51] uses a fuzzy system to adjust transmission power by maintaining dynamic topology control."

"This work predicts future RSSI variations using the grey model, and the transmission power is determined using a fuzzy inference system."

"(1) The authors have conducted experiments in three different environments, and the collected data is used to model the transmission power."

"(2) The impact of the interferences and transmission power level on the communication quality is evaluated by varying distances and the data set built from the experiment."

"(3) Classification-based power level prediction approach is proposed and evaluated with linear and significantly non-linear models using a data set built from natural environments."

Proposed System

"The communication quality among low power nodes is affected by the surrounding environmental factors such as environmental changes, background obstacles, the distance between sender and receiver, etc Real-time experiments for investigating the impact of these parameters are conducted in the present work."

"The parameters selected for investigation are RSSI and LQI to measure the quality of communication."

"The investigation experimented with barriers to show the impact of obstacles on communication quality."

"The prediction is made in the nodes and computed for the efficient transmission power for the given LQI and RSSI."

Experimental Results and Evaluations

"From the plots, it is clear that sender and receiver power levels influence the link quality and RSSI values of communication."

"Indoor communication of a distance up to 15 m is enough to set a minimum power level for the sender and receiver for quality communication."

"Outdoor, this minimum power combination can cover up to 20 m. Another inference from this experimental data is that if we wanted to change sender power dynamically, the power has to be set based on the receiver power since the LQI and RSSI rely on the transmission power-receiver power pair."

"The data sample contains the features such as distance between the sender and receiver, link quality indicator, RSSI, and receiver power level."

"The model is analyzed with linear classifiers such as linear regression and Naive Bayes classifiers, and non-linear classifiers support vector machine with rbf kernel and KNN to choose the best power prediction model for accuracy, model size, and prediction speed."

Conclusion
"A real-time investigation is conducted to analyze the impact of transmission power on communication quality."

"The parameters such as receiver power, transmit power, RSSI, LQI, and communication distance are considered in the present work to model the transmit power."

"The predicted power level maintains the quality of service parameters of wireless communication."

"The present work can be extended by installing dynamic power level prediction in network layer routing, and further investigations can be conducted to analyze communication quality."

PECDF-CMRP: A Power-Efficient Compressive Data Fusion and Cluster-Based Multi-Hop Relay-Assisted Routing Protocol for IoT Sensor Networks [52]

This is a machine-generated summary of:

Agbulu, G. Pius; Kumar, G. Joselin Retnar; Juliet, Vimala A.; Hassan, S. Adurrashid: PECDF-CMRP: A Power-Efficient Compressive Data Fusion and Cluster-Based Multi-hop Relay-Assisted Routing Protocol for IoT Sensor Networks [52]

Published in: Wireless Personal Communications (2022)

Link to original: https://doi.org/10.1007/s11277-022-09905-6

If you want to cite the papers, please refer to the original.

For technical reasons we could not place the page where the original quote is coming from.

Abstract-Summary
"Wireless sensor network (WSN) is the predominantly used technology in building real-time event monitoring and dissemination frameworks for the internet of things (IoT) applications."

"WSN being power-hungry, fragile, and highly susceptible to loss of connectivity, it becomes vital to strengthen the network's communication reliability and energy efficiency."

"A power-efficient compressive data fusion and cluster-based multi-hop relay-assisted protocol (PECDF-CMRP) for IoT sensor networks is proposed."

"Relays are chosen from the K cells to establish cooperative networks for data broadcast from the event sectors to the central gateway."

"The relay allocation is formulated as a maximization problem regarding the node's energy and path loss."

Introduction

"In multi-level clustered WSN structures with stationary gateways, head nodes closer to the gateways are constrained to drain their energy earlier, as they operate as relays to other higher layers."

"In the current literature, this relay cooperation in clustered WSNs has been validated to improve the node lifecycles and strengthen the global data reception rates of the networks [53]."

"Driven by these problems as stated above, we propose a power-efficient compressive data fusion and cluster-based multi-hop relay-assisted routing protocol (PECDF-CMRP) IoT sensor network."

"This protocol aims to resolve the relaying issues of multi-layered cluster-based sensor networks with stationary gateways by optimizing the global energy saving, lifetime, data reception rates and delay performances."

"Our contributions in this paper are outlined as follows: We propose a power-efficient compressive data fusion and cluster-based multi-hop relay-assisted routing protocol (PECDF-CMRP) for IoT sensor networks."

Related Work

"Wang and others in [54] proposed the energy-efficient compressive sensing-based clustering routing (EECSR) protocol to resolve hitches of Hot spot and power consumption in clustered WSNs."

"In the proposed model, the relay allocation is framed as an NP-hard problem in terms of the node's distance and residual energy and resolved using gradient descent (GD) cost minimization scheme."

"In [55], Kaiyan and Wang propose a relay assignment strategy for cooperative rateless-codes based EH WSNs wherein the cooperator nodes harvest energy from the RF signals diffused from a source."

"The proposed relaying solution assigns relay nodes from one of its adjoining nodes based on the collected levels of vitality of the nodes and applies piezoelectrical (EH) energy harvesting at the node devices to reinforce the energy efficiency at the nodes."

"In order attain further optimized performances, a new power-efficient compressive data fusion, and cluster-based multi-hop relay-assisted routing protocol (PECDF-CMRP) for IoT sensor networks is proposed and experimented in this paper."

System Model and Communication Protocol

"The network is formed within a 2D target area $A=(X, Y)m^2$, with some resource-constrained nodes and one BS."

"The nodes are arbitrary dispersed across the monitoring area, while the BS is set at the field extreme."

"It is assumed that a cell of cluster covers N/K dimension of the sensing area; where, N is the over-all nodes, K denote the total cell built over the network."

"Such nodes can be subdivided into smaller and more K-groups to secure the network against hot-pot overheads."

Node Communication Protocol

"To evade collisions and enhance the bandwidth usage at cluster-level, MAC-based TDMA is applied to coordinate the communication between the ON and CH nodes in each cell."

"During transmission, the ON's employs their prearranged channel-slots to disseminate their observations to the CH nodes."

"The both levels of communication as per, ON-to-CH and CH-RN is defined upon event occurrences."

"The CH of the event source serves as the coordinator at both levels, to allocate slots to the K members and convey preamble notices to the nearest relay terminal, declaring needs for pending transmissions."

"This helps to prepare each relay for pending cooperation, while the CH receives the observations from the members."

"Through these protocols, the communication between ON-to-CH and CH-RN are defined."

Proposed Routing Protocol Design Phases

"This section details the design methods of the proposed PECDF-CMRP, which comprises the three-phases namely: Network initialization and grouping phase; Relay node allocation phase; Data fusion and cooperative transmission phase."

"Uneven-clustering with K-means as per CH nodes with P probability is based on the number of K-cells and network density."

"The relay nodes are meant to support the heads report their fused messages to the BS exploiting multi-hop cooperative communication."

"After estimating GA for the nodes under consideration, the BS ascertains the optimal weights for selecting the appropriate relays."

"When a CH or relay node acquires a join notice (JN), it scrutinizes the request message for cooperative membership fitness by checking the signal strength and distance and of the sender."

"The CH node of the event cell as well, conveys a pre-transmission preamble notification to the nearest relay member using non-persistent CSMA."

Performance Evaluation

"PECDF-CMRP slightly outperforms EECSR by just 5.4%, however, it considerably outperformed EDC and TEEN by about 14.5% and 22.73%, respectively."

"Our proposed PECDF-CMRP optimizes the communication performances on the following grounds: Firstly, In EDC, number of CH nodes are isolated due to irregularities in the clusters formed during events and thereby they dissipate their energy timelier."

"In TEEN also, CH nodes work for other upper layers as relays and as result, they deplete their energy faster."

"In EECSR, more energy is expended during the CH selection stage since the nodes are enforced to flood messages about their position and energy to choose their heads."

"In PECDF-CMRP the nodes are effectively grouped into K unequal cells taking accounts factors of Hotspots, and energy dissipation at the grouping and data gathering stages."

Conclusions

"A power-efficient compressive data fusion and cluster-based multi-hop relay-assisted routing protocol (PECDF-CMRP) for IoT sensor networks was proposed and experimented .The proposed PECDF-CMRP exploits the full advantages wavelet-based data fusion and multi-hop relaying schemes."

"Phase, relay nodes are chosen from the different K cells to establish cooperative relay diversity across the sensor network."

"The relay node preferences was formulated as a maximization problem regarding path loss factors, and residual energy metrics."

"During the last stage, the relay nodes exploit amplify and forward (AF) policies to cooperatively to convey the data fused from the K event sector to the central BS."

Discovery of Random Delay Distribution Based on Complex WSNs Clock Synchronization [56]

This is a machine-generated summary of:

Wang, Xin; Yi, Deerkun; Zhang, Zengping: Discovery of Random Delay Distribution Based on Complex WSNs Clock Synchronization [56]

Published in: Wireless Personal Communications (2021)

Link to original: https://doi.org/10.1007/s11277-021-08340-3

If you want to cite the papers, please refer to the original.

For technical reasons we could not place the page where the original quote is coming from.

Abstract-Summary

"With the rapid development, Wireless Sensor Networks (WSNs) technology has been widely applied in many fields, such as military intelligence detection, medical condition detection, urban facility construction and smart home, etc Clock synchronization is one of the key technologies in WSNs, precise node time synchronization is critical for many networked control applications in WSNs."

"Once the clock of sensor nodes loses synchronization, the information collected by nodes will lose timeliness and value for users."

"The main purpose of clock synchronization is to control the interval time that between sending and receiving timestamps of wireless sensor nodes within acceptable clock synchronization accuracy."

"In ZigBee wireless sensor network, through theoretical model and data experimental analysis, we got the result that the random delay distributions of the uplink and downlink obey normal distribution when the sensor nodes exchange some information."

Introduction

"WSNs is a kind of network form formed by the free organization and combination of tens of thousands of sensor nodes with computing and data processing capabilities through wireless communication technology."

"In WSNs, clock synchronization is the premise of many applications, such as node location, energy saving, data fusion and so on."

"In order to realize the coordination of the whole network, it is necessary to synchronize the clock of the sensor node every other period of time."

"Clock synchronization technology of WSNs requires each node to exchange timestamps so as to synchronize the time of the whole network."

"In the clock synchronization algorithm, the network delay will have a relatively large impact, so many researchers assume the network delay as different distribution according to different conditions when calculating the clock synchronization algorithm, so as to realize the algorithm."

Problem Statement

"Messages may be arbitrarily delayed due to the deterministic and uncertain time delay in the process of message transmission."

"In the clock synchronization technology of WSNs, sensor nodes constantly exchange their clock information through packet transmission, But in fact, In the time recording interval before transmission and after receiving, there are many different link delays in radio signal transmission, which may be far greater than the clock synchronization accuracy we need."

"The transmission time, channel access time, interrupt processing time and receiving time are all uncertain delay sources, and the time range is usually between 5 us and 500 ms."

"Access times are uncertain based on current network traffic. (3) Transmission time Transmission time is the time it takes to transmit a message in the physical layer. (4) Propagation time Propagation time refers to the time of transmitting a message from the sender to the receiver in a wireless channel."

The Experimental Measurements

"Our experiment is based on the implementation of ZigBee network, setting up two nodes."

"They are the coordinator and the terminal node respectively, thus establishing a pattern of bidirectional information exchange in pairs."

"When sampling the timestamp information, we set the distance between the coordinator and the terminal node as 3.737 m, and sampled a set of data every second."

"After programming the coordinator node and the terminal node, we use the IAR tool to burn and write them into the module respectively, and then power the two nodes to communicate and exchange timestamp information with each other."

"All collected data is printed from the serial port of the terminal node and stored in the serial port receiving data file. (The serial port receiving data file here is a TXT file written in C#)."

Conclusions

"The random delay distribution of wireless sensor networks will greatly affect the network design of clock synchronization algorithm."

"It is very important to determine the random delay distribution in the study of clock synchronization in wireless sensor networks."

"We find the distribution of random delay between nodes of wireless sensor networks based on ZigBee protocol."

"Our theoretical model and several experimental results show that the random delay distribution between sensor nodes in ZigBee network obeys the normal distribution."

A Cooperative Protocol for Pervasive Underwater Acoustic Networks [57]

This is a machine-generated summary of:

Cerqueira, Lucas S.; Vieira, Alex B.; Vieira, Luiz F. M.; Vieira, Marcos A. M.; Nacif, José A. M.: A cooperative protocol for pervasive underwater acoustic networks [57]

Published in: Wireless Networks (2021)

Link to original: https://doi.org/10.1007/s11276-021-02550-0

If you want to cite the papers, please refer to the original.

For technical reasons we could not place the page where the original quote is coming from.

Abstract-Summary

"The novel Underwater Wireless Sensor Network (UWSN) can contribute to monitor and explore aquatic environments."

"Acoustic communication has a large packet error rate and low throughput."

"A large number of solutions to improve aquatic communication refers to routing protocols, medium access control protocols, and designing acoustic modems."

"Cooperative communication explores the broadcast nature of wireless transmission and enhances its performance."

"Cooperative communication has not been fully explored in UWSNs."

"We present COPPER, a Cooperative Protocol for Pervasive Underwater Acoustic Networks."

"COPPER exploits the broadcast nature of wireless communication and, sensor nodes that are idle can operate as a relay, enhancing communication by space diversity."

Introduction

"Jang and others [58] present an underwater sensor node that is capable of harvesting energy from underwater acoustic signals using piezoelectric interfaces and communicates by modulating the piezoelectric impedance."

"These devices contain several sensors, to collect several aquatic environment characteristics, such as water temperature, local pressure, water salinity, etc Once a sensor node collects underwater environment information, it sends the data to a central node—in most cases using an ad-hoc network—which will pre-process and group data of all sensor nodes it receives, and then send to a far computational center—in most cases using a satellite link or high power radio—where data will be further processed."

"A COoperative Protocol for PERvasive Underwater Acoustic Networks (COPPER) is presented."

"Differently from our previous work [59, 60], where we have introduced the basic concepts of our cooperative communication protocol, in this work, we also explore the distance between underwater nodes and the sink node to properly schedule data transmission."

Related Work

"Some works propose a relay-aided protocol on the physical layer."

"Vajapeyam and others [61] proposed a protocol where relays work using an amplify-and-forward scheme."

"They have also proposed an amplify-and-forward based protocol, namely Wave Cooperative."

"Javaid [62] proposes RBCRP: a region-based cooperative routing protocol based on the amplify-and-forward technique over Rayleigh fading channels in underwater wireless sensor networks."

"They propose an amplify-and-forward (AF) relay protocol."

"Ahmad and others [63] also use, at the relay, an amplify-and-forward method, to perform cooperative communication."

"Their proposal also explores the existence of physical layer diversity to perform cooperative communication and routing at the network layer."

"In their work, they properly select the mobile relay to cooperate taking into account the distance between nodes (i.e., the link distance) and residual energy each transmitting node has."

"A cooperative ARQ scheme that works together with a MAC-based handshake protocol has been proposed by Kim and others [64]."

Wireless Underwater Network Formulation

"We consider a wireless underwater network with n nodes, including the sink node."

"Sensor nodes contain several sensors, to collect several aquatic environment characteristics, such as water temperature, local pressure, water salinity, etc Once a sensor node collects underwater environment information, it sends the data to the sink node."

"It will pre-process and group data of all sensor nodes information, and then send them to a far computational center—in most cases using a satellite link or high power radio–, where data will be further processed."

"The sink node gathers sensor nodes data and sends it to an external network where data will be further processed."

"Systems that monitor oceans and coral reefs [65, 66] usually deploy underwater sensor nodes that capture water properties like temperature and salinity."

"These sensor nodes transmit data to a central node, usually deployed in a buoy."

Cooperative Communication Proposal

"If a node receives a correct packet that is addressed to the sink, this node saves this content in a particular buffer that stores only cooperation messages."

"If the sink node has properly received the packet, the sensor node may send its new packet (if it has data to transmit)."

"Data period, the sink correctly receives the nodes 1 and 2 cooperative messages."

"Asynchronous COPPER presumes that when a sensor node overhears a NACK frame, it will randomly pick, from its cooperative buffer, one of the messages that had failed before."

"As expected, the sink node sends a NACK message indicating that the frames from sensor nodes 3 and 4 were not properly received."

"Data transmission will be correctly aligned on the sink node but, a given sensor node may overhear two messages at the same time."

Evaluating COPPER

"When there are a great number of available sensor nodes to cooperate, when the protocol does not synchronize the cooperation process, various nodes may retransmit a packet (i.e., the same packet)."

"At lower network loads, COPPER-A also has a better flow rate (goodput) when compared to other protocols."

"As occurred to the packet error rate (and for similar reasons), COPPER-S has better performance, in terms of goodput, when we increase the network load."

"Of low network workload, COPPER-S and Scheduled COPPER (COPPER-SE) protocols show better performance (PER and goodput) and consume practically as much energy as the non-cooperative protocol."

"The scheduled protocol, in turn, transmits fewer cooperative packets when compared to COPPER-A and COPPER-S. This happens because in advancing the transmission of a packet, it can collide with another transmission from other nodes."

Conclusions

"We have presented COPPER, a new MAC protocol for underwater wireless sensor networks that take advantage of cooperative communication."

"COPPER also uses selective repeat ARQ schemes to recover from errors and improve communication."

"We have also studied three distinct versions of COPPER: the synchronous COPPER, the asynchronous COPPER, and the data a scheduled COPPER which explores the distance between underwater nodes and the sink node to properly schedule data transmission."

"The synchronous COPPER is able to improve the UWSN goodput by 17% and decrease the packet error rate by 65%, consuming less than 1% more energy."

Cross-Layer Design Based N-ary Huffman Coding for Performance Analysis of DSDV Routing Protocol in MANETs [67]

This is a machine-generated summary of:

Mehta, Ridhima: Cross-Layer Design Based N-ary Huffman Coding for Performance Analysis of DSDV Routing Protocol in MANETs [67]

Published in: Wireless Personal Communications (2022)

Link to original: https://doi.org/10.1007/s11277-022-09770-3

If you want to cite the papers, please refer to the original.

For technical reasons we could not place the page where the original quote is coming from.

Abstract-Summary

"The implementation of the optimal Huffman coding technique for achieving proficient data compression, lower information redundancy and minimal utilization of the scarce bandwidth is essential to handle the effective processing of massive network data."

"We employ cross-layer design technique among the data link, network and presentation layers of the traditional protocol stack to develop the improved DSDV routing protocol based on n-ary Huffman coding procedure."

"Diversified simulation settings are employed with deviating network size to measure the multitude of essential wireless network characteristics incorporating the average delay, packet error rate, packet delivery fraction, data rate, ping loss rate, entropy rate, and reception cache hit."

"Beyond, we significantly compare the performance of our developed cross-layer coded mobile routing model with several previous algorithms to validate its enhanced feasibility and superiority in terms of crucial network operation metrics such as throughput, packet drop rate, mean delay and packet delivery ratio."

Introduction

"Since a MANET has no established infrastructure and is multi-hop in nature, every node in the network functions as a router and takes part in discovery and maintenance of routes to other nodes in the network."

"Routing in MANETs is a challenging issue since it is characterized by mobile computing nodes, open error-prone medium, dynamic network partitioning, low capacity wireless links, infrastructure less ad-hoc environment, frequent link failures, and multi-hop wireless connectivity."

"Over the last 10 years, various MANET routing protocols have been proposed by network researchers and designed primarily to improve the MANET performance with respect to establishing correct and efficient routes between a pair of nodes for packet delivery [68]."

"We utilize the widely applicable proactive destination-sequenced distance-vector (DSDV) routing mechanism for hop-by-hop transmission of data packets through the ad-hoc network assisted with up-to-date topological view."

Background

"We introduce the preliminary concepts of the DSDV routing mechanism and cross-layer design implementation for ad-hoc networks."

"Contrary to the layered architecture in which a layer does not share information about its state with any other layer, the basic purpose of cross-layer design is to actively exploit relevant knowledge from multiple layers to attain performance gains in terms of throughput, average end-to-end delay, etc The system performance of next-generation wireless networks will be optimized by cross-layer design between physical, MAC and higher-layer protocols."

"To deal with the unique challenges of wireless ad-hoc networks and to optimally utilize the limited resources of the resource-constrained mobile nodes, the concept of cross-layer design is employed."

"There are six different types of cross-layer design implementations typically deployed in wireless network architectures."

Related Work

"Work in [69] estimated and compared the performance of the proactive Optimized Link State Routing protocol (OLSR), reactive Ad-hoc On-Demand Distance Vector Routing (AODV), and reactive Dynamic Source Routing (DSR) applied to the practical motorway surveillance system with infrastructure-less freeway mobility scenarios."

"In another preceding work [70], we deployed cross-layer design among four layers of the network protocol stack and applied adaptive Huffman coding technique to node persistence probabilities in wireless ad-hoc network."

"AODV and DSR routing schemes are examined in realistic vehicular ad-hoc network scenario with discrete-time urban mobility simulation modeling [71]."

"The authors in [72] analyzed and compared the performance of AODV, DSR, DSDV and OLSR routing protocols in MANET with the similar simulation environment based on the pause time and network density variables."

"We analyze the DSDV routing performance under various optimal coding implementations and IPv4 network protocol operations in miscellaneous simulation environments."

System Model

"The packet successful transmission probability from the data link layer is fed to the DSDV routing protocol at the network layer through the upward information flow implementation of the cross-layer design."

"The packet success probability from the data link layer is used by the Huffman coding algorithm at the presentation layer for data coding and compression."

"The symbols connected with each leaf node in the individual trees constitute the optimal codes conforming to the probability of packet successful transmission for various nodes in the ad-hoc network."

"Observe that as the number of digits utilized in the Huffman coding procedure increases, the average number of symbols assigned to individual leaf nodes representing the probability of packet successful delivery in ad-hoc networks decreases."

"An ad-hoc wireless network is modeled as a directed graphical structure comprised of non-stationary nodes interconnected by a set of wireless data links."

Simulation Setup Environment

"OMNeT + + discrete-event simulation framework coupled with the Qtenv set up is utilized to conduct the miscellaneous analytical experiments [73]."

"Multiple Network Description (NED) files are loaded and various NED parameters are adaptively estimated for assessment of optimal performance metrics associated with the presented coding-based DSDV routing methodology."

"Based on the flexible Eclipse platform [74], OMNeT + + Integrated Development Environment (IDE) is deployed for conducting discrete-event simulations."

"This modular simulation environment incorporates the Mobility framework (MF) [75] to reinforce the generalized execution of the mobile wireless networks by development of advanced networking protocols within the simulation tool."

"The proposed model development is implemented and simulated in the Qtenv environment through the interactive graphical runtime interface."

"This modeling scenario deploys the 32-bit logical addresses to facilitate the wireless host identification followed by the expected data routing with best-effort delivery mechanism over the underlying network to reach the desired destination node in an optimum manner."

Algorithm of the Proposed Routing Scheme

"The source code of the omnetpp.ini file, i.e., the simulation initialization file for editing and configuration of the general network parameters, and the code in the.ned (Network Description) file are modified and the relevant network attributes are set to the corresponding optimal values associated with the cross-layer information exchange."

"The following is the source code of the DsdvNetwork.ned file."

"We provide the C + + development code for the design and simulation of the proposed system model in OMNeT + + IDE."

"The following is the source code implemented in the omnetpp.ini file associated with the DSDV routing protocol."

Simulation Results and Analysis

"For dynamic IPv4 protocol, the optimal quaternary coding algorithm applied to the proposed system model obtains 56.27% and 14.08% lower redundancy measure in contrast to the binary and ternary coding schemes, respectively."

"The highest network data rate achieved over the wireless communication channel with binary Huffman coding correlated with the IPv4 routing methodology is estimated to be 326.67 Mbps."

"For generalized n-ary Huffman coding mechanism, the multi IPv4 protocol outperforms the primary IPv4 and dynamic IPv4 routing schemes in terms of minimal required data rate for information transmission, and coding attributes comprising of the optimal codeword length, mitigated redundancy, and improved Huffman encoding efficiency."

"Our modified DSDV protocol based on the integrated framework of coding optimization and adaptive cross-layer information exchange is compared via simulations with the thirteen earlier works existing in literature in terms of significant network performance attributes including throughput, packet loss rate, data delivery ratio and delay metrics."

Conclusion

"We show that the cross-layer optimization technique employing three layers of the protocol stack can be used to model and implement optimal data coding-based DSDV routing protocol in MANETs."

"The effects of these attributes is surveyed for disparate Huffman coding algorithms executed at the presentation layer and various IPv4 routing procedures implemented at the network layer."

"These protocols utilize the packet success probability from the data link layer to carry out the enhanced routing and coding schemes in the proposed DSDV routing method."

"As the number of wireless nodes in the network increases, the dynamic IPv4 routing algorithm outperforms the other two data forwarding schemes with respect to improving the ping loss ratio by upto 66.03%."

"Multi IPv4 methodology executed with the quaternary coding scheme indicates minimal redundancy performance by more than 20%, minimal data traffic rate by upto 52.2%, and enhanced coding efficiency by around 58.5%."

Cross-Layer Protocol for WSN-Assisted IoT Smart Farming Applications Using Nature Inspired Algorithm [76]

This is a machine-generated summary of:

Mahajan, Hemant B.; Badarla, Anil: Cross-Layer Protocol for WSN-Assisted IoT Smart Farming Applications Using Nature Inspired Algorithm [76]

Published in: Wireless Personal Communications (2021)

Link to original: https://doi.org/10.1007/s11277-021-08866-6

If you want to cite the papers, please refer to the original.

For technical reasons we could not place the page where the original quote is coming from.

Abstract-Summary

"The functionality of IoT depends on the Wireless Sensor Networks (WSNs) design."

"The IoT-empowered WSNs received the researcher's attention for the Smart Farming (SF) applications."

"As the nodes are resource restrained, the process of periodic farm data gathering and multi-hop delivery needs to be effective in terms of Quality of Service (QoS) and energy-efficiency of information transmission by reflecting the long-distance transmission difficulties of SF applications."

"To enhance the network lifetime substantially of densely deployed WSN for periodically monitoring of farm conditions, we propose a novel Nature-Inspired algorithm-based Cross-layer Clustering (NICC) protocol."

"We design NICC to find a reasonably better solution for clustering and routing in SF applications."

"NICC explores the idea of a nature-inspired optimization algorithm called Bacterial Foraging Optimization (BFO) with optimal fitness function, which models the trade-off among the energy efficiency and optimal data transmission."

"We design a BFO algorithm to select the optimal sensor node for clustering and routing problems based on cross-layer parameters-based fitness value computation."

Introduction

"Smart Farming (SF) is one of the vital applications of WSN-assisted IoT to automate the agriculture method to increase farm potency and conserve resources like energy, water, etc , in minimum cost requirement."

"Such evolutionary algorithms have been presented to solve the various optimization problems of different domains [77–80]."

"The optimization algorithm such as PSO, Ant Colony Optimization (ACO), GA, etc recently adopted for solving the optimal CH selection and routing problems in WSNs."

"We propose the BFO algorithm using an efficient fitness function for the problems like clustering and routing of WSNs-assisted SF applications."

"We design the Nature-Inspired algorithm-based Cross-layer Clustering (NICC) protocol using the BFO with novel fitness function in this paper."

"The sensor nodes are assessed and selected based on a cross-layer fitness function optimization problem."

Related Work

"The new clustering convention proposed for energy utilization decrease in IoT empowered WSNs in [81]."

"In [82], the FEEC-IIR clustering convention proposed depends on FEEC (Fuzzy-based Energy-Efficient Clustering) and IIR (Immune-Inspired Routing) strategies for WSN-helped IoT applications."

"The clustering convention proposed with the principle center around ideal CH determination by using the distance with other sensor hubs and the current status of energy in [83]."

"The tale clustering procedure proposed for heterogeneous WSNs in [84] consolidating meanings of the CH directing tree and CH revolution to adjust the energy utilization of the hub."

"The current clustering techniques concentrated on the energy efficiency for WSNs/or WSN-assisted IoT forms using techniques like fuzzy logic, PSO, ACO, etc However, the trade-offs among various performance metrics for WSN-assisted SF applications using these protocols is a challenging research problem."

Preliminaries

"Alongside this, proposed framework model having a few suspicions, for example, Network comprises of on-area sensor nodes and two Gateway/Sink/BS nodes sent inverse sides to address the significant distance correspondence issues."

"All sensor nodes are homogenous and haphazardly sent across the homestead area."

"All sensor nodes are static and having an extraordinary ID along energy imperatives."

"We utilized the main request radio model to appraise the energy utilization of every sensor node."

"Reduction of energy consumption may not always solve a higher data transmission delay problem, hence we consider the reduction in data transmission delay by selecting the sensor node with higher Received Signal Strength Indicator (RSSI) and minimum queue congestion."

"To achieve the performance trade-off, we propose the cross-layer parameters for sensor nodes evaluation during the process of NICC clustering and routing."

NICC Methodology

"The NICC protocol proposed to enhance the network lifetime with minimum data transmission delay based on cross-layer parameters of each sensor node during BFO fitness value evaluation."

"The stable and optimal CH selection or data forwarder node selection is achieved by the cross-layer parameter computation and its evaluation using BFO."

"In each cluster, the BFO algorithm functions such as bacterium initialization (all sensor nodes in the cluster), cross-layer probability-based fitness value computation of each bacterium, then the chemotaxis initialization as optimization function for CH selection with subsequent steps like swarming and reproduction, elimination, and dispersal, and finally the selection of optimal bacterium as CH."

"Both clustering and routing algorithms are based on the computation of cross-layer probability values to select more reliable and stable nodes."

"This section presents the steps of the BFO algorithm used for optimal CH selection or next-hop selection to form the cluster or route respectively in NICC protocol."

Simulation Results

"NICC scheme related with existing routing protocols in terms of various performance metrics such as average energy consumption, network lifetime, packet delivery ratio (PDR), communication overhead, and communication cost main in three categories such as energy efficiency, QoS efficiency, and computational efficiency."

"The BFO algorithm had designed using the cross-layer probability value as the fitness function in the NICC protocol that ensures more stable and reliable CH selection and route formation for data transmission."

"The proposed BFO-based cross-layer fitness function includes not only the energy consumption and communication distance but also considers the link quality and network congestions."

"The BFOCHR and PSO-UFC protocols used the BFO and PSO algorithms respectively for clustering and routing protocols of WSNs; however, they used only two parameters for fitness computation in terms of residual energy and geographical distance."

"The BFO-based clustering and information transmission algorithms of NICC apply cross-layer boundaries which show an extraordinary effect on delay outcome contrasted along with different conventions."

Conclusion and Future Work

"We proposed a novel routing protocol for precision farming applications aiming to achieve long-distance communications and scalability."

"NICC is a cluster-based WSN protocol for IoT applications."

"Data transmission is another challenge for such networks, therefore, we applied the similar approach of BFO for optimal relays selection in NICC protocol."

"The BFO-based clustering and routing in NICC protocol had to achieve energy balancing, network lifetime enhancement, QoS improvement, minimum communication delay, jitter, and overhead."

Ensuring Reliability of High-priority Data Transport Using Expected Congestion Shortfall Prediction in Wireless Sensor Networks [85]

This is a machine-generated summary of:

Ganesan, Sangeetha; Muthuswamy, Vijayalakshmi: Ensuring reliability of high-priority data transport using expected congestion shortfall prediction in wireless sensor networks [85]

Published in: Wireless Networks (2021)

Link to original: https://doi.org/10.1007/s11276-021-02792-y

If you want to cite the papers, please refer to the original.

For technical reasons we could not place the page where the original quote is coming from.

Abstract-Summary

"Sending critical data to sink without any data loss is a challenging task in wireless sensor nodes."

"Critical data has to reach sink node without any loss upon event driven triggers."

"A Reliable High priority data transport using Congestion Prediction (RHpCP) technique is proposed in this paper to handle critical data exchange."

"There are two types of precedence data reaching sink node namely high priority or critical data and low priority data sensed from the environment."

"The proposed RHpCP method is compared with the existing techniques of priority data-based congestion prediction using a Network Simulator."

"RHpCP is found to yield greater high priority data to reach sink node in contrast with existing traffic prediction techniques."

Introduction

"Heavy traffic condition in nodes leading to congestion is the prime factor that causes packet loss in the network."

"We consider eliminating congestion in high priority packets by foreseeing congestion occurrence thereby boosting greater emergency data to reach sink node."

"This simple worst-case prediction is carried over to ECS mechanism for routing highly sensitive data to sink node."

"End Tail dropping of packets is another important issue when routing data from source to sink nodes in network [86]."

"In order to construct a promising delivery [87] of highly considered information, there must be an increased fairness level for sensitive data and a moderate fairness level for low priority data to reach sink node."

"A prior forecast of traffic jamming conditions among nodes can be monitored and packets can be routed to sink without packet losses."

Related Works

"This arises the need to find a congestion-free node path to transport high priority data without dropping packets when compared with less prioritized data."

"If incoming data packets from source node is within Qmin buffer size of receiving node, then there is no congestion occurrence in network."

"This congestion state may recursively build up along with unwanted retransmissions resulting in more packet drops including greater loss of critical data, delays and node deaths thereby affecting the network throughput."

"Congestion control in a network is a global design applied among all sensor nodes to take independent decisions to route packets from intermediate and sink nodes."

"The proposed work focusses on routing data packets based on congestion control decisions carried over individual nodes."

"Congestion related parameters among network nodes play a major role in routing data packets to sink collector [88]."

Proposed RHpCP Mechanism

"In the first level, ECS estimation is performed to find the less traffic node to sink in high priority packet flows as a result of greater event triggers."

"The number of packets transferred, confidence level and packet loss are monitored in sink node during the completion of a whole data transmission cycle."

"Both high and low priority data involves choosing next hop based on least congested neighbours and the incoming packets are marked with high or low priority flag setting."

"The less traffic accommodated path is chosen and is replaced for the congested node for both high and low precedence data sensed."

"For low priority data packets, an individual estimation of CI is calculated and the best node is selected in next hop to avoid packet loss."

"The least ECS valued node is selected for next hop from present node to transmit HP data to sink."

ECS Estimation in Sink Node

"This packet loss rate is multiplied with congestion level fraction to obtain the expected congestion shortfall (ECS) at a current node."

"Expected congestion shortfall (ECS) is the average of worst (100-α)% packet losses occurring during tail transmissions in a node."

"Values for node n2 as $\mu_{n2} = 0.2$, $\sigma_{n2} = 2.7$ and $CL_{n2} = 0.24$, then for 98% confidence level for congestion occurrence in n2 is given as, For node n3, $\mu_{n3} = 0.4$, $\sigma_{n3} = 1.8$ and $CL_{n3} = 0.5$ for 95% confidence level For node n4, $\mu_{n4} = 0.08$, $\sigma_{n4} = 4.2$ and $CL_{n4} = 0.75$ for 92% confidence level, For node n5, $\mu_{n5} = 0.06$, $\sigma_{n5} = 3.6$ and $CL_{n5} = 0.44$ for 94% confidence level, The predicted ECS values indicate the maximum congestion shortfall that can be expected for a confidence level in each transmission round."

Low-Priority Data Handling

"The average end-to-end delay D is calculated as 14 ms or 0.0014 s. Given if NP is 150 packets/s, BU is 50 bps, B_T is 500 bps with a N_T value of 200 for 25th node during data transmission."

"The CI value is calculated as follows: Therefore, CI value yields 3.76 for the 25th node in network at a time interval t. Lower CI value marks less congestion at the node whereas higher CI value marks a highly congested node in network."

"This CI factor is calculated only when low-priority data is transported across the network."

"For incoming high priority data, normal transmission is enabled using randomized queue swapping between source nodes and hop count values."

"Low priority data is also given an improved fairness level for less important packets."

Simulation Results

"In PMUI, packet loss rates of high priority data reach nearly 0, 0.015 for medium priority and 0.025 for low priority data at the end of 100 s. Delay caused in HP data is lower than LP data followed by MP data such as 0.002 s, 0.004 s and 0.003 s for 100 s time interval."

"The proposed RHpCP tops with 96.5% packet delivery rate followed by EERET, FROG-MAC, TACCP, Priority based Data Reporting and PMUI at the end of 100 s. TACCP has a higher control overhead followed by RHpCP."

"The time taken for HP data to reach sink node is 5 s to 20 s on average with first node death in network creating energy hole is 1250th round on exposure to node congestion."

"Efficiency is achieved by expanding minimal number of congestion weighted paths to reach sink node for transmitting both critical and less priority data."

Conclusion and Future Enhancement

"The proposed RHpCP technique enables highly-sensitive data to reach sink node without any disruption and also handles transmission of low priority data by local congestion computation."

"There is a randomized transmission from the sender node for high priority data packets so that same node path is not followed each time to reach sink."

"Further ECS computation does the congestion risk prediction among sensor nodes to choose next optimal node for high priority packet transportation."

"More randomized queue formation for high priority data packets can be done for scheduling packets to sink."

An analytical Survey of Attack Scenario Parameters on the Techniques of Attack Mitigation in WSN [89]

This is a machine-generated summary of:

Ávila, Karen; Sanmartin, Paul; Jabba, Daladier; Gómez, Javier: An analytical Survey of Attack Scenario Parameters on the Techniques of Attack Mitigation in WSN [89]

Published in: Wireless Personal Communications (2021)
Link to original: https://doi.org/10.1007/s11277-021-09107-6

If you want to cite the papers, please refer to the original.

For technical reasons we could not place the page where the original quote is coming from.

Abstract-Summary

"This work reviews 93 recent proposals in which different solutions were formulated for the different attacks in WSN in the network layer; in total, 139 references were considered."

"These attacks are mainly Sybil, wormhole, sinkhole, and selective forwarding."

"The main goal of this contribution is to present the evaluation metrics used in the state of the art to mitigate the Sybil, wormhole, sinkhole, and selective forwarding attacks and show the network topologies used in each of these proposals."

Introduction

"Wireless sensors networks (WSN) have been very successful in the last decade; this is due to the many different areas in which they can be used and its impact on the Internet of Things paradigm (IoT)."

"A wireless sensor network becomes one of the most important ingredients of IoT applications [90]."

"WSN has a vulnerability that is an inherent part of the system and relates to the low processing capacity of the nodes or sensors [91–93], this shortcoming makes it difficult to implement algorithms that could potentially increase network security [91]."

"Most of the time, the sink node has a higher processing capacity than the rest of the nodes in the network."

"Wireless sensor networks are the base for the IoT paradigm."

"The IoT paradigm is expected to focus more and more on cybercriminals because more and more devices are connected to the Internet, which is a challenge for experts and security companies."

WSN and IoT Architecture

"According to [94], a sensor network needs five layers: physical, data link, network, transport, and application layer."

"These new three layers are used to manage the network and make the sensors work together, increasing the overall of the network [94]."

"The related user interfaces and services are always based on the characteristics of the applications as such, that is an intelligent transport system, environmental monitoring, remote medical system, etc The perception layer in IoT consists of data capture and its format, grouping the link and physical layers of WSN."

"The network layer includes all the technologies and mechanisms used to transmit captured data, which groups the network and transport layers in WSN."

"The application layer in IoT is the data processing and the applications that make it possible to visualize them, which corresponds to the application layer in WSN."

Applications Based on WSN and IoT

"Many areas can benefit from solutions based on the internet of things, in this review we include the areas of health and care at home, shopping, logistics and transportation, cities and smart homes."

"Health and home care are some of the most important areas for IoT applications since it is directly related to the lives of people."

"The authors of [95] propose a prediction scheme for the location of articles, based on current and previous locations, in which the objective is to locate products in the shopping store."

"In [96] authors design a smart home automation system turning a customary home to a smart home for accessing and controlling devices and appliances remotely, using Android based Smartphone applications."

Security Threats

"Security controls in WSN should be designed taking into account each of the network components."

"The security in a system must be considered from the beginning, taking into account security along all the other network components."

"Maintaining data confidentiality in WSN is important, but maintaining a secure architecture and topology is also necessary."

"One key security goal of WSN is to ensure data security."

"Traditional wireless networks do not have as many limitations as WSN, which is why the implementation of security mechanisms is not an easy task [97], and many authors ignore this aspect in their proposals."

"The wireless sensor network, as compared to traditional computer networks, has certain limitations that are highlighted in [98], including: Limited resources: in many cases, the implementation of a secure mechanism requires the availability of certain resources, such as memory and power, however, these resources are limited in a wireless sensor network."

Attacks on the Network Layer

"The transmission medium is one of the reasons why wireless networks are vulnerable to certain types of attacks [99]."

"If a wireless network is composed of nodes with low processing capacity and useful lifetime, this network may be compromised."

"This is why WSN are even more vulnerable to attacks than traditional wireless networks."

"We analyze different attacks (in the network layer) that can occur in wireless sensor networks as well as the network topologies being proposed during the attacks."

"This attack infiltrates the network of 2 or more illegitimate nodes which send messages to each other using low latency and high bandwidth channel, which becomes attractive to legitimate nodes, causing them to use a compromised transmission medium."

"The goal of a sinkhole attack is to attract network traffic to a specific node, this is achieved by altering the routing tables of the network."

Network Topologies

"The analysis presented in the following section includes the network topologies used in the revised proposals."

"The topology of a network is a graphic description of how the nodes of the network are connected."

"With the technological advancements, it is common to find mixed topologies generated by MANET or VANET networks in which nodes must have the intelligence to redistribute their topology given their constant movement."

Analysis of Proposals

"We analyze the proposals that mitigate each one of the attacks evaluated in this revision."

"The analysis identifies aspects such as the topology implemented, if the proposal evaluates the detection of intruders or if the proposal detects or prevents the attack, as well as the metrics used for the evaluation of each of proposal."

"Only the proposals that mitigate the sybil attack were analyzed."

"The most used metric to evaluate the proposals that mitigate the sybil attack is the Accuracy Detection/Detection Rate, which is used by 17 proposals."

"The most used metric that mitigates the wormhole attack is the PDR used by 16 proposals."

"Only the proposals that mitigate the sinkhole attack were analyzed."

"The most used metric to mitigate the sinkhole attack is the PDR, which is used by 8 proposals."

Future Work

"Future work is expected to identify vulnerabilities in wireless sensor networks inherited from the Internet of Things technology."

"These vulnerabilities can generate attacks in the network layer allowing access to confidential information of unauthorized people."

"Identifying these attacks in the network layer would lead us to the implementation of strategies that can mitigate that."

Conclusions

"It is important to highlight this type of analysis because of its benefits, disadvantages, and security problems that IoT inherited from WSN."

"This work also highlights the importance of including security mechanisms from the very beginning of the design of the topology of the network, these mechanisms cannot be treated in an exclusive manner."

"As a concluding remark, it is clear that it is time to give more importance to the area of security in the networks of sensors and the internet of things given their recent growth and the applications emerging around this paradigm."

Novel Distributed Dynamic Backbone-Based Flooding in Unstructured Networks [100]

This is a machine-generated summary of:

Saeedvand, Saeed; Aghdasi, Hadi S.; Mohammad Khanli, Leili: Novel Distributed Dynamic Backbone-based Flooding in Unstructured Networks [100]

Published in: Peer-to-Peer Networking and Applications (2019)

Link to original: https://doi.org/10.1007/s12083-019-00817-0

If you want to cite the papers, please refer to the original.

For technical reasons we could not place the page where the original quote is coming from.

Abstract-Summary

"The primary method for resource discovery on the unstructured networks is flooding a query on the network."

"All existing flooding algorithms for unstructured networks generate almost high additional duplicated queries."

"This paper, therefore, proposes a novel flexible Distributed Dynamic backbone-based Flooding (DDBF) algorithm for distributed unstructured networks."

"This paper explores Grid middleware, Peer-to-Peer (P2P) paradigm, and cloud networks resource discovery requirements and it proposes flooding algorithm based on the P2P networks using simulation."

"To evaluate and prove DDBF algorithm we, first, evaluated it on four fixed network topologies along with two different query flooder distributions, then we evaluated it with one dynamic network topology."

Introduction

"For information discovery in structured networks, there is no need to flood a query on the network."

"Structured networks are quite widespread in the research literature based on the different network structures such as grid middleware or cloud networks, but because of the dynamic nature (nodes enter and leave the network rapidly) most of the networks, it is hard to use such networks widely [101].Unstructured networks: These types of networks don't use centralized directory nor any accurate control over the network topology or file placement [102]."

"All of the unstructured networks use resource sharing systems, where resource discovery is an inevitable challenge."

"In the unstructured networks the current query flooding algorithms are generating a high number of duplicated queries, which causes large loads on the network."

"For flooding query in unstructured networks, different algorithms have been proposed."

"DDBF algorithm proposes a reliable approach to create a novel flexible backbone on the dynamic unstructured networks for flooding queries."

Literature Review

"That the original Gnutella algorithm [103] uses a simply constrained flooding approach for search, which generates too much traffic on the network."

"Qin Lv and others in [104] proposed a query flooding algorithm based on multiple random walks that resolves queries almost as quickly as Gnutella's flooding method while reducing the network traffic by two orders of magnitude in many cases."

"It's known that GA is an evolutionary algorithm that in unstructured networks if adjacent peers' changes rapidly, the algorithm cannot act real-time enough for flood queries properly."

"Proposed algorithms don't ensure connectivity in different situations; for instance, because of wireless sensor network nature, they don't have any strategy to add some new peers to the network after connected dominating set creation."

"In [105] authors presented a protocol for 3D P2P overlay over mobile ad hoc networks (MANETs) in which each peer runs a distributed algorithm to exploits a 3D-overlay."

Problem Description

"Even with this duplicate suppression, the number of duplicate messages in flooding algorithms can be excessive, and the problem worsens as the TTL (Time-To-Live) increases."

"These problems mean that flooding incurs considerable message processing overhead for each query, increase the load on each node as the network expands, and increase the query rate [106]."

"In the latest research which named lightweight path flooding scheme, one efficient approach for decreasing the repeated query overhead is proposed in [107]."

"In the lightweight flooding scheme, a node does a flooding on the query that includes the list of the adjacent peers' addresses."

"Lightweight path flooding scheme was compared with earlier state-of-the-art works, and it demonstrated higher performance by decreasing the number of repeated queries significantly."

"We provide a reliably distributed algorithm for activating only some of the nodes as a backbone to decrease the number of the duplicated queries."

Proposed Distributed Dynamic backbone-based Flooding Algorithm

"If a node is active and leaves the network, it becomes inactive and only proposed distributed rules can make it active again."

"At backbone creation phase; first of all, all existing nodes on the network follow four rules to be in the active or inactive mode, then assurance of connectivity step starts to ensure the connectivity."

"All nodes on the network can forward a new query but just activated nodes are able to flood queries to all of its adjacent nodes."

"When a new node is added to the network, it follows proposed rules to be in one of the active or inactive modes (then it enters the query flooding phase)."

"In the activation rules step, nodes performing proposed rules to be in one of the active or inactive modes."

"It stays inactive and sends a message to its single adjacent node to activate it (sends $AAM(N_j)$); otherwise, it goes to the second rule."

Performance Evaluation

"The number of involved nodes for flooding on the network has been compared in which results are flooding averages per both uniform and Zipf-like query flooder distributions."

"DDBF algorithm decreased the number of the redundant queries in a high number of nodes on the network about 49% compared to flooding scheme and 27% compared to Lightweight path flooding algorithm in all topologies."

"For dynamic network topology, we propose the results of two important metrics as a number of engaged nodes on the flooding and number of queries on the network."

"In the results obtained for dynamic network topology, average improvement percentage in terms of the number of engaged nodes for flooding on the network is 42% for DDBF in comparison to the TF, and in comparison to the lightweight algorithm is 24%."

Conclusion

"Networks' performance can be influenced by using an efficient algorithm for flooding queries."

"In the DDBF algorithm, we proposed a distributed algorithm whose nodes are able to create a reliable backbone for flooding in the cloud network."

"In the results obtained for dynamic network topology, in terms of the reduced engaged nodes at DDBF is 42% in comparison to the TF, and 24% in comparison to the lightweight algorithm."

"In terms of the reduced flooded queries at DDBF in comparison to the TF is 67%, and in comparison to the lightweight algorithm is 34%."

"In all different topologies DDBF achieves better network performance in comparison to existing state-of-the-art algorithms."

A Hybrid Computing Approach to Improve Convergence Time for Scalable Network [108]

This is a machine-generated summary of:

Devikar, R N; Patil, D V; Prakash, V Chandra: A hybrid computing approach to improve convergence time for scalable network [108]

Published in: Sādhanā (2020)

Link to original: https://doi.org/10.1007/s12046-020-01454-7

If you want to cite the papers, please refer to the original.

For technical reasons we could not place the page where the original quote is coming from.

Abstract-Summary

"The minimum route advertisement interval (MRAI) timers are mostly used by network operators to reduce the issues occurring at the time of increasing convergence time."

"Many researchers have been working on variation in MRAI timer and effect of it on scalability and network convergence."

"Keeping the value of MRAI timers optimum results in reducing the issue of slow convergence for the scalable network."

"With the static value of MRAI timer i.e., 30 s, FAPSO is a suitable algorithm that gives the optimal value of convergence time for the scalable network."

Introduction

"Convergence time of a network increases because of failure and flapping in the network."

"As there are multiple paths available in the network, so BGP requires more time to go through all the paths."

"Convergence time is the amount of time needed for a network to reroute the packets after a node or route failure."

"Convergence time is defined as the time required for all the routers available in the network to become updated with new, disappeared routes, or changed routes."

"Convergence optimization can be done by correctly using the minimum route advertisement interval (MRAI) timer for various network scenarios."

"The convergence time depends upon the network scenarios (topology) which were shown in the simulation."

"The topologies were manipulated by including a large number of autonomous systems (ASs) for checking whether the convergence time deviates in respect of network size or not."

Literature Survey

"Labovitz et al [109, 110] examined that BGP is an inter-domain routing protocol which takes a significant amount of time to converge because of different internet policy and topology."

"The Aiabdulkreem et al [111] proposed a modified path exploration technique i.e., fight or flight response technique to reduce the convergence time."

"BGP Churn Aggregation (CAGG) technique proposed by Wang et al [112] tries to reduce BGP path Exploration by gathering several AS routes in one path, without any bad impact on convergence time."

"Devikar et al [113] developed an approach in addition to load balancing for improving network convergence time and to reducing the congestion."

"In this 3 routers are available from source to destination and BGP keepalive time set as 60 s. So the convergence time for this scenario is calculated as 180 s. Moubarak et al [114] proposed a BGP-NCM (BGP Novel Control Mechanism) for newly updated path."

Research Methodology

"The convergence time can be maximally reduced by keeping the value of MRAI timer optimal without increasing the number of update messages."

"The proposed approach incorporates the fuzzy-based approach (filtering approach) with PSO to find the optimal value of the MRAI timer."

"In the implementation we have taken MRAI values from 5–30 s for different scenarios (number of nodes) range from 50–100."

"The algorithm identifies the best value is between 1–2 s as an optimal MRAI timer, and it will reduce the convergence time by around 9%."

"By making the value of MRAI timer variable we can improve the performance related to packet losses, number of updates, and convergence time for the scalable network."

"The Algorithm 1 is used to generate dynamic MRAI timer which improves the convergence time of a network."

Experiment Setup and Result Analysis

"From the experimental result, it has been observed that the value of convergence time and optimum MRAI are proportional to MRAI timer."

"The convergence time is calculated for different MRAI values from 5–30 s for 50–100 nodes."

"Result also shows that as we increase the nodes from 60 to 100 there is a small change in the value of convergence time."

"This happens because, for a lower value of MRAI, the routes are discovered rapidly after links or node failure which results in a reduction in packet losses."

"Analysis it has been observed that with an increase in the number of nodes, we are getting low PDR value for default MRAI."

"From the experimentation, it has been observed that FAPSO is having lower convergence time so that all the node are converged with updated information."

"Experimental results show that with an increase in the number of nodes gradually it increases the convergence time."

Conclusion

"The proposed work optimised the MRAI timer by using the FAPSO technique and compares the results with FMRAI technique."

"From the experimental result, it has been observed that FAPSO technique gives better optimised MRAI timer than FMRAI."

"As we have optimised the MRAI timer, it reduces the convergence time by around 9%–10%, required to reach the updated information in the scalable network."

"The proposed FAPSO technique is also used to improve different parameters like packet delivery ratio, scalability, network delay and throughput for different network scenarios i.e., the number of nodes."

Performance Evaluation of Scheduling Approaches for Wireless Sensor Networks [115]

This is a machine-generated summary of:

Al-Ghamdi, Bandar; Ayaida, Marwane; Fouchal, Hacène: Performance Evaluation of Scheduling Approaches for Wireless Sensor Networks [115]

Published in: Wireless Personal Communications (2019)

Link to original: https://doi.org/10.1007/s11277-019-06775-3

If you want to cite the papers, please refer to the original.

For technical reasons we could not place the page where the original quote is coming from.

Abstract-Summary

"Many research teams work on various issues as communication protocols, task scheduling, physical architecture, etc Scheduling tasks in such environments is an important issue to consider."

"In order to respect these conditions, adapted scheduling algorithms are proposed in this study."

"We consider that a WSN is composed of clusters of sensors."

"We propose three different scheduling procedures: semi-dynamic algorithm (for each cluster we have a fixed planning but its sensors will have a dynamic slot repartition), dynamic algorithm (the sensor clusters will have a dynamic time slot repartition as all sensors) and high-priority dynamic (in addition to have dynamic scheduling, we consider tasks with different priorities)."

"Of the use of scheduling methods, the communication between sensors is managed to be efficient and targets to avoid conflicting packets."

"The simulation results show that our proposals reduce the latency and improve the packet delivery."

Introduction

"Most of research studies focus on efficient power consumption using smart communication protocols or proposing specific scheduling algorithms or splitting the network into clusters [116]."

"Each sensor node works within its period of time."

"These time periods are scheduled on the Personal Area Network Coordinator (PANC)."

"The periodic time of each sensor node could be predefined (fixed) or could vary."

"It is important to find an efficient algorithm that can schedule each time slot properly."

Related Works

"They have proposed a slot assignment mechanism for a distributed centralized applications based on tree networks of ZigBee."

"They perform their scheduling process based on the time division approach."

"This paper works mainly on the Time Division Cluster Scheduling (TDCS) that tunes the configuration of IEEE 802.15.4/ZigBee beacon-enabled cluster-tree WSNs."

"[117], authors also have done their scheduling approach based on the time division approach as well."

"Our proposals are based on the first approach of the IEEE 802.15.4/ZigBee standard."

"Static approach has a number of drawbacks, (1) the WSN applications need a dynamic and an adaptive scheduling procedures whereas the static one cannot handle the mobility and to be adaptive with the network traffic. (2) Wasting time with unused packet interval time in static mode."

The Scheduling Algorithms

"Only the PANC has the eligibility to assign addresses and time slots to its sensor nodes in the network."

"Solution, is a priority scheduling algorithm which is based on the dynamic approach."

"That the time slots will not be provided only upon the node's request but also based on the schedule availability on the PANC."

"If the requested time slots is greater than the pre-defined slots to a particular node, the PANC will assign only the scheduled slots."

"After the end of scheduling, the PANC sends a beacon in order to synchronize all the nodes with the new time slots assignments."

"The PANC in the dynamic part will synchronize its child nodes based on time slots availability."

"The PANC decides the priority for each event based on the First Ready First Serve (FRFS) algorithm with the ability of changing every duty-cycle according to the nodes activities, i.e. some nodes do not have data to send."

Performance Evaluation

"We compare them with the static simulation model."

"Our simulation models are designed to support a cluster-tree (multi-hop) self-organizing network based on ZigBee/IEEE 802.15.4 protocol in beacon-enabled mode which runs at 2.4 GHz with a data rate of 250 Kbps."

"The feasibility of our work is done by comparing each simulation model with the static model."

"The Semi-Dynamic scheduling model also shows an acceptable improvement by 2% compared to the Static model."

"To illustrate, we presented here the results obtained for the node 5, which is considered as prioritized sensor in the High-Priority Dynamic Scheduling Algorithm example."

"This explains why the throughput is improved in the High-Priority Dynamic scheduling algorithm compared to the Dynamic scheduling algorithm."

"The priority scheduling model improves the network performances in terms of the total traffic successfully received/sent through the network."

Conclusion

"We have proposed three scheduling approaches called Semi-Dynamic, Dynamic and High Priority Dynamic Scheduling in Beacon Enabled Mode for Wireless Sensors Networks."

"These approaches can perform both static and auto generation (dynamic) of the source address, that are assigned by the PAN coordinator to its children, which are accepted to be working as routers."

"In the other hand, dynamic nodes can send their data through their parents (routers) to the PAN Coordinator."

"It is shown from this work that the dynamic assignment, i.e. based on the data ready to be sent, usually gives high performance than other approaches."

"If we consider a node as prioritized one, all the data sent by this node will be considered as priority events."

"The scheduling algorithm should be able to manage these different types of data which is not the case now."

Design Systematic Wireless Inventory Trackers with Prolonged Lifetime and Low Energy Consumption in Future 6G Network [118]

This is a machine-generated summary of:

Meenakshi, N.; Jaber, Mustafa Musa; Pradhan, Rahul; Kamruzzaman, M. M.; Maragatham, T.; Ramamoorthi, Jaya Subalakshmi; Murugesan, Mohanraj: Design systematic wireless inventory trackers with prolonged lifetime and low energy consumption in future 6G network [118]

Published in: Wireless Networks (2023)

Link to original: https://doi.org/10.1007/s11276-023-03325-5

If you want to cite the papers, please refer to the original.

For technical reasons we could not place the page where the original quote is coming from.

Abstract-Summary

“Integrating heterogeneous IoT networks with the existing ones, mainly cellular networks, is in great demand.”

“This study focuses on Heterogeneous IoT networks over 6G networks with ultra-dense deployment using Mobile edge computing and software-defined wide-area network (MEC/SDWAN).”

“This software-defined wide-area network is used for designing systematic Wireless Inventory Trackers (WIT).”

“The heterogeneous network consists of interconnected nodes and links of different types with a software-defined Wide Area Network(SD-WAN).”

“To address the heterogeneous network system capacity issue, the HIoT and Ultra-Dense deployment are employed for making the promising technologies in the 6G to manage them, enhancing the spatial reuse through flexible, intensive deployment.”

“This procedure helps to improve the prolonged lifetime of up to 2.5 h with low energy consumption of up to (3%) on efficient communications in the 6G cellular network.”

Introduction

“Article, the heterogeneous IoT(HIoT) Networks over 6G networks with ultra-dense deployment using Mobile Edge Computing software-defined wide-area network (MEC/SDWAN) for making a design of systematic wireless inventory trackers (WIT) are established [119].”

“This technology provides several benefits, including improved accuracy by eliminating manual data entry, increased efficiency in managing inventory levels, access to real-time data for informed decision-making, better customer service through rapid product fulfilment, and cost savings by reducing inventory carrying costs and avoiding stockouts."

"The main contributions of the study: i. The design of systematic wireless inventory trackers(WIT) using heterogeneous IoT(HIoT) networks over 6G computing and MEC/SDWAN for improving the prolonged lifetime and low energy consumption for efficient communication in 6G cellular networks is proposed."

"Heterogenous IoT network over 6G computing and MEC/SDWAN benefits in a concise perspective for improving the prolonged lifetime and low energy consumption for efficient communication."

Literature Review

"NodeB, which stands for "Node Base Station," is used in 3G wireless communication networks and manages the wireless communication between mobile devices and the core network."

"ENodeB, or "enhanced Node Base Station," is employed in 4G Long-Term Evolution (LTE) wireless communication networks and is responsible for providing radio access to mobile devices."

"The sensor is converted to digital values to maintain the data to be secure and to make the formation of the system to produce the managing factors to follow the progressive for the systematic elaboration of the IoT networks."

"This local receiver is designed by the local wireless and the inventory system to form in the wide-area network for the heterogeneous computing software to mention the network to perform the telecommunication in the signal for the heterogeneous for the travelling and the formation of the system in the wide-area network for the database and the cloud in the formation of the HTTP server for the end-user."

Result Analysis

"The heterogeneous distributed IoT system includes wireless nodes organized into wireless sub-networks."

"The NS4-programmable data plane simulation is a P4-driven network simulator supporting simulation of P4-enabled networks to address the problems existing in traditional sim emulators."

"The NS4-programmable data plane simulation is used for an internal P4 pipeline to process packets as P4 programs are defined and enable queues and buffers in a pipeline to model the behaviours of real P4 devices."

"The proposed system is to construct a design of a systematic wireless inventory tracker(WIT) using a heterogeneous IoT network over 6G computing and MEC/SDWAN() for improving the prolonged lifetime and low energy consumption for efficient communication in a 6G cellular network."

"The comparison result is based on mobile edge computing or software-defined wide-area network and heterogeneous distributed IoT system Nguyen and others [120]."

"Edge computing is the practice of capturing, storing, processing, and analyzing data near the client, and the heterogeneous distributed IoT system is various sub-systems."

Conclusion

"The observational results express that the proposed framework constructing a design of systematic Wireless Inventory trackers (WIT) using heterogeneous IoT(HIoT) networks over 6G Computing and MEC/SDWAN for improving the prolonged lifetime and low energy consumption for efficient communication in 6G cellular network has achieved a high precision-recall measure as compared to the existing technique."

"The total energy consumption regarding the proposed technique's number of iterations is achieved."

"The lifetime and energy consumption of the heterogeneous IoT (HIoT) networks over 6G networks is not that much efficient in the proposed model."

"In the future, the work can be strengthened in such a way that heterogeneous IoT (HIoT) networks over 6G networks with Ultra-dense deployment using Mobile edge /SDWAN computing to improve the prolonged lifetime of up to 2.5 h with low energy consumption of up to (3%) on efficient communications in the 6G cellular network."

An Improved Emergency Response Routing Protocol for Internet of Things [121]

This is a machine-generated summary of:

Araghipour, Amirhossein; Mostafavi, Seyedakbar: An Improved Emergency Response Routing Protocol for Internet of Things [121]

Published in: Wireless Personal Communications (2021)

Link to original: https://doi.org/10.1007/s11277-021-09187-4

If you want to cite the papers, please refer to the original.

For technical reasons we could not place the page where the original quote is coming from.

Abstract-Summary

"Due to the limited resources of Internet of Things (IoT) nodes, routing protocols for these networks should be designed in such a way that not only reduce the packet loss rate and transmission delay, but also to utilize the network resources efficiently."

"The ERGID protocol has been specifically designed to reduce end-to-end delay for emergency data."

"A novel emergency response routing protocol based on the ERGID is proposed in this paper to remedy the disadvantage of ERGID protocol in guaranteeing delay for emergency data."

"The proposed protocol meets the requirements of reliability and energy efficiency for the IoT. To achieve these goals, a prioritization mechanism is designed for emergency data that leads to more resource allocation to emergency packets and thus, reduction in transmission delay of this data."

Introduction

"The routing protocols designed for IoT should consider these limitations so that data can be transmitted securely with minimal energy consumption and minimal network latency."

"Given the large-scale and expanding scope of IoT networks, the routing protocol should have a mechanism to deliver the emergency data immediately with respect to the remaining energy of the nodes and identify the best route by estimating the delay of the routes and prioritizing the emergency packets [122]."

"The proposed protocol is more focused on the timely delivery of emergency data and its primary design goal is to reduce packet loss and end-to-end delay for emergency data comparing with state-of-art routing protocols."

"The main contributions of this paper are summarized as follows: A novel IoT routing protocol is designed for IoT networks is designed that, given the type of packets transmitted from the network nodes, prioritize packets so that the emergency data reach their destination on time."

Related Work

"ERGID is a routing protocol which aims to reliably transmit emergency data with minimum delay and to provide traffic load balancing in the network while taking the nodes residual energy into consideration."

"The authors in Khaleghnasab and others [123] proposed an improved version of AOMDV routing protocol which includes the energy level of nodes and signal-to-noise ratio in the packets and utilizes the gray theory to select multiple routes for concurrent data transmission."

"A trust-based routing protocol based on RPL is proposed in Selvaraj and others [124] which forms a community of trusty nodes to forward data packets."

"A has three choices to reach S by passing through B, C, or D. The calculated delay of each path are 113+6 (119), 110+8 (118), and 115+5 (120) and A can select the shortest one, which is 118, as its shortest distance to S. Residual Energy Probability Choice (REPC) REPC mechanism is focused on nodes remaining energy for route selection process to balance the network load."

The Improved Emergency Response Routing Protocol

"This type of control messages keeps the network status information updated in each node."

"This message is broadcast in the network to notify the other nodes that they should not forward the normal packets to this node."

"When node i receives an epath message from node j, it omits node j from its next hop candidate set for normal packet transmission."

"The second phase is control message generation which includes the operation of sharing the network status information among the nodes."

"In this phase, the nodes update their information using the received control messages from their neighbors."

"When network setup phase is completed, the IoT nodes start to generate control messages."

"An epath message is generated when a node is forwarding an emergency packet."

"When a node receives one of response or notification messages, it must update its information about the network status."

Performance Evaluation

"The performance of our proposed routing protocol is evaluated by comparing it with ERGID regarding three main routing metrics including packet loss, end-to-end delay, and energy consumption."

"In all scenarios, packet loss for emergency data is reduced in the proposed protocol."

"The best performance is achieved in transmission rate 50 kb/s. In this scenario, emergency packet loss rate in our protocol is 49.3% less than in ERGID."

"In the next scenario, the packet loss rate for normal data are evaluated for ERGID and the proposed protocol."

"As it is shown, end-to-end delay for emergency packets in ERGID is higher than proposed protocol which means that emergency data in the proposed protocol reach sooner to the destination than ERGID."

"The average end-to-end delay for emergency packets is 227 ms in our proposed protocol which shows reduction of 60% reduction compared to ERGID performance."

Conclusions

"Several routing protocols for emergency data transmission in Iot networks have been proposed, among which ERGID has shown better performance."

"We proposed a routing protocol based on ERGID, which applies a packet prioritization mechanism through reserving a path for delivering emergency packets."

"If an IoT node generates an emergency data, it will be forwarded to the destination with less packet loss and transmission delay."

"The simulation results in different network scenarios shows that the proposed protocol reduces the packet loss rate and end-to-end delay by 35% and 60%, respectively, comparing with ERGID."

A Hybrid Fuzzy-Genetic Algorithm for Performance Optimization of Cyber Physical Wireless Body Area Networks [125]

This is a machine-generated summary of:

Choudhary, Amit; Nizamuddin, M.; Sachan, Vibhav Kumar: A Hybrid Fuzzy-Genetic Algorithm for Performance Optimization of Cyber Physical Wireless Body Area Networks [125]

Published in: International Journal of Fuzzy Systems (2019)

Link to original: https://doi.org/10.1007/s40815-019-00751-6

If you want to cite the papers, please refer to the original.

For technical reasons we could not place the page where the original quote is coming from.

Abstract-Summary

"The use of fuzzy decision-making in datapath selection extends the sensor network lifetime with a uniform distribution of routing load among network nodes."

"Fuzzy-logic based routing protocols are mostly designed for general wireless sensor networks (WSN)."

"This paper presents a fuzzy-logic-based clustering protocol for data routing in WBANs."

"Nodes are grouped into clusters and cluster head nodes are selected through a Fuzzy-Genetic Algorithm termed as EB-f_g-MADM."

"EB-f_g-MADM makes an assessment of dual attributes of each cluster node in terms of node residual energy and CH selection cost."

"CH selection cost of a node is the forecasted value of network energy consumption if the node acts as a cluster head."

"EB-f_g-MADM utilizes a fuzzy-TOPSIS function which makes a quantitative comparison of cluster nodes and selects the cluster head node possessing the aforementioned attributes closest to their ideally desired values."

"EB-f_g-MADM provides enhanced network lifetime with a uniform distribution of routing load."

Introduction

"The requirement of high QoS performance with the constraint of limited node power demands an energy-efficient data routing protocol for WBAN systems."

"The first attribute, CH selection cost of a particular node is calculated by assuming the node as next cluster head and forecasting the value of network energy loss if it is used for executing the clustering-based data routing operation as cluster head."

"EB-f_g-MADM estimates the CH selection cost values of each candidate cluster member node using the first-order radio model for node transmission energy consumption [126]."

"For an optimum cluster head node, its CH selection cost needs to be minimized while the value of Residual energy attribute is desired to be maximized."

"After having an assessment of node attribute values, EB-f_g-MADM decides the relative importance of two node attributes for cluster head selection."

"EB-f_g-MADM applies the node attribute values and attribute weights to a fuzzy logic-based TOPSIS function for cluster head selection."

Literature Survey and Motivation

"The consideration of suitable node attributes helps in achieving the desired objectives of selection process like energy-efficient data routing, improved network throughput and latency etc The final step utilizes a Multi-Attribute Decision Making (MADM) algorithm for the final selection of optimum cluster head node [127, 128]."

"Fuzzy logic based MADM approaches The fuzzy logic based MADM approaches transform the attribute values into fuzzy linguistic grades and adequately deals with the vagueness of cluster head selection process [129]."

"Xie and others proposed CRT2FLACO protocol which utilized Ant colony optimization in conjunction with type-2 Mamdani fuzzy inference-based MADM approach for CH node selection [130]."

"The proposed EB-f_g-MADM algorithm performs model-based estimation of node CH selection cost attribute value."

"The fuzzy-MADM-based cluster head selection mechanisms offer a uniform distribution of cluster head load among network nodes resulting in a substantial increase in network lifetime as compared to conventional MADM approaches [127]."

WBAN System Model

"These sensor nodes are allowed to transmit their data packets directly to the sink using a single-hop LOS communication channel."

"Proposed work adopts first-order radio model for the estimation of node energy expended by a node in data transmission [126, 131]."

"According to first-order radio model, node transmission energy consumption for relaying a data packet (W number of bits) to a distance 'D' can be estimated as follows: The same model also accounts for node energy consumption in receiving a data packet of 'W' bits as follows: Using the same model, node energy consumed in aggregating 'W' number of bits can be accounted as follows: Here, $E_{Tx\text{-}elect}$, $E_{Rx\text{-}elect}$, and E_{Amp} denote per bit node energy consumption in running node transmitter, receiver, and amplifier circuits, respectively."

"In the proposed technique, node processing energy losses are ignored as they are negligible as compared to data transmission [94]."

Proposed WBAN Routing Protocol

"To select an appropriate CH node, EB-f_g-MADM assesses following attributes of each candidate cluster member node: i. CH selection cost of the node, $E_L(i)$; $\forall$ $i \in N$. ii."

"Step-1: Estimation of CH selection cost of each node Each candidate cluster member node estimates its CH selection cost attribute value."

"Step 3: Genetic Algorithm (GA) based weight optimization In this step, the sink decides the relative importance of three node attributes for cluster head selection by assigning certain weights to them such as W_L for E_L attribute and W_{CH} for E_R attribute."

"Genetic Algorithm optimizes the two node attribute weights in such a manner so that the selected cluster head needs minimum transmission power for communicating its data to sink."

"After receiving the node attribute values, sink performs Fuzzy-TOPSIS based cluster head selection process and Genetic Algorithm based attribute weight optimization process."

Simulation and Analysis of Proposed Protocol

"Network lifetime is defined as the total number of transmission rounds until last network node dies out."

"Simulated performance suggests a network stability period of 8420 rounds for the proposed protocol (EB-f_g-MADM) while for DSCB, EERP, and M-ATTEMPT protocols stability periods are 7938, 5315 and 2551, respectively."

"Proposed protocol offers an elongated network lifetime around 14,410 transmission rounds as compared to existing protocols."

"The proposed protocol offers elongated network stability lifetime periods as compared to existing protocols."

"It proves that the proposed protocol uniformly distributes the CH load across network nodes."

"Proposed protocol (EB-f_g-MADM) also adopts loss-less data compression technique which removes node redundant data from transmission."

"Network throughput can be measured by the number of effectively delivered data packets to the sink node."

"EB-f_g-MADM protocol successfully delivers a sum of 5.6×10^4 packets from source to sink node in 15,000 transmission rounds."

Conclusions and Future Work

"An energy efficient data routing protocol is of paramount importance for WBAN because bio-sensor nodes consume a considerable amount of assigned node energy for transmitting data to other nodes."

"A clustering-based protocol provides an energy-efficient data routing scheme with affordable data loss rates and network latency."

"Simulation results verified that the EB-f_g-MADM protocol offers an elongated network lifetime and stability period of 14,410 and 8420 transmission rounds, respectively, as compared to existing classical MADM-based routing protocols."

"The proposed protocol offers a consistent distribution of routing load among network nodes which results in elongated stability period and network lifetime."

"EB-f_g-MADM protocol utilizes a loss-less data compression technique for intra-cluster data communication which removes the node redundant data from the transmission."

"Our future research directions are focused on designing the WBAN routing protocol which overcomes the problem of cross-channel interference and provides the desired QoS performance in a multiple WBAN scenario."

Energy Efficient and Secure Healthcare Data Transmission in the Internet of Medical Things Network [132]

This is a machine-generated summary of:

Mondal, Sanjoy; Ghosh, Indrakshi; Das, Abhishek: Energy efficient and secure healthcare data transmission in the internet of medical things network [132]

Published in: Microsystem Technologies (2022)

Link to original: https://doi.org/10.1007/s00542-022-05398-2

If you want to cite the papers, please refer to the original.

For technical reasons we could not place the page where the original quote is coming from.

Abstract-Summary

"In the monitoring system, Biomedical Sensor Nodes (BSN) are used to sense the physiological parameters of the patients and transmitted to the coordinator or Base Station (BS) via the Internet of Medical Things (IoMT) frameworks for further analysis."

"A key research problem is to reduce energy consumption and communication costs."

"Patient data, transmitted over the insecure channel may lead to security threats."

"We propose an energy-efficient and secure health monitoring system to reduce the energy consumption and communication cost of the monitoring system."

"It also provides secured data transmission by incorporating a Cipher Block Chaining (CBC) based lightweight security mechanism."

"Our proposed algorithms prove to be a feasible solution for securing and extending the IoMT network lifetime."

Introduction

"Way to advancements in IoT network infrastructure, it has turned out to be possible to screen several patients through the IoMT network and the framework is known as a group health monitoring (GHM) [133]."

"The contributions of the proposed IoMT-based healthcare monitoring frameworks are highlighted below. (i) Designed an End-to-end secure and energy-efficient e-health monitoring framework for a group of individuals based on IoMT frameworks. (ii) Krill-herd Optimization based efficient leader node (LN) selection strategy is developed with sub-optimal cost value. (iii) In the case of a large-scale network, a distributed LN selection algorithm is designed. (iv) For sensitive medical information transmission, a lightweight cryptographic method is proposed to provide a secure data transmission. (v) To evaluate the performance of the proposed algorithms we conduct a series of experiments."

Related Works

"While the Temperature Aware procedures work to keep the Temperature intensity of the Sensors, network speed is a crucial concern in addition to energy efficiency."

"An energy-efficient and reliable routing protocol (EERP) for WBAN is proposed by Ali et al. [134]."

"Another cost-effective and energy-efficient routing protocol is proposed by Kiran and others based on path link factors, residual energy, and link reliability."

"An EH-RCB mechanism that combines a powerful and energy-efficient harvested focused routing algorithm was presented by Ullah et al. [135]."

"Anwar et al. [136] collectively systemized an energy-aware link efficient routing approach for WBAN (ELR-W)."

"Energy and link-aware efficiency is used to select the next-hop data routing."

"By taking the usage of the Critical data routing (CDR) protocol, Sagar et al. [137] volunteered the energy-aware WBAN for Health monitoring."

"An energy-balanced alternative-aware routing algorithm (SEAR) is proposed for WBAN in Jiasong et al. [138]."

Network and System Model

"In this IoMT application, IoT devices are randomly placed in regions of interest (ROIs)."

"Each IoT node has a fixed detection and transmission range."

"Each IoT node can communicate with other IoT devices and base stations directly or in a multi-hop fashion."

"IoT nodes are first organized into a certain number of clusters, and in each cluster, a unique node known as a leader node (LN) oversees gathering information from those other network nodes and communicating with the BS."

"All IoT nodes have a fixed initial power, and power consumption is a big issue during communication."

"At the time of foraging, the Krill's swarm is communicating with another member of the swarm and this is the basis of the KHO algorithm."

Proposed Method

"We suggest two different routing mechanisms: a distributed routing a system that operates in the peer-to-peer mode without the involvement of a central node or BS, and a centralized routing strategy in which BS collects network data from all the nodes."

"The CN nodes are divided into several groups (called a cluster) and within each group, a leader node (LN) collects data from its member nodes and transmits the whole data packet to the BS."

"The BS calculates variables such as node residual energy, neighbor count, average node distance, etc using network data from all CNs."

"Residual energy of sensor Node: In this data communication scheme, the leader node transmits the aggregated data to the next-hop node or the BS."

"It consumes more energy than the member nodes because the leader node is involved in an additional job like aggregation, processing, and data transmission to the next hop or BS."

Performance Evaluation and Result Analysis

"In each experiment, the communication range between each node is adjusted to 80 m. To evaluate the proposed algorithms we use several evaluation metrics defined below: Alive Nodes: Alive nodes indicate the number of nodes alive in the network."

"The higher number of alive nodes increases the performance of the network."

"Network lifetime: Network lifetime describes the number of rounds until sensor nodes are died out."

"FND, HND and LND.FND: FND: The FND specifies the number of rounds before the network's first node fails."

"HND:The HND specifies the number of rounds until 50% of the network's nodes are dead."

"LND: The LND specifies the number of rounds until the network's nodes are destroyed."

"It can be observed that the network delay increases in both cases, with the increasing network size (i.e., node numbers) and with the increasing of many rounds."

Statistical Significance

"To prove the statistical significance of the simulation results, we have conducted a pair-wise statistical t-test [139]."

"Test, a certain number of alive nodes (NoA) are taken into consideration for both the existing and proposed algorithms in every round, and they create a pair known as an observation pair in each round."

"Another pair can be forms for EERP and KCLSA, SEAR, DLSA, EERP, DLSA."

"It can be concluded with confidence that our suggested methods outperform SEAR and EERP."

Conclusion

"Due to the limited capabilities of IoT nodes, energy-efficient data transmission technologies are required."

"We propose an IoMT-based secure, and energy-efficient e-healthcare architecture that reduces the energy consumption of IoT networks."

"Statistical research showed that the KCLSA and DLSA are more energy-efficient and secure than other studies, with reduced network latency, in part because patient data is transported over the unsecured Internet via several access points in such database examinations."

"The proposed framework must also enhance network security and energy consumption for Inter-WBAN data transmission."

Comparative Analysis of Energy Efficient Routing Protocols with Optimization in WSN [140]

This is a machine-generated summary of:

Sreedevi, Pogula; Venkateswarlu, S.: Comparative analysis of energy efficient routing protocols with optimization in WSN [140]

Published in: International Journal on Interactive Design and Manufacturing (IJIDeM) (2022)

Link to original: https://doi.org/10.1007/s12008-022-00958-2

If you want to cite the papers, please refer to the original.

For technical reasons we could not place the page where the original quote is coming from.

Abstract-Summary

"In various industries such as industry, automation, health, military etc , Wireless Sensor Network (WSN's) has been developed."

"In order to boost the network life, increase power efficiency and resolve the problems of fault tolerance in wireless sensor networks, we reviewed numerous literature surveys concerning energy optimize routing protocols."

"The objectives achieve lower consumption of electricity, increase packet delivery rates and assess SINK's optimal location in a wireless sensor system."

Introduction

"Collaboration between sensor nodes is essential to minimize total data transferred to the processing center to monitor the redundant data flow through a network."

"Collaboration (i.e. the correlation between nodes located spatially to reduce the total data transmitted to a processing center) in the nodes capable of compensating for each other's deficits, as the sensor nodes, includes battery-powered devices with restricted sensors and limited compute capabilities to monitor the redundant data flow in a network."

"WSNs should operate with optimal energy to increase the lifetime of the sensor nodes, simultaneously ensuring network connectivity and availability."

"Because of the scarcity of energy in WSNs, energy optimization is needed to minimize the energy consumed by the sensor nodes to prolong network lifetime."

"Energy efficiency must be considered in every aspect of network design and operation, for both operation of the individual sensor nodes and communication of the overall network."

Literature Review

"The proposed algorithm is executed by comparing existing methods such as EASR (energy aware sink relocation), and Tree Cluster data gathering algorithms

(TCBDGA) in different efficiency metrics such as network life, number of nodes and communication range."

"The literature survey examines WSN energy efficient cluster protocols, such as DEACP, Low Duty Cycles Energy Efficient (ELDC), Threshold Sensitive Energy Efficient Sensor Network Protocol, and Sensor Clustering Protocols (TEEN),Hill Climbing Particle Swarm Optimization—Monte Carlo Localization (HCPSO-MCL), a Memetic Algorithm—Topology and Orchestration Specification for Cloud Applications (MA-TOSCA), an ant-colony optimization algorithm, etc We have investigated these existing techniques and observed that they have their own certain limitations like secure routing, reliability, energy-efficiency, routing overhead and network lifetime."

"Based on the investigations on these proposed techniques such as RE-TOPSIS, CCR protocol, energy-efficient zone-based dual sink protocol, and a new centralized fuzzy based clustering, we have observed some limitations like determining optimal path and improving of network lifetime."

Conclusion

"We examine different literature studies focused in the wireless sensor network energy-efficient cluster routing protocols in order to address the energy consumption, network existence and reliability disadvantages."

"We have set the objectives and discuss the algorithms of FTOR-ModPSO, PSO-GA, and fuzzy optimal clustering method."

"The techniques simulate the outcome and performance with regard to energy efficiency, network existence and the transmission rate of packets."

An Asynchronous Anycast Cross-Layer Protocol for WSN Suited to Noisy Environments [141]

This is a machine-generated summary of:

Heimfarth, Tales; Giacomin, João Carlos; de Oliveira Schneider, Bruno: An Asynchronous Anycast Cross-layer Protocol for WSN Suited to Noisy Environments [141]

Published in: Wireless Personal Communications (2021)

Link to original: https://doi.org/10.1007/s11277-021-08505-0

If you want to cite the papers, please refer to the original.

For technical reasons we could not place the page where the original quote is coming from.

Abstract-Summary

"This paper presents AGA-MAC+: an asynchronous anycast protocol for wireless sensor networks."

"Latency is a problem for asynchronous MAC due to sleep-delay, since each relay must wait for the receiver to wake-up."

"The first FCS member to wake up becomes the next relay, reducing latency."

"This new protocol is an improvement over AGA-MAC, our previous work, which considers data size to determine the number of FCS members."

"AGA-MAC+ was compared with two other well-known protocols, for different packet sizes and SINR values."

"Comparing energy spent in the communication process, AGA-MAC+ presented a performance gain of at least 5%, in 88% of the cases."

"When latency is assessed, AGA-MAC+ outperforms both protocols for all scenarios, presenting an improvement of at least 18% in 83% of the tests."

Introduction

"[142], we proposed AGA-MAC, an anycast cross-layer protocol that takes advantage from geographic routing and preamble sampling for low latency communication in duty cycled networks."

"Differently from previous works, our protocol selects a target area for FCS members based on the packet size in relation to the duty cycle for reducing the sleep-delay."

"Differently from the previous approach, our protocol tunes MAC parameters based on the packet size and on the accessed channel state, aiming at reducing total latency time."

"The contributions of this work are: to present a novel mathematical model that takes in consideration packet size and channel state for selecting protocol parameters; to present a concrete cross-layer protocol which uses this model to handle WSN communication in duty cycled networks with efficiency, even in the presence of error-prone communication channels."

Related Work

"The sender node precedes data frame with a preamble that should be as long as the cycle time, the interval between two wake ups of any node."

"While in B-MAC each data transmission between nodes takes place after a cycle time interval, X-MAC shortens this interval to 50% of cycle time on average."

"The inherent redundancy of the network is explored since a sender can select one node among the neighbor group to forward the message."

"The neighborhood of a sender node is modeled as a circle of radius R. Anycast protocols achieve even more latency reduction per-hop than X-MAC."

"Like AGA-MAC, it is an anycast cross-layer protocol for asynchronous wireless sensor networks."

"The BSC-MAC protocol [143] approaches energy conservation by coordinating duty cycles of each node according to the routes used by the network."

"We propose an alternative approach to consider transmission errors applied to a latency optimized duty cycling protocol."

The AGA-MAC+ Protocol

"The second parameter is the minimum signal-to-interference-plus-noise ratio (SINR) required for the data message to be received error-free with a pre-established probability."

"Although node 3 may promote message forwarding toward the destination, this advance is below the threshold determined by the sender."

"Node 4 could be included in the FCS if a smaller SINR were accepted by the sender."

"The larger the FCS, the faster the sender receives an answer to a preamble and the lower the propagation delay of the data message."

"The first member of the FCS to wake up sends a CTS, causing the sender to interrupt the preamble sequence and sending the data message."

"The reduction in threshold reduces the average advance of messages, increasing the necessary number of hops to reach the destination node."

Mathematical Model

"In order to estimate the latency to send a message from the Source node S to the Destination D, the node S needs to calculate the distance the message will travel (distance(S, D)) and its average propagation speed s.The average propagation speed of the message from source to destination node is the same average speed per hop, being calculated by the ratio between the average advance per hop and the mean time spent in each hop:Initially we calculate the average time that one node spends to send the data message one hop forward."

"The maximum number of preambles a sender node transmits in a cycle is calculated as the ratio between Cycle Time and Preamble plus eACK duration:A greedy rule for FCS composition is to select the nodes closest to the destination position (D), in order to enlarge advance towards D, reducing the number of hops spent to take the message from source to destination."

AGA-MAC+ State Machine

"When an eACK is received from a FCS member, transition 7 takes the node to sending message state where it transmits data packet."

"After sending the message, the sender node goes to waiting ACK state (transition 8) in order to wait for an ACK packet."

"Upon ACK is received, the sender node goes to carrier sense state through transition 9 and communication ends."

"If the acknowledgement packet is not received after an amount of time, the sender node goes to carrier sense state through transition 9 and programs another attempt to send the message."

"During carrier sense operation, if a preamble transmission is sensed, the node goes to receiving preamble state through transition 10."

"When data packet is correctly received, transition 20 drives the node to sending ACK state, where receiver transmits an acknowledgement message to sender node."

Results

"When the range of GeRaF and X-MAC were limited to 33 m, which is nearer to the original protocol without limitation (if the preamble is received without errors, the eACK packet is transmitted), our protocol presented an even better relative result."

"Large packets have a very increased latency due to retransmissions to both X-MAC and GeRaF protocols."

"This happens since small packets are less susceptible to transmission errors and the range was limited to 27 m. For this HARSH environment, we do not present results with the limitation of 33 m since no packet could be delivered by both X-MAC and GeRaF protocol with this setting."

"Experiments performed demonstrated that the novel protocol introduced in this work can deliver data packets with a lower latency than GeRaF and X-MAC for most of the test scenarios."

Conclusion

"Asynchronous MAC protocols for WSN rely on preamble sampling in order to achieve rendezvous between a node transmitting a packet and the next relay."

"The AGA-MAC+ is an advance of the original protocol since it considers, besides the packet size, the quality of the communication channel in order to parameterize the protocol."

"Further, our protocol was compared with the X-MAC [144] and GeRaF [145] protocols for different packet sizes and channel qualities."

"For the X-MAC and GeRaF protocols, two different minimum SINR for answering the preamble were established, in order to make a more fair comparison, since without limit middle/large packets may never be relayed."

"For large packet sizes, both X-MAC and GeRaF presented a big latency when compared to our protocol due to the necessity of several re-transmissions."

"Other preamble-sampling protocols that do not take in account the quality of channel and packet size will suffer from the same drawback when transmitting large packets."

NEWTR: A Multipath Routing for Next Hop Destination in Internet of Things with Artificial Recurrent Neural Network (RNN) [146]

This is a machine-generated summary of:

Sumathi, A. C.; Javadpour, Amir; Pinto, Pedro; Sangaiah, Arun Kumar; Zhang, Weizhe; Mahmoodi Khaniabadi, Shadi: NEWTR: a multipath routing for next hop destination in internet of things with artificial recurrent neural network (RNN) [146]

Published in: International Journal of Machine Learning and Cybernetics (2022)

Link to original: https://doi.org/10.1007/s13042-022-01568-w

Copyright of the summarized publication:

If you want to cite the papers, please refer to the original.

For technical reasons we could not place the page where the original quote is coming from.

Abstract-Summary

"WSN nodes are battery powered, therefore energy management is a key factor for long live network."

"One way to prolong lifetime of network is to utilize routing protocols to manage energy consumption."

"To have an energy efficient protocol in environment interactions, we can apply ZigBee protocols."

"Artificial Intelligence Interactions routing methods, Tree Routing (TR) that acts in the tree network topology is considered a simple routing protocol with low overhead for ZigBee."

"We present an artificial Intelligence Tree Routing based on RNN and ZigBee protocol in IoT environment."

"Simulation results show that NEWTR improve the network lifetime by 5.549% and decreases the energy consumption (EC) of the network by 5.817% as compared with AODV routing protocol."

Introduction

"Different applications in sensor networks require high accuracy in design, architectural selection, and also determining the placement of the nodes in network."

"Nodes are different in terms of sensors, energy requirements and ability to process in Heterogeneous networks."

"In Wireless Sensor Networks Transmit data between nodes should be done in several steps by the central node, so that it would not be necessary for each node to have a wide range of transmission and sufficient energy."

"Apart from the number of neighbors, destination of a node to another is considered and causing lower power consumption in such limited network, included directly."

"The Key of Contribution of this research are: A Multipath Routing for Next Hop Destination in IoT based on ZigBee Protocol Drawbacks Control traffic arriving at routers with using artificial recurrent neural network in IoT for next hop destination Moreover the WSNs features, importance and necessities are explained."

Literature Review

"Routers have the accessibility to network routing related data, and in result they have the ability to conduct data packets among several networks."

"Destination address, Adjacent routers that by using them, the possibility of obtaining information on remote networks would be provided, Existence paths to the whole remote networks, The best path to each remote networks, The maintenance method and observing routing information."

"Routers define the best path to pass data packets, based on existence information in routing tables."

"Using a multiple-routing algorithm is one the common methods in sensor network which the data transmission is done through intermediate nodes."

"Tree Routing (TR), which for tree topology is used, is considered as simple and efficient routing algorithm compared with other network routing protocols."

"Routings algorithm, Tree Routing (TR) which acts in the tree network topology is considered as simple routing protocol with low overhead for ZigBee."

Proposed Method

"This is an optional parameter that varies according to network conditions dist: the distance to the neighboring node nb: the number of neighboring nodes for neighboring node (the next hop)."

"The parameter calculates the distance between neighboring nodes based on the number of steps in the path."

"We count all neighboring nodes together including parent and child nodes, after considering the next hop selection that can minimize routing costs."

"Since the router in the tree algorithm paths is to upwards to parent node or to the bottom to the child node, the number of remaining steps to destination is calculated according to tree topology."

"For improving routing we will use Deep learning neural network to control traffic arriving at routers."

"The traffic packets are routed to less crowded routers by training an LSTM network on each router."

Simulation Results

"A unique set of protocols has been discussed in this paper to evaluate the performance of the proposed routing protocol against ESTR and AODV based on their average end-to-end delays, packet delivery ratios (PDR), routing control overhead and energy consumption."

"In this simulation, AODV and ESTR routing protocols are compared based on their packet delivery ratios and throughputs."

"This simulation compares the throughput of our proposed algorithm with AODV and ESTR routing protocols."

"In this simulation our proposed algorithm is compared based on Routing Control Overhead with AODV routing protocol and ESTR routing protocol."

"To obtain routing control overhead of proposed method in data entry in NS2 parameter related to node, a counter is set that sends packet with specified byte length with different data rate (bytes per second) 2000 times."

"In this simulation our proposed algorithm is compared based on energy consumption with AODV routing protocol and ESTR routing protocol."

Conclusion and Future Work

"A tree pattern is used to select optimal routes using TR algorithm-based ZigBee routing protocol-based neighbor node counts."

“A proposed algorithm increases the throughput and reduces latency in wireless sensor networks based on the IoT and ZigBee protocol.”

“The routing algorithms perform dynamic routing based on the number of nodes surrounding the target node.”

“A superior performance was also demonstrated for the proposed algorithm in distributed networks.”

“As the number of guiding nodes grows, network costs will increase.”

“Implementing a localization algorithm for node locating operations is also included in the cost.”

“Developing an energy-constrained wireless sensor network means designing algorithms that minimize the use of energy.”

“TR is not the last algorithm based on ZigBee protocol.”

On-Chip and Distributed Dynamic Parallelism for Task-Based Hardware Accelerators [147]

This is a machine-generated summary of:

Heinz, Carsten; Koch, Andreas: On-Chip and Distributed Dynamic Parallelism for Task-based Hardware Accelerators [147]

Published in: Journal of Signal Processing Systems (2022)

Link to original: https://doi.org/10.1007/s11265-022-01759-2

If you want to cite the papers, please refer to the original.

For technical reasons we could not place the page where the original quote is coming from.

Abstract-Summary

“TaPaSCo relies on the host to manage task parallelism and perform the actual task launches.”

"For more expressive parallel programming patterns, such as pipelines of task farms, the round trips from the hardware accelerators back to the host for launching child tasks, especially when exploiting data-dependent execution times, quickly add up."

"The major contribution of this work is the addition of on-chip task scheduling and launching capabilities to TaPaSCo."

"This enables not only low-latency dynamic task parallelism, it also encompasses the efficient on-chip exchange of parameter values and task results between parent and child accelerator tasks."

Introduction

"FPGAs have become widely available as accelerators in computing systems."

"With highly-compute intensive domains such as AI/ML, not only does the complexity of the individual accelerators grow, but the computing demand is so large that the workload needs to be distributed across many compute nodes."

"There is a need to efficiently load-balance a high-rate of task launches between the nodes of a distributed system, and quickly communicate task parameters and results among nodes."

"This new feature enables low-latency interactions directly between processing elements, even across the network in distributed systems, without the need for host involvement."

"The same mechanisms can be used even across a network with multiple FPGAs, with integrated load-balancing facilities ensuring an even use of processing elements across the nodes."

Heterogeneous Computing Architecture

"TaPaSCo realizes hardware thread pools, each having a set number of PEs to perform the same task."

"In its initial version, TaPaSCo employed a software runtime to dispatch a task to a suitable, currently idle PE."

"The task dispatch (finding a suitable idle PE) and the launch, including the transfer of task parameters and the collection of results to/from the selected PE, is now handled on-chip."

"This off-loading of the task dispatch decouples the software application on the host side from the PEs on the FPGA."

"We present Cascabel 2, which extends the prior version by now allowing the PEs themselves to autonomously launch new tasks without the need for host interaction."

"The main goal of direct on-chip task launches is to reduce the latency, resulting in task launches with both low latency and high throughput."

Implementation

"To the regular TaPaSCo interfaces for PE control and interrupt-based signalling, two AXI4 streams are used to enable dynamic parallelism: A 512bit stream, shown in green, flows from the PEs to the Cascabel 2 unit, and carries new launch requests, including child task parameters."

"A second 64bit stream, shown in red, flows from the the Cascabel 2 unit back to the PEs and transports the task result, which is generally a single scalar value."

"Cascabel 2 supports this by allowing a child task to skip its parent task when returning results, and instead provide its result to its grandparent task."

"Further processing is dependent on the selected action: it either forwards the result to the (grand-)parent, waits for additional return values, or issues a new merge task using the Cascabel 2 unit."

"For the Cascabel 2 functions, it provides methods to launch tasks and to receive results."

Evaluation

"For evaluating this, we use the on-chip launch interface and measure the required clock cycles from writing the task launch command for an immediately returning (NOP) task to the Cascabel 2 launch-command stream, up to when the parent task receives the result value from the child task."

"Hardware measurements have shown that fewer than 100 clock cycles (at the network clock domain running at 322MHz) are actually required for the operation within the FPGA logic to receive a packet and send out the result."

"If a block is visible at the timestamp of the snapshot, a new on-chip task launch of the NTT kernel is performed."

"The host software fetches the header from the memory on the FPGA and, if the header signals a visible block, it launches the execution on the NTT hardware module."

Related Work

"To much prior work on HLS that focuses on parallelism within a PE [148, 149], our focus is parallelism across PEs and the required infrastructure to support this."

"One recent example is ParallelXL [150], which also aims for dynamic on-chip parallelism: Their PEs are grouped into tiles, which are attached to two NoCs to perform work-stealing scheduling and argument/task routing."

"We believe that our simple n-to-1 and 1-to-n streaming interconnects will scale better than ParallelXL's more expensive NoCs, and still allow performance gains even for highly scheduling intensive workloads, as demonstrated by our task-excessive Fibonacci example."

"ParallelXL lacks advanced features such as hardware support for merge/reduce operations and has more limited customizability (e.g., omitting the result interconnect on PEs for void child tasks)."

"Our work does not provide an automated way to generate full clusters, but by incorporating the TaPaSCo framework, the kernel composition on individual FPGAs can easily be replicated."

Conclusion and Future Work

"It extends the scheduling capabilities of prior work (e.g., barriers) with new mechanisms for performing inter-task reduction operations and optimized result passing."

"Cascabel 2 enables low-latency task launches both locally on-chip, as well as in a distributed system over the network."

"With all its proven advantages, for some applications, the task-based programming model currently at the heart of TaPaSCo is not the optimal one."

"We are currently working to combine task-based reconfigurable computing with self-scheduling streaming operations for use in data-flow applications."

JOSP: Joint Optimization of Flow Path Scheduling and Virtual Network Function Placement for Delay-Sensitive Applications [151]

This is a machine-generated summary of:

Lyu, Qing; Zhou, Yonghang; Fan, Qilin; Lyu, Yongqiang; Zheng, Xi; Xu, Guangquan; Li, Jun: JOSP: Joint Optimization of Flow Path Scheduling and Virtual Network Function Placement for Delay-Sensitive Applications [151]

Published in: Mobile Networks and Applications (2022)

Link to original: https://doi.org/10.1007/s11036-021-01868-5

If you want to cite the papers, please refer to the original.

For technical reasons we could not place the page where the original quote is coming from.

Abstract-Summary

"As the service requests are highly dynamic and different services require distinct bandwidth occupation amount and time, how to schedule the paths of flows and place VNFs efficiently to guarantee the performances of network applications and maximize the utilization of the underlying network is a challenging problem."

"We present a joint optimization approach of flow path scheduling and VNF placement, named JOSP, which explores the best utilization of bandwidth from two different aspects to reduce the network delay."

"We first present a delay scaling strategy that adds the penalty to the link bandwidth occupation that may cause congestion in accordance with the network placement locations."

"We present a reinforcement learning framework and use both the flow path delay and the network function-related delay to calculate the reward of placing VNFs adaptively."

Introduction

"The influence of bandwidth occupation amount and time should be further explored to improve the network utilization; ii) Learning-based methods only considered the flow path delay in the reward calculation while ignoring the function-related delay which is also a significant indicator for deciding the placement locations; iii) The

VNF placement has an impact on the path scheduling as the flow needs to traverse to the locations of VNFs to enforce the service chain on the path."

"We develop a framework named Joint Optimization of flow path Scheduling and virtual network function Placement (JOSP) to reduce network delay."

"Unlike traditional works that consider the flow path scheduling and VNF placement separately, JOSP combines them together as these two problems are correlated with each other."

"The main contributions of the paper are listed as follows: We investigate the problem of flow path scheduling and VNF placement jointly and formulate it as an integer linear programming to minimize the delay."

Related Work

"Flow path scheduling and VNF placement have been studied in intensively in recent years."

"Besides, as the bandwidth of the network is limited, not all flows can be scheduled to desired paths."

"A method called fast path assignment [152] was provided to minimize the influence of the bandwidth occupation of the bottleneck links in the scheduling of flow paths, but the overall bandwidth utilization was not considered."

"The influence of the bandwidth occupation amount and time of the current flow on subsequent flows were not considered in the path scheduling in these works."

"These works only used flow path delay to calculate the reward, ignoring the network function-related delay."

"Our approach in this paper is differentiated by solving the flow path scheduling and VNF placement problems all together."

"Besides, a learning-based method which utilizes both the flow path delay and the network function-related delay to improve the VNF placement is proposed."

Network Model and Problem Formulation

"The flow in the network firstly needs to traverse to the ingress node f_{in}, then the network nodes placed with the network functions in the service chain in sequence, and lastly the egress node f_{eg}."

"The flow path delay varies to the changes of the delay between two connected nodes, the service chain specifications and the placement locations of the network functions."

"To reduce the placement cost, the number of instances of a type of network function that placed to a node should not exceed 1, which is represented as Service Chain Constraint."

"The flow f needs to traverse the nodes placed with the network functions in the service chain in sequence."

"If the k th network function in the service chain is the t th type of network function in the network function set, there should be a node placed with the network function, which is represented as Function Type Constraint."

Design of JOSP

"A reinforcement learning method is presented to decide the placement locations of VNFs in JOSP, where both of the flow path delay and network function-related delay are used to update the reward and persistently improve the placement performance by adjusting the placement locations of VNFs."

"The link bandwidth occupation in the path scheduling for current flow will further influence the path selection for other flows and even bring extra congestion and increase the total delay."

"The unreasonable longtime occupation of the link bandwidth and processing resource are avoided when scheduling the path of long flow to the Cool Track."

"The VNF placement scheme needs to be carefully considered to facilitate the flow path scheduling so that the service chain functionalities and delay requirements from flows are satisfied."

"The flow path delay and the network function-related delay are feedback to calculate the reward to the agent to help improve the placement."

Evaluations

"The DES strategy imposes the penalty of occupation of the link bandwidth that has the possibility to cause congestion in the flow path scheduling to reduce the delay."

"The short flow would be scheduled to path with the least delay, it will quickly release the bandwidth of the links on the flow path after finishing traversing and let the subsequent flows use the links to reduce the overall delay."

"JOSP uses the flow path delay as well as the network function-related delay to calculate the reward for the agent to adjust the placement locations of the VNFs."

"This is because JOSP schedules the flows to more available links to avoid the congestion based on the link utilization situation and the bandwidth occupation requirement, thus reducing the delay caused by waiting for the link bandwidth to be released."

Conclusion

"DES adds the penalty to the occupation of the link bandwidth that has the possibility to generate congestion and LSFD differentiates the long flow and short flow in the path selection to explore the best utilization of the bandwidth to reduce the delay."

"Empowered with the achieved flow path delay and network function-related delay to update the reward for the learning agent, JOSP adjusts the placement locations of the VNFs."

"In our future work, we plan to add the bandwidth occupation time of the flows as a feature to be learned without human intervention and use distributed learning-based methods to enhance the placement performance."

Security-Aware Multi-Cloud Service Composition by Exploiting Rough Sets and Fuzzy FCA [153]

This is a machine-generated summary of:

Lahmar, Fatma; Mezni, Haithem: Security-aware multi-cloud service composition by exploiting rough sets and fuzzy FCA [153]

Published in: Soft Computing (2021)

Link to original: https://doi.org/10.1007/s00500-020-05519-x

If you want to cite the papers, please refer to the original.

For technical reasons we could not place the page where the original quote is coming from.

Abstract-Summary

"With the emergence of cloud computing and to better satisfy users' complex requirements in front of the huge number of cloud services, these latter may be combined while considering the virtualized environment's constraints, including quality of services, security policies, resources availability, interoperability, etc. As composing services from multiple clouds was proved to be more beneficial than relying on services from one single cloud, the new approaches are now spanning multiple clouds."

"We propose a security-aware multi-cloud service composition approach using fuzzy formal concept analysis (fuzzy FCA) and rough set theory (RS), which are two techniques with a strong mathematical background."

"To guarantee a high security level of the hosting clouds and the selected services, we exploit the fuzzy relations of fuzzy FCA and the approximation of RS."

Introduction

"This latter is an evolving approach that reuses existing services or process fragments, to generate a composite one and to increase the number of applications in the cloud environment [154]."

"Most of the existing MCSC approaches concentrated on reducing the number of used clouds [154, 155], reducing the inter-cloud communication cost [156], etc. But, they neglected other important multi-cloud issues, such as the lack of load balancing among cloud servers [154, 155, 157], the high execution time [156, 158, 159], etc. MCSC approaches also do not specify the environmental constraints and the security concerns when dealing with different dimensions and modalities of cloud service composition."

"Existing approaches have tackled security issues when combining services hosted in the same cloud."

"We propose in this work a security-aware MCSC approach that allows combining the safest services from the most trusted clouds."

"We combine fuzzy FCA with rough set theory [160] to define algorithms that approximate the user request, extract the most secure cloud combination, and return a high-quality and safe service composition."

Related Work

"Combinatorial-based approaches consist in selecting the clouds that offer the highest number of services in order to find the best services in a short execution time, whereas in clustering-based approaches, grouping the most pertinent services or clouds according to their common features allows keeping only those able to satisfy the user requirements, that is to say reducing the search space to a specific cluster and, thereafter, minimizing the composition time."

"Despite that meta-heuristics-based approaches [161] achieve service composition from several clouds, they suffer from a high time complexity and inefficient load balancing between cloud servers."

"Security aspects as a remarkable composition criteria in untrusted cloud environments are neglected in simple services-based approaches [155, 157, 161, 162], which makes the consumers' data at stake."

Preliminaries and Basic Concepts

"We adopt fuzzy formal concept analysis (Fuzzy FCA) and rough set theory (RS) which have been successfully applied in several domains."

"As for the theory of rough sets, it is adopted in order to approximate the user request when the needed services are not found or the selected services do not completely satisfy the security requirements."

"As defined in Pawlak [160], the lower approximation is the largest set in the extent/intent part whose objects/attributes entirely satisfy the user request, whereas the upper approximation deals with the smallest set in the extent/intent part."

"The notion of approximation is naturally used in fuzzy FCA, given that each formal concept extracted from the formal context is a definable set of objects and a definable set of attributes."

"While looking for the requested objects (services/clouds in our case), if none of the existing fuzzy formal concepts is able to satisfy the user request, this latter is called a non-definable set."

Problem Formulation

"In order to measure the security level, we assign to each service/cloud a score that represents the satisfaction degree of security policies."

"The security has two meanings: to what level the selected service meets the user's security needs and how trusted are the clouds involved in the service composition."

"The security level for each service and cloud will be determined depending on their adopted security policies."

"In SaaS services, the cloud provider may give some control to SaaS developers to manage the security needs of their cloud applications on top of PaaS platforms."

"Since each cloud layer has its specific security aspects, incorporating these latter into the MCSC process depends on the type of service composition."

"This means that a better security level requires a smaller membership value (close to 0) denoting the satisfaction degree of this policy by a cloud provider."

"Data leakage and loss of service are also negative security policies."

Proposed Approach

"The motivation behind using fuzzy FCA is transforming the insecure MCE into a secure search space, by filtering the clouds which do not satisfy the required security policies."

"The advantage of using fuzzy FCA, in this case, is to facilitate the selection of clouds and services, by grouping them according to their common satisfied security policies."

"Fuzzy FCA allows us to find the combination of most secure clouds and to run through their corresponding cloud lattices in order to select the safest services."

"By secure cloud combination, we mean the set of clouds that satisfy at best the user's security policies."

"If the Intent part of a fuzzy formal concept is able to satisfy the required security policies, then the candidate cloud combinations will be extracted from the concept's Extent."

"The user request, including security requirements, is satisfied in case the needed services are found in the best cloud combination."

Complexity Study

"Secure clouds extraction In this step, the first algorithm is responsible for approximating the required security policies."

"Based on the resulted concepts from Algorithm 1, the second algorithm in this step represents the cloud combinations' extraction process."

"Each combination extracted from the lower approximation concepts must contain a cloud from each fuzzy concept (lines 5–6)."

"Selection of secure cloud combination In this step, Algorithm 3 is responsible for calculating the score of each cloud combination outputted by Algorithm 2."

"The complexity of Algorithm 4 is calculated according to the number of combined clouds (line 4)."

"In the worst case, each policy will be found in one single concept's Intent (line 6)."

Experiments

"This is the reason behind the increased size of each cloud combination including that of the optimal combination, because the number of Intents that satisfy the user request tends to the number of required security policies."

"The high sizes of cloud combinations are the reason behind the decrease in the score values without approximation, compared to the scores generated by our approach."

"The security policies approximation and the bottom-up parsing approach are the reason behind reducing the number of hosting clouds."

"In all test cases, our approach requires a very reduced composition time (<1 s) to find the optimal cloud combination and, thereafter, extract the most secure services."

"In the worst cases (when we have 5 and 10 clouds and a density equals to 20% and 30%, respectively), the size of the optimal cloud combination generated by our approach is surpassed by that of MCSC-FCA."

Conclusion and Future Works

"The approach proposed in this paper consists of selecting and composing services in a multi-cloud setting, while taking into account the security level of each cloud zone and service."

"The first technique was used to ensure precision, by generating fuzzy relations between clouds/services and security policies."

"The secure clouds will serve as a trusted search space, from which the safest services will be composed to satisfy the user's security requirements."

"We only checked if the existing clouds adopt a set of security policies, but we did not ensure if they also verify them or not."

A Traffic Intensive Virtual Channels Allocation Scheme in Network-on-Chip [163]

This is a machine-generated summary of:

Joshi, Bansidhar; Thakur, Manish K.: A Traffic Intensive Virtual Channels Allocation Scheme in Network-on-Chip [163]

Published in: Arabian Journal for Science and Engineering (2022)

Link to original: https://doi.org/10.1007/s13369-022-07191-9

If you want to cite the papers, please refer to the original.

For technical reasons we could not place the page where the original quote is coming from.

Abstract-Summary

"While transitioning from source to destination IP cores, allocation of the packets (flits) to the next router's available VCs is done as per an allocation strategy."

"An ideal VC allocation determines an effective usage of resources, which typically results in an increasing number of packets received at the destination(s) and can minimize the per-hop delay as a performance enhancement."

"This paper introduces an artificial neural networks (ANN)-based VC allocation technique that prioritizes the most appropriate VCs for allocation and provides all-round performance improvisation."

"The proposed scheme is compared with the state-of-the-art First-Come-First-Serve (FCFS)-based VC allocation under two distinct routing strategies."

"Conducted experiments under various diverse traffic scenarios for the proposed scheme demonstrate an upsurge of up to 5.75% in the number of packets received at destinations, while also incurring up to 24.4% lesser network delay as compared to the FCFS-based VC allocation scheme."

Introduction

"The VC-based flow control technique requires each packet to gain exclusive access to an available virtual channel in a preferred port of downstream router before this packet's flits can actually start traversing."

"The VC allocation stage assigns available VC buffers of downstream neighbor routers to the awaiting header flits at the current upstream router."

"The previous section, virtual channels allow flits from different sources to be packed together to minimize the use of valuable buffer space while maintaining maximum bandwidth between routers and the channel, thus reducing the power required to maintain the state of these buffers."

"The VC allocation stage of router micro-architecture decides the clock cycle time, and hence the allocation algorithm needs to be optimized enough to prevent consequent drops in throughput and also reduce network's latency and power consumption."

Background Work

"In [164], Smai and others elucidated that allocation of virtual channels is usually performed on First-Come-First-Served (FCFS) basis; i.e., after route computation stage, any of the VCs available in the ports of a downstream router can be assigned to any of the requesting packet's header flit, regardless of its priority over other packets."

"The model is compared with a conventional FCFS-based VC allocator and optimization is achieved while performing extensive trade-off between network's latency and throughput."

"Michelogiannakis and others have presented a packet chaining approach in [165] to maximize allocator's matching efficiency and hence overall network performance of networks having small/single size flits and short cycle time."

"While the above cited contributions focused toward performance enhancement in NoC-based system, most of these still make the trade-off the network's throughput with overall latency to ensure the effective allocation of the virtual channels."

"Considering that performance of the NoC-based system will enhance further with effectively allocated virtual channels, this paper presents an artificial neural network-based scheme for effective allocation of virtual channels."

Artificial Neural Network-Based VC Allocation

"In the proposed scheme, we assign a conventional neural network to every router, which then facilitates the allocation of virtual channels in neighboring routers to the flits residing in various virtual channels at the current router."

"Upon receiving the tileID (IP core from where the flits generated currently) and the flitSRC (the virtual channels that are holding the flits in current cycle), a forward pass is performed in the neural network, predicting which virtual channels these flits should be allocated."

"Softmax squashes all the probabilities of the classes generated by the forward pass and define the absolute value with which all the virtual channels should be reserved for the upstream router's flits."

"For training the neural network, the prediction list post-forward pass is compared to the output of the default allocation method (FCFS in our case), and the mean squared error (MSE) is calculated between the two predictions."

Evaluation

"For the specified mesh configuration featuring heavy bursty traffic flow throughout the simulation cycles, the number of received flits for the proposed ANN-based VC allocation scheme has achieved an enhancement of up to 1.7% and 5.1% under the DyAD and OE routing schemes, respectively, over FCFS-based VC allocation scheme."

"For overall latency as another performance parameter, the proposed VC allocation scheme performs similar to the traditional FCFS-based allocation under DyAD routing, while achieving up to 9.8% less delay for OE routing at higher injection rates."

"With a minuscule improvement of up to 1.8% in terms of the number of flits received at destinations, the proposed allocation scheme performs as better as the conventional FCFS-based allocation scheme for both off-time interval cases in DYAD and OE routing schemes."

Conclusion and Future Work

"It is safe to infer from the assumption of application-specific traffic being not high in its entropy, that upon receiving a particular input, those learned weights allow for an approximate solution to VC allocation in each router."

"We have used a perceptron model for the neural network in this paper, which has proven to perform optimal VC allocation of comparable standards to that of the industry standard."

"An extensive comparison under various diverse traffic models and routing schemes of the proposed VC allocation strategy to the state-of-the-art FCFS-based VC allocation scheme shows a 5.75% increase in the number of packets received at destinations as well as a reduction in the network delay of up to 24.4%."

"We will thus explore these models further and try to bring an optimal resource allocation strategy to our future research contributions."

Bibliography

1. Hassan, S., & Rufai, A. (2023). Modified dropping-random early detection (MD-RED): A modified algorithm for controlling network congestion. *International Journal of Information Technology*. https://doi.org/10.1007/s41870-023-01201-1
2. Thiruchelvi, G., & Raja, J. (2008). A survey on active queue management mechanisms. *International Journal of Computer Science and Network Security, 8*(12), 130–145.
3. Dana, A., & Malekloo, A. (2010). Performance comparison between active and passive queue management. *International Journal of Computer Science Issues, 7*(5), 13–17.
4. Chydzinski, A. (2022). Impact of the dropping function on clustering of packet losses. *Sensors, 22*(20), 1–16. https://doi.org/10.3390/s22207878
5. Patel S, Karmeshu (2019) A new modified dropping for congested networks. Wireless Personal Communications 104(1):37–55. https://doi.org/10.1007/s11277-018-6007-8
6. Bie, Y., Li, Z., Hu, Z., & Chen, J. (2022). Queue management algorithm for satellite network based on traffic prediction. *IEEE Access, 10*, 54313–54324. https://doi.org/10.1109/ACCESS.2022.3163519
7. Singh, M., & Singh, M. P. (2022). Congestion avoidance with source location privacy using octopus-based dynamic routing protocol in WSN. *Wireless Networks*. https://doi.org/10.1007/s11276-022-03165-9
8. Lightfoot, L., Li, Y., & Ren, J. (2016). STaR: Design and quantitative measurement of source-location privacy for wireless sensor networks. *Security and Communication Networks, 9*(3), 220–228.
9. Berdibek, A., & Saginbekov, S. (2019). A routing protocol for source location privacy in wireless sensor networks with multiple sources. In *Proceedings of the 15th ACM international symposium on QoS and security for wireless and mobile networks* (pp. 93–99).
10. Baroutis, N., & Younis, M. (2017). Load-conscious maximization of base-station location privacy in wireless sensor networks. *Computer Networks, 124*, 126–139.
11. Panda, N., & Supriya, M. (2022). Efficient data transmission using trusted third party in smart home environments. *EURASIP Journal on Wireless Communications and Networking*. https://doi.org/10.1186/s13638-022-02200-9
12. Yoo, H., Shim, M., Kim, D., & Kim, K. H. (2010). GLOBAL: A gradient-based routing protocol for load-balancing in large-scale wireless sensor networks with multiple sinks. In *The IEEE symposium on Computers and Communications (IEEE)* (pp. 556–562).
13. Al-Turjman, F., Mostarda, L., Ever, E., Darwish, A., & Khalil, N. S. (2019). Network experience scheduling and routing approach for big data transmission in the Internet of Things. *IEEE Access, 7*, 14501–14512.
14. Durand, A., Ngoko, Y., & Cerin, C. (2017). Distributed and in-situ machine learning for smart-homes and buildings: Application to alarm sounds detection. In *2017 IEEE International Parallel and Distributed Processing Symposium Workshops (IPDPSW)* (pp. 429–432). IEEE.
15. Kamgueu, P. O., Nataf, E., Ndié, T. D., & Festor, Q. (2013). *Energy-based routing metric for RPL*. (Doctoral dissertation). INRIA.
16. Joshi, M., & Kaur, B. (2015). Web integrated smart home infrastructure using internet of things. *International Journal of Engineering Research and Generic Science, 3*(6), 153–158.
17. Sousa, N., Sobral, J. V., Rodrigues, J. J., Rabêlo, R. A., & Solic, P. (2017). ERAOF: A new RPL protocol objective function for Internet of Things applications. In *2017 2nd International Multidisciplinary Conference on Computer and Energy Science (SpliTech)* (pp. 1–5). IEEE.
18. Musaddiq, A., Zikria, Y. B., & Kim, S. W. (2020). Routing protocol for low-power and lossy networks for heterogeneous traffic network. *EURASIP Journal on Wireless Communications and Networking, 2020*(1), 1–23.
19. Shirmarz, A., & Ghaffari, A. (2021). Taxonomy of controller placement problem (CPP) optimization in Software Defined Network (SDN): A survey. *Journal of Ambient Intelligence and Humanized Computing*. https://doi.org/10.1007/s12652-020-02754-w

20. Kumari, A., Member, S., Sairam, A. S., & Member, S. (2019). A survey of controller placement problem in Software Defined Networks. *Networking and Internet Architecture*. https://doi.org/10.1109/ACCESS.2019.2893283
21. Wang, G., Zhao, Y., Huang, J., & Wang, W. (2017). The controller placement problem in software defined networking: A survey. *IEEE Network, 31*, 21–27. https://doi.org/10.1109/MNET.2017.1600182
22. Casado, M., Freedman, M. J., Pettit, J., et al. (2007). Ethane: Taking control of the enterprise. In *Proceedings of the conference on Applications, technologies, architectures, and protocols for computer communcations—SIGCOMM '07* (pp. 1–12).
23. Pan, J., Paul, S., & Jain, R. (2011). A survey of the research on future internet architectures. *Communications Magazine IEEE, 49*, 26–36. https://doi.org/10.1109/MCOM.2011.5936152
24. Feamster, N., Rexford, J., & Zegura, E. (2014). The road to SDN: An intellectual history of programmable networks. *ACM Sigcomm Computer Communication Review, 44*, 87–98. https://doi.org/10.1145/2602204.2602219
25. Shirmarz, A., & Ghaffari, A. (2019). An adaptive greedy flow routing algorithm for performance improvement in software-defined network. *International Journal of Numerical Modelling: Electronic Networks, Devices and Fields*. https://doi.org/10.1002/jnm.2676
26. Shirmarz, A., & Ghaffari, A. (2020a). Performance issues and solutions in SDN-based data center: A survey. *The Journal of Supercomputing*. https://doi.org/10.1007/s11227-020-03180-7
27. Shirmarz, A., & Ghaffari, A. (2020b). An autonomic Software Defined Network (SDN) architecture with performance improvement considering. *Journal of Information Systems and Telecommunication, 8*, 1–9.
28. Masoudi, R., & Ghaffari, A. (2016). Software defined networks: A survey. *Journal of Network and Computer Applications, 67*, 1–25.
29. Farhady, H., Lee, H., & Nakao, A. (2015). Software-defined networking: A survey. *Computer Network, 81*, 79–95.
30. Hakiri, A., Gokhale, A., Berthou, P., et al. (2014). Software-defined networking: Challenges and research opportunities for future internet. *Computer Networks, 75*, 453–471.
31. Heller, B., Sherwood, R., & McKeown, N. (2012). The controller placement problem. In *Proceedings of the first workshop on hot topics in software defined networks* (pp. 7–12). ACM.
32. Bouzghiba, S., Dahmouni, H., Rachdi, A., & Garcia, J.-M. (2017). Towards an autonomic approach for software defined networks: An overview. In *Advances in ubiquitous networking* (Vol. 2, pp. 149–161). Springer.
33. Hu, Y., Wendong, W., Gong, X., et al. (2013). Reliability-aware controller placement for software-defined networks. In *2013 IFIP/IEEE International Symposium on Integrated Network Management (IM 2013)* (pp. 672–675). IEEE.
34. Hu, Y., Wang, W., Gong, X., et al. (2014b). On reliability-optimized controller placement for software-defined networks. *China Communications, 11*, 38–54. https://doi.org/10.1109/CC.2014.6821736
35. Ramasamy, M., & Pawar, S. (2018). Pareto-optimal multi-controller placement in software defined network. In *2018 3rd International Conference for Convergence in Technology* (pp. 1–7).
36. Liu, S., Wang, H., Yi, S., & Zhu, F. (2015). NCPSO: A solution of the controller placement problem in software defined networks. In *International conference on algorithms and architectures for parallel processing* (pp. 213–225). Springer.
37. Zhang, B., Wang, X., Ma, L., & Huang, M. (2016). Optimal controller placement problem in internet-oriented software defined network. In *2016 International Conference on Cyber-Enabled Distributed Computing and Knowledge Discovery (CyberC)* (pp. 481–488). IEEE.
38. Perrot, N., & Reynaud, T. (2016). Optimal placement of controllers in a resilient SDN architecture. In *2016 12th International Conference on the Design of Reliable Communication Networks (DRCN)* (pp. 145–151). IEEE.

39. Liu, Y., Gu, H., Yu, X., & Zhou, J. (2018). Dynamic SDN controller placement in elastic optical datacenter networks. In *2018 Asia Communications and Photonics Conference (ACP)* (pp. 1–3). IEEE.
40. Chinnasamy, P., Udgata, S. K., Lalitha, K., & Jeevanantham, A. (2020). Multi-objective based deployment of throwboxes in Delay Tolerant Networks for the Internet of Things environment. *Evolutionary Intelligence*. https://doi.org/10.1007/s12065-020-00474-w
41. Ibrahim, M., Hanbali, A., & Nain, P. (2007). Delay and resource analysis in MANETs in presence of throwboxes. *Performance Evaluation, 64*(9), 933–947.
42. Zhao, W., Ammar, M., & Zegura, E. (2004). A message ferrying approach for data delivery in sparse mobile ad hoc networks. In *Proceedings of ACM MobiHoc* (pp. 187–198). ACM Press.
43. Bin Tariq, M. M., Ammar, M., & Zegura, E. (2006). Message ferry route design for sparse ad hoc networks with mobile nodes. In *Proceedings of the 7th ACM international symposium on mobile ad hoc networking and computing (MobiHoc '06), Florence, Italy* (pp. 37–48).
44. Li, F., Yin, Z., Cheng, Y., & Wang, Y. (2016). Optimization problems in throwbox-assisted delay tolerant networks: which throwboxes to activate? How many active ones i need? *IEEE Transactions on Computers, 65*(5), 1663–1670.
45. Javvaji, G., & Udgata, S. K. (2020). Soft computing approach for multi-objective task allocation problem in wireless sensor network. *Evolutionary Intelligence*. https://doi.org/10.1007/s12065-020-00412-w
46. Chai, Z., & Liang, S. (2020). A node-priority based large-scale overlapping community detection using evolutionary multi-objective optimization. *Evolutionary Intelligence, 13*, 59–68.
47. Basu, M. (2008). Dynamic economic emission dispatch using nondominated sorting genetic algorithm-II. *International Journal of Electrical Power & Energy Systems, 30*, 140–149. https://doi.org/10.1016/j.ijepes.2007.06.009
48. Hu, C., Liu, Y., Tao, F., Lu, Y., & Wang, Z. (2020). Minimum transmission delay algorithm for multi-hop nodes in Internet of Things based on symmetric algorithm. Journal of Ambient Intelligence and Humanized. *Computing*. https://doi.org/10.1007/s12652-020-02603-w
49. Qiu, T., Jie, L., & Si, W. (2017). A data-driven robustness algorithm for the internet of things in smart cities. *IEEE Communications Magazine, 55*(12), 18–23.
50. Vidhya, S. S., Senthilkumar, M., & Anantha Narayanan, V. (2022). An empirical investigation based quality of service aware transmission power prediction in low power networks. *Sādhanā*. https://doi.org/10.1007/s12046-022-01982-4
51. Zhang, J., Chen, J., & Sun, Y. (2009). Transmission power adjustment of wireless sensor networks using fuzzy control algorithm. *Wireless Communications and Mobile Computing, 9*, 805–818.
52. Agbulu, G. P., Kumar, G. J. R., Juliet, V. A., & Hassan, S. A. (2022). PECDF-CMRP: A power-efficient compressive data fusion and cluster-based multi-hop relay-assisted routing protocol for IoT sensor networks. *Wireless Personal Communications*. https://doi.org/10.1007/s11277-022-09905-6
53. Agbulu, G. P., Joselin, G. R. K., & Juliet, A. V. (2020). A lifetime-enhancing cooperative data gathering and relaying algorithm for cluster-based wireless sensor networks. *International Journal of Distributed Sensor Networks*. https://doi.org/10.1177/1550147719900111
54. Raja Basha, A. (2022). A review on wireless sensor networks: Routing. *Wireless Personal Communications*. https://doi.org/10.1007/s11277-022-09583-4
55. Zhu, K., & Wang, F. (2018). Relay selection in energy harvesting cooperative networks with rateless codes. *AIP Conference Proceedings, 1955*, 040109.
56. Wang, X., Yi, D., & Zhang, Z. (2021). Discovery of random delay distribution based on complex WSNs clock synchronization. *Wireless Personal Communications*. https://doi.org/10.1007/s11277-021-08340-3
57. Cerqueira, L. S., Vieira, A. B., Vieira, L. F. M., Vieira, M. A. M., & Nacif, J. A. M. (2021). A cooperative protocol for pervasive underwater acoustic networks. *Wireless Networks*. https://doi.org/10.1007/s11276-021-02550-0

58. Jang, J., & Adib, F. (2019). Underwater backscatter networking. In *Proceedings of the ACM special interest group on data communication, SIGCOMM '19* (pp. 187–199). New York, NY: ACM.
59. Cerqueira, L. S., Vieira, A. B., Vieira, L. F., Vieira, M. A., & Nacif, J. A. (2018). Copper: Increasing underwater sensor network performance through nodes cooperation. In *2018 IEEE symposium on computers and communications (ISCC)* (pp. 00322–00327). IEEE.
60. Cerqueira, L. S., Vieira, A. B., Vieira, L. F., Vieira, M. A., & Nacif, J. A. M. (2017). Cooperative protocol for medium access control in underwater acoustic sensor networks. In *Magnetic communications: From theory to practice* (pp. 151–168). CRC Press.
61. Vajapeyam, M., Vedantam, S., Mitra, U., Preisig, J. C., & Stojanovic, M. (2008). Distributed space-time cooperative schemes for underwater acoustic communications. *IEEE Journal of Oceanic Engineering, 33*(4), 489–501.
62. Javaid, N., Hussain, S., Ahmad, A., Imran, M., Khan, A., & Guizani, M. (2017). Region based cooperative routing in underwater wireless sensor networks. *Journal of Network and Computer Applications, 92*, 31–41.
63. Ahmad, A., Ahmed, S., Imran, M., Alam, M., Niaz, I. A., & Javaid, N. (2017). On energy efficiency in underwater wireless sensor networks with cooperative routing. *Annals of Telecommunications, 72*(3–4), 173–188.
64. Kim, H.-W., & Cho, H.-S. (2016). A cooperative arq-based mac protocol for underwater wireless sensor networks. In *Proceedings of the 11th ACM international conference on underwater networks & systems*.
65. Acar, G., & Adams, A. (2006). Acmenet: An underwater acoustic sensor network protocol for real-time environmental monitoring in coastal areas. *IEE Proceedings-Radar, Sonar and Navigation, 153*(4), 365–380.
66. Bainbridge, S., Eggeling, D., & Page, G. (2011). Lessons from the field-two years of deploying operational wireless sensor networks on the great barrier reef. *Sensors, 11*(7), 6842–6855.
67. Mehta, R. (2022). Cross-layer design based N-ary Huffman coding for performance analysis of DSDV routing protocol in MANETs. *Wireless Personal Communications*. https://doi.org/10.1007/s11277-022-09770-3
68. Taneja, S., & Kush, A. (2010). A survey of routing protocols in mobile Ad Hoc networks. *International Journal of Innovation, Management and Technology, 1*(3).
69. Hassnawi, L. A., Ahmad, R. B., Yahya, A., Aljunid, S. A., & Elshaikh, M. (2012). Performance analysis of various routing protocols for motorway surveillance system cameras' network. *International Journal of Computer Science Issues, 9*(2), 7–21.
70. Mehta, R. (2020). Optimal Huffman coding performance of ad-hoc networks based on cross-layer design. *Journal of Information Science and Engineering, 36*(6), 1375–1386. https://doi.org/10.6688/JISE.202011_36(6).0015
71. Malik, S., & Sahu, P. K. (2019). A comparative study on routing protocols for VANETs. *Heliyon, 5*(8), E02340. https://doi.org/10.1016/j.heliyon.2019.e02340
72. Shankar, A., & Chelle, L. (2016). Performance comparison of AODV, DSR, DSDV and OLSR MANET routing protocols. *International Journal of Engineering Research & Technology (IJERT), 5*(10), 218–223.
73. OMNeT++ (2018). *The omnet++ homepage*. Retrieved from http://www.omnetpp.org
74. Varga, A. (2001). The OMNeT++ discrete event simulation system. In *Proceedings of the European simulation multiconference*.
75. *Mobility framework (MF) for simulating wireless and mobile networks using OMNeT++*. Retrieved from http://mobility-fw.sourceforge.net/.
76. Mahajan, H. B., & Badarla, A. (2021). Cross-layer protocol for WSN-assisted IoT smart farming applications using nature inspired algorithm. *Wireless Personal Communications*. https://doi.org/10.1007/s11277-021-08866-6
77. Muruganathan, S. D., Ma, D. C. F., Bhasin, R. I., & Fapojuwo, A. O. (2005). A centralized energy-efficient routing protocol for wireless sensor networks. *IEEE Communications Magazine, 43*(3), S8–S13. https://doi.org/10.1109/mcom.2005.1404592

78. Ammari, H. M., & Das, S. K. (2008). A trade-off between energy and delay in data dissemination for wireless sensor networks using transmission range slicing. *Computer Communications, 31*(9), 1687–1704.
79. Calhoun, B. H., Daly, D. C., Verma, N., Finchelstein, D. F., Wentzloff, D. D., Wang, A., Cho, S. H., & Chandrakasan, A. P. (2005). Design considerations for ultra-low energy wireless microsensor nodes. *IEEE Transactions on Computers, 54*(6), 727–740. https://doi.org/10.1109/tc.2005.98
80. Potdar, V., Sharif, A., & Chang, E. (2009). Wireless sensor networks: A survey. In *2009 International conference on advanced information networking and applications workshops*. https://doi.org/10.1109/waina.2009.192
81. Wang, Z., Qin, X., & Liu, B. (2018). An energy-efficient clustering routing algorithm for WSN-assisted IoT. In *2018 IEEE Wireless Communications and Networking Conference (WCNC)*. https://doi.org/10.1109/wcnc.2018.8377171
82. Preeth, S. K. S. L., Dhanalakshmi, R., Kumar, R., & Shakeel, P. M. (2018). An adaptive fuzzy rule based energy efficient clustering and immune-inspired routing protocol for WSN-assisted IoT system. *Journal of Ambient Intelligence and Humanized Computing*. https://doi.org/10.1007/s12652-018-1154-z
83. Saranraj, G., Selvamani, K., & Kanagachidambaresan, G. (2019). Optimal energy-efficient cluster head selection (OEECHS) for Wireless Sensor Network. *Journal of the Institution of Engineers (India): Series B*. https://doi.org/10.1007/s40031-019-00390-3
84. Micheletti, M., Mostarda, L., & Navarra, A. (2019). CER-CH: Combining election and routing amongst cluster heads in heterogeneous WSNs. *IEEE Access, 7*, 125481–125493. https://doi.org/10.1109/access.2019.2938619
85. Ganesan, S., & Muthuswamy, V. (2021). Ensuring reliability of high-priority data transport using expected congestion shortfall prediction in wireless sensor networks. *Wireless Networks*. https://doi.org/10.1007/s11276-021-02792-y
86. Koley, I., & Samanta, T. (2018). Mobile sink based data collection for energy efficient coordination in wireless sensor network using cooperative game model. *Telecommunication Systems, 71*(3), 377–396.
87. Li, J., Zhang, L., Guo, L., Ren, Q., & Guo, Y. (2016). Multi-path reliable routing with pipeline schedule in wireless sensor networks. In Q. Yang, W. Yu, & Y. Challal (Eds.), *Wireless algorithms, systems, and applications, WASA 2016* (Lecture notes in computer Science) (Vol. 9798).
88. Lee, J., & Jung, I. (2010). Adaptive-compression based congestion control technique for wireless sensor networks. *Sensors, 10*, 2919–2945.
89. Ávila, K., Sanmartin, P., Jabba, D., & Gómez, J. (2021). An analytical survey of attack scenario parameters on the techniques of attack mitigation in WSN. *Wireless Personal Communications*. https://doi.org/10.1007/s11277-021-09107-6
90. Park, J., Gofman, M., Wu, F., & Choi, Y.-H. (2016). Challenges of wireless sensor networks for internet of thing applications. *International Journal of Distributed Sensor Networks, 12*(8), 1550147716665506.
91. Khan, S., Lloret, J., Song, H., & Du, Q. (2017). Qos based cooperative communications and security mechanisms for ad hoc sensor networks. *Journal of Sensors*.
92. Pongle, P., & Chavan, G. (2015). A survey: Attacks on rpl and 6lowpan in iot. In *International conference on pervasive computing (ICPC)* (pp. 1–6).
93. Wang, T., Zhang, G., Yang, X., & Vajdi, A. (2016). A trusted and energy efficient approach for cluster-based wireless sensor networks. *International Journal of Distributed Sensor Networks, 12*(4), 3815834.
94. Akyildiz, I. F., Su, W., Sankarasubramaniam, Y., & Cayirci, E. (2002). Wireless sensor networks: A survey. *Computer Networks, 38*(4), 393–422.
95. Rezazadeh, J., Sandrasegaran, K., & Kong, X. (2018). A location-based smart shopping system with iot technology. In *2018 IEEE 4th World Forum on Internet of Things (WF-IoT)* (pp. 748–753).

96. Govindraj, V., Sathiyanarayanan, M., & Abubakar, B. Customary homes to smart homes using internet of things (iot) and mobile application. In *2017 International conference on smart technologies for smart nation (SmartTechCon)* (pp. 1059–1063).
97. Krishnan, S., & Chaporkar, P. (2017). Stochastic approximation based on-line algorithm for fairness in multi-rate wireless LANs. *Wireless Networks, 23*(5), 1563–1574.
98. Walters, J. P., Liang, Z., Shi, W., & Chaudhary, V. (2007). *Wireless sensor network security: A survey*. Auerbach Publications.
99. Kulkarni, G., Shelk, R., Gaikwad, K., Solanke, V., Gujar, S., & Khatawkar, P. (2013). Wireless sensor network security threats. In *Communication and Computing (ARTCom 2013), Fifth International Conference on Advances in Recent Technologies in, IET* (pp. 131–135).
100. Saeedvand, S., Aghdasi, H. S., & Mohammad Khanli, L. (2019). Novel distributed dynamic backbone-based flooding in unstructured networks. *Peer-to-Peer Networking and Applications*. https://doi.org/10.1007/s12083-019-00817-0
101. Madan, M., & Mathur, M. (2014). Cloud network management model—A novel approach to manage cloud traffic. *International Journal of Cloud Computing and Services Architecture, 4*(5). https://doi.org/10.5121/ijccsa.2014.4502
102. Daswani S, Fisk A (2002) *Gnutella UDP extension for scalable searches (GUESS)*
103. The gnutella protocol specification v0.4. Retrieved 01 April, 2008 from http://rfc-gnutella.sourceforge.net/developer/stable
104. Lv, Q., Cao, P., Cohen, E., Li, K., & Shenker, S. (2002). Search and replication in unstructured peer-to-peer networks. In *ICS* (pp. 84–95). ACM.
105. Abid, S. A., Othman, M., & Shah, N. (2014). 3D P2P overlay over MANETs. *Computer Networks, 64*, 89–111. https://doi.org/10.1016/j.comnet.2014.02.006
106. Hughes, D., Coulson, G., & Walkerdine, J. (2005). Free riding on Gnutella revisited: The bell tolls? *IEEE Distributed Systems Online, 6*(6). https://doi.org/10.1109/MDSO.2005.31
107. Kang, S., Kim, T., Jeon, H., Lee, W., & Kang, S. (2015). A healthcare information sharing scheme in distributed cloud networks. *Cluster Computing*, 1–6. https://doi.org/10.1007/s10586-015-0488-y
108. Devikar, R. N., Patil, D. V., & Prakash, C. (2020). A hybrid computing approach to improve convergence time for scalable network. *Sādhanā*. https://doi.org/10.1007/s12046-020-01454-7
109. Labovitz C, Malan G R and Jahanian F 1999 Origins of Internet routing instability. In *Eighteenth annual joint conference of the IEEE computer and communications societies proceedings*
110. Labovitz, C., Ahuja, A., Wattenhofer, R., & Venkatachary, S. (2001). The impact of Internet policy and topology on delayed routing convergence. In *IEEE INFOCOM* (pp. 537–546).
111. Aiabdulkreem E A, Al-Raweshidy H S and Abbod M F 2014 Using a fight-or-flight mechanism to reduce BGP convergence time. In *International conference on communications and networking (ComNet)*.
112. Wang X, Bonaventure O & Zhu P (2011). Stabilizing BGP routing without harming convergence. In *IEEE conference on computer communications workshops (INFOCOM WKSHPS)* (pp. 840–845)
113. Devikar, R. N., Patil, D. V., & Chandra Prakash, V. (2016). A mathematical approach to improve the network performance. *Indian Journal of Science and Technology, 9*, 1–8.
114. Moubarak, M. T., Elbayoumy, A. D., & Megahed, M. H. (2018). Design and implementation of BGP novel control mechanism (BGP-NCM) based on network performance parameters. *Ain Shams Engineering Journal, 9*, 2079–2091.
115. Al-Ghamdi, B., Ayaida, M., & Fouchal, H. (2019). Performance evaluation of scheduling approaches for wireless sensor networks. *Wireless Personal Communications*. https://doi.org/10.1007/s11277-019-06775-3
116. Al-Ghamdi, B., Ayaida, M., Fouchal, H., & Baala, H. (2014). A slot assignment for wireless body area networks. In *2014 IEEE symposium on computers and communications (ISCC)*.

117. Pantazis, N. A., Vergados, D. J., Vergados, D. D., & Douligeris, C. (2005). Energy efficiency in wireless sensor networks using sleep mode TDMA scheduling. *Ad Hoc Networks, 7*(5), 322–343.
118. Meenakshi, N., Jaber, M. M., Pradhan, R., Kamruzzaman, M. M., Maragatham, T., Ramamoorthi, J. S., & Murugesan, M. (2023). Design systematic wireless inventory trackers with prolonged lifetime and low energy consumption in future 6G network. *Wireless Networks*. https://doi.org/10.1007/s11276-023-03325-5
119. Anpalagan, A., Ejaz, W., Sharma, S. K., de Costa, D. B., Jo, M., & Kim, J. (2021). Guest editorial special issue on green communication and computing technologies for 6G networks. *IEEE Transactions on Green Communications and Networking, 5*(4), 1653–1656.
120. Nguyen, D. C., Pathirana, P. N., Ding, M., & Seneviratne, A. (2021). Deep reinforcement learning for collaborative offloading in heterogeneous edge networks. In *2021 IEEE/ACM 21st international symposium on cluster, cloud and internet computing (CCGrid)* (pp. 297–303). IEEE.
121. Araghipour, A., & Mostafavi, S. (2021). An improved emergency response routing protocol for Internet of Things. *Wireless Personal Communications*. https://doi.org/10.1007/s11277-021-09187-4
122. Thibaud, M., Chi, H., Zhou, W., & Piramuthu, S. (2018). Internet of Things (IoT) in high-risk Environment, health and safety (EHS) industries: A comprehensive review. *Decision Support Systems, 108*, 79.
123. Khaleghnasab, R., Bagherifard, K., Nejatian, S., Parvin, H., & Ravaei, B. (2020). A new energy-efficient multipath routing in internet of things based on gray theory. *International Journal of Information Technology & Decision Making, 19*(06), 1581. https://doi.org/10.1142/S0219622020500388
124. Selvaraj, S., Thangarajan, R., & Saravanan, M. (2022). Trust based and optimized RPL routing in social internet of things network. In S. Shakya, R. Bestak, R. Palanisamy, & K. A. Kamel (Eds.), *Mobile computing and sustainable informatics* (pp. 513–529). Springer.
125. Choudhary, A., Nizamuddin, M., & Sachan, V. K. (2019). A hybrid fuzzy-genetic algorithm for performance optimization of cyber physical wireless body area networks. *International Journal of Fuzzy Systems*. https://doi.org/10.1007/s40815-019-00751-6
126. Heinzelman, W.R., Chandrakasan, A., Balakrishnan, H. (2000) Energy-efficient communication protocol for wireless microsensor networks. In *Proceedings of the 33rd Hawaii international conference on system sciences—(HICSS '00), Washington DC, USA*, 2000 (Vol. 2, pp. 10), IEEE Computer Society. https://doi.org/10.1109/hicss.2000.926982
127. Ayati, M., Ghayyoumi, M. H., & Keshavarz-Mohammadiyan, A. (2018). A fuzzy three-level clustering method for lifetime improvement of wireless sensor networks. *Annales des Telecommunications, 73*(7–8), 535–546. https://doi.org/10.1007/s12243-018-0631-x
128. Junior, F. R. L., Osiro, L., & Carpinetti, L. C. R. (2014). A comparison between Fuzzy AHP and Fuzzy TOPSIS methods to supplier selection. *Applied Soft Computing, 21*, 194–209. https://doi.org/10.1016/j.asoc.2014.03.014. (ISSN 1568-4946).
129. Kim, Y. H., Ahn, S. C., & Kwon, W. H. (2000). Computational complexity of general fuzzy logic control and its simplification for a loop controller. *Fuzzy Sets and Systems, 111*(2), 215–224. https://doi.org/10.1016/S0165-0114(97)00409-0. (ISSN 0165-114).
130. Xie, W. X., Zhang, Q. Y., Sun, Z. M., et al. (2015). A clustering routing protocol for WSN based on type-2 fuzzy logic and ant colony optimization. *Wireless Personal Communications, 84*, 1165. https://doi.org/10.1007/s11277-015-2682-x
131. Ahmed, S., Javaid, N., Yousaf, S., Ahmad, A., Sandhu, M. M., Imran, M., Khan, Z. A., & Alrajeh, N. (2015). Co-LAEEBA: Cooperative link aware and energy efficient protocol for wireless body area networks. *Computers in Human Behavior, 51*(1), 1205–1215. https://doi.org/10.1016/j.chb.2014.12.051
132. Mondal, S., Ghosh, I., & Das, A. (2022). Energy efficient and secure healthcare data transmission in the internet of medical things network. *Microsystem Technologies*. https://doi.org/10.1007/s00542-022-05398-2

133. Wei, K., Zhang, L., Guo, Y., & Jiang, X. (2020). Health monitoring based on internet of medical things: Architecture, enabling technologies, and applications. *IEEE Access, 8*, 27468–27478.
134. Ali, R., Hussain, K., Soomro, A., Awan, J., Khan, S., Arain, T. H., & Zafar, H. (2018). An energy efficient routing protocol for wireless body area sensor networks. *Wireless Personal Communications, 99*(1443–1454), 04.
135. Ullah, Z., Ahmed, I., Ali, T., Ahmad, N., Niaz, F., & Cao, Y. (2019). Robust and efficient energy harvested-aware routing protocol with clustering approach in body area networks. *IEEE Access, 7*, 33906–33921.
136. Anwar, M., Abdullah, H., Altameem, A., Qureshi, K., Masud, F., Cao, Y., & Kharel, R. (2018). Green communication for wireless body area networks: Energy aware link efficient routing approach. *Sensors, 18*(09), 1–17.
137. Sagar, A., Singh, S., & Kumar, A. (2020). Energy-aware wban for health monitoring using critical data routing (cdr). *Wireless Personal Communications, 112*, 273–302.
138. Jiasong, M., Liu, X., & Yi, X. (2019). Simplified energy-balanced alternative-aware routing algorithm for wireless body area networks. IEEE. *Access, 7*, 108295–108303., 07.
139. Sharma, J. K. (2014). *Fundamentals of business statistics* (2nd ed.). Vikash Publishing House.
140. Sreedevi, P., & Venkateswarlu, S. (2022). Comparative analysis of energy efficient routing protocols with optimization in WSN. *International Journal on Interactive Design and Manufacturing, IJIDeM*. https://doi.org/10.1007/s12008-022-00958-2
141. Heimfarth, T., Giacomin, J. C., & Schneider, B. d. O. (2021). An asynchronous anycast cross-layer protocol for WSN suited to noisy environments. *Wireless Personal Communications*. https://doi.org/10.1007/s11277-021-08505-0
142. Heimfarth, T., Giacomin, J., & De Araujo, J. (2015). Aga-MAC: Adaptive geographic anycast mac protocol for wireless sensor networks. In *2015 IEEE 29th international conference on advanced information networking and applications (AINA)* (pp. 373–381). https://doi.org/10.1109/AINA.2015.209
143. Toklu, S., & Ayhan Erdem, O. (2014). BSC-MAC: Energy efficiency in wireless sensor networks with base station control. *Computer Networks, 59*, 91–100. https://doi.org/10.1016/j.bjp.2013.12.012
144. Buettner, M., Yee, G. V., Anderson, E., & Han, R. (2006). X-MAC: A short preamble mac protocol for duty-cycled wireless sensor networks. In *Proceedings of the 4th international conference on Embedded networked sensor systems* (pp. 307–320). ACM, SenSys'06. https://doi.org/10.1145/1182807.1182838
145. Zorzi, M., & Rao, R. (2003). Geographic random forwarding (GeRaF) for ad hoc and sensor networks: Multihop performance. *IEEE Transactions on Mobile Computing, 2*(4), 337–348. https://doi.org/10.1109/TMC.2003.1255648
146. Sumathi, A. C., Javadpour, A., Pinto, P., Sangaiah, A. K., Zhang, W., & Mahmoodi Khaniabadi, S. (2022). NEWTR: A multipath routing for next hop destination in internet of things with artificial recurrent neural network (RNN). *International Journal of Machine Learning and Cybernetics*. https://doi.org/10.1007/s13042-022-01568-w
147. Heinz, C., & Koch, A. (2022). On-chip and distributed dynamic parallelism for task-based hardware accelerators. *Journal of Signal Processing Systems*. https://doi.org/10.1007/s11265-022-01759-2
148. Canis, A., Choi, J., Aldham, M., Zhang, V., Kammoona, A., Anderson, J. H., Brown, S., & Czajkowski, T. (2011). LegUp: High-level synthesis for FPGA-based processor/accelerator systems. In *Proceedings of the 19th ACM/SIGDA international symposium on Field programmable gate arrays* (pp. 33–36).
149. Prabhakar, R., Koeplinger, D., Brown, K. J., Lee, H., De Sa, C., Kozyrakis, C., & Olukotun, K. (2016). Generating configurable hardware from parallel patterns. *ACM SIGPLAN Notices, 51*(4), 651–665.

150. Chen, T., Srinath, S., Batten, C., & Suh, G. E. (2018). An architectural framework for accelerating dynamic parallel algorithms on reconfigurable hardware. In *2018 51st Annual IEEE/ACM International Symposium on Microarchitecture (MICRO)* (pp. 55–67). IEEE.
151. Lyu, Q., Zhou, Y., Fan, Q., Lyu, Y., Zheng, X., Xu, G., & Li, J. (2022). JOSP: Joint optimization of flow path scheduling and virtual network function placement for delay-sensitive applications. *Mobile Networks and Applications*. https://doi.org/10.1007/s11036-021-01868-5
152. Lyu, Q., & Zhu, H. (2019). Optimizing network latency with fast path assignment for incoming flows. *International Journal of Computational Engineering and Information Technology, 13*(9), 465–471.
153. Lahmar, F., & Mezni, H. (2021). Security-aware multi-cloud service composition by exploiting rough sets and fuzzy FCA. *Soft Computing*. https://doi.org/10.1007/s00500-020-05519-x
154. Kurdi, H., Al-Anazi, A., Campbell, C., & Al Faries, A. (2015). A combinatorial optimization algorithm for multiple cloud service composition. *Computers and Electrical Engineering, 42*, 107–113.
155. Zou, G., Chen, Y., Yang, Y., Huang, R., & Xu, Y. (2010). AI planning and combinatorial optimization for web service composition in cloud computing. In *Proceedings of the international conference on cloud computing and virtualization* (pp. 1–8).
156. Mezni, H., & Sellami, M. (2017). Multi-cloud service composition using formal concept analysis. *Journal of Systems and Software, 134*, 138–152.
157. Zhang, F., Hwang, K., Khan, S. U., & Malluhi, Q. M. (2015a). Skyline discovery and composition of multi-cloud mashup services. *IEEE Transactions on Services Computing, 9*(1), 72–83.
158. Dou, W., Zhang, X., Liu, J., & Chen, J. (2013). Hiresome-ii: Towards privacy-aware cross-cloud service composition for big data applications. *IEEE Transactions on Parallel and Distributed Systems, 26*(2), 455–466.
159. Wang, D., Yang, Y., & Mi, Z. (2015). A genetic-based approach to web service composition in geo-distributed cloud environment. *Computers and Electrical Engineering, 43*, 129–141.
160. Pawlak, Z. (1982). Rough sets. *Int J Parallel Program, 11*(5), 341–356.
161. Yu, Q., Chen, L., & Li, B. (2015). Ant colony optimization applied to web service compositions in cloud computing. *Computers and Electrical Engineering, 41*, 18–27.
162. Gutierrez-Garcia, J. O., & Sim, K. M. (2013). Agent-based cloud service composition. *Applied Intelligence, 38*(3), 436–464.
163. Joshi, B., & Thakur, M. K. (2022). A traffic intensive virtual channels allocation scheme in network-on-chip. *Arabian Journal for Science and Engineering*. https://doi.org/10.1007/s13369-022-07191-9
164. Smai, A.-H., & Thorelli, L.-E. (1998). Distance-based flow control in wormhole networks. In *Parallel computer routing and communication* (pp. 189–202).
165. Michelogiannakis, G., Jiang, N., Becker, D., & Dally, W. (2011). Packet chaining: Efficient single-cycle allocation for on-chip networks. *IEEE Computer Architecture Letters, 10*(2), 33–36.

Chapter 7
Performance Optimization for Wireless Broadband Access

Milind Pande, Anand J. Kulkarni, and Apoorva S. Shastri

Introduction by the Editor

This chapter discusses strategies for improving the performance of Wireless Broadband Access. This includes Multi-user Scheduling Algorithms for non-identical channel distribution, LTE Cellular Networks, optimal resource allocation in LTE, and performance comparison of various tunnelling techniques and routing protocols. In addition, it addresses optimal multipath routing for video transmission in VANETs, and transmission of audio over LTE packet-based wireless networks using wavelets. This chapter reviews different aspects involved in device-to-device (D2D) communication in cellular networks such as transmission mode selection, quality of experience (QoE) awareness in video streaming, traffic offloading techniques, mobility models, resource sharing, and power management schemes. It also shows a comparison of different resource allocation schemes for D2D networks Metaheuristic algorithms such as sleep scheduling algorithm (iSleep) and Whale Optimization Algorithm (i-WOA) for clustering in Vehicular Ad Hoc Networks are also introduced.

In device to device (D2D) communication, to maximize system throughput and spectral efficiency allocation of channel and power based on the mode selection. This maximization problem is called Joint mode selection, channel assignment, and power allocation (JMSCPA) problem. Such problem requires two step solutions also called bi-level optimization problem, which has high complexity and

M. Pande
Dr. Vishwanath Karad MIT World Peace University, Pune, Maharashtra, India
e-mail: milind.pande@mitwpu.edu.in

A. J. Kulkarni · A. S. Shastri (✉)
Institute of Artificial Intelligence, Dr. Vishwanath Karad MIT World Peace University, Pune, Maharashtra, India
e-mail: anand.j.kulkarni@mitwpu.edu.in

M. Pande et al. (eds.), *Optimization Methods in Mobile Communication Systems*, https://doi.org/10.1007/978-981-95-1810-4_7

computational time. Artificial bee colony optimization is successfully implemented to solve this problem.

In high user equipment density network, quality of video streaming is main concern. Dynamic adaptive streaming over HTTP (DASH) is used to adapt quality of video segments to network conditions. In those architecture video segments are cached progressively as requested by user equipment. Performance of DASH is improved by introducing one measure that is quality of experience, which gets improved by cellular resource allocation and caching of video segments and storage member assignment problem.

Cluster-based opportunistic traffic offloading scheme CBOT for device-to-device communication, that divided the network into heterogeneous clusters. A cluster head is chosen that provides the inter-cluster communication by using LTE connection via node and intra-cluster communication through Wi-Fi. To reduce energy consumption, the LTE traffic is diverted to Wi-Fi for intra-cluster communication.

Unpredictable device mobility may influence the service continuity of a D2D communication. RPGM (Reference Point Group Mobility), a group mobility model, use for D2D communication. It is suitable for modelling the D2D mobility and manages the relationship among mobile devices. Mobility model enables the D2D candidates to stay in proximity of each other so the D2D communication is maintained for a longer time so that better service continuity is performed. When D2D mobile terminals are in proximity to each other, they undergo a joint handover.

Multi-user scheduling is an important task to be performed in wireless communication networks to share the bandwidth resource effectively between large number of users. Multi-user scheduling algorithms is implemented over independent and non-identically distributed generalized fading channel and 3GPP long term evolution (LTE) networks. Various scenarios are considered such as (1) independent identical fading parameter, independent non identical average SNR channel conditions and (2) both fading parameter and average SNR are independent and non-identically distributed under absolute SNR (AS) scheduling and threshold SNR scheduling.

An iterative heuristic optimal resource allocation (HORA) algorithm and a chunk-based resource block allocation (CRBA) scheduling algorithm determine resource block (RB) allocation among users to satisfy the quality-of-service requirement. It performs RB and power allocation separately to reduce computational complexity. In the CRBA algorithm sets of RBs are allocated to groups of users keeping power constant to all users. The HORA algorithm assigns most RBs to users who have high-value signal to noise ratio and continues the RB allocation until it meets the QoS criteria of all users in consideration of the threshold value of the power budget.

To get a smooth communication between IPv6 and the existing IPv4 network, the transitions mechanisms are used to prevent this sudden change. This mechanism is called tunnelling. The transition mechanisms can be utilized to deploy and adopt the IPv6 network for direct connectivity with the existing IPv4 network. The payload is transmitted and received at tunnel destination endpoint, checked for protocol value, fragmented, and reassembled for the delivery of IPv6 packet to its final destination address. Some advantages of tunnelling are as follows: It Connects two separate networks/subnetworks. It Allow multiprotocol network over a single protocol backbone network. It allows communication between private IPv6 networks and public IPv4 networks.

In vehicular technology features of surrounding video and its usage is gaining importance. Which also include transmission of videos from one vehicle to other vehicles in the VANET to safeguard the vehicles from critical situations that occur in the highways. Moth Whale Optimization Algorithm (MWOA) determine the optimal multipath for the transmission of the videos from one vehicle to other in the VANET network.

Audio signal processing is the procedure used to analyse audio using different processing techniques. Pulse Code Modulation does not require compression in audio devices having large bandwidth, however compression is necessary for digital audio signal that has to be transmitted over VOIP link. Wavelets are more practical and dominant tool meant for analysing audio signals, as it exhibits multi-resolution property with a considerable decrease in the time complexity for removing resolution problems. Wavelet based wireless VOIP model can be utilize for audio data transmission.

The optimal operation of Wireless Sensor Network (WSN) requires a sustainable topology. A dead node may result in a breach of end-to-end connectivity in the network. Every sensor node has a typical operating temperature zone. iSleep enables the sensor nodes to auto-control the sleep state of the nodes and maintain network connectivity for a longer amount of time.

Intelligent Whale Optimizer algorithm (i-WOA) is a cluster optimization method. It uses probabilistic approach that aims to reduce processing time while forming a cluster, computational cost, network overhead, packet latency, and end-to-end delay between the vehicles. Once the cluster optimization is performed, other vehicles searching for CHs inform their areas agreeing on to the optimized vehicle specialist.

Machine Generated Summaries

Disclaimer: The summaries in this chapter were generated from Springer Nature publications using extractive AI auto-summarization: An extraction-based summarizer aims to identify the most important sentences of a text using an algorithm and uses those original sentences to create the auto-summary (unlike generative AI). As the constituted sentences are machine selected, they may not fully reflect the body of the work, so we strongly advise that the original content is read and cited. The auto generated summaries were curated by the editor to meet Springer Nature publication standards. To cite this content, please refer to the original papers.

Machine generated keywords: cellular, cellular network, protocol, scheduling, allocation, message, packet, qos, spectrum, resource, streaming, resource allocation, underlay, video, route.

An Adaptive Transmission Mode Selection Scheme for Cellular Underlaid D2D Communication [1]

This is a machine-generated summary of:

Sharma, Nitin; Khanolkar, Shailesh; Kumar, Shreyas Ramesh; Anpalagan, Alagan: An Adaptive Transmission Mode Selection Scheme for Cellular Underlaid D2D Communication [1]

Published in: International Journal of Wireless Information Networks (2022)

Link to original: https://doi.org/10.1007/s10776-022-00550-1

If you want to cite the papers, please refer to the original.

For technical reasons we could not place the page where the original quote is coming from.

Abstract-Summary

"We propose to use Artificial Bee Colony (ABC) optimization to solve the joint mode selection, channel assignment, and power allocation (JMSCPA) problem to maximize system throughput and spectral efficiency."

"JMSCPA is a problem where the allocation of channel and power depends on the mode selection."

"As bi-level optimization increases the complexity and computational time, we propose a modified version of single-level ABC algorithm aided with the adaptive transmission mode selection algorithm to allocate the cellular, reuse, and dedicated modes to the DUs along with channel and power allocation based on the network traffic load scenarios."

"A single variable, represented by the users (CUs and DUs) is used to allocate mode selection, and channel allocation to solve the JMSCPA problem, leading to a simpler solution with faster convergence, and significant reduction in the computational complexity which scales linearly with the number of users."

Introduction

"Inspired by the success of bio-inspired meta-heuristics in a variety of complex real-world optimization problems, we propose the use of Artificial Bee Colony (ABC) [2] optimization to solve the joint mode selection, channel assignment, and power allocation (JMSCPA) problem for underlay D2D communication."

"We propose a modified version of single-level ABC algorithm aided with the adaptive transmission mode selection (ATMS) algorithm to adaptively allocate the cellular, reuse, and dedicated modes to the DUs along with channel, and power allocation based on the network traffic load scenarios."

"We propose a modified version of single level ABC algorithm aided with the adaptive transmission mode selection (ATMS) algorithm to allocate the cellular, reuse, and dedicated modes to the DUs along with channel and power allocation based on the network traffic load scenarios."

Related Work

"The authors in [3] solved a joint mode selection, channel allocation, and power control problem in D2D communications underlaying cellular networks to maximize the overall system throughput while guaranteeing the SINR of D2D and cellular links."

"The authors in [4] solved a centralized resource allocation problem for DUs by allocating cellular, reuse, and dedicated mode in case of a two-tier cellular network, comprising a macro base station and femtocell to maximize the system throughput and spectral efficiency."

"The authors in [5] solved a throughput and spectrum efficiency maximization problem by formulating a joint mode selection, resource block allocation, and power control optimization problem."

"The evolutionary algorithm was used in [6] to solve the joint mode selection problem and resource allocation problem in D2D communication underlaying cellular networks, with the objective of maximizing the system sum-rate and guaranteed fairness in the allocation of the resources under the system constraints."

System Model

"This section describes the three modes used in this paper for D2D communication. D2D communication in cellular mode This mode is the same as normal cellular communication, and a pair of unused uplink and downlink channel is allocated to enable the DUs to communicate."

"DUs use orthogonal channels, and reuse of these channels by any other CUs or DUs is not allowed in this mode."

"Like the cellular mode, since there is no reuse and the channels are orthogonal, there is no interference in this mode."

"The reuse of the uplink channel allocated to the CU by the DU will lead to co-channel interference."

"The D2D communication takes place in the cellular mode, dedicated or reuse mode, depending on the channel condition and the traffic load allocated to the network."

"In the reuse mode, only the CUs uplink channel can be used by the DU as it is easier to control the uplink interference."

Problem Formulation

"The solution to the JMSCPA problem, obtained using the modified ABC algorithm, jointly allocates resources to the CUs and DUs under SINR, power, and QoS constraints."

"Constraint C2 restricts the number of DUs operating in the cellular mode to less than the number of unused uplink/downlink channels."

"Constraint C3 ensures that the number of channels used by DUs in the cellular and dedicated mode cannot be more than the number of unused channels."

"Constraints C8 and C9 ensure that SINR of the DUs and CUs are always above a predefined threshold, respectively."

ABC Optimization Aided Solution

"The pseudocode for generating the initial population to describe the process of adaptive transmission mode selection and channel allocation is given in Algorithm 1."

"The employed bee phase modifies the optimization variables to obtain better solutions."

"In the fourth step, the scout bee phase replaces solution which cannot be improved further with randomly produced solutions, generated as per the procedure followed in the initialization phase."

"The CUs and DUs are allocated to the channels, represented by 1 to C. The DUs are allocated cellular, reuse, and dedicated mode of operation depending on the network load and subject to SINR and power constraints."

"In the JMSCPA formulation, the employed bee phase involves modification in the mode selection, channel assigned, and the power allocated to the CU and DUs."

"The modified onlooker bee phase uses mutation and crossover operators of the DE algorithm to update the solutions generated in the employed bee phase."

Numerical Evaluation

"The ATMS-aided ABC algorithm was able to allocate either dedicated mode or reuse mode to all the DUs in all the 100 runs."

"The proposed ATMS-aided ABC algorithm consistently outperforms all the other algorithms as it allows more DUs in reuse mode compared to the other three algorithms, irrespective of load conditions."

"The selection of population size plays a significant role in the convergence of the solution obtained using the proposed ATMS aided ABC algorithm."

"It can be concluded from the above discussion that proposed ATMS-aided ABC algorithm converges faster, requires a lesser number of iterations and population size for convergence."

"The computational complexity of the proposed ATMS-aided ABC algorithm reduces significantly on account of the use of a single variable represented by the users (CUs and Dus), for performing joint mode selection and channel allocation for the JMSCPA problem."

Conclusion

"This paper proposed an ATMS-aided ABC algorithm to provide a solution to the JMSCPA optimization problem, with the objective of maximization of system throughput under SINR and power constraints for D2D communication underlaying cellular networks."

"We proposed a modified version of single-level ABC algorithm aided with the adaptive transmission mode selection (ATMS) algorithm to adaptively allocate the modes to the DUs along with channel, and power allocation based on the network traffic load scenarios."

"A single variable, represented by the users (CUs and DUs) is used to allocate mode selection and channel allocation to solve the JMSCPA problem, leading to a simpler solution with a faster convergence, and significantly reduction in the computational complexity."

"The proposed algorithm used a single variable to allocate mode selection and channel allocation to maximize the system throughput and spectral efficiency by allocating cellular, reuse, and dedicated mode of operation to the DUs depending on the traffic load conditions."

QoE Awareness in Progressive Caching and DASH-Based D2D Video Streaming in Cellular Networks [7]

This is a machine-generated summary of:

Al-Habashna, Ala'a; Wainer, Gabriel: QoE awareness in progressive caching and DASH-based D2D video streaming in cellular networks [7]

Published in: Wireless Networks (2019)

Link to original: https://doi.org/10.1007/s11276-019-02055-x

If you want to cite the papers, please refer to the original.

For technical reasons we could not place the page where the original quote is coming from.

Abstract-Summary

"We present an architecture to improve video streaming quality of experience (QoE) in cellular networks with high user equipment (UE) density."

"In the proposed architecture, video segments are progressively cached, as requested, in selected UEs called storage members (SMs)."

"We study the improvements achieved by the proposed architecture in terms of many video streaming QoE metrics."

"We improve the operation of the proposed architecture by introducing QoE awareness to both caching and distribution of video segments."

"We employ QoE awareness in three aspects of the proposed architecture; cellular resource allocation, caching of video segments, and SM-assignment optimization."

Introduction

"In [8], we propose an architecture to improve video streaming QoE in cellular networks under high traffic load."

"In [8], we evaluate the performance improvements achieved by DABAST in terms of many video streaming QoE metrics."

"Results show that DABAST significantly improves video streaming QoE for many users in the cell by enhancing all the measured QoE metrics."

"As these results show, despite the clear improvements achieved by DABAST for a significant portion of the video streams in the cell, a high number of video streams did not experience considerable improvement, and hence, did not receive video streaming service with good QoE, due to the repetitive video rebuffering."

"Making DABAST aware of video streaming QoE for users in the cell can further increase the QoE gains achieved by DABAST."

"We analyze the improvements achieved by each QoE-awareness technique in terms of video streaming QoE metrics."

Background and Related Work

"We present and evaluate DABAST; an architecture to improve the QoE of video streaming in cellular networks with high user density."

"In DABAST, SMs are assigned dynamically to requesting UEs on a segment-by-segment basis, depending on SMs availability and video bit rates of available segments to maximize video streaming QoE. Many video streaming QoE-aware approaches for resource allocation in conventional cellular networks (without D2D communication) have been proposed [9–12]."

"In [9], an adaptive video steaming QoE maximization approach was proposed for resource allocation in LTE networks, where the playout buffer levels are signaled from the UEs to the QoE optimizer at the BS."

"Our scheduling algorithm is proposed to operate in DABAST and considers not only the reported playout buffer length, but also an estimation of the transmitted video contents over both the cellular and the D2D channel."

The DABAST Architecture

"DABAST employs the cached and segmented download algorithms at the BS to provide progressive caching of video segments in SMs and to improve the delivery of video contents by providing D2D transmission of video segments from SMs to requesting UEs."

"When a UE requests a video file, the BS processes the request and responds as follows: Send To an SM (STSM): This case is employed to progressively cache video segments in the SMs of the clusters as requested."

"Send With Assistance (SWA): if requested segments of a video file are available in any of the SMs, the BS will ask the SMs to send the segments to the requesting UE over D2D links."

"In such cases, requested video segments are transmitted to the UEs from the distributed cache (when found) even if the requested video bit rate does not match that of cached segments."

DABAST with QoE-Aware Cellular Resource Allocation

"Our BB scheduling algorithm is proposed to operate in DABAST and considers not only the reported playout buffer length, but also an estimation of the transmitted video duration over both the cellular and the D2D channel (from all SMs)."

"We can see that the metric used for BB scheduling keeps the scheduler aware of the current playout buffer length at the UEs, and hence, avoid rebufferings by allocating more resources to users with imminent playout buffer stalling, and by saving the cellular resources for users who get their video segments through the BS only (over cellular resources)."

"We discuss how SMs are assigned the task of video segment transmission to requesting UEs in case of RR, PF, and BB cellular resource scheduling."

DABAST with HRC and SM-Assignment Optimization

"A video file can be considered popular enough for HRC if it is requested more than a certain number of times recently. 4. To avoid increasing the initial delay, the first few segments are always transmitted with a low video bit rate, even in HRC."

"We consider the problem of SM assignment in the case where there are many UEs requesting video segments that are cached with multiple video bit rates."

"An optimal SM-assignment approach, in this case, would maximize the achievable video bit rate in the cell, without causing a considerable increase in the number of rebufferings or in the initial delay."

"This can be achieved by assigning to each requesting UE the SM that is caching the requested segment with the highest video bit rate, which results in setting the binary variable of that combination, x_s, to 1."

Performance Evaluation of QoE-Aware DABAST

"This usually does not result in rebuffering because such UEs (playing videos cached in the SMs) have relatively higher playout buffer as most segments are delivered over the D2D channel."

"This reduction is expected as BB scheduling utilizes more of the cellular resources to help users who get their video segments exclusively through the BS, and hence fewer segments will be sent with higher video bit rate to users with higher playout buffer, i.e., users who get some of their segments over the D2D channel."

"Regarding BB, it is expected to have the least average number of rebufferings, as it does not employ HRC, which means no video segments are downloaded with high video bit rate."

"Thanks to BB scheduling, where cellular resources are allocated to UEs with low playout buffer, which distributes the extra delay caused by transmission of high video bit rate segments over all UEs in the cell sharing cellular resources."

Conclusion

"We present our proposed architecture for improving the QoE of video streaming over cellular networks with high user density."

"The architecture is called DASH-based BS-Assisted D2D video STreaming in cellular networks (DABAST)."

"We evaluate the performance of DABAST, through computer simulations, in terms of video streaming QoE metrics."

"Results show that DABAST achieves significant improvements in terms of QoE when compared to conventional video streaming over a cellular network, i.e., DASH streaming without collaborative D2D communication."

"We analyze the performance achieved by each QoE-awareness technique in terms of video streaming QoE metrics."

Clustering Based Opportunistic Traffic Offloading Technique for Device-to-Device Communication [13]

This is a machine-generated summary of:

Kumar, Prashant; Chauhan, Naveen; Kumar, Mohit; Awasthi, Lalit K.: Clustering based opportunistic traffic offloading technique for device-to-device communication [13]

Published in: International Journal of System Assurance Engineering and Management (2021)

Link to original: https://doi.org/10.1007/s13198-021-01136-5

If you want to cite the papers, please refer to the original.

For technical reasons we could not place the page where the original quote is coming from.

Abstract-Summary

"Authors proposed a new technique named as Clustering Based Opportunistic Traffic Offloading (CBOT), for Device-to-device communication."

"To improve the energy consumption and throughput of the system, authors have proposed cluster formation, cluster head selection, and rotation techniques."

"The simulation results show that CBOT improves network lifetime by achieving high energy efficiency and increased system performance by achieving higher throughput."

"Further, the performance of CBOT tested under opportunistic networks scenario and simulation results demonstrate that the proposed approach improves energy consumption and throughput in a better way than the existing approaches."

Introduction

"Cellular Device-to-device communication can be defined as direct communication between the devices i.e., two users having mobile devices can communicate without traversing or aid of the base station and underlying core network."

"Device-to-device communication is a promising technique that has been considered to provide a feasible solution to deploy the cellular network in remote areas, public safety applications, or in the areas where network infrastructure is damaged."

"The outband D2D communication uses an unlicensed spectrum, which requires an added interface in common cases wireless technologies like Wi-Fi, Bluetooth, ZigBee, Wi-Gig, ZigBee are used."

"Authors proposed an opportunistic traffic offloading scheme for Device-to-device communication based on clustering (CBOT)."

"Authors proposed CBOT, a clustering based opportunistic traffic offloading scheme for Device-to-device communication, which opportunistically uses LTE and Wi-Fi interfaces for communication."

Related Work

"Asadi and others [14], Liu and others [15], Gandotra and Jha [16] and Hematian and others [17] present surveys on Device-to-device communication in LTE and cellular networks."

"The proposed MCN architecture for wireless communications extends the services of the conventional one-hop cellular network by incorporating the flexibility of ad-hoc networks and support multi-hop relays."

"Choi and others [18] proposed a code-based discovery protocol for device-to-device (D2D) communications."

"Lei and others [19] present the study of direct communications between UEs in the LTE-A cellular networks."

"Asadi and Mancuso [20] proposed a protocol for D2D communication that used LTE and Wi-Fi direct interface."

"Arain and others [21] proposed CEECP, Clustering-based Energy Efficient, and Communication Protocol (CEECP) for road networks."

"Cao and others [22] proposed a cooperative device-to-device communication framework that introduces a cooperative relay technique for conventional underlay/overlay D2D communication."

"In Garrocho and others [23], the authors proposed a Wi-Fi middleware to support D2D communication in infrastructure mode."

Proposed Methodology

"If the NDREP sender is a cluster head, then UE starts pairing with the cluster and if NDREP is received from another UE, then a new cluster is formed."

"To elect the new CH, UEs share their SIGN value as defined in [2] which is calculated based on the signal strength and residual energy of UEs."

"The UE having a lower SIGN value will be the new CH."

"The new CH will broadcast information to the UEs of the cluster."

"After successful completion, of the "Attach" process, CH sends S-TMSI of its cluster UEs to the eNB."

"Upon receiving the S-TMSI of its cluster UEs, eNB allocates cluster id to CH."

"Upon receiving the NDREP the UE will acknowledge CH with its S-TMSI."

"CH broadcasts new UE information among other cluster members."

"In reply, each UE sends their SIGN value to the current CH."

Energy Consumption in CBOT

"In the proposed model, the UE will be in one of the four following modes: (1) cluster head (i.e. in LTE mode) in receiving or transmitting mode; (2) cluster head in idle mode; (3) cluster member (i.e. in Wi-Fi mode) in receiving or transmitting mode; (4) cluster member in idle mode."

"This model is extended by adding the probability of UE to be in active or idle mode and also considered the energy consumption in packet receiving."

"Energy consumption in Wi-Fi is analyzed in Garcia-Saavedra and others [24], which states that Wi-Fi requires power in packet transmission as well as in packet processing."

"The total energy consumed by UE_i can be obtained by adding the energy consumed in LTE and Wi-Fi modes by UE_i."

Evaluation of CBOT

"CBOT forms more stable clusters as compare to the ECO-M. Further, in ECO-M, clusters may need to reforms if the battery of CH is dead as there is no mechanism is defined for cluster head rotation."

"It can be observed that CBOT significantly achieves higher average throughput than the ECO-M approach, due to much efficient use of the resources."

"CBOT achieved better throughput than ECO-M because it always chooses the cluster head having the best channel quality determined the SIGN value."

"It can be observed that in CBOT, the average remaining energy per UE is 36% as compared to the 14% in the case of the ECO-M approach."

"CBOT produces low latency and consumes less energy compared with ECO-M. This is due to fact that CBOT selects a cluster head based upon SIGN value that consists of two parameters: signal strength and remaining energy of UEs."

"In CBOT devices consume less energy as compared to ECO-M."

Conclusion

"Authors have proposed a cluster-based opportunistic traffic offloading scheme CBOT for device-to-device communication, that divided the network into heterogeneous clusters."

"A cluster head is chosen that provides the inter-cluster communication by using LTE connection via eNB and intra-cluster communication through Wi-Fi."

"To reduce energy consumption, the LTE traffic is diverted to Wi-Fi for intra-cluster communication."

"As the future scope, the implementation and testing of proposed CBOT methodology under multiple cell environment can be performed."

Performance Evaluation of an Effective Mobility Model for D2D Communications [25]

This is a machine-generated summary of:

Ouali, Kaouthar; Kassar, Meriem; Nguyen, Thi Mai Trang; Sethom, Kaouthar; Kervella, Brigitte: Performance Evaluation of an Effective Mobility Model for D2D Communications [25]

Published in: Wireless Personal Communications (2020)

Link to original: https://doi.org/10.1007/s11277-020-07816-y

If you want to cite the papers, please refer to the original.

For technical reasons we could not place the page where the original quote is coming from.

Abstract-Summary

"Unpredictable device mobility may influence the service continuity of a D2D communication."

"The effects of the mobility model on D2D communication performances must be fully studied."

"We adapt the RPGM (Reference Point Group Mobility), a group mobility model, to the D2D communication."

"It is suitable for modelling the D2D mobility and manages the relationship among mobile devices."

Introduction

"The quality of service and connection between devices strongly depends on the relative position of D2D mobile devices which makes the accuracy of the mobility model for cellular network more critical."

"The effects of the mobility model on D2D communication performances must be fully studied especially with mobility and context-aware D2D communications where the candidates exploit context information about the traffic, users velocity and location to decide whether their traffic load can be transferred to D2D tier and which D2D users should be followed."

"D2D is opportunistic in nature; therefore our work insights on D2D mobiles having the same target (i.e. sharing resources to compute a common task in a context of a cloudlet or a smartphone can be set up as a personal hotspot or relay so that nearby devices in poor coverage can be served through the mobile's cellular connection) or on a mobile terminal fetching for a D2D partner to form a D2D group useful for content sharing."

Selected Mobility Models

"To comprehensively evaluate the effects of realistic mobility in a D2D study case and to validate the accuracy of our model, we consider both group and individual mobility models."

"These models have been carefully selected to capture mobility as appropriate for a D2D communication."

"Nabi and others [26] presented MoBAN, a mobility model for Wireless Body Area Networks (WBAN)."

"We believe that this model could be considered an attractive candidate for characterizing the effects of mobility in future D2D networks."

"The main idea of the Community Mobility Model (CMM) is that the users with strong social network ties in the real world tend to move in the same geographical space."

"Based on these social relationships, this model can be used to model D2D mobility patterns with the exact application scenario."

"We present in this paper a comprehensive configurable mobility model based on RPGM for evaluating D2D communications."

Adopted Mobility Model

"The first mobile that moves is selected as a leader and the other D2D candidate will be considered as a D2D follower."

"Every D2D candidate will try to maintain the connectivity with the reference node (i.e., leader) in the network during a certain duration of time."

"All D2D members will freely roam but within the safe zone of the reference node."

"The value of speed and traveled distance of the D2D leader follow the RWP model."

"The movement of the D2D follower is fully characterized by his leader, which makes the model more convenient to use."

"Ping-pong effect is defined as the unnecessary handover to the neighbouring eNodeB while D2D mobile node is returning to the source after a very short span of time."

"For the D2D follower, we suppose that the follower is behind his leader."

Performance Evaluation

"The proposed mobility model enables the D2D candidates to stay in proximity of each other so the D2D communication is maintained for a longer time compared to the MoBAN, RPGM, CMM, RWP, BM and GM mobility models so that better service continuity is performed."

"We continue by investigating how the group mobility models metrics translate into D2D communication performance indicators."

"When D2D mobile terminals are in proximity to each other, they undergo a joint handover."

"This is obvious since the D2D mobiles are both in the overlapping zone and able to trigger a handover to the neighboring cell."

"The second user has to wait to trigger a handover to the same target eNodeB. This waiting time is the ones taken to move to the overlapping area, different from one mobility model to another."

Conclusion

"We have investigated on how mobility impacts the D2D communication performances in LTE-A based network."

"We have deployed a simple yet efficient mobility model based on RPGM model to capture the D2D users mobility patterns."

"Comparing to various group mobility models, our proposed model has resolved the trade-off between delay-sensitive and delay-tolerant applications and has performed well in terms of both D2D mode ratio and handover latency."

"No real life D2D mobility traces for realistic application scenarios are available for research purposes."

Uplink Resource Sharing and Power Management Scheme for an Underlay D2D Communication [27]

This is a machine-generated summary of:

Mohamad, Noor Mohammed Vali; Nagarajan, Rajkumar; Muthukaruppan, Lakshmanan; Subramaniyam, Hariharan; Sendrayan, Nandakumar; Thangappa, Velmurugan; Manickavelu, Palanivelan: Uplink Resource Sharing and Power Management Scheme for an Underlay D2D Communication [27]

Published in: Wireless Personal Communications (2019)

Link to original: https://doi.org/10.1007/s11277-019-06747-7

If you want to cite the papers, please refer to the original.

For technical reasons we could not place the page where the original quote is coming from.

Abstract-Summary

"The interference to the cellular user equipment (CUE), due to D2D pairs is reduced with proper strategies on resource allocation and transmit power assignment to the D2D pairs."

"We propose a distance-based interference reduction and power management scheme to reduce the interference on CUEs and evolved NodeB (eNB) and thus improve the sum rate of D2D pairs."

"The power allocation is based on the D2D pairs distance to the CUEs and eNB, which is estimated based on Haversine formula."

"The system sum rate, total transmit power of D2D pairs and the battery life of D2D user equipment are analyzed and validated through simulation."

Introduction

"In [28], resource allocation has been done in D2D pairs using mixed integer programming formation and Greedy strategy to mitigate the interference and then power distribution algorithm is employed."

"Resources for the D2D pairs were allocated based on distance [29], CUE uplink transmit power was allocated based on fixed signal to noise ratio (SNR) and eNB selects the CUE which has the negligible interference to the D2D pairs."

"An iterative based resource allocation mechanism was discussed to improve the sum rate of D2D pairs."

"The second algorithm is power management scheme where the power is allocated to D2D pairs using the estimated distance from first algorithm."

"The system sum rate, total transmit power of D2D pairs and battery life of DUEs is analysed based on the proposed algorithms."

System Model

"That CUEs communicate through eNB where D2D pairs work over an undelay network."

"ENB and CUEs coordinate with known Channel Quality Indicator (CQI) and allocate the channels where the interference in the network is minimal."

"The channel allocated by the eNB to CUEs are orthogonal which are shared by any D2D pair within the cell."

"The primary motivation is to reduce the interference from CUEs to D2D pairs in order to provide reliable communication to D2D users."

Analytical Modelling of Proposed Algorithm

"The high transmission power of eNB causes significant interference to D2D receivers so it is difficult to maintain reliable D2D communication."

"The eNB allocates the resources to D2D pairs based on estimated distance."

"The urban macro cell propagation is considered for estimating the path loss and it is given by [30],where d is the distance between two UEs or between UE and eNB in km."

"In the uplink phase of an underlay cellular network, D2D pairs exhibits interference to the eNB, as it shares the same cellular resource."

"Based on the distance estimated from Algorithm I, the power allocation to D2D pairs is done."

"The expected battery life time of jth DUE having the transmit power of P_j and operating voltage V_o is given as Transmit power of UE is controlled to balance the battery life time and the rate of transmission."

Simulation Results

"The improvement in the sum rate of the proposed scheme compared to the Greedy Heuristic algorithm as a function of distance is summarized as: for maximum D2D distance per cell radius of 0.1, the improvement in sum rate is 10%, 9.5% and 8.6% approximately for CUE transmit power of 17 dBm, 20 dBm and 23 dBm."

"The average battery life time of DUE in hours (h) is obtained as, for maximum number of D2D pairs in the network is 5, the average battery life time of DUE is 204 h, 201 h, 197 h for CUE transmit power of 17 dBm, 20 dBm and 23 dBm respectively."

"If maximum number of D2D pairs in the network is 30, the average battery life time of DUE is 197 h, 192.5 h, 187.5 h for CUE transmit power of 17 dBm, 20 dBm and 23 dBm respectively."

Conclusion

"The results of the proposed scheme show significant improvement in sum rate, transmit power of D2D pairs when compared to the conventional Greedy Heuristic algorithm, in addition the battery life time of DUE is analyzed."

"The system sum rate of D2D pairs is analyzed as a function of distance between D2D to CUEs and eNB along with limited CUE transmit power."

"For the same transmit power of D2D pairs, the proposed algorithm witnessed better sum rate with increase in CUE transmit power and maximum D2D distance per cell radius."

"Greedy Heuristic and proposed algorithms almost have same D2D transmit power for different transmit power of CUE, the proposed algorithm outperforms Greedy Heuristic algorithm in terms of sum rate against the distance and number of D2D pairs."

A Comparative Study of Resource Allocation Schemes for D2D Networks Underlay Cellular Networks [31]

This is a machine-generated summary of:

Dhilipkumar, S.; Arunachalaperumal, C.; Thanigaivelu, K.: A Comparative Study of Resource Allocation Schemes for D2D Networks Underlay Cellular Networks [31]

Published in: Wireless Personal Communications (2019)

Link to original: https://doi.org/10.1007/s11277-019-06204-5

If you want to cite the papers, please refer to the original.

For technical reasons we could not place the page where the original quote is coming from.

Abstract-Summary

"Device-to-Device (D2D) communication is a promising take over in cellular communication."

"The concept of two tier network can be evolved to enlighten the cellular communication."

"Making each cellular user as transmission relays it would form a massive D2D network resulting in increased offloading of cellular traffic."

"The D2D scheme have practical challenges which needs to be met in order to keep its place primary in the next generation wireless communication."

"Various resource allocation scheme for D2D networks in cellular network are discussed."

"D2D assisted Cognitive radio networks [32] can be developed to utilise the spectrum opportunistically for next generation communication systems."

Introduction

"D2D is the direct communication between two cellular users with or without the control of BS (base station)/eNodeB (evolvedNodeB)."

"D2D communication was basically categorized into [14] Inband D2D and Outband D2D. Inband D2D is prominent for cellular communication because which utilize the licensed spectrum."

"Inband D2D communication is categorized into Underlay approach and Overlay approach based on spectrum sharing."

"In Underlay approach the licensed spectrum is shared between CU (cellular users) and D2D users and in the latter there is dedicated spectrum allotted for D2D and CU communication."

"In controlled mode eNodeB controls the D2D communication and in autonomous mode without the control of eNodeB, D2D users communicate each other."

"As the evolution of cellular communication is in rapid progress with that scarcity of spectrum resource is also growing rapidly."

"Cisco predicts the global mobile data traffic will increase up to 49 exabytes per month within next 5 years."

Mixed Integer Non-linear Programming Method

"Being the motivation for rich multimedia communication, the D2D have to face the interference problem during both uplink and downlink when resources are not allocated properly."

"The LTE system with RB (resource block with 0.5 ms) occupying 1 time slot uses OFDMA (orthogonal frequency division multiple access) for downlink and SC-FDMA (single carrier-frequency division multiple access) for uplink."

"The resource blocks are assigned for D2D communication for optimum throughput without affecting the existing primary network."

"The alternative resource allocation is formulated by considering the Channel Quality Indicator (CQI)."

"The eNodeB identify the UEs with higher channel quality indication through the channel gain information and share the resource blocks for D2D system."

"The interference mitigation is better in Heuristic resource allocation on comparison with the random allocation."

Distributed Resource Allocation Algorithm

"In [33] multi-user and multi relay D2D network is considered and a distributed resource allocation was formulated."

"The approach is based on the message passing technique in which UEs transmit and receive to/from relays in iterative manner."

"The resource allocation problem is mapped to the max-sum message passing (MP) problem and by which the messages are passed till a global allocation is obtained."

"As the distance increases the average achievable data rate increases since the distributed resource allocation ensures data rate throughput by using relays."

Downlink Resource Reuse Algorithm

"CUs have primary priority and their QoS should be protected but in the previous works [14, 34–36] the resource of single CU can be reused at one time and near-far effect causes degradation in downlink QoS. To overcome these issues a resource reuse design is formulated as generic utility maximization problem."

"To evaluate a system of hybrid (having both D2D and Device to CU communication) single-cell network with C orthogonal downlink CUs and L D2D Links are considered."

"It is also evident that the proposed algorithm has nearly 50% to 100% better sum data-rate."

"When the number of CUs increases, the sum rate of D2D link increases more than the previous case since, more CUs in the system would have more resources to reuse for D2D links; but it also comes with the constraint of limited power."

Game Theoretic Based Resource Allocation

"Game theory can be utilised in communication networks where resources are in demand for the users."

"In Song and others [37], evaluated the game theoretic models for the radio resource allocation problems in D2D communication."

"Unlike in [38] the multiple D2D network is termed as D2D local area network in [37] and resource allocation problem is treated as games where the devices compete with each other or cooperate with each other to resource reuse."

"The cross-tier interference (between D2D and CU) and co-tier interference (D2D user with other D2D user) and game theoretic approach is used to reduce both the types of interferences."

"Different models using game theory are formulated and evaluated such as Non co-operative control game [39], stackelberg-Type game model [40], Auction based model [41] for different applications such as power control, local radio resource allocation and global radio resource allocation."

"Pricing is proposed for the non-cooperative game theory for optimizing rate and power in [42]."

Interference Aware Graph Based Resource Allocation

"The resource sharing problem is modelled as interference aware graph based method."

"This method proves to be the best solution for optimal resource allocation with low computational complexity."

"The worst case is nothing but considering all the communication links experiencing all the virtual clusters and the complexity is calculated using the following equation where M and N represents the number of CUEs and D2D UEs."

"It matches only approachably with the optimal resource sharing scheme, the interference aware method has low complexity than the others."

Conclusion

"For such an inevitable requirement D2D communication technology is considered as the most reliable scheme."

"On analysing the various resource allocation schemes other than conventional cellular network communication the D2D systems provide better throughput and average data sum-rate."

"Distributed resource allocation method provides a normalized network throughput of 0.7 at the same time CU-D2D Heuristic algorithm provides better normalized throughput than the former as 0.8 when the number of CU-D2D links are 20."

"As we look into the data sum-rate, the resource reuse algorithm provides better rate than the random and heuristic algorithms."

Investigation on Multi-user Scheduling Algorithms Over Non-identical Channel Distribution and LTE Networks [43]

This is a machine-generated summary of:

Krithiga, S.; Bhaskar, Vidhyacharan; Malarvizhi, S.: Investigation on Multi-user Scheduling Algorithms Over Non-identical Channel Distribution and LTE Networks [43]

Published in: Wireless Personal Communications (2020)

Link to original: https://doi.org/10.1007/s11277-020-07782-5

If you want to cite the papers, please refer to the original.

For technical reasons we could not place the page where the original quote is coming from.

Abstract-Summary

"We investigate the performance of various multi-user scheduling algorithms over independent and non-identically distributed generalized fading channel and 3GPP long term evolution (LTE) networks."

"Novel expressions for average BER, average channel capacity and fairness among users are derived for various scenarios such as (1) independent identical fading parameter, independent non identical average SNR channel conditions and (2) both fading parameter and average SNR are independent and non-identically distributed under absolute SNR (AS) scheduling and threshold SNR scheduling."

"More over the performance of threshold SNR scheduling is compared with AS scheduling and other existing scheduling techniques to identify a suitable scheduling algorithm for LTE networks."

Introduction

"The opportunistic schedulers have greedy nature to allocate the resources to the users having better channel quality compared to the users with poor channel conditions."

"It becomes very crucial to implement multi-user scheduler in networks that aims at best performance and guarantees required level of service to all users."

"In Multi-User (MU) MIMO, a scheduler can be used to exploit both MU and antenna diversities for enhanced throughput, higher stability, lower delay or improved fairness among users."

"The authors in [44] proposed a novel joint mode selection and user scheduling algorithm to enhance network throughput for a cellular network with Full-Duplex (FD) Base Stations (BSs)."

"This motivates us to propose threshold SNR scheduling technique which improves system performance with better fairness among users while maintaining better performance in proposed system."

System Model

"We employ Absolute SNR (AS) scheduling in which the user with the maximum SNR is considered as the best user (b) and is allowed to transmit information."

"Threshold SNR algorithm allow the users having SNR greater than preselected threshold SNR to be transmitted."

"In case all the users have SNR less than the preselected threshold SNR, the user having the maximum SNR is allowed to transmit."

Numerical Results and Discussions

"It has been made clear that fairness among users is lower in i.ni.d, i.ni.d conditions in AS scheduling because of the reason that users having the maximum absolute SNR value will be allowed to transmit, whereas users with poor channel conditions will not have the priority."

"The threshold SNR scheduling provides better fairness among users compared to the i.ni.d case in AS scheduling due to the fact that not only the user with the maximum SNR, but all users having SNRs greater than the threshold SNR are allowed to transmit the signal."

"Further, it is noted that at BER = 10^{-2}, K = 500 users, AS scheduling technique has SNR gain of 1 and 2 dB compared to threshold SNR, scheduling method in [45] respectively."

"This result indicates that threshold SNR algorithm has better fairness among users compared to scheduling technique [46] and absolute SNR scheduling algorithm."

Conclusions

"We analyzed the average BER and average channel capacity performance of MIMO–OSTBC systems over α-μ fading with different scenarios, such as independent identically distributed and non-identically distributed fading parameters and average SNRs of the users under Absolute SNR scheduling."

"Simulation results showing the BER comparison of AS scheduling and Threshold SNR scheduling is obtained for a LTE system of 30 users with different SNR levels."

"Scheduling technique proposed in [45] provides good fairness among users but BER is high, thereby, system performance is degraded."

"Threshold SNR scheduling achieves better fairness among users compared to scheduling technique [46] with slight increase in BER compared to AS scheduling technique and reduced BER compared to [45]."

"It is made clear that threshold SNR scheduling can be a better choice with reduced BER while maintaining fairness among the users at the same time."

Adaptive Spectrum Aggregation Regimen for Downlink NR-gNodeB and Device to Device Systems in 5G [47]

This is a machine-generated summary of:

Priyadharsini, Naveena A.; Tamil Selvi, S.: Adaptive Spectrum Aggregation Regimen for Downlink NR-gNodeB and Device to Device Systems in 5G [47]

Published in: Wireless Personal Communications (2020)

Link to original: https://doi.org/10.1007/s11277-020-07944-5

If you want to cite the papers, please refer to the original.

For technical reasons we could not place the page where the original quote is coming from.

Abstract-Summary

"Device to device communication (D2D) and internet of things are anticipated to rule the globe in the upcoming days."

"The ever increasing demand for data traffic and throughput pose a unique venture in fifth generation (5G) industry, including limited spectrum and mass device connectivity."

"The proposed system signifies the specifications putforth by third generation partnership project for D2D communication underlying cellular network."

"Since the users of D2D and new radio g-NodeB apportion same spectrum, interference would be the resultant."

"To quell such interference effects and also to maintain high capacity and throughput, here two mechanisms, namely spectrum aggregation and scheduling are emphasized."

Introduction

"A days handheld devices are enriched with artificial intelligence, so they not only used for personal communication but also for multitasking such as live video streaming, home automation, health monitoring, on line computations, IoT etc Also the 5G communication system encourages device to device communication and smart applications."

"The upcoming 5G system relies on D2D communication and IoT with artificial intelligence and avoids the human interactions."

"To fuel the ever increasing spectral demand in 5G cellular system, millimeter wave (mm wave) spectrum has opted."

"It has benefits like, ability to penetrate materials like plastic, wall, etc. The NR is the global standard, proposed by 3GPP for cellular communication in 5G systems."

"During the design of 5G system, spectrum aggregation and scheduling of radio resources are the cutting edges to overcome such inconsolable effects."

Related Works

"The requirements are expected as more than 1 Gbps data rate in mobility, higher spectral efficiency, Wearable devices with artificial intelligence, low power battery consumption, World Wide Wireless Web (WWWW) access etc. In [48], they proposed a two tier cellular network."

"They concluded that this setup could be suitable for indoor communication system with high data rate and low latency application. Ibraheem et al. [49] explored two mechanisms such as Spectrum Aggregation (SA) and adaptive modulation coding (AMC) to enhance spectral efficiency in LTE-A network."

"They concluded that, the developed scheme protects the WiFi users very well than the conventional methods like listen before talk.Panagiotis et al. [50] focused on the cognitive radio network with the spectrum aggregation mechanisms."

"Based on the position of the nodes and the relaying protocols, the power was allocated to each component carriers and the best relay was selected. Li et al. [51] proposed an excellent resource allocation and power allocation policies for uplink D2D underlying 5G cellular networks."

Proposed System Architecture

"5G is formulated to operate in various platforms like D2D, IoT, M2M communication."

"In 5G, various applications operate in different frequency bands simultaneously."

"The ultimate objective is to establish uninterrupted reliable communication among the users."

"Further two major mechanisms such as spectrum aggregation and scheduling of radio resources are carried out for the improvement of data throughput, capacity, spectral efficiency and interference mitigation of the network."

Design Methodology

"All these induce the chances for interferences like co-channel interferences, Inter Symbol Interference (ISI), Inter Carrier Interference (ICI), Inter cell interferences and adjacent cell interferences."

"Since the system is working in the noisy environment, the received signal consists of both valid information and noise as well as interference components."

"The received signal is given as, where H denotes the fading channel response."

"Each user is capable to receive signals from multiple component carriers."

"Quality of all available radio channels are manipulated based on their SINR, data rate and average throughput on each resource block of all the component carriers."

"Average throughput of each user on every resource block within the past window of size τ in every component carrier is expressed as, These steps are followed iteratively until every user is prioritized for assignment of radio resources."

"From the throughput values, the fairness of the user is manipulated."

Experimental Results and Discussion

"The path loss parameter is derived with respect to attenuation, base station density and transmission distance under Rayleigh fading channel condition."

"System throughput depends on base station density and SINR."

"The GFDM signal is transmitted through the channel."

"The overall throughput depends on base station density and SINR."

"The SINR depends on the channel gain which depends on fading effects."

"This scheduler picks the channel with peak SINR and allocates it to the highest priority users."

"The experimental results clearly depicts that spectrum aggregated system has improved throughput than the system without aggregation mechanisms."

"The inclusive spectral efficiency of the network depends on its SINR values."

"The over all result shows that, 4×4 MIMO configuration declares maximum efficiency."

"The simulation result declares that the spectrum aggregated system operating in mm wave frequency band results with peak data rate."

Conclusion

"The proposed system investigates on D2D communication underlying cellular networks with spectrum aggregation in 5G system."

"Spectrum aggregation feature is embedded in NR technology to emphasis the capacity and system throughput."

"The spectrum aggregation system results with SER of 10^{-8} at 10 dB SINR."

"The overall data throughput of 98.97% is achieved for systems which support spectrum aggregation."

Small Cells Integration with the Macro-Cell Under LTE Cellular Networks and Potential Extension for 5G [52]

This is a machine-generated summary of:

Alsharif, Mohammed H.; Nordin, Rosdiadee; Shakir, Mohammed Mudhafer; Mohd Ramly, Athirah: Small Cells Integration with the Macro-Cell Under LTE Cellular Networks and Potential Extension for 5G [52]

Published in: Journal of Electrical Engineering & Technology (2019)

Link to original: https://doi.org/10.1007/s42835-019-00173-2

If you want to cite the papers, please refer to the original.

For technical reasons we could not place the page where the original quote is coming from.

Abstract-Summary

"Heterogeneous Networks (HetNet) requires the deployment of small cell networks that can co-exist with the existing macro-cell."

"This paper highlights three important aspects of the emerging small cell wireless networks."

"The third part of this study focuses on the future trends of small cell deployment in 5G and the critical technical preparation needs to be done to allow integration with the existing 4G network."

"The results show that the proposed integration of 5G small cells into existing macro 4G networks has improved both of the data rate (91.46%) and energy efficiency (98.66%)."

Introduction

"The main aim of the study is to propose a strategy as well as an analytical investigation of integrated an overlay small cells with the underlay macro-cell in LTE network."

"To propose a strategy to deploy an overlay small cells integrated with the underlay macro-cell in LTE network in high demand outdoor urban environments in order to a two different network objectives: (1) to provide a full radio coverage area with high QoS besides a high speed data rate, and (2) enhance energy efficiency."

"2 is that in this paper, it mainly introduces the comprehensiveness of the paper where it investigates and design technically on the deployment strategy LTE small cell network where the main motivations are to provide a full radio coverage area with high QoS besides a high-speed data rate and enhancement of energy efficiency."

Related Works

"Authors in [53] presented a fully distributed algorithm for sharing time access to small cell networks (SCNs) and multi-homing capabilities of macro cellular users to enhance and improvise the overall data rates of downlink and to fulfill the Quality-of-Service (QoS) throughput requirements."

"An extension of an urban ray-based model for the assessment of a 60 GHz outdoor small-cell network is proposed by the authors in [54]."

"The main focus of this study is to design a deployment strategy of an overlay small cells integrated with the underlay macro-cell in LTE network in high demand outdoor urban environments in order to a two different network objectives: (1) to provide a full radio coverage area with high QoS besides a high-speed data rate, and (2) enhance energy efficiency; where these two metrics are important performance indicators to assess users QoS and their impact on network performance and power consumption."

LTE Small Cell Network Architecture

"The successful deployment of small cell networks relies upon how a network operator can integrate small cell into the existing mobile access networks to provide seamless device-to-core network connectivity."

"The E-UTRAN consists of multiple evolved Node Bs (eNBs), which connect via the X2 interface to support handover with the EPC via the S1 interface for traffic and control purposes."

"In the context, the small cells that fulfill the following high-level criteria support the X2 interface, S1/X2 traffic ratio: (1) one macro cell overlaps several small cells in the same area; (2) as small cells, they are deployed and managed by operators; (3) they grant an open access to all users (of the same operator); (4) are characterized by a lower equipment and installation cost if compared to macro cells; and finally (5) mobility support (support of handover between macro and small cells or among small cells)."

System Model

"The 3GPP UMa-NLoS path loss model is expressed as a function that includes the frequency (fc) in GHz, macro-BS antenna height (hMBS) in meter, UE antenna height (hUE) in meter, average building height (hbl) in meter, street width (w) in meter, and the radius of the macro-cell (RMBS) in meter, as follows: UEs receive a signal from the SBS, so the path loss model of the SBS is considered as Line of Sight (LoS) fading channel."

"The path loss model is expressed as a function that includes the frequency (fc) in GHz, and the radius of the small cell (RSBS) in meter, as follows in the next formula, The cell coverage area in a cellular system is defined as the percentage of the area within a cell that has received a signal at a power above a given minimum, P_{min}."

Results and Discussion

"When 1/5 QPSK is used at the edge of the macro-cell (2 bits per symbol with coding rate 1/5), the data rate will be 6.72 Mbps for a single chain, and with 2 × 2 MIMO (2T, 2R), the data rate will be two times that of a single chain, i.e., 13.44 Mbps."

"While the MBS provides 13.44 Mbps under the same BW condition and number of antennas; which means that the performance of the network (data rate) has improved by 91.67% through deploying the small cell underlay macro-cell in LTE technology."

"The small cell provides a high SINR, due to the close distance between the BS and UE, which provides a high order of MCS to use with high data rate."

"High order modulations, such as 128-QAM and 256-QAM with small cell in 5G communication technology can achieve multi-gigabit data rates for mobile subscribers."

Conclusion

"Performance comparison of the deployment strategy of an overlay small cells at edge of the underlay macro-cell in LTE network shown that the small cell provides

161.28 Mbps data rate with 1.04 Mb/J energy efficiency at the cell edge with full radio coverage area and a strong radio signal, while macro-cell provides 13.44 Mbps with 13.93 kb/J at the cell edge under the same conditions of the BW and number of antennas."

"From the deployment of small cells underlay on the existing LTE macro-cell technology, the proposed small cells model has improved both of the data rate and energy efficiency by 91.46% and 98.66%, respectively."

"The 5G small cells will improve the data rate that expected to offer up to more than 20 Gbps with LOS, and energy efficiency up to 0.5 Gb/J."

Optimal Performance of Resource Allocation in LTE-A for Heterogeneous Cellular Network [55]

This is a machine-generated summary of:

Kukade, Shweta; Sutaone, Mukul; Patil, Rajendrakumar: Optimal performance of resource allocation in LTE-A for heterogeneous cellular network [55]

Published in: Wireless Networks (2021)

Link to original: https://doi.org/10.1007/s11276-021-02635-w

If you want to cite the papers, please refer to the original.

For technical reasons we could not place the page where the original quote is coming from.

Abstract-Summary

"We propose an iterative heuristic optimal resource allocation (HORA) algorithm and a chunk based resource block allocation (CRBA) scheduling algorithm to determine resource block (RB) allocation among users to satisfy the quality of service requirement."

"It performs RB and power allocation separately to reduce computational complexity."

"In the CRBA algorithm sets of RBs are allocated to groups of users keeping power constant to all users."

"User selection is performed based on channel conditions to improve throughput."

"RB usage and quality-of-service (QoS) constraint are considered for the scheduling algorithm."

"The HORA algorithm assigns most RBs to users who have high-value signal to noise ratio and continues the RB allocation until it meets the QoS criteria of all users in consideration of the threshold value of the power budget."

"Problems that arise during continuous resource allocation to the scheduled user are considered as APX-hard and NP-hard problems."

"An RB and power allocation optimization problem is formulated for the maximum data rate in the cellular network."

Introduction

"In order to achieve high spectral efficiency and throughput, LTE-A, adopts dynamically changing system resource parameters based on channel conditions, e.g., rate, power levels, transmit antenna selection, and modulation and coding schemes (MCS)."

"Another consideration is efficient data RB allocation based on the proposed optimized HORA and SRBA algorithms in time-varying channel conditions."

"During the allocation of resources, the following conditions can be considered [56]: Adjustment of transmit power and resource allocation concerning the channel link condition."

"A maximum data rate is decided with an adaptive modulation technique which depends on the quality of the link to ensure QoS. The hierarchical allocation method cannot optimize the user resource distribution equally for power and spectrum."

"The HORA algorithm decides RB allocation among the users along with the QoS requirement."

"RB usage, QoS and power allocation are the three—tuplet constraints considered for the allocation of RBs to the UE."

Related Work

"Efficient allocation of resources among the UEs under good channel conditions with a max–min and delay-fairness algorithm was presented in [57]."

"The author proposed the algorithm to serve the resources to these users to improve the throughput of the system."

"Dynamic user weight calculation was carried out in [58, 59] based on queuing delay, throughput, and resource allocation which was used by the utility function for RB allocations in the UL direction."

"Power allocation using a water-filling algorithm to groups of users was discussed in [60, 61] considering UE requirements."

"All of these works used different methods to allocate RBs to the UEs considering noise and diverse channel conditions."

"Our main contributions are as follows: To allocate RBs to the UEs efficiently in the UL direction, we proposed the HORA and CRBA algorithms."

LTE Physical Layer

"Depending on the size of the OFDM channel, it may have 6 RBs in 1.4 MHz and 100 RBs in a 20 MHz channel for transmission in a single time slot [62]."

"A resource element (RE) is a one subcarrier, for the duration of 1 OFDM symbol."

"Depending on the length of the CP, 6 or 7 OFDM symbols will fit into one slot."

"In the uplink, the physical uplink control channel (PUCCH) occupies one resource block at each end of the system bandwidth and does frequency hopping between the first and second slots within a subframe."

"The resource blocks between the PUCCH are used for the physical UL shared channel (PUSCH) and may do frequency hopping between slots [63]."

"The SRS occupies the last OFDM symbol in the second slot of a subframe."

System Model

"At the receiver the MMSE channel estimator is used considering its higher effectiveness in the ISI and noise reduction even though it is complex compared to the least-square (LS) channel estimators and zero-forcing (ZF) equalizer."

"The LS channel estimator amplifies noise at a low SNR, but it is minimally complex."

"The MMSE channel estimator has a high computational complexity but it provides a good compromise between improved noise and the elimination of ISI [64]."

"In the RB allocation, at most each RB is to be used at any instant of time between the user link and the BS and is considered and given by: The power level allocation on each RB has a discrete distribution."

"The consumed power at the BS can be represented by: The non-zero power level allocation on each RB at a given point of time is considered."

Proposed Algorithm

"In the HORA algorithm, we use a heuristic approach to decide the allocation of RBs amongst the users with the QoS requirement of every user."

"Users with the highest SNR are the allocated a large set of RBs and the highest power with good channel gain until the QoS requirements of the users are met."

"In CRBA, RBs and UEs are divided into different sets and then the RB allocation in the UL direction is performed to increase the throughput."

"Algorithm 2 shows the creation of the sets of UEs and RBs."

"The sets of UE, RB pair allocation are shown in Algorithm 3."

"The set of maximum RB groups are allocated with maximum SNR, UE group, and then the allocated RB group is subtracted from the available pool."

Link Adaption in LTE

"It changes system parameters based on actual channel quality by adapting to channel conditions."

"The DL radio channel quality is identified based on the CQI report."

"Based on the channel quality the MCS is selected to match the channel link and to ensure that the bit error rate (BER) will not exceed 10%."

"In the UL transmission, channel coding structure selection of the MCS process is similar to the DL and is under the control of the BS."

"In response to the CQI, the MCS and the channel coding rate are selected."

"A wideband CQI report recommends a single MCS value to the RBs available in the bandwidth, while a sub-band CQI report selects multiple modulation and coding scheme (MCS) values to different continuous RBs i.e., different adaption assigns different RBs in each subframe [63]."

Simulations and Results

"One of the benefits of the HORA and CRBA algorithms is the usage of the link adaption technique which increases throughput with an increasing number of users."

"In the HORA algorithm initially, the RBs are allocated to the UEs based on the channel condition."

"In the CRBA algorithm the highest chunks of RBs are allocated to the group of users whose channel condition is good."

"A fair allocation of RBs is possible in the CRBA algorithm even though the computational complexity is high."

"The largest chunk of RBs is allocated to the users who have the highest SNR but the algorithm complexity is high."

"In the proposed algorithm with good channel conditions the EVM is 0.1% and 0.015% in the CRBA and HORA algorithm respectively."

"In the HORA algorithm, the variable power as per the number of RBs for one user is allocated."

Conclusion

"We investigate the system throughput performance of optimal allocation of RBs to users using the proposed HORA and CRBA algorithms."

"In CRBA algorithm, a chunk of RBs is allocated to groups of users based on the channel quality."

"These algorithms can easily be applied to improve the network performance in the LTE uplink, since it considers all constraints in LTE-A. The optimal performance of both methods in terms of throughput and power efficiency is compared."

"It shows that RB allocation to users at its highest level can be possible with better channel quality."

"Simulation results exhibit that the proposed algorithm shows a better performance in terms of throughput."

"The throughput of the HORA algorithm is higher by 4.44% than the CRBA algorithm as the HORA algorithm is computationally more efficient compared to the CRBA."

Performance Comparison Between Different Tunneling Techniques Using Different Routing Protocols [65]

This is a machine-generated summary of:

Jain, Neha; Payal, Ashish: Performance Comparison Between Different Tunneling Techniques Using Different Routing Protocols [65]

Published in: Wireless Personal Communications (2021)

Link to original: https://doi.org/10.1007/s11277-021-09186-5

If you want to cite the papers, please refer to the original.

For technical reasons we could not place the page where the original quote is coming from.

Abstract-Summary

"Both IPv4 and IPv6 have different structures, and directly bringing IPv6 into the current IPv4 network does not work due to their compatibility issues."

"Dual-Stack, Tunneling, and Protocol Translator mechanisms are adopted to smooth the transition between IPv4 and IPv6 networks."

"The transitions allow us to communicate between both IPv4 and IPv6 networks together despite having their different structures."

"The proper transition mechanism has to be adopted for a complex network, and the in-depth study and analysis can help find the best transition technique."

"The 6to4 tunneling is configured for a real-time network using real-time applications—FTP and Email, and the performance of the transition between IPv4 and IPv6 addressing networks is evaluated."

"This novel work is helpful for researchers as it gives a detailed study to choose the appropriate tunneling technique with a mixture of proper routing protocols for efficient performance."

Introduction

"IPv6 also eliminates Network address translation (NAT), supports automatic autoconfiguration, and Internet Control Message Protocol version 6 (ICMPv6) for neighbor discovery [66]."

"To get a smooth communication between IPv6 and the existing IPv4 network, the transitions mechanisms are used to prevent this sudden change."

"The transition mechanisms allow us to deploy and adopt the IPv6 network for direct connectivity with the existing IPv4 network."

"The two commonly used methods in the transition process are encapsulating IPv6 addresses into IPv4 packets and sending them through the tunnel on the IPv4 network."

"It uses IPv4 addresses, which are no longer available and consumes more power and resources to handle two routing tables, thus increasing overall cost [67]. Tunneling—When similar IP networks need to communicate with each other, but a different IP network separates them, the virtual or logical path is established between them for a connection."

Tunneling

"The payload is transmitted and received at tunnel destination end-point, checked for protocol value, fragmented, and reassembled for the delivery of IPv6 packet to its final destination address [68]."

"Some advantages of tunneling are as follows—(1) Connects two separate networks/subnetworks (2) Allow multiprotocol network over a single protocol backbone network (3) Limited hop count network protocols are no longer a problem. (4) Allows communication between private IPv6 networks and public IPv4 networks."

"6to4—(Connection of IPv6 Domains via IPv4)—In this tunneling technique, IPv6 networks are connected through an intermediate IPv4 network using routers or tunnel end-points, and the packets are transferred between tunnel end-points."

"6over4—(Transmission of IPv6 over IPv4 domains without explicit tunnels)—This technique is used in multicast networks."

"ISATAP—(Transmission of IPv6 over IPv4 domains without explicit tunnels)—Inter-Site Automatic Tunnel Addressing Protocol—ISATAP is an automatic, point-to-point tunneling technique designed for Intrasite."

Network Analysis Methodology

"The network implementation requires proper planning and designing before its commencement."

"The analysis of the network is accomplished through seven stages in this paper."

"Each stage is vital for successful designing, simulation, result evaluation, and comparison."

Literature Review

"In 2015, A. Ahmed and others considered the IPv4, IPv6, and 6to4 tunneling techniques on a network, and ethernet delay, throughput, and packet loss parameters were evaluated."

"Year in 2016, Y. Sookun and V. Bassoon configured the dual-stack, automatic 6to4 tunneling, 6RD, ISATAP, and performance parameters evaluated were round trip time, throughput, packet loss, and CPU utilization."

"In 2018, K. El Khadiri and others configured the dual-stack, manual tunnel, 6to4 automatic tunnel, only IPv4 and IPv6 network for real-time applications—VoIP and video conferencing."

"In 2018, K. El Khadiri and others had used dual-stack, manual, automatic 6to4, native IPv4, and IPv6 tunneling techniques for video conferencing real-time applications on a network using RIP and RIPng routing protocol."

"That the dual-stack transition performed better than 6to4 tunnel brokers in all areas, while the 6to4 tunnel broker was better where there is no native IPv6 connectivity."

Simulation Network

"This paper proposes a network for 6to4 tunneling, and the network consists of Site A and Site B, which are connected in-between through the IPv4 backbone network."

"We have configured automatic and manual 6to4 tunneling networks for analysis."

"The 6to4 automatic and manual tunneling is configured individually, using RIP on the IPv4 network, RIPng on the IPv6 network, and similarly OSPF on the IPv4 network and OSPFv3 for IPv6 network."

"The automatic and manual tunnel is configured differently in the 6to4 network, and both are assigned with different interface addresses."

Results and Their Analysis

"For manual 6to4 RIP network, Router A and Router C have traffic dropped of 0.013619 packets/s and 0.030004 packets/s, while other objects have a drop of 0.001427 packets/s."

"For manual 6to4 OSPF network, Wkstn A, Email server, Router A, and Router C has a traffic drop of 0.013911 packets/s, 0.013244 packets/s, 0.018543 packets/s, and 0.034397 packets/s, and remaining have a drop of 0.001427 packets/s."

"In manual 6to4 for both RIP and OSPF network, Wkstn B and Router A has high traffic sent of 0.028577 packets/s and 0.035466 packets/s."

"In automatic 6to4 for both RIP and OSPF networks, it can be seen that Router A has higher traffic sent in both networks with a calculated value of 1.39003 packets/s and 1.580971 packets/s."

"In manual 6to4 for both RIP and OSPF network, it can be seen that Router C has higher traffic sent with the calculated value of 0.02929 packets/s and 0.035823 packets/s."

Conclusion and Future Work

"This paper has shown the automatic and manual 6to4 tunneling techniques using RIP/RIPng and OSPF/OSPFv3 routing protocol for two native end-to-end IPv6 networks connected through an intermediate IPv4 backbone network."

"The minimum IP traffic dropped obtained 0.389756 packets/s for automatic 6to4 RIP network, and the maximum IP traffic dropped obtained 0.456353 packets/s for manual 6to4 OSPF network."

"IPv6 Traffic Dropped—The manual 6to4 RIP and automatic 6to4 RIP network perform better than manual 6to4 OSPF and automatic 6to4 OSPF network."

"The minimum IPv6 traffic dropped obtained 0.049329 packets/s for manual 6to4 RIP network, and the maximum IPv6 traffic dropped obtained 0.128409 packets/s for automatic 6to4 OPSF network."

"OSPF Traffic Sent—The automatic 6to4 OSPF network sent OSPF traffic as 200.5539 bits/s, which is better than the manual 6to4 OSPF tunnel network."

Connecting Disjoint Nodes Through a UAV-Based Wireless Network for Bridging Communication Using IEEE 802.11 Protocols [69]

This is a machine-generated summary of:

Ullah, Hanif; Abu-Tair, Mamun; McClean, Sally; Nixon, Paddy; Parr, Gerard; Luo, Chunbo: Connecting Disjoint Nodes Through a UAV-Based Wireless Network for Bridging Communication Using IEEE 802.11 Protocols [69]

Published in: EURASIP Journal on Wireless Communications and Networking (2020)

Link to original: https://doi.org/10.1186/s13638-020-01727-z

If you want to cite the papers, please refer to the original.

For technical reasons we could not place the page where the original quote is coming from.

Abstract-Summary

"Cooperative aerial wireless networks composed of small unmanned aerial vehicles (UAVs) are easy and fast to deploy and provide on the fly communication facilities in situations where part of the communication infrastructure is destroyed and the survivors need to be rescued on emergency basis."

"We worked on such a cooperative aerial UAV-based wireless network to connect the two participating stations."

"The proposed method provides on the fly communication facilities to connect the two ground stations through a wireless access point (AP) mounted on a UAV using the IEEE 802.11a/b/g/n. We conducted our experiments both indoor and outdoor to investigate the performance of IEEE 802.11 protocol stack including a/b/g/n. We envisioned two different cases: line of sight (LoS) and non-line of sight (NLoS)."

"In LoS, we consider three different scenarios with respect to UAV altitude and performed the experiments at different altitudes to measure the performance and applicability of the proposed system in catastrophic situations and healthcare applications."

Introduction

"The main contribution of this paper is listed below: In this paper, we particularly consider a network where a single UAV will bridge communication between two ground stations through an AP mounted on a UAV using 802.11 a/b/g over 2.4 and 802.11n at both 2.4 and 5 GHz band."

"In LoS communication, we analyse three different scenarios with respect to UAV height from the ground stations: in scenario 1, we calculate the data rate, signal strength, and SNR between UAV and ground stations at 10 m height."

"In scenario 2, we calculate the same characteristics for the communication links between UAV and ground stations at a height of 15 m, while in scenario 3, we revise the same experiment at a height of 20 m to analyse the same performance metrics."

Related Work

"The authors mainly concentrated on providing a theoretical overview of the UAV-coverage area in an outdoor environment and to experimentally check the performance evaluation of the configured board both in lab as well as in real aerial deployment in order to study both the infrastructure mode and ad hoc operational mode of the IEEE 802.11."

"The authors concluded their experiments with respect to different antenna orientations and summarised how poor the results could be in terms of throughput and RSS if the right antenna orientation is not deployed on the UAV [70]."

"The authors addressed only the issue of UAV-to-UAV communication, while the communication between UAV and ground nodes was not considered which may prevent the use of the tool in real-time applications."

"In [71], the authors performed real-time experiments to exemplify air-to-ground wireless channels between UAV and ground users over a range of frequencies including 900 and 1800 MHz (cellular), and 5 GHz (WiFi) with respect to LoS and NLoS scenarios."

Experimental Setup

"Experimentation is carried out in an outdoor environment with a single UAV acting as an AP for communication bridge between two ground stations operating as client and server, respectively."

"Our testbed consists of two ground stations, i.e. a client machine and a server machine, and both of these machines are connected to each other through an AP that is mounted on a UAV."

"The experiments are performed at 10, 15, and 20 m of UAV altitude from the ground stations using the IEEE 802.11 protocol stack including a/b/g/n at 2.4 GHz and 5 GHz(n) band in an outdoor environment for LoS scenario, while at a 5 m height with the same setup in an indoor environment for NLoS scenario."

"With its powerful dedicated WiFi signal carried by the 3DR link, it provides connectivity between UAV and solo App to exchange real-time data and videos between UAV, controller, and other ground stations [72]."

Results and Discussion

"The results are generated in terms of graphs for different metrics such as data rate, signal strength, and SNR at 10, 15, and 20 m of UAV altitude and at a distance of $\approx$15 m between the participating ground stations in an outdoor environment for LoS scenario, while at a 5 m height with the same distance between the stations in an indoor environment for NLoS scenario."

"For 11n at 2.4 GHz, the SNR is quite high at both 10 and 15 m altitude on both client and server machines, but at 20 m, the SNR is quite low and ranges from −98 to ∼−87 dBm on both ground stations, while the SNR for 802.11n at 5 GHz remains almost the same in all three cases for both client and server."

Conclusion and Future Work

"We have tested IEEE 802.11a/b/g at 2.4 GHz and IEEE 802.11n at both 2.4 and 5 GHz bands using a 20 MHz channel in both indoor (NLoS) and outdoor (LoS) environment to check the performance of communication links between UAV and ground stations connected through an air-lifted AP on a UAV in terms of data rate, SNR, and signal strength for both scenarios."

"Based on the SNR, IEEE 802.11b performs slightly better than the others followed by 802.11n at 2.4 GHz (for LoS scenario), while 802.11n at both 2.4 and 5 GHz performs well as compared to others in the case of NLoS. Moreover, in terms of signal strength, again 802.11b and 802.11n at both 2.4 and 5 GHz are slightly better than the other IEEE standards for both LoS and NLoS scenarios respectively."

Optimal Multipath Routing for Video Transmission in VANETs [73]

This is a machine-generated summary of:

More, Shivaprasad; Naik, Udaykumar: Optimal Multipath Routing for Video Transmission in VANETs [73]

Published in: Wireless Personal Communications (2020)

Link to original: https://doi.org/10.1007/s11277-020-07740-1

If you want to cite the papers, please refer to the original.

For technical reasons we could not place the page where the original quote is coming from.

Abstract-Summary

"A new algorithm, known as Moth Whale Optimization Algorithm (MWOA) is proposed in this paper to determine the optimal multipath for the transmission of the videos from one vehicle to other in the VANET network."

"The proposed algorithm is developed with the integration of Moth Search (MS) algorithm and Whale Optimization Algorithm (WOA)."

"The performance of the proposed Moth Whale Optimization Algorithm is estimated with the metrics, such as Packet delivery Ratio (PDR), Packet end-to-end delay, and the throughput."

"The Moth-Whale Optimization Algorithm produces the minimum end-to-end delay of 4.4427, and maximum Packet delivery ratio and throughput of 96.5135 and 94.2429, respectively, which shows the superiority of the method in effective video transmission."

Introduction

"Each category comprises of various routing protocols, and most of the precious research works in VANET concentrates either on data dissemination or routing protocol, but both the process is needed to be carried out depending on the data packet type that are received from the vehicles [74]."

"The main goal of this paper is develop an efficient algorithm for the transmission of videos from one vehicle to other vehicles in the VANET in order to safeguard the vehicles from critical situations that occur in the highways."

"The major contribution of the paper is: Proposed Moth-Whale Optimization Algorithm (MWOA) for transmission of videos in VANET: The proposed Moth-Whale Optimization Algorithm is developed with the combination of the Moth

Search (MS) algorithm and Whale Optimization Algorithm (WOA) for the selection of multipath in order to transmit the videos from one vehicle to other in the VANET."

Motivation of the Research

"The literature survey is carried out with eight papers as follows: Xie and others [75] designed a Multi-channel Error Recovery Video Streaming (MERVS) as this method was capable of minimizing the video transmission error to control the problem of message loss."

"Zaidi and others [76] developed the Enhanced User Datagram Protocol (EUDP) that recovered the erroneous sub-packets for retransmission, improved the quality of video at the receiver side, and the transmission of video streaming over VANET, but the performance of this method was very less due to the presence of burst errors."

"The major challenges of the work are: The problem of reduced quality video streaming in VANET is usually due to the packet loss and delay in transmission."

"In case of adaptive geographic routing scheme posing dynamic nature of urban VANETs, life time of the routes discovered are very limited and are not capable of transferring the packets of a video file of larger size."

System Model for VANET

"VANET are treated as a special instance of mobile ad hoc network (MANET) for the provision of V2V and V2I communication."

"The major application of the VANETs is to offer secure, safe, and enjoyable ride for the passengers, who drive on the road in vehicles."

"Each vehicle is provided with a special electronic device to make the vehicle to be connected to the VANET."

"The VANET system usually transmits the emergency signal to other vehicle if there arise any accident or sudden breaking in any of the vehicles."

"The most important improvement in the VANET architecture is the development of the driver assistant system."

Proposed Method of Multipath Routing in VANET

"The primary goal of this paper is to develop a routing protocol for achieving multipath routing in VANETs to transfer videos with the selection of an optimal path by proposing an optimization algorithm."

"The VANET is simulated and the number of routes to transmit videos to destination from the source is determined using the adaptive geographic routing scheme [77]."

"After encoding the video through HEVC, the packet transmission is done through the multipath that is selected optimally for the transmission of videos to the destination from the source."

"The compressed video produces bit streams that are given as input to the routing process to obtain the optimal path for transferring the video."

"Both the algorithms are combined to produce the proposed MWOA to obtain the optimal solution in an enhanced manner."

"The proposed MWOA is used for selecting the optimal route to transmit the video between the vehicles in order to enable safe and effective transmission."

Results and Discussion

"The end-to-end delay of the existing methods, such as Adaptive routing protocol, MERVS, MFO, WOA, GA, and the proposed MWOA at the number of round 50 is 31.8888, 30.6177, 29.8105, 28.0086, 27.6991, and 26.7782, respectively."

"The end-to-end delay produced by the methods, such as Adaptive routing protocol, MERVS, MFO, WOA, GA, and the proposed MWOA is 11.6208, 10.3551, 9.8796, 8.8639, 7.5901, and 6.8774, respectively, which shows that the proposed method produces less delay in transmission as compared to that of the existing methods."

Conclusion

"The Moth-Whale Optimization Algorithm is proposed in this paper for the transmission of videos from one vehicle to the other, so as to provide secure travel on road."

"The proposed Moth-Whale Optimization Algorithm is the integration of the Moth Search algorithm and Whale Optimization Algorithm."

"The performance of the proposed Moth Whale Optimization Algorithm is estimated with the metrics, such as Packet delivery Ratio, end-to-end delay, and the throughput."

"The maximum PDR and throughput of 96.5135 and 94.2429, respectively are produced using the proposed method, and proves the capability of transmitting the videos in an enhanced manner."

Transmission of Audio Over LTE Packet Based Wireless Networks Using Wavelets [78]

This is a machine-generated summary of:

Singh, Sarvjit; Gupta, Amit; Sohal, J. S.: Transmission of Audio over LTE Packet Based Wireless Networks Using Wavelets [78]

Published in: Wireless Personal Communications (2020)

Link to original: https://doi.org/10.1007/s11277-020-07059-x

If you want to cite the papers, please refer to the original.

For technical reasons we could not place the page where the original quote is coming from.

Abstract-Summary

"The main goal is to analyse the performance of audio data for wireless VOIP (Voice Over IP) proposed model utilizing wavelets."

"This treatise represents a performance study of wireless VOIP for recorded audio signal using different wavelet families such as Coiflet, Daubechies family (Db2, Db4, Db6, Db8, and Db10), Haar and Symlet."

"The recorded audio data was decomposed up to 4-levels by means of wavelet transform."

"From simulation results it is clear that wavelets are more practical and dominant tool meant for analyzing audio signals, as it exhibits multi-resolution property with a considerable decrease in the time complexity for removing resolution problems."

"It was concluded that Coiflet performs best at first and third decomposition levels, while Haar and Db8 (Daubechies 8) performs the best among the other wavelets at second and fourth decomposition levels respectively."

Introduction

"Audio signal processing is the procedure used to analyse audio signals (such as music signals, voice signals and speech signals) using different processing techniques."

"Pulse Code Modulation does not require compression in audio devices (e.g., in CD players) having large bandwidth, however compression is necessary for digital audio signal that has to be transmitted over VOIP link [79]."

"The wavelet theory has been establishing vast field for generally stimulating developments in engineering fields not limited to signal processing, image processing, audio (such as music signals, voice signals and speech signals) processing, communication, and mathematics."

"The advantages of time frequency resolution and localization in wavelets makes it favourable tool for audio signals transmitted via wireless voice over IP applications [80–83]."

"The present objective of the research study is to analyse and retain the quality of audio signal during wireless VOIP communication by utilizing wavelet transform."

Literature Review

"Long time ago [84, 85] classified rate control schemes into three categories i.e. equation-based mechanisms, window-based mechanisms and additive increase, multiplicative decrease (AIMD) mechanisms."

"Equation-based rate control popularly known as rate control scheme for wired networks, also termed as TCP-Friendly Rate Control (TFRC)."

"To opt a consistent depiction of TCP throughput; in TFRC the transmitter makes use of an equation characterizing the allowed transmission rate of a TCP connection as a function of packet loss and Round Trip Time and packet loss rate, further its transmission rate is adjusted on the basis of measured parameters."

"The research work based on TCP protocols were investigated by [86] while the work utilizing UDP were studied by [87–90]. Further the audio data has been compressed by the means of specific codecs as described in [91–93], Author [94] presented IP based delays that includes RTT variation, packet loss ratio and difference between application-level RTT and network level RTT."

"Another author [95] also investigated the different audio codecs on the basis of packet loss."

Methodology

"The main purpose of this study is to analyse the performance of wireless VOIP intended for audio data by utilization of WT (Wavelet Transform)."

"For simulation, an audio data is under consideration for transmission by use of different levels of wavelet decomposition."

"This model can be used towards enhancement of QOS in terms of performance parameters evaluated for transmitted audio data via Voice over IP by use of different wavelet decomposition levels as stated above."

"In the proposed wavelet based wireless VOIP model, the most important scheme is to use different wavelets with different decomposition levels for the analysis of performance."

"After selection of wavelet and its decomposition level for an audio data, a process of encoding and packetization are done."

"During communication of audio data from transmitter to receiver there happens to be time imbalance for the received packets."

Analysis of Variance (ANOVA) Statistical Technique

"Using such a series of two samples test for hypothesis testing is invalid for the following reason."

"When we perform each two samples test at 5% (confidence) level of significance, there is 95% probability that we shall accept H_0 when two populations mean are equal."

"We do not prefer t test in hypothesis testing for more than two means; instead we select a procedure that tests the equivalence of mean."

"The appropriate procedure is a single factor analysis of variance, often abbreviated as ANOVA, namely the F-test."

"If there is significant difference, there is provision to go for post hoc-multiple comparison test."

"Post hoc-multiple comparison tests enable us to find the range in response and test the significance of difference between means in pair as well."

Feature Extraction for Performance Analysis

"The main objective of this wavelet based wireless VOIP model is to get hold of improved the QOS in the recovered audio data."

"This wavelet based VOIP QOS also depends on transmission capabilities i.e. delay and the packet loss."

"The RMSE is expressed as square root of MSE between the two audio signals i.e. original audio data and reconstructed audio data; Normalization of the RMSE enables the contrast among system datasets or system models with different scales."

Results and Discussions

"Prior to implement embedded hardware, data processing becomes significant to compute the results for attaining the correct and actual audio reconstructed signals (so as to analyse the QOS in terms of performance) for transmitted audio data via Voice over IP by use of different wavelet decomposition levels."

"A MATLAB® based simulation code and (GUI) was designed for analysis purpose for proposed model at different decomposition levels."

"The results realized from present proposed wavelet based wireless VOIP model assumes that a large value of precision is necessary for the analysis purpose so as to extract usable features from non-stationary audio signals."

"Analysis of proposed VOIP model, without using wavelet is done on the recorded audio data, whereas in second case distinct wavelets as intimated above were chosen to study the recorded audio data."

"To study the behaviour of wavelet based wireless VOIP model two network transmission parameters Delay and Packet Loss were evaluated."

Conclusion

"The objective of the present work was to investigate the audio data transmission through wavelet based wireless VOIP model at different decomposition levels."

"To achieve the objective different features were analysed to evaluate the quality of the reconstructed audio signal at the destination, Thereafter the level of accuracy of recorded data was examined by using one way ANOVA statistical technique."

"Since the calculated F-Value is 3.68 which is greater than the critical value ($f_c = 2.95$) at 0.05 level of significance, which shows that the samples of SNR produced statistical significant difference among the four groups (i.e. four decomposition levels)."

"By utilizing higher order mathematics additional wavelet families can be resembled for present work as well as for video data."

A Reliable Link-Adaptive Position-Based Routing Protocol for Flying Ad Hoc Network [96]

This is a machine-generated summary of:

Usman, Qamar; Chughtai, Omer; Nawaz, Nadia; Kaleem, Zeeshan; Khaliq, Kishwer Abdul; Nguyen, Long D.: A Reliable Link-Adaptive Position-Based Routing Protocol for Flying ad hoc Network [96]

Published in: Mobile Networks and Applications (2021)

Link to original: https://doi.org/10.1007/s11036-021-01758-w

If you want to cite the papers, please refer to the original.

For technical reasons we could not place the page where the original quote is coming from.

Abstract-Summary

"Conventional routing protocols of FANET generally use a minimum hop count criterion to find the best route between source and destination, which results in lower latency with the consideration that there is single source/destination network environment."

"In a network with multiple sources, the minimum hop count routing criterion along with the 1-Hop HELLO messages broadcasted by each node in the network may deteriorate the network performance in terms of high End-to-End (ETE) delay and decrease in the lifetime of the network."

"RLPR enhances the network performance by selecting those relay nodes which are in the forwarding zone and whose geographic movement is towards the destination."

"RLPR selects the next hop with better energy level and uses signal strength and relative speed of the nodes to achieve high connectivity-level."

"Based on the performance evaluation performed in the Network simulator (ns-2.35), it has been analysed that RLPR outperforms the Robust and reliable predictive based routing (RARP) and Ad hoc on-demand distance vector (AODV) protocols in different scenarios."

"The results show that RLPR achieves a 33% reduction in control messages overhead as compared to RARP and 45% reduction as compared to AODV."

"RLPR shows a 55% improvement in the lifetime of the network as compared to RARP and 65% as compared to AODV."

Introduction

"Each and every device in an infrastructure-based wireless network is directly connected to a central entity; and if a device wants to communicate with another device in the network, it needs to communicate through a central node, that is generally known as an access point."

"The cooperation among UAVs in multi-hop communication can increase the performance of the entire network."

"FANET requires a routing protocol that rapidly accepts topology variations in the network and may able to forward the traffic reliably with reduced ETE delay and with a smaller number of control messages in the network."

"The following list shows the major contribution of this research work: A reliable link-adaptive position-based routing protocol is designed for FANET that can work in a highly dynamic environment and enhances the network performance in terms of high connectivity-level with a minimum control overhead."

Literature Review

"Topology-based routing protocols need topological information of nodes and the associated links along with the orientation to create and maintain the routing path."

"In these types of routing protocols, nodes periodically exchange HELLO messages or 1-hop messages based on the vicinity in the network with the neighboring nodes."

"Due to the maintenance of the routing information at each node, overhead as signaling, increases in FANET because of the high mobility and therefore, proactive routing is not suitable for such networks."

"To solve various problems of topology-based routing protocol, geographical-based routing is proposed, in which the routing path is established by predicting the positions of the nodes in the network."

"Routing decision on each and every node is constructed based on the location of the destination node and then all forwarding nodes are required to select the next hop, which is nearer to the destination to forward the data packet."

Proposed Reliable Link-Adaptive Position-Based Routing Protocol for FANET

"Node 1 broadcast the message in the network upon receiving the message each node first computes their energy if their energy is greater than define threshold level then nodes computes forwarding angle with respect to node 1."

"When a source node needs to discover a route, it broadcasts a message in the network; only those nodes are allowed to rebroadcast that are towards the destination and must be the part of the front relative list of the predecessor node."

"Steps involved in 1-Hop HELLO broadcast message are as follows: Each node in the network sends Hello message with its 1-Hop Neighbors if its energy is greater than threshold level."

"When the nodes who have stared their timer receive an RLRQ packet for the same path based on the source, destination, broadcast, and sequence number identifiers, it discards the packet, otherwise it rebroadcast the RLRQ message when its timer expires."

Simulation and Results

"It has been observed that the control message overhead increases with the increase in the number of nodes in the network using RLPR, RARP, and AODV routing protocols."

"It has been observed that the control message overhead increases with the increase in the number of sources in the network using RLPR, RARP, and AODV routing protocols."

"RARP and AODV allow all the nodes in the network to rebroadcast the messages/route-requests."

"In RARP and in AODV, chances of collision increase, as both of these protocols allow all the nodes to rebroadcast the message in the network, and as a result, nodes consume more energy that significantly decreases the network lifetime."

"In the case of RARP and AODV, the connectivity level decreases that results in an increase in the packet loss and thus the source nodes ultimately rebroadcast the message in the network."

Conclusion and Future Work

"Performance evaluation of RLPR was carried out by comparing with RARP and conventional AODV routing protocol and simulation was run in different scenarios."

"The results showed that RLPR increased the network life time and minimized the number of control message that was used for path discovery between source and destination."

"With different number of nodes in the network the control messages were reduced up to 33% by RLPR as compared to RARP and 45% when compared with AODV and life time of the network was enhanced by 55% when compared with RARP and 65% when compared with AODV."

"When the number of sources in the network were increased, RLPR had also shown the reduction in control messages, 49% as compared to RARP and 57% when compared with AODV and life time of the network was enhanced by 21% when compared with RARP and enhanced by 35% when compared with AODV."

An Intelligent Probabilistic Whale Optimization Algorithm (i-WOA) for Clustering in Vehicular Ad Hoc Networks [97]

This is a machine-generated summary of:

Husnain, Ghassan; Anwar, Shahzad: An Intelligent Probabilistic Whale Optimization Algorithm (i-WOA) for Clustering in Vehicular Ad Hoc Networks [97]

Published in: International Journal of Wireless Information Networks (2022)

Link to original: https://doi.org/10.1007/s10776-022-00555-w

If you want to cite the papers, please refer to the original.

For technical reasons we could not place the page where the original quote is coming from.

Abstract-Summary

"An intelligent probability-based bio-inspired Whale Optimization Algorithm for clustering in VANETs (i-WOA) has been proposed considering communication range, the number of nodes (density), velocity, route on the highway during the process of cluster formation for vehicular communication by incorporating fitness function probability thus minimizing the randomness."

"The results were compared with already established methods and demonstrate that the proposed i-WOA method produces an optimal number of cluster heads (CHs) in various scenarios, for instance, communication ranges, network size, and node density."

"The results exhibit a 75% (regression analysis) improvement in cluster optimization for VANETs with application in ITS, consequently reducing communication cost and routing overhead hence increasing network lifetime."

Introduction

"Keeping in view all these limitations, a new network VANETs emerges by considering the above-mentioned limitations and to increase reliability and mobile connectivity where every vehicle is connected either directly with each other or with the roadside unit and access points [98]."

"Vehicle to Roadside Communication (VRC), another kind of ITS application offers by organization vehicles that use unicast mode for security messages transmission."

"There is a range of vehicular communication infotainment applications that upgrade the traveler encounter and offer instructive types of assistance, for example, fuel stations, climate data, crisis administrations, intelligent correspondence, Internet access, etc. It is of great importance to developing a system that could send and receive messages between the vehicles directly with High-Quality of Service (QoS) [99], and its media services [100]."

"Vehicle to Vehicle (V2V) has gained an advantage by eliminating roadside units and enables vehicles to share information directly thus introducing Dedicated Short-Range Communication (DSRC) for VANETs [101]."

Literature Review

"In [102], the authors proposed an intelligent-based clustering algorithm in VANETs (IBCAV) that improves direction-finding by incorporating inter-layered methods, traffic flow system as well as the AI-based system designed for selecting a cluster by considering the cluster size, network density, and velocity of CNs."

"This algorithm pursues the cluster life and the perfect CH number as a chief factor of network stability through utilizing Moth Flame Optimization (MFO)."

"In [103] the authors proposed a synchronized self-motivated direction-finding optimization method employing ACO and PSO algorithms that help the central decision-making routing system by taking into account the three parameters that include the length of the route, road quality, and mobility congestion."

"In [104], the authors proposed a cluster optimization algorithm for VANETs based on whale optimization by tuning several parameters such as communication range, number of nodes, network size, and load balancer results in an optimized number of clusters thus efficiently allocating the network resources and increasing lifetime."

Materials and Methodology

"The challenge intended is to propose an intelligent cluster optimization method (i-WOA) by incorporating a probabilistic approach when initializing a random vehicle on the highway, increasing the number of vehicles by considering processing time and computational cost that consolidates all the challenges and issues of ITS."

"To optimize the number of clusters in VANETs, an intelligent Whale Optimizer using a probabilistic approach has been developed that aims to reduce processing time while forming a cluster, computational cost, network overhead, packet latency, and end-to-end delay between the vehicles."

"Searching of vehicle, formation of the cluster, and selection of CH is numerically illustrated for performing the optimization of clustering."

"Once the cluster optimization is performed, other vehicles searching for CHs inform their areas agreeing on to the optimized vehicle specialist as appeared in [104]."

Results and Discussion
"The results obtained from the developed intelligent WOA for clustering in VANETs (i-WOA) have been compared with established bio-inspired clustering methods i.e.; Ant Lion Optimizer (ALO) [105], WOA [104], and Grey Wolf Optimizer (GWO) [106]."

"It could be observed here that developed i-WOA produces a considerably fewer number of clusters as compared to the other two methods."

"i-WOA has outclassed other methods which means that the developed method has more tendency to balance the number of nodes (load) in a cluster than other established methods."

"An interesting relationship has been observed between the transmission range and the number of clusters, as the transmission range increases, i-WOA produces more optimized clusters."

"The theory states that with each rise in the grid size a lesser increase in the number of clusters will lead to the optimum results."

Conclusion
"The developed method minimizes network randomness by obtaining an optimized number of clusters by incorporating probability-based function."

"The proposed optimization method performs superior to the ALO, GWO, and WOA in terms of CHs, whereas changing communication range, network approximation, and the number of vehicles."

"This study reduces network overhead, computational cost, end-to-end delay amongst the vehicles, and randomness while initializing the network nodes thus producing optimal number of clusters."

"A study on dynamic network topology for cluster optimization employing Harris–Hawk's optimization method is in progress."

An Effective Practical Method for Narrowband Wireless Mesh Networks Performance [107]

This is a machine-generated summary of:

Rattal, Salma; Ar-Reyouchi, El Miloud: An effective practical method for narrowband wireless mesh networks performance [107]

Published in: SN Applied Sciences (2019)

Link to original: https://doi.org/10.1007/s42452-019-1595-9

If you want to cite the papers, please refer to the original.

For technical reasons we could not place the page where the original quote is coming from.

Abstract-Summary

"This study presents a new practical measurement method for analyzing narrowband wireless mesh networks (NWMNs) application performance."

"It investigates a scenario for packet transmissions in NWMNs, leveraging wireless interconnection between two devices."

"The measurement results obtained, with the mentioned critical operating parameters, provide a quicker and clearer idea as well as conclusive evidence for overall network performance."

Introduction

"Narrowband wireless mesh networks (NWMNs) [108] are emerging technologies that provide connectivity to control, manage, supervise, and offer more capacity to devices in critical applications."

"The paper presents an experimental analysis of a narrowband wireless mesh network, using one-hop forwarding time, payload bitrate, message size, modulation, ACK, channel spacing, payload bitrate, and maximum transmission unit (MTU) as evaluation parameters."

"In wireless mesh networks, the communication system operates on frequencies between 30 MHz and 1 GHz using channel separations up to 50 kHz, which has been elected and adapted to various radio packet switching networks."

"We use various critical parameters [e.g., one-hop forwarding time, payload bitrate depending on message size, modulation, forward error correction (FEC), ACK, payload bitrate, and maximum transmission unit (MTU)] to measure simple router/bridge radio network performance."

Related Works

"They analyze the effects of Acknowledge and Forward Error Correction on the practical transmission performance of different message sizes."

"The authors in [109] experimentally demonstrated that the data size has a great effect on forwarding time, which is increased when the packet length increases."

"The authors in [110] have studied the impact of message size on the average delay in a wireless real-time application based on packet switching."

"They have also clearly illustrated experimentally the relationship between modulation type, channel spacing, forwarding time, and data length based on average message size."

Technical Parameters

"The binary information, transmitted by a channel, has a relationship with packet transmission time and packet size, which affects network performance."

"The payload bitrate of custom IP packets transmitted by routers on the radio channel includes 28 B of IP packet overhead, 20 B of IP header, and 8 B of user datagram protocol (UDP) header."

"On a data communication link, the packet size is in bytes, and the MTU is the largest packet size that can be transmitted over a network."

"The mission of the application implementation is to integrate the network analyzer by illustrating the effect of packet size, carrier spacing, and modulation rate on the performance of wireless mesh networks through experimental results."

"Since the overhead of the packet radio protocol is fixed, the length of the user data measurably affects the bitrate of the payload."

Systems Model

"A practical illustration of the router configuration transmitting a single packet from IoT's main radio modems to IoT RTU."

"Packets coming from the SCADA center are transmitted to internet networks."

"The IoT main radio modem is duplex, and the transmission carrier is on all the time while the SCADA center sends and receives the packets."

"The IoT SCADA center sends requests one-by-one to RTU1 employing internet and IoT main radio modems."

"That the transfer time on and the processing time are zero, only the time of transmission of the packets on the medium intervenes to determine the performance."

"A repeater is intended to relay all received packets through the radio channel, but it is also used as a simple device by offering a suitable and practical level of communication and spectrum efficiency in the small or medium wireless network size."

Results

"Execution forwarding time measurement is done with varying CS to transmit a single packet to quantify the effect of NB on message size."

"A bridge mode can perform a transmission time based on the average message size in NWMNs."

"Bridge mode can improve forwarding time depending on message size and fixing CS."

"Comparison results demonstrate that in router or bridge modes when increasing message size and CS, the forwarding time and the modulation rate increase while payload bitrate decreases."

"The authors analyze the effect of message size and channel spacing on the resulting bitrates, forwarding time, and modulation rate through experimental results."

"The practical method described in this paper quickly generalize the analysis of Narrowband wireless mesh networks application performance using the robust design metric using only one-hop forwarding time, payload bitrate, message size, modulation, ACK, channel spacing, and maximum transmission unit (MTU), as evaluation parameters."

Conclusions

"This paper illustrated the focusing effect of the various packet sizes, CSs, and modulation rates over a wireless mesh network."

"Experimental results proved the efficacy of the proposed method for a quick performance overview, based on several basic parameters, providing a quick and easy idea of the possible bitrate and forwarding time for the operating mode of the router and bridge network performance."

"Based on the proposed method, we can see the effect of packet length, spacing channel, and modulation rate on the resulting time or transmission time."

"The model's router/bridge mode based on the mentioned parameters provides an effective method to estimate network performance under QoS constraint."

"We are now examining the effectiveness of other novel methods for different wireless networks to facilitate the analysis of their performance."

iSleep: Thermal Entropy Aware Intelligent Sleep Scheduling Algorithm for Wireless Sensor Network [111]

This is a machine-generated summary of:

Banerjee, Partha Sarathi; Mandal, Satyendra Nath; De, Debashis; Maiti, Biswajit: iSleep: thermal entropy aware intelligent sleep scheduling algorithm for wireless sensor network [111]

Published in: Microsystem Technologies (2019)

Link to original: https://doi.org/10.1007/s00542-019-04706-7

If you want to cite the papers, please refer to the original.

For technical reasons we could not place the page where the original quote is coming from.

Abstract-Summary

"The optimal operation of Wireless Sensor Network (WSN) requires a sustainable topology."

"A dead node may result in a breach of end-to-end connectivity in the network."

"Every sensor node has a typical operating temperature zone."

"An intelligent sleep scheduling algorithm 'iSleep' is proposed based on available neighboring nodes, associated network entropy, and traffic flow pattern."

"iSleep enables the sensor nodes to auto-control the sleep state of the nodes and maintain network connectivity for a longer amount of time."

"It is found that the proposed algorithm outperforms the existing scheduling algorithms in terms of the lifetime of a network for both the energy models."

Introduction

"Research and industrial applications show that the functions of Wireless Sensor Networks are considerably affected by the onboard temperature of the nodes."

"Change in clock drift affects the latency of the nodes, while the quality of wireless links degrades with an increase of temperature of the sensors and a quantitative account is given in [112]."

"Exposure to sunlight, diurnal and seasonal variation of temperature, moisture content of air and others affect the usual functioning of the sensor nodes [113–116]."

"All these nodes are within each other's transmission range and experience highly different temperatures."

"Added to this, the temperature of a sensor node increases due to Joule heat generation in the electronic or electrical circuitry of the nodes."

"The variation may be attributed to the difference in the work schedule of the two nodes and shows the temperature heterogeneity of the network."

Related Works

"There is a plethora of techniques proposed by eminent authors towards efficient sleep-scheduling of sensor nodes."

"The authors have proposed a temperature adaptive power model where the sensor nodes intelligently switch to different current levels in order to control the transmission range and thereby the connectivity and coverage of the network according to the temperature sensed by the onboard sensor."

"On careful study of the works of the literature, it is evident that most of the sleep-scheduling techniques are based on energy-consumption, connectivity, coverage, etc. Temperature aware efficient sleep scheduling strategy is required to manage the thermal disturbances generated within the sensor nodes."

"We propose a temperature aware auto-controlled sleep scheduling algorithm for nodes in Wireless Sensor Network."

"Nodes can sense temperature and its own neighborhood and smartly decide about the respective sleep-awake strategy by itself with the help of the proposed algorithm i-Sleep."

Contributions

"Temperature adaptive sleep scheduling reduces overall energy consumption in a multihop wireless sensor network."

"Reduced energy consumption prolongs the lifetime of the network."

"The network consists of intelligent sensor motes that can make their radio ON/OFF depending on the average energy consumption by the motes."

"The schedule of sleep of the nodes is tuned with their respective relative entropy so that the best possible node is found for hibernation and the network is minimally disturbed. (b) The neighborhood status of nodes plays a pivotal role in maintaining the end-to-end connectivity and other QoS parameters of the network."

"The neighborhood status has been quantified and combined with relative entropy to find an optimal candidate node to enter the sleep state. (c) Temperature aware intelligent sleep scheduling algorithm has been tested in the Cooja simulator on InstantContiki-2.7 with Z1 and Sky motes."

System Model Preliminaries and Definitions

"WSN is assumed to resemble a Random Geometric Graph [117] which means that the connectivity between any two nodes in the network is only a function of the transmission range of the sensor nodes."

"Every sensor node contains three functional modules: sensing module, analog pre-conditioning, and data conversion circuitry, and the third one is a digital signal processing (DSP) unit with a radio link."

"In a particular session, the energy consumption of different nodes will be different and it depends on their position in the network and the schedule of work they are to perform."

"The node with the maximum value of relative entropy will be allowed to enter sleep state, but with due concern to the status of its neighborhood and the amount of traffic to be handled, so that the network never gets disconnected."

Proposed Algorithm

"An optimal sleep-scheduling algorithm is proposed in ALGORITHM 1: (iSleep) where the node with maximum relative-entropy will go to sleep-state followed by reorganization of the nodes among themselves in order to maintain connectivity by finding an alternative path to the destination using neighbor nodes."

"This probability is calculated by all nodes for every session (line 5)."

"Temperature based node entropy and network entropy is calculated (line 7 and 8)."

"The relative contribution of every node to the network entropy is found (line 9)."

"A node having higher relative entropy will tend to get into a sleep state only if the connectivity is minimally disturbed."

Simulation and Results

"Added to this, due to continuous engagement, the onboard temperature of a node increases to its upper limit of rating and may lead it to the non-functioning state and affects network connectivity."

"Some of the nodes may reach the upper bound of working temperature and stop functioning and result in an unstable network with increased entropy."

"Threshold entropy seems to relax with an increased number of nodes as congested network leverages lesser distances between node-pairs and hence less energy consumption for communication."

"A 6LoWPAN (IPv6 over Low -Power Wireless Personal Area Networks) network is considered to measure the energy consumption per node per message."

"Every node records its power consumption in different states at a regular interval from as mentioned in the code."

"SOPC algorithm schedules the operation of nodes in such a way that the neighborhood of a node remains consistent and the power consumption remains uniform throughout the lifetime of the node."

Conclusion and Future Scope

"Sensor nodes in WSN are subjected to a faster rise in temperature due to overusage in the network when it is placed centrally."

"Temperature driven random shutting down of a node may lead to an abrupt disconnection of the network."

"An efficient sleep scheduling algorithm proposed in this article ensures this by auto-selecting the sleep state of a node, minimally disturbing the connectivity of a network."

"If these are included in the proposed sleep scheduling algorithm it should better represent the simulation of a real network."

Impact of Secondary User Block on the TCP Throughput in Cognitive Radio Sensor Networks [118]

This is a machine-generated summary of:

Hassani, Mohammad Mehdi; Berangi, Reza: Impact of Secondary User Block on the TCP Throughput in Cognitive Radio Sensor Networks [118]

Published in: Wireless Personal Communications (2019)

Link to original: https://doi.org/10.1007/s11277-019-06677-4

If you want to cite the papers, please refer to the original.

For technical reasons we could not place the page where the original quote is coming from.

Abstract-Summary

"Two main contributions are provided on the modeling of SBL and throughput evaluation of transport layer protocol for CRSN."

"A new congestion control algorithm is proposed to distinguish between actual congestion from the wrong congestion due to the SBL by considering the dynamic nature of CRSN."

"The obtained results through proposed model are verified using the COGNS framework based on NS2, which is a simulation framework for cognitive radio sensor networks."

"The proposed algorithm is compared with some of the well-known transport protocol TFRC-CR, OHTP and TCP Reno."

Introduction

"Cognitive radio network (CRN) is a wireless network that its unlicensed users, referred as the secondary users (SUs), have cognitive radio (CR) capabilities to exploit the vacant channel opportunistically."

"A CRSN is a distributed network of wireless cognitive radio sensor nodes, which sense an event signal and collaboratively communicate their readings over dynamically available spectrum bands in a multi-hop manner."

"Packets may be lost due to congestion or noisy wireless channels in wireless networks, which misled TCP to shape the traffic and reduce the throughput rather than optimizing it."

"When a PU occupies the channel, the Secondary User (SU) leaves the channel (spectrum handoff), and as a result, the route is broken and some packets are lost."

"We propose a new rate-based transport protocol called TCP over CRSN (T-CRSN) for cognitive radio sensor networks."

Related Work

"Simulation based performance evaluation shows that sensing time has a significant effect on TCP throughput."

"TCP performance is investigated over cognitive radio ad-hoc network (CRAHN) in [119] using an extension of NS2 to evaluate their CRAHN model that supports SUs spectrum management operations."

"The cross-layer designs based on the CR-related operations (such as spectrum sensing, decision, and handoff) are considered by many researchers [120–127] to enhance transport layer performance in terms of throughput and delay."

"Researchers believe that instead of changing TCP and using alternative transport layer protocols, wireless networks must be optimized to enhance TCP performance."

"Optimization of lower layer parameters like spectrum sensing, physical layer modulation, datalink frame size and coding scheme are considered to enhance TCP throughput."

"TFRC-CR [128] is proposed as the first equation based transport protocol based on the TCP friendly rate control for cognitive radio network."

CRSN Architecture

"A CR sensor node must enable to sense spectrum."

"That sink node is equipped with unlimited power and with more than one transceiver to send and receive data simultaneously."

"The CRSN node has a radio transceiver unit that enables to sense spectrum and dynamically adapts its parameters such as transmission power, carrier frequency and modulation [129]."

"CRSN node cannot concurrently send/receive the packet at the spectrum sensing time because of half-duplex communication ability."

"Packet loss due to buffer overflow and sending many data packets to the next hop are other results of long sensing duration which will mostly degrade TCP performance in the term of throughput in CRSN."

TCP Challenges in CRSN

"When a PU activity is detected, CRSN stops the SU from transmitting and coordinates the use of new channel."

"Sending data packet by SUs is blocked either when there is no free channel after sensing or by primary user presence on the channel."

"(3, 1) denotes a model with four available channels, and there are three SUs and one PU where occupy four channels in CRSN."

"P_{block} is a state that the SU is blocked either due to primary user activities or lack of free channel."

"It is the blocking probability due to: (1) One PU returns to the channel and wants to send packets. (2) All of the N channels are occupied and there are no vacant channels."

"On the other hands, when all of the three channels are filled with one SU and two Pus (state (1, 2)), if an SU wants to transmit on a channel with the rate of λ_s, because all channels are occupied then SU traffic locates in the block situation."

Analysis of the SBL Probability

"By increasing of channel numbers, the SU blocking probability rate is decreased in each curve."

"This reduction is due to the fact that by increasing number of channels, SU's chance to occupy the channel and transmission of data is increased."

"By increasing the SU arrival rates, $P_{blocking}$ is increased per each N channels model intuitively, when SU's traffic is increased, the competition between SUs to access the vacant channel is increased too, hence more SUs are blocked."

"With the increasing number of channels, the SU chance is opportunistically enhanced to access channel dynamically."

"They can occupy channels more times and block probability rate is decreased."

The Proposed Congestion Control Model

"A cross-layer approach is used to enhance TCP throughput of CRSN in the new model, which the SU can use channels when the PU is not active."

"If TCP could detect the difference between packet loss and timeout situation due to the primary user arrival from congestion loss, TCP throughput would not be Degraded."

"TCP adjusts the transmission rates by considering the difference between congestion loss and packet loss due to PU arrival."

"If a primary user activity confirms in the FCC database, it moves to Wait state, otherwise, congestion is detected and it moves to congestion state."

"When the packet loss is due to the primary activity, hence the current TCP throughput is stored."

"If the PU activity is detected, current TCP throughput is saved and the secondary user must find another vacant channel and move to it."

Simulation Results and Discussions

"The number of channels is proportional to the value of throughput."

"The SBL due to PU activity misleads TCP Reno to detect is as a congestion loss."

"The throughput degradation occurs at the beginning of the PU activity due to the long time it takes TFRC-CR to reduce the rate."

"The throughput values of the OHTP are almost close to our proposed model."

"The OHTP is comparatively close to the proposed model among all the protocols but it takes long time to switches between rate based and window based transmission."

"The proposed model considers shortest PU activity period by calculating PU arrival rate in DTMC model."

"In this zone, the proposed congestion control experiences lower throughput than OHTP."

"It affects throughput value in the first 18 s of simulation time."

Conclusion

"We introduced SBL as a new type of packet loss by considering the impact of two factors: (1) PU activities, (2) SU blocking."

"In case of primary user detection during the SU, the connection is blocked."

"For distinguishing between SBL and packet loss due to congestion, the Sus dynamic spectrum access is modelled by the discrete-time Markov chain (DTMC)."

"It could detect the difference between packet losses due to the primary user arrival from congestion loss."

"Given the requirements of CRSNs, (1) obtaining a discrete-time Markov chain for the total throughput of secondary users together with the primary uses and (2) investigating the SUs mobility on the TCP throughput could be added to this model as future work."

Trade-Off Between Spectral Efficiency and Normalized Energy in Ad-Hoc Wireless Networks [130]

This is a machine-generated summary of:

Mehta, Ridhima: Trade-off between spectral efficiency and normalized energy in Ad-hoc wireless networks [130]

Published in: Wireless Networks (2021)

Link to original: https://doi.org/10.1007/s11276-021-02610-5

If you want to cite the papers, please refer to the original.

For technical reasons we could not place the page where the original quote is coming from.

Abstract-Summary

"The available bandwidth spectrum should be utilized in the most efficient and controlled way for practical economic growth of industries in conformity with the evolving demand of energy for powering wireless devices."

"In the proposed optimization scheme, energy management and limited spectrum sharing are distributively implemented for handling quality of service provisioning and potential resource allocation across multiple nodes with inherent operational capability deficiencies."

"We employ cross-layer information exchange and convex optimization techniques to simultaneously achieve more efficient radio spectrum usage and optimal energy consumption in ad-hoc wireless networks with distributed scenarios."

"We consider the time-invariant additive white Gaussian noise channel and the time-varying Rayleigh fading channel to study the trade-off between the two network design objectives of achieving improved spectrum efficiency and minimizing the energy overhead in data routing paradigm."

"The effectiveness of the proposed optimization model is substantiated through the simulation comparison results with the existing schemes in terms of various key performance parameters such as throughput, energy efficiency, and average bit errors."

Introduction

"As the current wireless systems are progressively moving towards packet switching network deployment with statistical multiplexing architecture, there has been tremendous demand for increase in data rates to support modern QoS applications over future wireless networks."

"Wireless static nodes in these networks are deployed in an ad-hoc mode and can dynamically allocate the radio signal power to their associated outgoing links for data packet transmission."

"The design of cooperative ad-hoc wireless networks with multi-hop relaying and distributed allocation of limited radio spectrum is implemented in conjunction with the available energy optimization technique."

"These layers respectively deal with the diverse networking issues of data-rate control among multiple-commodity flows, medium access time allocation and power control operation."

"The next section presents the related work on the efficient utilization of restrained resources through the optimized system design in various widely employed wireless networks conducted over the past few years."

Review of Literature

"In the proposed work, we present a distributed optimization framework to perform cross-layer design in ad-hoc wireless communications technology for realizing the improved energy efficiency and optimal spectrum usage."

"Mehta and Lobiyal [131], the three network design objectives of higher application throughput, lower delay and energy-efficient real-time communication are jointly optimized through cross-layer problem formulation."

"This work did not consider consolidated optimization of energy and spectral resource utilization with multiple heterogeneous wireless network limitations of channel errors, interference-aware link scheduling and temporal restraints using cross-layer methodology, as is accomplished in our proposed system design."

"Despite these previous works, our optimization model aims at achieving dual performance objectives of optimal radio spectrum assignment by maximizing the network capacity and minimizing the energy consumption of each individual link for data transmission over the allotted time slot."

System Model

"In this channel admission control mechanism, a node with data to transmit contends for the medium access and senses the channel status for a specific interval."

"The wireless channel is logically partitioned into multiple smaller time slots, that are exclusively allocated to a particular node willing to transmit the data through the network."

"The distributed network scenarios are considered in which data packets modeling the bursty traffic are traversed through multiple hops to reach the intended destination node."

"During this assigned slotted interval, the node is permitted to transmit at a dynamic data-rate while ensuring efficient channel utilization."

"The fading phenomenon, perceived as a time-fluctuating data filtering operation on the transmitted signal, occurs owing to multipath propagation of signal transmission between the transmitter and the receiver nodes."

"Further, we consider that the wireless channel does not change significantly over all time instants due to high correlation between subsequent radio signals transmitted across the network."

Problem Formulation

"The developed optimization problem is implemented as the following bi-objective maximization problem subject to the underlying network operational constraints."

"The remaining sets of constraints are convex functions of the optimization variables."

"The above optimization problem is a convex optimization problem since it consists of maximizing a concave combination of objective functions subject to convex and affine constraint sets."

"Considering the first objective of maximizing the spectral efficiency, it is expressed as the sum of logarithmic functions of link transmission powers."

Simulation Results and Discussion

"The trade-off behaviour between the radio spectrum resource allocation and the normalized energy utilization for both the wireless channel models can be discriminated by reasonably higher values of both the contemplated criteria for Rayleigh fading networks."

"On an average, it can be computationally accounted that the objective function value for the developed optimization model for Rayleigh fading wireless network is higher in magnitude by a factor of 1.345 as compared to the AWGN wireless network."

"The proposed system model implemented with the Rayleigh fading wireless channel exhibits 33.87% higher energy demand as compared with the primal AWGN channel type."

"Compared with the similar works [132–135] proposed previously, our unified bi-optimization work to accomplish the enhanced capacity and minimal energy utilization for practical data delivery across multiple network sessions retrieves the improved performance of ideal radio spectrum allocation and effective energy management in dynamic wireless networks."

Conclusion
"The simulation trade-off curves between normalized energy consumption and radio spectrum allocation are plotted for both AWGN and Rayleigh fading optimization models."

"It was computationally established that the joint objective function value for Rayleigh fading optimization model is higher in magnitude by around 34.45% than the corresponding AWGN wireless model."

"For the investigated network of size 25, our proposed AWGN model demonstrated higher throughput by up to 80.06% and energy efficiency by around 71.18% than the other existing methods."

"The Rayleigh fading channel model has been proven to outperform the considered previous works by exhibiting 87.416% higher throughput and 44.93% greater energy efficiency."

Scheduling and Resources Allocation in Network Traffic Using Multiobjective, Multiuser Joint Traffic Engineering [136]

This is a machine-generated summary of:

Latif, Ghazanfar; Saravanakumar, N.; Alghazo, Jaafar; Bhuvaneswari, P.; Shankar, K.; Butt, Muhammad O.: Scheduling and resources allocation in network traffic using multiobjective, multiuser joint traffic engineering [136]

Published in: Wireless Networks (2020)

Link to original: https://doi.org/10.1007/s11276-020-02407-y

If you want to cite the papers, please refer to the original.

For technical reasons we could not place the page where the original quote is coming from.

Abstract-Summary
"Reciprocal optimality is a desirable characteristic from the end user perspective, for both Best Effort and Quality of Service (QoS) datagrams delivery within a network."

"It is also aimed to formulate and evaluate enhanced scheduling algorithm for packets delivery, focusing on maximal user satisfaction and minimal networking operation costs."

"Mathematical optimization techniques have been applied to Application, Transport and Network layers of Open System Interconnection model."

"Novel elucidation of optimal packets switching, traffic engineering, congestion avoidance and QoS improvements are discussed."

Introduction

"For the network operator there is certain cost of effort to provide a particular throughput (bit rate) to each user on a link, keeping in view the overall capacity of the link."

"Bandwidth is regarded as the maximum possible bits transfer rate for a particular single path (link), measured as bits/s. However, throughput is the actual bit transfer rate on a particular link, also measured as bits/s. Scheduling can be regarded as the allocation of a particular link to a particular user."

"This paper targets traffic engineering related issues such as packets routing and link resources scheduling for wired and wireless digital communication networks."

"Each link is considered to have bandwidth greater or equal to the actual throughputs the users (nodes) will enjoy once routing is completed and all the nodes (users) packets flow is scheduled among the L links at a particular point of time."

Literature Review

"Notable researches, which contributed mainly to the area are works of Sally Floyd, Van Jacobson (Random Early Detection Congestion avoidance algorithm), Frank Kelly (very start of congestion control theory), Shenker S. (utility function introduction) [137–139]."

"Decentralization approach for congestion control is proposed in most recent works to enhance the networking performance."

"Packets losses due to data rate mismatches have to be avoided, in order to induce reliability in the network."

"The optimization algorithm becomes more effective when considered jointly for congestion control and routing to satisfy both performance metrics (users' objectives) and network robustness (operators' objectives) as described in the works of [140, 141]."

"The network traffic (packets) can be categorized into below types: Elastic traffic without critical timing and latency dependence) Soft or Firm Real Time (RT) performance gradually drops with delay increase Hard Real Time (HRT) guarantees sufficient performance up to a well-defined very small delay."

Proposed Traffic Engineering Scheme

"The link resources scheduling based routing would be depicted as a mathematical model comprising of: Objective function It includes traffic engineering penalty function, end user utility function, operator cost function Constraints set Routing matrix (R), link load has not to exceed links capacity (C) Variables Source, Link rates, etc. The scheme followed in this work is to construct the utility function with regards to end user application type i.e. for file transfer and sharing end user would be more satisfied with higher throughput, while for real time video streaming/gaming end user satisfaction would increase as low queuing and propagation delays are experienced."

"The packets at incoming links (N) may vary in length, priority, source and destination addresses."

"The queue lengths of the transmitted packets include varied lengths, priority, as well as source and the destination address."

Evaluation of the Proposed Traffic Engineering Problem

"Approach is based on Pattern Search algorithm and solver supplied with TCE Matlab Optimization Toolbox function patternsearch."

"The obtained outcomes for both approaches Sequential Quadratic Programming (SQP) and Pattern Search Algorithm include below parameters [142]: Current Function value/Best Function Value The value of Objective Function at the current point."

"We can deduce that SQP certainly produced more accurate results as can be seen from the final function value, preset solution vector relative change tolerances and optimization process completion."

"The objective function # 1 with linear link utilization and similar satisfaction degree for the N users is most accurate and fastest in optimization."

"It shall be noted that the JRAFI index is bound between the interval 0 to 1, and the higher value of JRAFI index means more accurately optimized objective function i.e. more balanced traffic."

Outcomes Analysis

"As, lower the link capacities induce difficulty to optimize such problem and the lower the optimal bandwidths become."

"For both of the applied techniques i.e. SQP and Pattern Search Solver, it is observed that the later approach was not able to provide accurate optimal results."

"The optimization function settings for Pattern Search solver method shall be modified in order to increase the accuracy and change tolerance."

"First approach based on SQP algorithm is implemented in TCE Matlab using builtin fmincon function, and is significantly faster and less computationally complex than the second approach implemented with pattern search TCE Matlab builtin routine."

Conclusion

"Only part of the delay sources have been taken into account within the optimization problem formulated and solved in this work, processing time and energy constraints might be added to the utility function as the continuation of this research."

"The defined and solved problem is utilizable also in optimal paths in graph problems in adhoc networks (wired and wireless)."

"The simulation in this work covers Pareto Efficiency, however there exist other efficiency principles and conditions, such as Bayesian, Kuldor-Hicks, which might be incorporated into the basic optimization problem to reflect some hidden or advanced performance characteristics."

"Another viable addition to this work can be done by introducing the Stochastic Objective Functions Optimization, where the networking traffic is modeled using probability theory."

Mobility Management Using Dynamic Cross-Layer Signaling in Heterogeneous Wireless Networks [143]

This is a machine-generated summary of:

Al Emam, Fatma A.; Nasr, Mohamed E.; Kishk, Sherif E.: Mobility management using dynamic cross-layer signaling in heterogeneous wireless networks [143]

Published in: Wireless Networks (2019)

Link to original: https://doi.org/10.1007/s11276-019-02162-9

If you want to cite the papers, please refer to the original.

For technical reasons we could not place the page where the original quote is coming from.

Abstract-Summary

"Vertical handover cross-layer designs lack automatic adaption with different applications."

"The upper layers in a cross-layer design need to be aware of the events in the lower layers, so that sessions will not be interrupted during vertical handover."

"Challenges of the cross-layer designs were summarized."

"Design architecture of cross-layer include a message exchange between modules from different layers."

Introduction

"There is a need for a cross layer tuned to the needs of applications and network conditions."

"The adaptive transport layer protocol should suit architectural heterogeneities for roaming mobile nodes (MNs) [144]."

"HO decisions are based on lower layer events and do not consider applications' requirements."

"Mobility management protocols enable changing the point of network attachment based on a decision taken."

"Trigger messages (TRG) provide signaling needed for application adaptation in heterogeneous network mobility [145]."

"If the transport layer is notified using triggers, it will compensate for data-rate fluctuations and minimizes impact of mobility."

"Vertical handover decision schemes don't take the demands of running applications into consideration, according to the new network profile."

"Performance significantly improved by cross-layer triggers: Receiver reports allows adjusting the transmission rate when MN changed its PoA. Reduce the impact of trigger messages propagation in the network."

Handover Decision Algorithm

"The overall score of candidate networks is computed as the weighted sum of the attribute values."

"RSS_{th} threshold values are chosen to satisfy the requirements of the active applications."

"These threshold values RSS_{th} are suggested that give the best handoff delay during scenario iterations (such as RSS_{th} of −85/−72 dBm in 802.11 WLAN, RSS_{th} of −119/−75 dBm in UMTS, RSS_{th} of −102/−85 dBm in 802.16WIMAX)."

"RSS_{th} can be made adaptable to delay threshold."

"When a delay threshold is 40 ms RSS_{th} is −87.5 dBm."

Cross Layer

"It builds upon the triggering architecture and allows mobility management protocols to get required information for HO decision."

"These triggers are used to adjust the behaviors of other protocols to mobility events e.g., Transport protocols implementation could provide triggers to the client."

"This framework produce new triggers for services at different protocol stack layers in support of context-aware communications."

"The Interlayer signaling makes use of IP data packets to deliver the cross layer information with no need to use a dedicated message protocol."

"Getting timely information from other layers makes upper layers adapt to mobility-incurred events in advance."

"A mobility management protocol can take advantage of link up/down triggers and crossing a threshold indication event."

"Application layer can get MIH events or decision of handover algorithm to adapt its operation more efficiently Through TRG."

Media Type TCP and RTP

"The sender adjusts its sending rate relying on receiver reports (RR) messages and RTCP messages."

"Changing the transmission rate of the sender according to receiver reports about the characteristics of the new link improves performance for vertical handovers."

"Post-handover adaptation time of TCP connections was considered to realize seamless handoff in heterogeneous networks by adjusting the window size adaptively [146]."

"The sender starts adjusting its streaming rate in heterogeneous networks during handoff [147]."

"New adaptive, reliable transport protocol where the packet is adjusted to adapt to abrupt variation in the wireless link conditions [148]."

"The sender gets notified of the event and controls its streaming rate upon receiving HN message."

"Dummy packets are transmitted to the MN via the new AP to estimate new network characteristics and report feedback via a signaling message."

"Dummy packets are used to probe the available bandwidth for the streaming server to adjust its sending rate accordingly."

Rate Controller Module "Quality Adaptation"

"The rate switching feedback trigger takes values from 0 to 31."

"DN trigger is sent to the server before association response message from next PoA. The Link Going Up/Down event messages can be generated before VHO occurrence so the server react accordingly in advance."

"Expected VHO occurrence time should be synchronized with the rate switch request to the streaming server."

"DN/UN feedback messages should arrive at the proper time to reflect the new rate before VHO occurrence to reduce packet losses."

"Rate switching is triggered by an UPDATE message from MN with predefined quality threshold limit (ranging from 0 to 31)."

"Based on the MN feedback requests (up/down), the server dynamically selects the most suitable Q for video stream as described in algorithm 1."

"As a server receives triggers from MN it switches into handover state."

Simulation Environment

"Dummy packets are sent to IF2 to calculate appropriate streaming rate while IF1 keeps receiving data."

"HN event notifies the sender about handoff occurrence then it adjusts streaming rate appropriately using dummy packets."

"The highest throughput is achieved thanks to the dummy segments that inform the sender of available bandwidth within a single RTT and accordingly increase its sending rate."

"The proposed design handoff period becomes shorter to probe for bandwidth availability with varying speed from 10 to 100 km/h. Throughput decreases as the mobile node speed increase."

"The behavior of proposed scheme is illustrated as follows: 1.t0 $\leq$ t < t1: The MN transmits an ACK when it enters the overlapping area in response to received dummy packet.2.t0 + RTT $\leq$ t < t1 + RTT: sending rate is adjusted in response to dummy packets."

Conclusion

"A cross-layer framework designed for the efficient delivery of services in heterogeneous wireless networks."

"The key idea is to imminent notify the server with these events to probe for the characteristics of the new wireless network using dummy packets."

"The performance of the proposed framework is evaluated for both TCP RTP-based applications with adequate enhancements for each."

"Dummy packets makes estimation of the available bandwidth more accurate."

Bibliography

1. Sharma, N., Khanolkar, S., Kumar, S. R., & Anpalagan, A. (2022). An adaptive transmission mode selection scheme for cellular underlaid D2D communication. *International Journal of Wireless Information Networks*. https://doi.org/10.1007/s10776-022-00550-1
2. Karaboga, B., & Basturk, D. (2007). A powerful and efficient algorithm for numerical function optimization: Artificial bee colony (abc) algorithm. *Journal of Global Optimization, 39*(3), 459–471.
3. Yu, G., Xu, L., Feng, D., Yin, R., Li, G. Y., & Jiang, Y. (2014). Joint mode selection and resource allocation for device-to-device communications. *IEEE Transactions on Communications, 62*(11), 3814–3824.
4. Huang, Y., Nasir, A. A., Durrani, S., & Zhou, X. (2016). Mode selection, resource allocation, and power control for d2d-enabled two-tier cellular network. *IEEE Transactions on Communications, 64*(8), 3534–3547.
5. Du, Y., Zhang, W., Wang, S., Xia, J., & Mohammad, H. A. (2021). Joint resource allocation and mode selection for device-to-device communication underlying cellular networks. *IEEE Access, 9*, 29020–29031.
6. Pang, H., Wang, P., Wang, X., Liu, F., & Ngoc, N. (2013). Joint mode selection and resource allocation using evolutionary algorithm for device-to-device communication underlaying cellular networks. *Journal of Communications, 8*, 751–757.
7. Al-Habashna, A.'a., & Wainer, G. (2019). QoE awareness in progressive caching and DASH-based D2D video streaming in cellular networks. *Wireless Networks*. https://doi.org/10.1007/s11276-019-02055-x
8. Al-Habashna, A., Wainer, G., & Fernandes, S. (2017). Improving video streaming over cellular networks with DASH-based device-to-device streaming. In *2017 International Symposium on Performance Evaluation of Computer and Telecommunication Systems* (pp. 468–475). IEEE.
9. El, E. A., Schroeder, D., Steinbach, E., et al. (2015). QoE-based traffic and resource management for adaptive HTTP video delivery in LTE. *IEEE Transactions on Circuits and Systems for Video Technology, 25*, 988–1001.
10. Cicalo, S., Changuel, N., Tralli, V., et al. (2016). Improving QoE and fairness in HTTP adaptive streaming over LTE network. *IEEE Transactions on Circuits and Systems for Video Technology, 26*, 2284–2298.
11. Ghalut, T., Larijani, H., & Shahrabi, A. (2016). QoE-aware optimization of video stream downlink scheduling over LTE networks using RNNs and genetic algorithm. *Procedia Computer Science, 94*, 232–239.
12. Lee, G., Kim, H., Cho, Y., & Lee, S.-H. (2014). QoE-aware scheduling for sigmoid optimization in wireless networks. *IEEE Communications Letters, 18*, 1995–1998.
13. Kumar, P., Chauhan, N., Kumar, M., & Awasthi, L. K. (2021). Clustering based opportunistic traffic offloading technique for device-to-device communication. *International Journal of System Assurance Engineering and Management*. https://doi.org/10.1007/s13198-021-01136-5
14. Asadi, A., Wang, Q., & Mancuso, V. (2014). A survey on device-to-device communication in cellular networks. *IEEE Communications Surveys & Tutorials, 16*(4), 1801–1819.
15. Liu, J., Kato, N., Ma, J., & Kadowaki, N. (2014). Device-to-device communication in LTE-advanced networks: A survey. *IEEE Communications Surveys & Tutorials, 17*, 1923–1940. https://doi.org/10.1109/COMST.2014.2375934
16. Gandotra, P., & Jha, R. K. (2016). Device-to-device communication in cellular networks: A survey. *Journal of Network and Computer Applications*. https://doi.org/10.1016/j.jnca.2016.06.004
17. Hematian, A., Yu, W., Lu, C., Griffith, D., & Golmie, N. (2017). Towards clustering-based device-to-device communications for supporting applications. *ACM SIGAPP Applied Computing Review, 17*, 35–48. https://doi.org/10.1145/3090058.3090063

18. Choi, K. W., Lee, H., & Chang, S. C. (2014). Discovering mobile applications in device-to-device communications: Hash function-based approach. *IEEE Transactions on Mobile Computing, 15*, 336–339.
19. Lei, L., Zhong, Z., Lin, C., & Shen, X. (2012). Operator controlled device-to-device communications in LTE-advanced networks. *IEEE Wireless Communications, 19*(3), 96–104.
20. Asadi, A., & Mancuso, V. (2013). WiFi direct and LTE D2D in action. In *IFIP wireless days (WD)* (pp. 1–8).
21. Arain, Q. A., Uqaili, M. A., Deng, Z., Memon, I., Jiao, J., Shaikh, M. A., Zubedi, A., Ashraf, A., & Arain, U. A. (2017). Clustering based energy efficient and communication protocol for multiple mix-zones over road networks. *Wireless Personal Communications, 95*, 411–528. https://doi.org/10.1007/s11277-016-3900-x
22. Cao, Y., Jiang, T., & Wang, C. (2015). Cooperative device-to-device communications in cellular networks. *IEEE Wireless Communications*. https://doi.org/10.1109/MWC.2015.7143335
23. Garrocho, C. T. B., da Silva, M. J., & Oliveira, R. A. R. (2020). D2D pervasive communication system with out-of-band control autonomous to 5G networks: Project and evaluation of a middleware for networking and content exchange to D2D communication without human interaction. *Wireless Networks, 26*, 373–386. https://doi.org/10.1007/s11276-018-1820-2
24. Garcia-Saavedra, A., Serrano, P., Banchs, A., & Bianchi, G. (2012). Energy consumption anatomy of 802.11 devices and its implication on modeling and design. In *8th International Conference on Emerging Networking Experiments and Technologies* (pp. 169–180). https://doi.org/10.1145/2413176.2413197
25. Ouali, K., Kassar, M., Nguyen, T. M. T., Sethom, K., & Kervella, B. (2020). Performance evaluation of an effective mobility model for D2D communications. *Wireless Personal Communications*. https://doi.org/10.1007/s11277-020-07816-y
26. Nabi, M., Geilen, M., & Basten, T. (2011). Moban: A configurable mobility model for wireless body area networks. In *Proceedings of the 4th International ICST Conference on Simulation Tools and Techniques* (pp. 168–177).
27. Mohamad, N. M. V., Nagarajan, R., Muthukaruppan, L., Subramaniyam, H., Sendrayan, N., Thangappa, V., & Manickavelu, P. (2019). Uplink resource sharing and power management scheme for an underlay D2D communication. *Wireless Personal Communications*. https://doi.org/10.1007/s11277-019-06747-7
28. Zhao, W., & Wang, S. (2015). Resource sharing scheme for device-to-device communication underlaying cellular networks. *IEEE Transactions on Communications, 63*(12), 4838–4848.
29. Duong, Q., Shin, Y., & Shin, O.-S. (2015). Distance-based resource allocation scheme for device-to-device communications underlaying cellular networks. *AEU-International Journal of Electronics and Communications, 69*(10), 1437–1444.
30. GPP TR 36.814V9.0.0 (2010-03). (2010). 3rd Generation partnership project; technical specification group radio access network; evolved universal terrestrial radio access (E-UTRA); further advancements for E-UTRA physical layer aspects (Release 9).
31. Dhilipkumar, S., Arunachalaperumal, C., & Thanigaivelu, K. (2019). A comparative study of resource allocation schemes for D2D networks underlay cellular networks. *Wireless Personal Communications*. https://doi.org/10.1007/s11277-019-06204-5
32. Ahmad, M., et al. (2018). Joint user selection, mode assignment, and power allocation in cognitive radio-assisted D2D networks. *IET Communication, 12*(10), 1207–1214.
33. Hasan, M., & Hossain, E. (2014). Distributed resource allocation for relay-aided device-to-device communication: A message passing approach. *IEEE Transactions on Wireless Communications, 13*(11), 6326–6341.
34. Yu, C. H., Doppler, K., Ribeiro, C. B., & Tirkkonen, O. (2011). Resource sharing optimization for device-to-device communication underlaying cellular networks. *IEEE Transactions on Wireless Communications, 10*(8), 2752–2763.
35. Min, H., Lee, J., Park, S., & Hong, D. (2011). Capacity enhancement using an interference limited area for device-to-device uplink underlaying cellular networks. *IEEE Transactions on Wireless Communications, 10*(12), 3995–4000.

36. Feng, D., Lu, L., Yuan-Wu, Y., Li, G. Y., Feng, G., & Li, S. (2013). Device-todevice communications underlaying cellular networks. *IEEE Transactions on Communications, 61*(8), 3541–3551.
37. Song, L., Niyato, D., Han, Z., & Hossain, E. (2011). Game-theoretic resource allocation methods for device-to-device communication. *IEEE Transactions on Wireless Communications, 14*, 1284–1536.
38. Zhu, D., Jiaheng Wang, A., & Swindlehurst, L. (2014). Downlink resource reuse for device-to-device communications underlaying cellular networks. *IEEE Signal Processing Letters, 21*(5), 531–534.
39. Wang, F., et al. (2013). Energy-aware resource allocation for device-to-device underlay communication. In *IEEE international Conference Communications (ICC), Budapest.*
40. Wang, F., et al. (2013). Joint scheduling and resource allocation for device-to-device underlay communication. In *Proceedings of IEEE Wireless Communications Networking Conference (WCNC), Shanghai.*
41. Xu, C., et al. (2012). Resource allocation using a reverse iterative combinatorial auction for device-to-device underlay cellular networks. In *IEEE Global Communication Conference (Globecom), Los Angeles.*
42. Musku, M., et al. (2005). Joint rate and power control with pricing. In *IEEE GLOBECOM'05* (pp. 3466–3470).
43. Krithiga, S., Bhaskar, V., & Malarvizhi, S. (2020). Investigation on multi-user scheduling algorithms over non-Identical channel distribution and LTE networks. *Wireless Personal Communications*. https://doi.org/10.1007/s11277-020-07782-5
44. Yu, G., Wen, D., & Qu, F. (2016). Joint user scheduling and channel allocation for cellular networks with full duplex base stations. *IET Communications, 10*(5), 479–486.
45. Bechira, N., Nasreddine, M., Mahmoud, A., Walid, H., & Sofien, M. (2014). Novel scheduling algorithm for 3GPP downlink LTE cellular network. *Procedia Computer Science, 40*, 116–122.
46. Noliya, A., & Kumar, S. (2020). Performance analysis of resource scheduling techniques in homogeneous and heterogeneous small cell LTE-A networks. *Wireless Personal Communications, 112*(4), 2393–2422.
47. Priyadharsini, N. A., & Tamil Selvi, S. (2020). Adaptive spectrum aggregation regimen for downlink NR-gNodeB and device to device systems in 5G. *Wireless Personal Communications*. https://doi.org/10.1007/s11277-020-07944-5
48. Tehrani, M. N., Uysal, M., & Yanikomeroglu, H. (2014). Device to device communication in 5G cellular networks: Challenges, solutions and future directions. *IEEE Communication Magazine, 52*, 86–92.
49. Ibraheem, S., Mahamod, I., & Nordin, R. (2013). Downlink spectral efficiency evaluation with carrier aggregation in LTE-Advanced system employing Adaptive Modulation and Coding schemes. In *IEEE 11th Malaysia International Conference on Communications (MICC)* (pp. 98–103).
50. Panagiotis, D. D., Koralia, N. P., Muhaidat, S., et al. (2016). Carrier aggregation for cooperative cognitive radio networks. *IEEE Transactions on Vehicular Technology, 66*(7), 5904–5918.
51. Li, R., Hong, P., Xue, K., Zhang, M., & Yang, T. (2020). Energy-efficient resource allocation for high-rate underlay d2d communications with statistical csi: A one-to-many strategy. *IEEE Transactions on Vehicular Technology, 69*(4), 4006–4018.
52. Alsharif, M. H., Nordin, R., Shakir, M. M., & Mohd Ramly, A. (2019). Small cells integration with the macro-cell under LTE cellular networks and potential extension for 5G. *Journal of Electrical Engineering & Technology*. https://doi.org/10.1007/s42835-019-00173-2
53. Walid, A., Sabir, E., Kobbane, A., Taleb, T., & Koutbi, M. E. (2016). Exploiting multi-homing in hyper dense LTE small-cells deployments. In *2016 IEEE Wireless Communications and Networking Conference, Doha* (pp. 1–6).

54. Charbonnier, R., Aslam, M. Z., Corre, Y., & Lostanlen, Y. (2017). Mixing deterministic and stochastic propagation for assessing mmwave small-cell networks. In *11th European Conference on Antennas and Propagation (EUCAP), Paris* (pp. 136–140).
55. Kukade, S., Sutaone, M., & Patil, R. (2021). Optimal performance of resource allocation in LTE-A for heterogeneous cellular network. *Wireless Networks*. https://doi.org/10.1007/s11276-021-02635-w
56. Monikandan, B. S., Sivasubramanian, A., Babu, S. P. K., Venkatesan, G. P., & Arunachalaperumal, C. (2020). Channel aware optimized proportional fair scheduler for LTE downlink. *Peer-to-Peer Networking and Applications, 13*, 1–10.
57. Kabir, M. T., & Masouros, C. (2020). Delay-constrained beamforming and resource allocation in full duplex systems. *IEEE Transactions on Vehicular Technology, 69*(3), 3476–3480.
58. Chaudhuri, S., Baig, I., & Das, D. (2015). Utility based QoS aware uplink scheduler scheme for LTE small cell network. In *In 2015 IEEE International Conference on Communications (ICC)* (pp. 3149–3154). IEEE.
59. Thayammal, M. C., & Mary Linda, M. (2019). Utility-based optimal resource allocation in LTE-A networks by hybrid ACO-TS with MFA scheme. *The Computer Journal, 62*(6), 931–942.
60. Wu, X., Ma, Z., Chen, X., Labeau, F., & Han, S. (2019). Energy efficiency-aware joint resource allocation and power allocation in multi-user beamforming. *IEEE Transactions on Vehicular Technology, 68*(5), 4824–4833.
61. Lim, J. H., Ahmad, B., Jusoh, M., & Sabapathy, T. (2017). Deployment of resource allocation and power control schemes in long term evolution advanced (LTE-A) hybrid network. *Wireless Personal Communications, 96*(2), 2943–2963.
62. Baker, M. (2009). LTE-Advanced physical layer. In *Proc. IMT-Advanced Evaluation Workshop, 3GPP, Beijing* (pp. 1–48).
63. Wang, M., Zhang, J., Ren, B., Yang, W., Zou, J., Hua, M., & You, X. (2015). The evolution of LTE physical layer control channels. *IEEE Communications Surveys and Tutorials, 18*(2), 1336–1354.
64. Razavi, S. M., & Ratnarajah, T. (2016). Adaptive LS- and MMSE-based beamformer design for multiuser MIMO interference channels. *IEEE Transactions on Vehicular Technology, 65*(1), 132–144.
65. Jain, N., & Payal, A. (2021). Performance comparison between different tunneling techniques using different routing protocols. *Wireless Personal Communications*. https://doi.org/10.1007/s11277-021-09186-5
66. Albkerat, A., & Issac, B. (2014). Analysis of IPv6 transition technologies. arXiv Prepr. arXiv1410.2013.
67. Sookun, Y., & Bassoo, V. (2016). Performance analysis of IPv4/IPv6 transition techniques. *IEEE International Conference on Emerging Technologies and Innovative Business Practices for the Transformation of Societies (EmergiTech), 2016*, 188–193.
68. Carpenter, B., & Moore, K. (2001). Connection of IPv6 domains via IPv4 clouds. RFC 3056.
69. Ullah, H., Abu-Tair, M., McClean, S., Nixon, P., Parr, G., & Luo, C. (2020). Connecting disjoint nodes through a UAV-based wireless network for bridging communication using IEEE 802.11 protocols. *EURASIP Journal on Wireless Communications and Networking*. https://doi.org/10.1186/s13638-020-01727-z
70. Yanmaz, E., Kuschnig, R., & Bettstetter, C. (2011). Channel measurements over 802.11a-based UAV-to-ground links. In *2011 IEEE GLOBECOM Workshops (GC Wkshps)* (pp. 1280–1284). https://doi.org/10.1109/glocomw.2011.6162389
71. Shi, Y., Enami, R., Wensowitch, J., & Camp, J. (2018). Measurement-based characterization of LOS and NLOS drone-to-ground channels. In *2018 IEEE Wireless Communications and Networking Conference (WCNC)* (pp. 1–6). https://doi.org/10.1109/wcnc.2018.8377104
72. 3D-Robotics (3DR), Solo user manual V9, Â© 2015 3D Robotics Inc. (2017). https://3dr.com/support/articles/2083-96893/usermanual/. Retrieved Mar 2019.

73. More, S., & Naik, U. (2020). Optimal multipath routing for video transmission in VANETs. *Wireless Personal Communications*. https://doi.org/10.1007/s11277-020-07740-1
74. Mchergui, A., Moulahi, T., Alaya, B., & Nasri, S. (2017). A survey and comparative study of QoS aware broadcasting techniques in VANET. *Telecommunication Systems, 66*(2), 253–281.
75. Xie, H., Boukerche, A., & Loureiro, A. A. F. (2015). MERVS: A novel multi-channel error recovery video streaming protocol for vehicle ad-hoc networks. *IEEE Transactions on Vehicular Technology*, 14.
76. Zaidi, S., Bitam, S., & Mellouk, A. (2016). Enhanced adaptive sub-packet forward error correction mechanism for video streaming in VANET. In *2016 IEEE Global Communications Conference (GLOBECOM)* (pp. 1–6).
77. Salkuyeh, M. A., & Abolhassani, B. (2016). An adaptive multipath geographic routing for video transmission in Urban VANETs. *IEEE Transactions on Intelligent Transportation Systems*, 1–10.
78. Singh, S., Gupta, A., & Sohal, J. S. (2020). Transmission of audio over LTE packet based wireless networks using wavelets. *Wireless Personal Communications*. https://doi.org/10.1007/s11277-020-07059-x
79. Monro. (2004). *Audio compression*. US Patent App. 10(473), 649.
80. Singh, S., & Sohal, J. S. (2016). Wavelets based wireless VOIP and its future scenario, ICAET. In *MATEC Web of Conferences*. https://doi.org/10.1051/matecconf/20165701013
81. Singh, S., & Sohal, J. S. (2017). Performance analysis of data transmission over wireless VOIP using wavelets. *Indian Journal of Science and Technology*. https://doi.org/10.17485/ijst/2017/v10i18/112805
82. Singh, S., & Sohal, J. S. (2017). Wavelet based image transmission analysis for wireless VOIP. *International Journal on Recent and Innovation Trends in Computing and Communication, 5*(5), 622–627.
83. Singh, S., Gupta, A., Sohal, J. S., & Veer, K. (2018). Audio transmission over wavelet based wireless VOIP. In *Proceedings of Soft Computing: Theories and Applications (SoCTA)*. Dr. BR Ambedkar National Institute of Technology.
84. Floyd, S., & Fall, K. (1999). Promoting the use of end-to-end congestion control in the internet. *IEEE/ACM Transactions on Networking, 4*, 1–16.
85. Floyd, S., Handley, M., Padhye, J., & Widmer, J. (2000). Equation-based congestion control for unicast applications. In *Proceedings of ACM SIGCOMM'2000, Stockholm, Sweden* (pp. 43–56).
86. Paxson, V. (1999). End-to-end internet packet dynamics. *IEEE/ACM Transactions on Networking, 3*, 277–292.
87. Borella, M., Swider, D., Uludag, S., & Brewster, G. (1998). Internet packet loss: Measurement and implications for end-to-end QoS. In *Proceedings, Architectural and OS Support for Multimedia Applications/Flexible Communication Systems/Wireless Networks and Mobile Computing* (pp. 3–12).
88. Arai, M., Chiba, A., & Iwasaki, K. (1999). Measurement and modeling of burst packet losses in internet end-to-end communications. In *Pacific Rim International Symposium on Dependable Computing (PRDC)* (pp. 260–267).
89. Bolot, J. (1993). End-to-end packet delay and loss behaviour in the internet. *ACM SIGCOMM Computer Communication Review, 23*, 289–298.
90. Maxemchuk, N. F., & Lo, S. (1997). Measurement and interpretation of voice traffic on the Internet. In *IEEE Proceedings of communications, 1997. ICC'97 Montreal, Towards the Knowledge Millennium* (Vol. 1, pp. 500–507).
91. Bolot, J., & Vega-Garcia, A. (1996). Control mechanisms for packet audio in the internet. In *Proceedings IEEE Infocom, San Francisco, CA* (pp. 232–239).
92. Measurement and interpretation of voice traffic on the internet. In *Proceedings ICC, Montreal, Canada* (pp. 1–291).
93. Brosh, E., Baset, S. A., Rubenstein, D., & Schulzrinne, H. (2008). The delay-friendliness of TCP. *ACM SIGMETRICS Performance Evaluation Review, 36*(1), 49–60.

94. Mohamed, S., Cervantes-Perez, F., & Afifi, H. (2001). Integrating network measurements and speech quality subjective scores for control purposes. In *IEEE INFOCOM* (Vol. 2, pp. 641–649).
95. Hagsand, O., Hanson, K., & Marsh, I. (1999). Measuring internet telephony quality: Where are we today? In *Globecomm Telecommunications Conference—Globecomm'99* (Vol. 3, pp. 1838–1842).
96. Usman, Q., Chughtai, O., Nawaz, N., Kaleem, Z., Khaliq, K. A., & Nguyen, L. D. (2021). A reliable link-adaptive position-based routing protocol for flying ad hoc network. *Mobile Networks and Applications*. https://doi.org/10.1007/s11036-021-01758-w
97. Husnain, G., & Anwar, S. (2022). An intelligent probabilistic whale optimization algorithm (i-WOA) for clustering in vehicular ad hoc networks. *International Journal of Wireless Information Networks*. https://doi.org/10.1007/s10776-022-00555-w
98. Nazib, R. A., & Moh, S. (2020). Routing protocols for unmanned aerial vehicle-aided vehicular ad hoc networks: A survey. *IEEE Access, 8*, 77535–77560. https://doi.org/10.1109/ACCESS.2020.2989790
99. Shafiq, Z., Haseeb Zafar, M., & Qazi, A. B. (2018). QoS in vehicular ad hoc networks—A survey article info. *Journal of Information Communication Technology and Robotic Applications*. http://jictra.com.pk/index.php/jictra/article/view/81. Retrieved 6 Apr 2021
100. Senouci, O., Zibouda, A., & Harous, S. (2017). Survey: Routing protocols in vehicular ad hoc networks. In *ACM International Conference Proceeding Series* (pp. 1–6). https://doi.org/10.1145/3231830.3231838
101. Hu, L., Wang, H., & Zhao, Y. (2020). Performance analysis of DSRC-based vehicular safety communication in imperfect channels. *IEEE Access, 8*, 107399–107408. https://doi.org/10.1109/ACCESS.2020.3000534
102. Harrabi, S., Ben Jaafar, I., & Ghedira, K. (2017). Message dissemination in vehicular networks on the basis of agent technology. *Wireless Personal Communication, 96*(4), 6129–6146. https://doi.org/10.1007/s11277-017-4467-x
103. Chowdhary, N., & Kaur, P. D. (2018). Dynamic route optimization using nature-inspired algorithms in IoV. *Smart Innovation Systems and Technologies, 79*, 495–504. https://doi.org/10.1007/978-981-10-5828-8_47
104. Husnain, G., & Anwar, S. (2021). An intelligent cluster optimization algorithm based on Whale Optimization Algorithm for VANETs (WOACNET). *PLoS One, 16*(4), e0250271. https://doi.org/10.1371/JOURNAL.PONE.0250271
105. Yao, P., & Wang, H. (2017). Dynamic Adaptive Ant Lion Optimizer applied to route planning for unmanned aerial vehicle. *Soft Computing, 21*(18), 5475–5488. https://doi.org/10.1007/s00500-016-2138-6
106. Fahad, M., et al. (2018). Grey wolf optimization based clustering algorithm for vehicular ad hoc networks. *Computers and Electrical Engineering, 70*, 853–870. https://doi.org/10.1016/j.compeleceng.2018.01.002
107. Rattal, S., & Ar-Reyouchi, E. M. (2019). An effective practical method for narrowband wireless mesh networks performance. *SN Applied Sciences*. https://doi.org/10.1007/s42452-019-1595-9
108. Chatei, Y., Ghoumid, K., & Ar-Reyouchi, E. M. (2017). Narrow-band channel spacing frequencies metric in one-hop wireless mesh networks. In *2nd International Conference on Communication and Electronics Systems (ICCES)* (pp. 296–301). https://doi.org/10.1109/cesys.2017.8321284
109. Ar-Reyouchi, E. M., Chatei, Y., Ghoumid, K., & Lichioui. (2015). The powerful combined effect of forward error correction and automatic repeat reQuest to improve the reliability in the wireless communications. In *IEEE International Conference on Computer Science, Las Vegas, Nevada, USA* (pp. 691–696). https://doi.org/10.1109/csci.2015.39
110. Chatei, Y., Hammouti, M., Ghadban, A., Ar-Reyouchi, E. M., Mekaoui, S., & Ghoumid, K. (2018). Impact of message size on the average delay in narrowband wireless links. In *Proceedings of the 1st International Conference on Electronic Engineering and Renewable*

Energy. ICEERE 2018, 15–17 April 2018, Saidia, Morocco, 1 Jan 2019 (pp. 51–57). https://doi.org/10.1007/978-981-13-1405-6_8
111. Banerjee, P. S., Mandal, S. N., De, D., & Maiti, B. (2019). iSleep: Thermal entropy aware intelligent sleep scheduling algorithm for wireless sensor network. *Microsystem Technologies*. https://doi.org/10.1007/s00542-019-04706-7
112. Bannister, K., Giorgetti, G., & Gupta, S. K. (2008). Wireless sensor networking for hot applications: Effects of temperature on signal strength, data collection and localization. In *Proceedings of the 5th Workshop on Embedded Networked Sensors (HotEmNets' 08)*.
113. Hati, S., Dey, P., & De, D. (2019). WLAN based energy efficient smart city design. *Microsystem Technologies, 25*(5), 1599–1612.
114. Boano, C. A., Zúñiga, M., Brown, J., Roedig, U., Keppitiyagama, C., & Römer, K. (2014). Templab: A testbed infrastructure to study the impact of temperature on wireless sensor networks. In *Proceedings of the 13th International Symposium on Information Processing in Sensor Networks* (pp. 95–106). IEEE Press.
115. Beutel, J., Buchli, B., Ferrari, F., Keller, M., Zimmerling, M., & Thiele, L. (2011). X-sense: sensing in extreme environments. In *2011 Design, Automation and Test in Europe* (pp. 1–6). IEEE Press.
116. Sun, J., & Cardell-Oliver, R. (2006). *An experimental evaluation of temporal characteristics of communication links in outdoor sensor networks* (pp. 73–77). Association for Computing Machinery (ACM).
117. Penrose, M. (2003). *Random geometric graphs* (Vol. 5). Oxford University Press.
118. Hassani, M. M., & Berangi, R. (2019). Impact of secondary user block on the TCP throughput in cognitive radio sensor networks. *Wireless Personal Communications*. https://doi.org/10.1007/s11277-019-06677-4
119. Di Felice, M., Chowdhury, K. R., & Bononi, L. (2009). Modeling and performance evaluation of transmission control protocol over cognitive radio ad hoc networks. In *Proceedings of the ACM Conference on Modeling, Analysis and Simulation of Wireless and Mobile Systems* (pp. 4–12). https://doi.org/10.1145/1641804.1641809
120. Amjad, M. F., Aslam, B., & Zou, C. (2013). Transparent cross-layer solutions for throughput boost in cognitive radio networks. In *Proceedings of IEEE CCNC 2013* (pp. 580–586). https://doi.org/10.1109/CCNC.2013.6488502
121. Wang, J., Huang, A., Wang, W., Zhang, Z., & Lau, V. K. N. (2014). On the transmission opportunity and TCP throughput in cognitive radio networks. *International Journal of Communication Systems, 27*(2), 303–321. https://doi.org/10.1002/dac.2362
122. Li, G., Hu, Z., Zhang, G., Zhao, L., Li, W., & Tian, H. (2011). Cross-layer design for energy efficiency of TCP traffic in cognitive radio networks. In *Proceedings of IEEE VTC fall* (pp. 1–5).
123. Wang, X., Sun, X., Zhao, C., & Zhou, Z. (2010). TCP-CReno-TCP enhancement using cross-layer for cognitive radio networks. In *Proceedings of IET AIAI* (pp. 37–40). https://doi.org/10.1109/icbnmt.2010.5705147
124. Kumar, A., & Shin, K. G. (2012). DSASync: Managing end-to-end connections in dynamic spectrum access wireless LANs. *IEEE/ACM Transactions on Networking, 20*(4), 1068–1081. https://doi.org/10.1109/TNET.2011.2178264
125. Luo, C., Yu, F. R., Ji, H., & Leung, V. C. (2010). Cross-layer design for TCP performance improvement in cognitive radio networks. *IEEE Transactions on Vehicular Technology, 59*(5), 2485–2495. https://doi.org/10.1109/tvt.2010.2041802
126. Wang, J., Huang, A., & Wang, W. (2012). TCP throughput enhancement for cognitive radio networks through lower-layer configurations. In *Proceedings of IEEE PIMRC* (pp. 1424–1429). https://doi.org/10.1109/pimrc.2012.6362571
127. Khalife, H., Conan, V., Leguay, J., & Spyropoulos, T. (2013). Point to multipoint transport in multichannel wireless environments. In *Proceedings of IEEE WCNC* (pp. 1404–1409). https://doi.org/10.1109/wcnc.2013.6554769

128. Al-Ali, A. K., & Chowdhury, K. R. (2013). TFRC-CR: An equation-based transport protocol for cognitive radio networks. *Elsevier Ad Hoc Networks, 11*(6), 1836–1847. https://doi.org/10.1016/j.adhoc.2013.04.007
129. Akan, O. B., Karli, O. B., & Ergul, O. (2009). Cognitive radio sensor networks. *IEEE Network, 23*(4), 34–40.
130. Mehta, R. (2021). Trade-off between spectral efficiency and normalized energy in Ad-hoc wireless networks. *Wireless Networks*. https://doi.org/10.1007/s11276-021-02610-5
131. Mehta, R., & Lobiyal, D. K. (2019). Performance modeling and optimization of multiple-objective cross-layer design in multi-flow ad-hoc networks. *International Journal of Communication Systems, 32*, e3861. https://doi.org/10.1002/dac.3861
132. Chen, X., Hu, R. Q., Wu, G., & Li, Q. C. (2015). Tradeoff between energy efficiency and spectral efficiency in a delay constrained wireless system. *Wireless Communications and Mobile Computing, 15*, 1945–1956. https://doi.org/10.1002/wcm.2469
133. Yu, W., Musavian, L., & Ni, Q. (2015). Multi-carrier link-layer energy efficiency and effective capacity tradeoff. In *2015 IEEE International Conference on Communication Workshop (ICCW), London, UK* (pp. 2763–2768). https://doi.org/10.1109/ICCW.2015.7247597
134. Xiang, L., Chen, H., & Zhao, F. (2017). Area spectral efficiency and energy efficiency tradeoff in ultradense heterogeneous networks. *Wireless Communications and Mobile Computing, 2017*, 4390197. https://doi.org/10.1155/2017/4390197
135. Maharazu, M., Hanapi, Z. M., & Alrashah, M. A. (2021). Energy and spectral efficiency balancing algorithm for energy saving in LTE downlinks. *Symmetry, 13*(211), 19. https://doi.org/10.3390/sym13020211
136. Latif, G., Saravanakumar, N., Alghazo, J., Bhuvaneswari, P., Shankar, K., & Butt, M. O. (2020). Scheduling and resources allocation in network traffic using multiobjective, multiuser joint traffic engineering. *Wireless Networks*. https://doi.org/10.1007/s11276-020-02407-y
137. Floyd, S., & Jacobson, V. (1993). Random early gateway for congestion avoidance. *IEEE/ACM Transactions on Networking, 1*(4), 397–413. https://doi.org/10.1109/90.251892
138. Gibbens, R. J., & Kelly, F. P. (1999). Resource pricing and the evolution of congestion control. *Automatica, 35*(12), 1969–1985.
139. Shenker, S. (1995). Fundamental design issues for the future Internet. *IEEE Journal on Selected Areas in Communications, 13*(7), 1176–1188.
140. He, J., Chiang, M., & Rexford, J. (2007). Towards robust multi-layer traffic engineering: Optimization of congestion control and routing. *IEEE Journal on Selected Areas in Communications, 25*(5), 868–880.
141. He, J. (2008). Rethinking traffic management: Design of optimizable networks. Ph.D. dissertation, Electrical Engineering Dept., Princeton, New Jersey.
142. Abramson, M. A. (2002). Pattern search algorithms for mixed variable general constrained optimization problems, Ph.D. thesis, Department of Computational and Applied Mathematics, Rice University, Houston, TX, 2002.
143. Al Emam, F. A., Nasr, M. E., & Kishk, S. E. (2019). Mobility management using dynamic cross-layer signaling in heterogeneous wireless networks. *Wireless Networks*. https://doi.org/10.1007/s11276-019-02162-9
144. Yuan, Z., & Muntean, G. (2013). A prioritized adaptive scheme for multimedia services over IEEE 802.11 WLANs. *IEEE Transactions on Network and Service Management, 10*(4), 340–355.
145. Mansouri, W., Zarai, F., Mnif, K., & Kamoun, L. (2017). Cross layer architecture with integrated MIH in heterogeneous. *Computer Networks, 127*, 126–137.
146. Wang, J., Meng, S., Sun, J., & Quo, Z. (2016). A general PID-based rate adaptation approach for TCP-based live streaming over mobile networks. In *IEEE International Conference on Multimedia and Expo (ICME)* (pp. 1–6).
147. Evensen, K., Kaspar, D., Griwodz, C., Halvorsen, P., Hansen, A., et al. (2011). Improving the performance of quality-adaptive video streaming over multiple heterogeneous access

networks. In *Proceedings of the Second Annual ACM Conference on Multimedia Systems* (pp. 57–68).

148. Qi, Q., Cao, Y., Li, T., Zhu, X., & Wang, J. (2010). Soft handover mechanism based on RTP parallel transmission for mobile IPTV services. *IEEE Transactions on Consumer Electronics, 56*(4), 2276–2281.

Chapter 8
Optimization Techniques for Optimizing the Next Generation Radio Network Parameters

Milind Pande, Anand J. Kulkarni, and Apoorva S. Shastri

Introduction by the Editor

For low latency, efficient channel modelling, improved error detection, and correction high-speed wireless connectivity optimization of radio network parameters is of foremost importance. This chapter is written after taking several techniques into consideration spanning from Routing constraints in the device-to-device communication for beyond IoT 5G networks: a review, Mixed Uplink, Downlink Channel Allocation, and Power Allocation Schemes for 5G Networks and Radio resource management: approaches and implementations from 4G to 5G and beyond, Deep Reinforcement Learning Aided Cell Outage Compensation Framework in 5G Cloud Radio Access Networks, Optimization of Spreading Code Using Modified Differential Evolution for Wireless Communication along with Intensive Benchmarking of D2D communication over 5G cellular networks: prototype, integrated features, challenges, and main applications tech which are based on 5G technology and the future it withholds with it. Along with these, techniques based on Bio inspired algorithms such as Improved Particle Swarm Optimization Algorithm for UAV Base Station Placement, Pattern Correction in Planar Antenna Array using Multi-Objective Particle Swarm Optimization and Continuous Excitation and optimal compact time-modulated circular antenna array synthesis using krill herd optimization are instituted in this chapter.

M. Pande
Dr. Vishwanath Karad MIT World Peace University, Pune, Maharashtra, India
e-mail: milind.pande@mitwpu.edu.in

A. J. Kulkarni · A. S. Shastri (✉)
Institute of Artificial Intelligence, Dr. Vishwanath Karad MIT World Peace University, Pune, Maharashtra, India
e-mail: anand.j.kulkarni@mitwpu.edu.in

M. Pande et al. (eds.), *Optimization Methods in Mobile Communication Systems*, https://doi.org/10.1007/978-981-95-1810-4_8

Routing is fundamental in any wireless sensor network to ensure fast and reliable data transfer, Ineffective routing results in flapping and low Quality of Service (QoS), D2D communication provides reliable communication as in this communication allows device to be connected without partial involvement of cellular network. In wireless networks, routing protocols can be divided into three categories: proactive, reactive, and hybrid routing protocols. Devices that use proactive routing protocols must share updated routing information tables. Devices with reactive routing protocols do not retain previous routing data; nevertheless, they only change network routing information when communication is required. Hybrid Routing Protocol combines the best futures of the first two categories. The Zone Routing Protocol (ZRP) is one of the hybrid routing protocols in which every network node proactively maintaining routing information about its routing zone, while reactively acquiring routes to destinations beyond the routing zone.

In D2D communication cellular user experience a degradation in quality of service due to interference of uplink and downlink results in reduced throughput. To maximize system throughput interference mitigation scheme incorporated in D2D communication. There are three subparts of this scheme first part is subdivision concept which virtually divide cell into outer centre and inner regions. In second step, based on particle swarm optimization power is allocated in mid-cell region. In third part resources are allocated in cell subdivisions. In the Mixed uplink and downlink channel allocation and power allocation problem, overall network throughput is maximized by efficiently assigning both the UL and DL channels and setting power for the cell users and D2D transmitters. UL and DL resource allocation and power allocation problem is divided into two sub-problems. In the first sub-problem, assignment of UL and DL channels to the centre and outer cell region users and in the second sub-problem, power allocation scheme take place for mid-cell region users.

Networking standards of 5G demands efficient management of interference and spectrum while considering fairness and Quality of Service (QoS). Radio resource management (RRM) schemes that evolved during LTE/LTE-A network environment period will remain relevant for 5G, however, these schemes must become more intelligent and adaptive for future as features and requirements of network and users will be diverse and highly demanding. Radio resource management is key for satisfying the standard requirements of 5G environment, as it demands efficient management of interference and spectrum while considering fairness and Quality of Service (QoS) requirements as well. There are few main functions of RRM in LTE/LTE-A cellular technologies such as packet scheduling, resource allocation, link adaptation and control, radio admission and removal and handover management which can also efficiently work with 5G.

Cloud Radio Access Networks (C-RAN) with cloud BBUs (Base Band Units) pool architecture and distributed RRHs (Remote Radio Heads) can provide the ubiquitous services. Failure at RRH lead to a significant drop in network performance. DRL based framework handle cell outages in C-RAN and maximize energy efficiency and connectivity of RRH while improving quality of service for each compensation user. In this framework compensation users are allocated to adjacent

RRHs by using the K-means clustering algorithm. Secondly, DQN is used to find the antenna down tilt and the power allocated to compensation users.

Evolving 5G cellular warless network are assumed to provide higher data rates, lower energy consumption and lower end to end latency.D2D in one of the keys to enhance performance of 5G. The D2D communication technology underlying mobile network provides numerous benefits such as raising overloaded cellular network efficiency to reduce latency and minimize consumed energy. But there are vital challenges D2D technology faces such as handover, energy consumption, interference, resource allocation, and security.

In cellular network, base station might get out of service due to some fault or natural distastes. Solution to this problem is to send low altitude drones equipped with small cellular BS. Mixed integer linear programming formulation solve problem of Finding the minimum number of drone base stations (DBS) and their optimal location. An improved PSO-based algorithm is proposed to jointly optimize DBSs' locations and find the minimum number of DBSs. As in the original PSO algorithm, the particles are randomly distributed in the initialization phase and a K-means-based clustering method is employed to generate the positions of the first-generation particles (DBSs). In addition, a custom communication protocol is presented for data exchange between the user's equipment (UE) and the network controller. Multi-objective particle swarm optimization (MOPSO) and continuous excitation distribution in the range of 0 to 2 are used for correction of the failed element in 5×5 planar antenna array for ISM band application by substituting role of non-operational elements to remaining operational elements. It was found that up to 72% element failure at the various position of the array can be effectively recovered for desired sidelobe level and beamwidth first null (FNBW).

Optimal far-field radiation pattern synthesis by using time modulation technique is carried out for a compact circular antenna array (CAA) design using a stochastic optimization method called krill herd (KH) optimization. A compact antenna array is achieved by reducing the inter-element spacing between the array elements, which leads to the miniaturization of the size of the array. In time modulation technique, 'time' has the most significant role. It is considered the fourth-dimensional parameter, which plays the role of an additional degree of freedom to achieve an improved radiation pattern. The time-modulated antenna array (TMAA) radiates at various harmonic frequencies for the periodic switching time sequence. The KH algorithm is applied here to find out the optimum sets of switching on-time duration for each array element and to determine the optimum inter-element separation between two array elements of the 20- and 36-element time-modulated circular antenna array (TMCAA).

Machine Generated Summaries

Disclaimer: The summaries in this chapter were generated from Springer Nature publications using extractive AI auto-summarization: An extraction-based summarizer aims to identify the most important sentences of a text using an algorithm and uses those original sentences to create the auto-summary (unlike generative AI). As the constituted sentences are machine selected, they may not fully reflect the body of the work, so we strongly advise that the original content is read and cited. The auto generated summaries were curated by the editor to meet Springer Nature publication standards. To cite this content, please refer to the original papers.

Machine generated keywords: image, device, optimization, cellular, edge, pattern, management, cellular network, antenna array, placement, content, element, array, discovery, code.

An Improved Particle Swarm Optimization Algorithm for UAV Base Station Placement [1]

This is a machine-generated summary of:

Pasandideh, Faezeh; E. Rodriguez Cesen, Fabricio; Henrique Morgan Pereira, Pedro; Esteve Rothenberg, Christian; Pignaton de Freitas, Edison: An Improved Particle Swarm Optimization Algorithm for UAV Base Station Placement [1]

Published in: Wireless Personal Communications (2023)

Link to original: https://doi.org/10.1007/s11277-023-10334-2

If you want to cite the papers, please refer to the original.

For technical reasons we could not place the page where the original quote is coming from.

Abstract-Summary

"One of the promising solutions to fix this situation is to send low altitude drones equipped with a small cellular BS (DBSs) to the target locations."

"Finding the minimum number of DBSs and their optimal locations are highly challenging issues."

"A Mixed-Integer Non-Linear Programming formulation is provided, in which the DBSs' location and the proper number of DBSs are jointly determined."

"An improved PSO-based algorithm is proposed to jointly optimize DBSs' locations and find the minimum number of DBSs."

"As in the original PSO algorithm, the particles are randomly distributed in the initialization phase and a K-means-based clustering method is employed to generate the positions of the first-generation particles (DBSs)."

"The acquired results show that the proposed solution based on the integration of PSO and K-means algorithms provides a low packet loss and latency."

Introduction

"The drone-mounted BSs (DBSs) provide cellular networks with vital communication links and make available temporary coverage for the users in circumstances of connection shortage."

"The above-mentioned literature have shown the network performance improvement is achievable by employing DBSs or UAV-BSs in cellular networks."

"They manage the DBSs or UAV-BSs in an uncoordinated manner which suffers from lacked global network information and centralized decision making."

"The main contributions of this paper can be summarized as follows: The DBS Placement Problem is formulated as a Mixed-Integer Non-Linear Programming (MINLP) and then solved; A coverage modeling module and traffic modeling module consisting of a novel queuing modeling is developed within the controller; A PSO-based algorithm is proposed to jointly optimize DBSs' locations and to find the minimum number of DBSs; The unsupervised ML clustering technique called K-means is employed in the PSO algorithm that predicts the initial value of the number of DBSs more intelligently; and A custom communication protocol (CCP) developed for exchanging the UE's main data between UEs and the network controller."

Related Work

"In [2], authors propose a method to solve the UAV-BS placement which includes a classical Branch and Bound algorithm with relaxation induced neighborhood search (RINS) and a heuristic algorithm to the UAV-BS placement and user association problem in linear time."

"In [3], authors take advantage of a Genetic Algorithm (GA) to solve the optimization problem in which the UAV-BS radius coverage and its horizontal position are considered as a gene in the GA model."

"After finding the optimal location of the UAV-BS, the authors reduce the resource allocation problem to a Multidimensional Binary Knapsack Problem (MBKP) that can be solved in polynomial time."

"In [4], authors solve the deployment problem of multiple UAVs by adopting a centralized multi-agent Q-learning algorithm."

"Authors utilize the modified alternating direction method of multipliers (ADMM) and the successive convex optimization (SCO) algorithms to make the 3D UAV deployment problem and user association and scheduling blocks more tractable."

System Model

"This paper only considers the data transmission between users and the DBSs (access link) for the sake of simplicity."

"DBS deployment affects the users coverage area as well as the reliability of the Air to Ground (AtG) communication links."

"The probability of having a LoS connection between user and DBS depends on different factors, such as building density, user's location, and DBS elevation angle between the user and DBS."

"The probability of having a LoS and NLOS connections between user j and the DBS i are calculated according to [5]."

"The initial number of required DBSs (NDBS) to start serving a set of users in an area can be defined as the total number of users in the network divided by the maximum number of users that each DBS covers."

Proposed Solution

"The network is composed of a number NDBS of DBSs each that can support a maximum number of users M. Both NDBS and M are known by every DBS and the controller, such as the traffic request (packet/s) for UE within the network."

"PSO algorithm optimizes the placement problem of DBSs by using the sets of candidate DBS positions called particles and moving the swarm of particles, which represent potential solutions, around in the search-space according to simple mathematical formulae over the particle's position, velocity, and cost value."

"If it is not minimized, it is realized that the number of DBSs is not sufficient, then this number is increased by one and perform the steps from scratch. (a) Improved population initialization phase in the original PSO algorithm, the particles are distributed randomly in the initialization phase."

"A K-means based clustering method can be proposed as an initialization method to generate the positions of the first-generation particles (DBSs)."

Evaluation

"The remote API acts as a controller, calculating the best positions for the DBSs, providing mobility and movement models, then sending all the information to the CoppeliaSim to verify if DBSs are located at the best places, the maximum number of users are covered and so forth."

"This section discusses the simulation results obtained from the characterization of the proposed algorithm including, the rate of packet loss, latency, bandwidth, and the number of covered and uncovered users."

"Since, 3 or 5 DBSs are not enough to serve 150 users, the proposed algorithm tries to divide the number of available DBSs, between all the users, which is not going to be able to fully cover all the users at the same time."

Conclusion

"This paper proposed an improved PSO-based placement algorithm to find the minimum number of DBSs and their optimal locations."

"An initialization approach that estimates the initial value of the number of DBSs is provided by employing a K-means clustering-based scheme in the PSO algorithm to improve the performance."

"To extend the work, in the future we will propose more other existing evolutionary computation algorithms such as Ant colony optimization (ACO) and Genetic Algorithm (GA) combining with the K-means algorithm or other advanced clustering algorithms such as Density-based spatial clustering of applications with noise (DBSCAN) to solve the placement problem."

Routing Constraints in the Device-to-Device Communication for Beyond IoT 5G Networks: A Review [6]

This is a machine-generated summary of:

Malathy, S.; Jayarajan, P.; Hindia, M. H. D. Nour; Tilwari, Valmik; Dimyati, Kaharudin; Noordin, Kamarul Ariffin; Amiri, I. S.: Routing constraints in the device-to-device communication for beyond IoT 5G networks: a review [6]

Published in: Wireless Networks (2021)

Link to original: https://doi.org/10.1007/s11276-021-02641-y

If you want to cite the papers, please refer to the original.

For technical reasons we could not place the page where the original quote is coming from.

Abstract-Summary

"Routing is fundamental in any wireless network for path selection, which provides the most effective way that legitimizes the data to be transmitted from a source to a destination device."

"Device-to-Device communications (D2D) is a technology that allows the devices to be connected without or partial involvement of the conventional cellular network."

"With the dynamic nature of D2D technology, the routing approach act as a principal architecture that essential to be implemented in every niche D2D aspect."

"If wrong routing decisions are made in D2D communication, the QoS performance would be worse than the conventional cellular network."

"This paper present the state of the art of fundamentals, recent progress, current challenges, future directions, and potential routing applications for D2D and Beyond IoT 5G Networks."

"This review will also act as a guide and reference for future researchers and scientists to explore and integrate the routing technique in D2D communication and Beyond IoT 5G Networks."

Introduction
"In the D2D communications B5G wireless network, routing is crucial to ensure the data is efficiently sent to the destination device (end-user)."

"The selection of an optimum path from the source to destination devices in the D2D communication over IoT B5G cellular networks is limited due to the network instability, data packets traffic congestion, energy resource constraints, and link quality of the devices."

"One of the limitations of the D2D communication over IoT B5G cellular networks is the data packets traffic congestion issue that takes place when there is an excessive amount of data packets flow are injected on a single or particular device, and that specific mobile device carries most of the network data packet traffic."

"In the D2D communication over IoT B5G cellular networks, energy resources and link quality of the device play an important role in providing a seamless connection with the end-users."

Overview of D2D Communication
"D2D communication technology is one of the essential components of the future IoT B5G mobile network."

"During times of calamitous events, for example, an earthquake, floods, tsunami, and etc., established cellular communication infrastructures may become dysfunctional owing to failure or damage; in such a scenario, the D2D communications based IoT B5G network technology becomes the most suitable mode to offer swift deployment and self-manage communication with the peer-users."

"The following section describes the importance of routing in the D2D communication over IoT B5G Cellular Networks."

"The next section illustrates the challenges that occur during route selection in the D2D communication over IoT B5G cellular networks."

The Importance of Routing in D2D Communication
"To implement D2D communications in the B5G network, resource allocation and relay selection need to be considered for providing better QoS performances [7–10]."

"The mobile devices in the D2D communications have the ability to established links among themselves without or partiality involve in the cellular network infrastructure, in which all mobile devices act as a host as well as an intermediates device for other devices to build network infrastructure."

"Based on the D2D communication, all mobile devices in the B5G IoT network interact with each other and act as relay devices, ensuring the reliable and efficient operation of the entire network."

"In the D2D communication over IoT B5G network, the governing body is absent, and the devices are moving unpredictably; hence, it is a very difficult task to keep the network deployed in this scenario."

Types of Routing Protocols
"In wireless networks, routing protocols can be divided into three categories: proactive, reactive, and hybrid routing protocols."

"Devices that use proactive routing protocols must share updated routing information tables."

"Devices with reactive routing protocols do not retain previous routing data; nevertheless, they only change network routing information when communication is required."

"Based on the reactive routing protocol, the source device only searches for a route when it needs to connect to the target device."

"The process of initiating a reactive protocol routing discovery involves all devices exchanging topology information with each other."

"Reactive protocols take time to collect, maintain, and analyze the topological information about the network; taking this to account, it increases the average delay in data transmission from the source to target devices [11–15]."

"If the target device is not situated within the location of the nearest neighbour, it uses the reactive routing protocol."

Routing Key Challenges in D2D Communication

"Challenges that occur during route selection in the D2D communication over IoT 5G networks such as energy resource utilization, network, and route instability, data packets traffic congestion, and link quality constraints of the devices are discussed as follows: The mobile device in the D2D communications network changes its position unpredictably; thus, the topology of the network also changes accordingly."

"In order to maintain network and route stability in the D2D communication over IoT 5G networks, several studies have been conducted by researchers are discussed as follow: In the article [16], the authors proposed an Energy-aware Multipath Routing Protocol (EMRP), which clarifies information about the physical and medium access control layers to evaluate the energy resource and link bandwidth availability of devices that enable it to choose the optimum path."

"For link instabilities, limited energy resource constraint, and device movement of dynamic topological changing networks, the article [17] proposed a Multipath-ChaMeLeon (M-CML) routing scheme to boost the overall network performance metrics."

The Future Direction of Routing in D2D Communication over IoT B5G Networks

"In order to fulfil the growing demand for peer-to-peer communication, the use of a D2D communication B5G network with efficient and reliable routing is one of the key solutions to deliver better QoS network performance to the users."

"It also brings various technical challenges like substantial computation cost, modulation scheme selection for communication mode, and noise control, which need to be considered for optimal route and network performance."

"D2D communication is one of the pivotal technologies in the B5G wireless network."

"In D2D communication, devices share the same spectrum resource link with cellular and mobile devices links, resulting in interference among the spectrum-sharing links, thereby degrading the entire network performance."

Conclusion

"This paper has reviewed the current state of art of routing and its applications of the D2D communications and Beyond IoT 5G Networks."

"The routing key challenges in D2D communications have been highlighted and discussed, while the future direction of routing in D2D communication over IoT B5G networks insight has been provided."

"Whereas the energy resource and link quality constraints are the key selection criteria that need to be critically considered when designing and implementing a routing protocol for D2D communication over IoT B5G networks."

"The selection of efficient and reliable relay devices in the path selection procedure remains one of the most imperative and crucial challenges for reliable data packets transmission in the D2D communication over IoT B5G networks."

Mixed Uplink, Downlink Channel Allocation and Power Allocation Schemes for 5G Networks [18]

This is a machine-generated summary of:

Dubey, Rishav; Mishra, Pavan Kumar; Pandey, Sudhakar: Mixed Uplink, Downlink Channel Allocation and Power Allocation Schemes for 5G Networks [18]

Published in: Wireless Personal Communications (2020)

Link to original: https://doi.org/10.1007/s11277-020-07148-x

If you want to cite the papers, please refer to the original.

For technical reasons we could not place the page where the original quote is coming from.

Abstract-Summary

"Without an efficient interference management scheme in place, cellular users experience a degradation in the quality of service, which makes interference mitigation a major deployment challenge."

"We have put forward an efficient interference mitigation scheme in order to maximize the system throughput."

"We introduce a cell sub-division concept under which we virtually subdivide the cell into three regions, namely center, mid and outer regions."

"We formulate the power allocation problem as a robust particle swarm optimization algorithm for the mid-cell region."

"This enables the scheme in handling the uncertainties in the allocated power, thus reaching a robust optimal solution."

"The third part consists of a resource allocation scheme built around a cell subdivision concept."

Introduction
"In [19] A mode selection and resource allocation scheme based on particle swarm optimization is proposed for downlink scenario, where users switch their mode (CU or D2D mode) based on the proximity."

"The proposed scheme is hybrid; it only deals with channel allocation for a restricted number of users with fixed transmit power."

"We propose a channel allocation scheme as well as a power allocation scheme in the 5G networks."

"Unlike the previous studies, we eliminate the need for perfect CSI by mapping the users into regions where direct mapping of resources can be carried out, thus significantly reducing the computation cost. We propose power allocation schemes for UL and DL scenarios respectively by formulating the robust PSO based scheme by considering the uncertainties caused due to the stochastic nature of the input parameters."

Related Work
"Resource allocation and power control schemes are addressed as the major approaches to alleviate the interference in D2D underlay cellular networks."

"In [20], the authors proposed a local search based resource allocation scheme for maximizing the network throughput."

"In [21] authors, introduced a joint mode selection, channel allocation, and power control scheme to maximize the network throughput."

"In [22] authors proposed a resource allocation scheme that aims for longer battery life at each user."

"In [23], the authors proposed a channel allocation scheme for minimizing the interference and also maximizing the overall network throughput."

"In [24], the authors proposed a scheme aiming that maximizing the spectrum efficiency of per user in the network by minimum power consumption."

"In [25], the authors proposed a genetic algorithm-based joint resource allocation and user matching scheme (GAAM)."

"The authors proposed a scheme for multi-sharing resource allocation with theoretical analysis that gives optimal solutions."

System Model
"The D2D pairs can reuse the either UL or DL channels of available CUs."

"We allow one D2D pair to reuse the channel of only one CUs."

"D2D pairs reuse the DL resources of CUs, then the victim of interference are CUs and D2D receivers."

"This interference is produced when D2D user reuses the DL resources of CUs."

"The volume of interference depends on transmit power of D2D-transmitter and eNB along with channel gain between the D2D transmitter to cellular user and eNB to D2D receivers."

"D2D pairs reuse the UL resources of CUs, then the victim of interference are eNB and D2D receivers."

"This interference is produced when D2D user reuse the UL resources of CUs."

"The volume of interference is depend on transmit power of D2D-transmitter and CUs along with channel gain between the D2D transmitter to eNB and CUs to D2D receivers."

Problem Formulation
"In the Mixed UL and DL channel allocation and power allocation problem, we aim to maximize the overall network throughput by efficiently assigning both the UL and DL channels and setting power for the CUs and D2D transmitters."

"In the first sub-problem, we assign the UL and DL channels to the center and outer cell region users and in the second sub-problem, power allocation scheme take place for mid-cell region users."

"We assume for the mid-region of the cell, D2D users reuse the UL or DL resource blocks of CUs using the proposed channel allocation scheme."

"The mathematical formulation of the UL and DL channel allocation problem in the center and outer cell region is given as follows: S.t."

Up-Link and Down-Link Resource Allocation and Power Allocation Schemes
"We propose UL and DL channel allocation scheme for center cell and outer cell region and R PSO based power allocation scheme for the mid-cell region."

"This eliminates the need for gaining the perfect CSI as the resources can be directly mapped from the center to outer region with least computation."

"Although PSO offers the optimality in computing the global maximum it somehow is not able to take into consideration because of the uncertainty involved with the changing inputs."

"In order to solve the problem of handling the uncertainty in the input, we employ robust PSO which considers the uncertainty in the input of the parameters."

"In the initialization phase, we pick the CU and D2D transmit power and for which we assign an upper bound and lower bound values in order to search for an optimal solution."

Performance Evaluation
"We compare the proposed uplink and downlink resource allocation and robust PSO-based power allocation scheme with three schemes mentioned in the literature."

"In the first scenario, we simulate the proposed power allocation scheme in the uplink resource domain while in the second scenario we consider the downlink resource domain."

"We put forward a comparison between the overall throughput of proposed MRUPA and MRDPA scheme with the above-mentioned schemes."

"We draw a comparison between the throughput attained by the proposed MRUPA and MRDPA scheme with the related works."

"Although the throughput achieved by the proposed MRUPA and MRDPA schemes outperforms the throughput attained by the previous works, observing the fact that most of the previous works take into consideration that at most two D2D pairs share the same resource with one CU, we thus formulate that the number of D2D pairs is at most twice the number of CU users."

Conclusion and Future Work

"We have proposed MRDPA and MRUPA schemes in order to carry out power allocation and channel allocation with an aim of interference mitigation and throughput maximization."

"The scheme subdivides the cell into three subparts, center, mid and outer region."

"For the mid-cell users, we devise a power control mechanism to mitigate the interference."

"Through extensive simulations under different scenarios and parameter settings, we have shown that our scheme achieves better interference mitigation and thus higher throughput."

Optimization of Spreading Code Using Modified Differential Evolution for Wireless Communication [26]

This is a machine-generated summary of:

Kashyap, Katyayani; Pathak, Sunil; Yadav, Narender Singh: Optimization of Spreading Code Using Modified Differential Evolution for Wireless Communication [26]

Published in: Wireless Personal Communications (2021)

Link to original: https://doi.org/10.1007/s11277-021-08949-4

If you want to cite the papers, please refer to the original.

For technical reasons we could not place the page where the original quote is coming from.

Abstract-Summary

"Optimization gives the best one among all the solutions and the spread spectrum is used to minimize distortion, hence optimized spreading code is very important in wireless communication."

"This study shows a comparison of two recent modified techniques of differential evolution (DE) with the help of engineering design problems and application of that modified DE, considering the better one, in wireless communication to optimize logistic map based spreading code."

"The performance of the optimized spreading code is also evaluated with the help of bit error rate (BER) by applying it in static and dynamic direct sequence spread spectrum system (DSSS)."

"The performance of the proposed technique (optimized dynamic logistic map code based DSSS) displays better results than non-optimized spreading code and orthogonal spreading code considering BER."

Introduction
"Optimized chaotic code plays a very important role in wireless communication techniques [27, 28]."

"Further, the deployment of modified DE in a chaotic system applying in wireless communication and other engineering applications like gas transmission compressor design, optimal capacity of gas transmission capacity and design of gear train etc. To optimize chaos code Differential Evolutionary Algorithm can be considered because it has random search nature, fewer parameter setting, valid for high dimensional complex problems and have high performances."

"This paper shows a comparative study of two recent modified techniques of DE with the help of engineering design problems and application of that modified DE (considering the better one) in wireless communication to optimize logistic map based spreading code."

Literature Review
"A presentation is made after a comparison of a linear and none linear channels with static chaotic sequence in term of a bit error rate, computational time, mutual information and signal power for faded channel, evaluations of performance is done taking into consideration of different modulation schemes, which finally dictates the efficiency of generated code."

"Author tested the capacity performance and bit error rate by implementing eight transmitting and receiving antenna arrays in communication system based on ultra wide band in [29]."

"Authors of the paper [28] explains the literature review of differential evolution and also application in the field of wireless communication."

"Differential Evolution with seven different domain of applications in the field of wireless communication are identified in this paper."

"Authors of the paper [30] proposed IDE (improved version of DE) for ORPD (optimal reactive power dispatch) problem and also for benchmark functions."

The Proposed Method
"The mathematical explanation of differential evolution is as shown in the following."

"Mathematical expression for Modified DE (using exponential scale factor) The modified DE is differing from the basic DE as shown in the following steps."

"In this modified differential evolution algorithm, the scheme used is DE/best/1/bin which is elaborated in the following by mathematical terms."

"Step-2: In basic DE, F the scaling factor, CR the crossover rate are constant values, but in modified DE, F is replaced by taking exponential scaling factor f. In DE, using scaling factor (F) and crossover rate (CR), exploration and exploitation are balanced."

"During the stage of initial iteration, the value of f will be high, which helps in the search area in exploration, after drop of some iterations, exponentially the value of f will decrease, hence with reduced step size, the solution is now move and which enhances the exploitation skill of the algorithm."

Parameter Settings and Experimental Results

"DCPCX (Differential evolution with probabilistic parent centric crossover), DCPCX (Differential evolution with parent centric crossover) using three real time optimization problems with the help of Mean, Standard Deviation and number of function evaluations."

"Optimized version of method 1 is used in a DSSS."

"We can say from the analysis of the result that dynamic optimization gives better results than static optimization or static non-optimization."

"When we compare static optimization and orthogonal, orthogonal gives the better result than the static optimization."

Conclusion

"DE with enhanced parameters can play a vital role to solve the realistic optimization problems and application in wireless communication techniques."

"To optimize chaos code differential evolutionary algorithm can be considered because it has the property of random search, requires only fewer parameters setting, high performance and valid to complex optimization problems."

"From the outcomes of the proposed system, it is clear that the optimized chaos code using modified DE can be used as a spreading code which gives better results than non-optimized spreading code and orthogonal spreading code."

"The work explores the spin-offs that can be obtained from optimized link oriented chaos system supported wireless techniques."

A Bio-Inspired Frequency-Based Approach for Tailorable and Scalable Speckle Pattern Generation [31]

This is a machine-generated summary of:

Mathew, M.; Wisner, B.; Ridwan, S.; McCarthy, M.; Bartoli, I.; Kontsos, A.: A Bio-Inspired Frequency-Based Approach for Tailorable and Scalable Speckle Pattern Generation [31]

Published in: Experimental Mechanics (2020)

Link to original: https://doi.org/10.1007/s11340-020-00631-3

If you want to cite the papers, please refer to the original.

For technical reasons we could not place the page where the original quote is coming from.

Abstract-Summary

"Conventional speckling methods use non-reliable, non-repeatable patterning methodologies including spray paints and air brushing leading to increased systematic and random errors based on the user's experience."

"A novel bio-inspired speckle pattern development technique is introduced, leveraging spatial imaging parameters in addition to frequency characteristics of speckle patterns, enhancing the results obtained through DIC."

"This novel technique leverages gradient parameters in the frequency spectrum obtained from patterns fabricated using a bio-templating manufacturing technique."

"The analysis presented shows that optimized gradient features alongside tailored spatial characteristics reduce errors while increasing the usefulness of DIC results across the entire region of interest."

"Numerical error analysis shows that the optimized patterns result in significant reduction in root mean square error compared to conventional speckling methods."

"Physical experiments show the scalability of this optimized pattern to allow for varying working distances while consistently maintaining a lower error threshold compared to conventional speckling techniques."

Introduction

"It is therefore important to control the size of the speckles as well as to maintain the speckle pattern's gradient quality towards proper image registration in order to improve the measured displacement resolution across the given FOV."

"They formulate an inverse relationship to systematic and random bias errors [32] based on the image gradient quality derived from the imaged speckle pattern."

"Random errors stem from image noise and can vary with image noise level, subset size and speckle patterns."

"Brief introduction of the state-of-the-art, the authors utilize in this manuscript gradient features derived from images of bio-template patterns, along with speckle size and spatial arrangements to numerically generate speckle patterns that would provide enhanced DIC measurements for given camera parameters and choice of a FOV."

Experimental Methods

"Once imaged, these speckling parameters are regarded in terms of image intensities (gray scale values) and pixel distribution."

"Since these quality metrics are based on the image gradient information, imaging within a highly controlled environment requires an understanding of the controllable parameters and their effect on the quality of the speckle pattern imaged."

"The properties that make a speckle pattern have higher quality metric values (MIG and SSSIG) are ingrained in the grayscale intensity variations within the image itself and therefore provide a framework for further analysis."

"A 512 × 512-pixel image was generated with speckle diameter of 6 pixels providing a size-controlled speckle distribution."

"The inverse Fourier transform of these two properties provides an image that retains the gradient information of the high-quality bio-template image, while maintaining the speckle size of the diameter-controlled field."

Testing and Results

"With the shifted images created via the FFT technique described the authors could evaluate the effect of the generated pattern quality on the expected errors when this pattern is used in the DIC method."

"The quality of a pattern applied on a specimen is quantified in its undeformed stage to ensure optimal speckle to pixel ratio, spatial distribution and image gradients of the imaged specimen."

"That the systematic errors for the larger working distance (larger speckles and coarse paint) are significantly different with reduced errors for the generated pattern, while those for the shorter distance remain similar."

"The random errors favor the fine spray paint patterns applied for the shorter distance and the water transferred optimized generated pattern for the larger distance."

Discussion

"The fact is that a speckle pattern made by various methods or by different users may demonstrate different grayscale value distributions."

"In the investigation presented in this manuscript it was shown that a pattern developed with gradient information taken from a high-quality speckle pattern along with constraints on spatial speckle size characteristics can be utilized to reduce systematic and random errors during DIC measurements."

"The simulated experiment comparing different speckle patterns show the effect of consistent and appropriate speckle sizing and distribution along with improved image gradients used during DIC."

"This study also showed the impact of speckle sizing given imaging parameters and their effects on errors under real testing conditions."

"In the physical experiment the speckle patterns were generated using the same power spectrum profile."

"This further shows that the generation of controlled and consistent speckle patterns for a given set of parameters can limit the error thresholds that are caused by improper speckling."

Conclusions

"This investigation presented a method to generate optimized speckle patterns for use with the Digital Image Correlation method (DIC)."

"To address common issues associated with applying speckle patterns for full field deformation measurements using the DIC method, spatial and frequency-based analysis of available high quality DIC patterns was performed to extract technical characteristics that determine the overall quality of such patterns for DIC measurements."

"To evaluate the performance of the generated speckle patterns with this approach a numerical validation approach was considered which showed reduced total bias (RMSE) errors to other patterns."

Pattern Correction in Planar Antenna Array Using Multi-Objective Particle Swarm Optimization and Continuous Excitation [33]

This is a machine-generated summary of:

Sharma, Kanhaiya; Pandey, Ganga Prasad: Pattern Correction in Planar Antenna Array using Multi-Objective Particle Swarm Optimization and Continuous Excitation [33]

Published in: Iranian Journal of Science and Technology, Transactions of Electrical Engineering (2020)

Link to original: https://doi.org/10.1007/s40998-020-00378-6

If you want to cite the papers, please refer to the original.

For technical reasons we could not place the page where the original quote is coming from.

Abstract-Summary

"Multi-objective particle swarm optimization (MOPSO) and continuous excitation distribution in the range of 0 to 2 are used for correction of the failed element in 5 × 5 planar antenna array for ISM band application by substituting role of nonoperational elements to remaining operational elements."

Introduction

"In the case of active antenna array, it is possible to control excitation of individual element of antenna from remote location, so it is possible to restore the original beam pattern of antenna array by recalculating excitation values only functional elements of the array."

"In the existence of failed element, various optimization techniques are more reliable and effective method, used by antenna engineers to improve the performance of the antenna array [34–47]."

"Researchers have used various optimization techniques to synthesis antenna array for SLL suppression and overall performance improvement [37, 48–59] ."

"The second investigation is done to find out the limit of compensation and the performance of antenna array after the failure of elements at various positions of the array and compared the performance of array after failure of elements without and with optimized weights given by multi-objective PSO, using simulation process in CST microwave studio 2019 and MAT LAB 2018 software."

Multi-Objective Particle Swarm Optimization

"Particle swarm optimization is meta-heuristic search evolutionary population-based technique, with a limited number of parameters, easy to write code."

"Particle in MOPSO moves and shares information in a similar way like PSO toward global best with more than one objective to decide global or local best."

"In the swarm, all nondominant particles are gathered into a repository called external memory to choose their global best target, among members of this repository."

"For selecting local best, a domination based and probabilistic rules are applied."

Array Design

"To variation in phase, variation in amplitude can easily restore damage SLL and FNBW."

"Investigation, we considered only amplitude variation to restore damaged pattern."

"All elements are excited with amplitude value one and equal phase 0°."

"When all 25 elements are excited with an equal value of amplitude one in the array, the value of maximum SLL is −12.9543 dB, and FNBW is 48° with directivity 19.03 dBi and gain 18.74 dBi."

Investigation and Observation

"In order to compensate these single element failures, the excitations of other elements were obtained from MOPSO to reestablish the same HPBW, SLL, and FNBW."

"To recover desired SLL and FNBW, at various positions 16, 19, 23, 16, 21, 19, 20, and 20 operational elements adjusted their excitation values."

"Due to the failure of two elements at various positions, SLL increases from −12.9543 dB to higher level up to −11.1195 dB, and maximum variation of FNBW is 4°."

"It also shows the new optimized value of SLL, FNBW, HPBW, and total number of operational elements adjusting their excitation values for restoring damaged pattern by applying MOPSO technique."

"When the total 18 element failure occurs at different positions, both the objectives can be recovered by compensating excitation of just five of the operational elements."

Conclusion

"Meta-heuristic search evolutionary techniques have always been better than a quantitative approach to restore the pattern of the failed element in the array."

"Authors have investigated meta-heuristic search evolutionary technique, MOPSO, to restore faulty radiation pattern with two objectives SLL of −12.9543 dB and FNBW 48° on planar antenna array of size 5 × 5."

"We have found that pattern may be restored even if 72% of the elements in the array fail by updating excitation values of remaining operational elements within a range of 0 to 2, in the presence of a failed element."

"This approach may be applied for other planar antenna arrays in a modern communication system where self-healing is essential to save cost and time to offer uninterrupted communication to the end-users."

An Optimal Compact Time-Modulated Circular Antenna Array Synthesis Using Krill Herd Optimization [60]

This is a machine-generated summary of:

Das, Avishek; Mandal, Durbadal; Kar, Rajib: An optimal compact time-modulated circular antenna array synthesis using krill herd optimization [60]

Published in: Annals of Telecommunications (2020)

Link to original: https://doi.org/10.1007/s12243-020-00827-7

If you want to cite the papers, please refer to the original.

For technical reasons we could not place the page where the original quote is coming from.

Abstract-Summary

"An optimal far-field radiation pattern synthesis by using time modulation technique is carried out for a compact circular antenna array (CAA) design using a stochastic optimization method called krill herd (KH) optimization."

"A compact antenna array is achieved by reducing the inter-element spacing between the array elements, which leads to the miniaturization of the size of the array."

"An improved far-field radiation characteristic is achieved with a low side lobe level (SLL) and narrow first null beam width (FNBW)."

"In time modulation technique, 'time' has the most significant role."

"It is considered the fourth-dimensional parameter, which plays the role of an additional degree of freedom to achieve an improved radiation pattern."

"The time-modulated antenna array (TMAA) radiates at various harmonic frequencies for the periodic switching time sequence."

"It is applied here to find out the optimum sets of switching on-time duration for each array element and to determine the optimum inter-element separation between two array elements of the 20- and 36-element time-modulated circular antenna array (TMCAA)."

Introduction

"In time modulation technique, 'time' plays a significant role for its additional degree of freedom over the conventional antenna array and it is often called as the fourth-dimensional (4D) parameter for the improved far-field radiation pattern synthesis of time-modulated circular antenna array 'TMCAA'."

"A lot of research on circular antenna array (CAA) synthesis has been reported in [61–73], but a very few attempts were made for the design of a compact CAA using time modulation technique."

"The 20- and 36-element time-modulated circular antenna arrays are designed for the improvement of radiation characteristics and compactness by applying a metaheuristic optimization technique called KH optimization."

"'Time' is applied for the additional design liberation in uniformly excited TMCAA for the compact size of the array and for the reduction of SLL and FNBW to avoid the interference and to obtain high directivity, respectively."

Time-Modulated Circular Antenna Array

"$\Theta_0 = 90°$ and $\phi_0 = 90°$ are considered, though it does not affect the presented results because of isotropic array elements."

"In TMCAA, all the array elements are controlled by a high-frequency switch."

"Consider the operating frequency of TMCAA is f_0 Hz, and its corresponding time period is T_0."

"The switching on-time duration of each array element is τ_n ($0 \leq \tau_n \leq T_p$), where T_p is the modulation period."

"M is considered to be zero, as because only the fundamental/central frequency is considered here for the 20- and 36-element TMCAA designs."

"This is used for the design and calculation of the parameters of the 20- and 36-element TMCAA at the fundamental/central frequency."

Cost Function Formulation

"W_1, W_2, and W_3 are the weighting factors, whose values are 18, 18, and 12, respectively; θ_0 is the peak elevation angle of the main beam; θ_{ms1} is the maximum side lobe angle below the first null position of the main beam, whereas θ_{ms2} is the maximum side lobe angle above the first null position of the main beam."

"The FNBW in the CF is considered only when $FNBW_{Computed} > FNBW(\tau_p = 1)$ and is employed to improve the directivity pattern by narrowing the FNBW value."

"The second term of the CF plays a significant role in SLL reduction by placing the nulls on every peak of the side lobes in the radiation pattern."

"CF minimization causes the reduction of SLL and FNBW values as far as possible."

Krill Herd Optimization

"The adaptive genetic reproduction mechanism is also incorporated in the KH algorithm after the calculation of the motion to the krill swarm."

"The performance of the KH algorithm is improved by the crossover and mutation operation followed by the position updating operation to all individual krills in the search space."

"The herding of krill swarm is mathematically modelled in the KH algorithm and discussed in details in [74, 75]."

"The KH algorithm is mathematically modelled in the following steps: In the KH algorithm, the fitness value depends on the food centre and the highest density of krill swarm, which is mathematically modelled as the Lagrangian model: where N_i represents the motion induced by other krill individuals, F_i is the foraging motion, and D_i is the physical diffusion of the ith krill individual."

"The KH optimization is a nature-inspired metaheuristic algorithm inspired by the krill herding motion which is able to solve several multidimensional real problems with unknown search space."

Results and Discussions
"Far better results of SLL, FNBW, and directivity values are achieved by using the KH algorithm for the same antenna array structure design for the 36-element compact TMCAA designs of case II."

"All the algorithms (PSO, DE, and KH) employed in this study are executed 50 times to obtain the best results, and all the CF values achieved in every execution are employed in box and whisker plots."

"For the 20-element TMCAA (case I) design using the PSO algorithm, the median value of CF is 0.6411, and the range of variation of CF value is from 0.5140 to 0.7570."

"The above plots of convergences for the synthesis of the 20- and 36-element TMCAA where the CF values are plotted against the number of iteration confirm the outstanding performance of the KH algorithm over the PSO and DE algorithm–based results in terms of the accuracy and convergence speed."

Conclusion
"This paper illustrates the optimal far-field radiation pattern synthesis of TMCAA to obtain the minimum SLL and FNBW, maximum directivity, and miniaturization of the size of the array by using the krill herd optimization which is applied for the first time for time modulation technique."

"The results achieved by using KH confirm the superiority and the prospect of the algorithm to be used as a proficient optimizer for optimal far-field radiation pattern synthesis of compact TMCAA by using the set of optimal parameters, such as optimal switching time sequence, the inter-element separation between the array elements, and the number of antenna array elements in each ring."

"The method applied here for the synthesis of TMCAA is a generic one and can be used for the optimal far-field radiation pattern synthesis using different antenna array elements and geometrics."

An Energy Efficient OFDM–MIMO Systems for Multimedia Data Transmission Based on Hybrid Fuzzy Approach [76]

This is a machine-generated summary of:

Sharma, Preeti; Dhubkarya, Devi Charan: An Energy Efficient OFDM–MIMO Systems for Multimedia Data Transmission Based on Hybrid Fuzzy Approach [76]

Published in: Wireless Personal Communications (2020)

Link to original: https://doi.org/10.1007/s11277-020-07109-4

If you want to cite the papers, please refer to the original.

For technical reasons we could not place the page where the original quote is coming from.

Abstract-Summary

"In a 4G cellular communication system, energy efficiency and power consumption are the key parameters in evaluating and designing communication system in multimedia."

"An efficient model with high energy is proposed for multiple input multiple output orthogonal frequency division multiplexing (MIMO–OFDM) mobile multimedia systems with less power consumption."

"Efficient energy based optimized power allocation with hybrid Fuzzy Grey Wolf Optimization (GWO) algorithm is presented to reduce the power consumption of MIMO–OFDM mobile multimedia communication systems."

"Experimental results show that the proposed hybrid Fuzzy GWO algorithm can assure the required service with maximum efficiency in MIMO–OFDM."

Introduction

"To represent the data in multimedia it requires computational energy (energy used for processing the data that are used for transmission) large bandwidth and energy for communication (energy that is used for transmitting the information wirelessly) [77]."

"The Quality of Service (QoS)—such as transmission latency or bit error rate (BER)—and Quality of Multimedia Data (QoMD)—including image/video quality that is required during the multimedia communication changes based on the current multimedia service [78]."

"A novel algorithm namely, hybrid fuzzy GWO is employed to improve the energy and also reduce the power required for the entire transmission."

"This section motivates the hierarchical video transmission and it requires highly efficient power optimization algorithms."

"Proposed work yields minimum power consumption, delay and high energy efficiency since wireless communication with hybridized fuzzy Grey Wolf Optimization could provide real-time data for farmers to monitor, track, and automate their agricultural systems, resulting in increased profitability, efficiency, and safety."

Related Work

"Ge and others [79] proposed an energy-efficiency model for multiple-input-multiple-output orthogonal frequency-division multiplexing (MIMO–OFDM) with energy-efficient mobile multimedia communication systems using statistical QoS constraints."

"To achieve the efficient energy in multiple-input-multiple-output orthogonal frequency-division multiplexing (MIMO–OFDM), mobile multimedia communication system an Energy Efficiency Optimized Power Allocation (eeopa) algorithm was used."

"Chen and others [80] proposed a near-optimal power allocation scheme where the target expects to increase the Qos for sending the videos that were based on scalable video coding (SVC) over multi-input multi-output (MIMO) systems."

"Energy efficiency for the [81]th literature is 26% but in case of our current system, the range of energy efficiency is 93% since proposed work has 67% better performance than the existing work."

"Power consumption for the [82]st reference is 12.5 and for current work is 11.09, so the proposed work attains 14% better performance."

"Another method depend on efficient power optimization and compression algorithm."

System Model

"Through the feedback channels, the sender can obtain the CSI without any delay from the receiver."

"In MIMO–OFDM an average power constraint was created for several blocks."

"Energy-efficient can be optimized in MIMO–OFDM cellular communication system by average transmission power constraint and the power control operation was processed along with the sender to the system QoS constraint and feedback CSI."

Proposed Hybrid Fuzzy Grey Wolf Optimization Methodology

"The data is encoded using erasure coding."

"That provides the maximum of its important enhancement in compression efficiency to prior video coding standards."

"The process of decoding and encoding for video compression was shown in the above algorithm."

"Formerly video was read and then the encoding process was done for the video clip."

"While performing the encoding process the design of conversion was first done for the video and then the video was converted from RGB to YCbCr and then the QPSK (Quadrature phase shift keying) modulation was done for the video frames."

"Wyner–Ziv coding for video: applications to compression and error resilience."

"The video data was analysed and the optimization process was performed."

"The data with minimum power is allocated based on the proposed optimization algorithm."

Simulation Results

"Some of the parameters were calculated namely bit error rate, power consumption, energy efficiency, processing time, PSNR and MSE."

"This method provide 20 dB better BER performance than the Hybrid optical and wireless MIMO Channel Estimation algorithm HCEA [83] and Selected Mapping Discrete wavelet transform (SLM-DWT) [84] obtain 25 dB less than the proposed method."

"We will show the performance gains for Fuzzy-GWO power optimization."

"Compared to MMO, our proposed method produces 1×10^{-3} better performance and in the case of DCSCE, 2×10^{-3}."

"Throughout the proposed method, power is optimized and it is allocated with less energy efficiency."

"While comparing with all other optimization algorithms, the proposed Fuzzy-GWO yields better performance."

"Analysis we can describe that the following suggestions: We compare the BER performance and verify that the above analysis of the proposed MIMO–OFDM with each having one transmitter and receiver, QPSK and different system parameters."

Conclusion

"The proposed work has been used in video transmission in wireless systems since it naturally merges the multiple video data with high energy efficiency in the wireless channel."

"Proposed work yields minimum power consumption, delay and high energy efficiency since wireless communication with hybridized fuzzy Grey Wolf Optimization could provide real-time data for farmers to monitor, track, and automate their agricultural systems, resulting in increased profitability, efficiency, and safety."

"A novel algorithm namely, hybrid fuzzy GWO is employed to improve the energy and also reduce the power required for the entire transmission."

"Comparing with other methodology, our proposed methods achieve improved performance in power and energy."

"From the analysis, power minimization has to been done with various optimization algorithms."

"Our proposed work has the maximum ability to minimize power consumption."

Deadline-Aware Cache Placement Scheme Using Fuzzy Reinforcement Learning in Device-to-Device Mobile Edge Networks [85]

This is a machine-generated summary of:

Somesula, Manoj Kumar; Kotte, Anusha; Annadanam, Sudarshan Chakravarthy; Mothku, Sai Krishna: Deadline-Aware Cache Placement Scheme Using Fuzzy Reinforcement Learning in Device-to-Device Mobile Edge Networks [85]

Published in: Mobile Networks and Applications (2022)

Link to original: https://doi.org/10.1007/s11036-022-02010-9

If you want to cite the papers, please refer to the original.

For technical reasons we could not place the page where the original quote is coming from.

Abstract-Summary
"Caching the most likely to be requested content at the mobile devices in a cooperative manner can facilitate direct content delivery without fetching content from the remote content server and thus alleviate the user-perceived latency, reduce the burden on backhaul and minimize the duplicated content transmissions."

"To content popularity, it is also essential to consider the users' dynamic behaviour for real-time applications, which can further improve the communication chances between user devices, leading to efficient content service time."

"We study an essential issue: caching content by taking advantage of user mobility and the randomness of user interaction time."

"In a cooperative caching problem, we consider a realistic scenario with user devices moving at various velocities."

"We formulate the cache placement problem as maximization of saved delay with capacity and deadline constraints by considering the contact duration and inter-contact time among the user devices."

Introduction
"Mobile edge caching (MEC) is a prominent technique that utilizes the edge nodes (i.e., base stations and mobile devices) as caching nodes to bring the contents near users, thus alleviating the core network burden and enhancing user Quality of Experience (QoE) [86]."

"To providing content near to users, caching at mobile devices even reduces network latency [87]."

"The user's mobility, short contact time, and inter-contact time make it challenging for mobile devices to cache content proactively."

"Real-time content caching on mobile devices improves user experience while maintaining a tolerable response latency."

"The contributions of this work are as follows: Design an Integer linear programming problem for cooperative content caching problem: maximization of saved download delay subject to constraints, namely cache capacity and deadline of the content in D2D enabled mobile edge networks."

Related Work
"Li and others [88] have investigated a distributed caching mechanism to select appropriate user nodes and allocate content at selected nodes in cellular networks using D2D communications."

"Zhao and others [89] have presented user mobility and interest prediction based cache placement and cache-aware replacement mechanisms to maximize the utility in IoT networks."

"In [90], the authors presented a network slicing mechanism to efficiently handle heterogeneous resources in a vehicular network with assured QoS. Zhang and others [91] have investigated the user interest request prediction based mobility aware D2D caching mechanism to maximize the opportunistic cache utility."

"Sun and others [92] have investigated a caching mechanism in a D2D environment by considering the user mobility to minimize the network latency."

"Jiang and others [93] have formulated the content caching in D2D networks as multi-armed bandit problem and presented two Q-learning based multi agent learning mechanisms."

"This inspires us to utilise machine learning methodologies to solve the exciting problem of designing an efficient content placement mechanism in a cache-enabled D2D network while considering the contact duration and inter-contact time."

System Model and Problem Formulation

"neighbour user: If the user requested content is not available at the local storage, then the content can be obtained from the nearby users in the communication range based on the availability of the content at neighbouring users."

"base station: If the requested content does not exist with any users in the communication range, the content would be served from the associated base station by fetching the content from the central server through a backhaul link."

"The mobile users in contact may exchange the content based on the contact time and inter contact time."

"If the requesting user obtains l segments from the users in contact, then the remaining content segments will be retrieved from the associated base station."

"We need to cache entire content at multiple users to serve the user request since the user is in mobility."

Fuzzy Reinforcement Learning Mechanism

"We designed a reinforcement learning mechanism to solve the proposed problem where the agent learns an optimal caching scheme to cache the content at the user devices in the mobile edge network."

"We present a fuzzy logic mechanism for computing mobile devices weights depending on the storage capacity, average contact time, number of contacts with other users in the network, and mobile devices' data transfer rate."

"Devices encountered: Number of devices encountered in the mobile user path is dented as E. Contact time: The average contact time of the mobile devise is denoted as T. Transmission rate: The amount of content transmitted in a contact time is denoted as R. The fuzzy logic system includes the following stages: fuzzification, rule base, inference process and defuzzification."

"The estimated Q-values of SARSA are updated using learning rate α at each time step t, as follows: where a_f denotes the action obtained from the max-weight policy (Algorithm 2)."

Performance Evaluation

"In scenario 2, the number of contents is 600, the deadline considered is 800 s, the Zipf parameter is 0.6, the cache capacity is 0.2 GB, and the number of user devices varies from 10 to 50 with step size 10."

"In scenario 3, the number of users is 20, the deadline considered is 800 s, the Zipf parameter is 0.6, the cache capacity is 0.2 GB, and the number of contents varies from 200 to 1000 with step size 200."

"The number of contents is 600, the deadline considered is 800 s, the Zipf parameter is 0.6, the cache capacity is 0.2 GB and the number of user devices varies from 10 to 50 with step size 10."

"The number of users is 20, the deadline considered is 800 s, the Zipf parameter is 0.6, the cache capacity is 0.2 GB and the number of contents varies from 200 to 1000 with step size 200."

Conclusion

"The cooperative cache placement problem has been analyzed in device-to-device mobile edge networks by placing the content to maximize the saved delay with deadline and capacity constraints."

"Since the proposed problem is NP-hard, we designed a caching scheme for large-sized D2D enabled mobile edge networks by integrating an on-policy reinforcement learning mechanism with fuzzy logic to enhance the long term reward and speed up the learning process."

"It has been demonstrated that exploiting user cooperation, content deadlines, and randomness of user interaction information results in a considerable performance gain in D2D enabled mobile edge networks."

Presenting an Optimal Energy-Aware Locating Structure Using the Internet of Things and Device-to-Device Communications on Smartphones [94]

This is a machine-generated summary of:

Rostami, Ali Shokouhi; Mohanna, Farahnaz; Keshavarz, Hengameh: Presenting an Optimal Energy-Aware Locating Structure Using the Internet of Things and Device-to-Device Communications on Smartphones [94]

Published in: Wireless Personal Communications (2021)

Link to original: https://doi.org/10.1007/s11277-021-08114-x

If you want to cite the papers, please refer to the original.

For technical reasons we could not place the page where the original quote is coming from.

Abstract-Summary

"Device-To-Device (D2D) communications as a novel technology for improving the capacity and reducing power consumption in wireless networks have been taken into account."

"To solve the above-mentioned problem, in this paper, a new approach was presented for the localization of D2D communications on smartphones, which was based on the Dynamic Packet Transport and Multimedia Presentation Tracking, protocols as well as the distortion rate."

"Due to this new approach, the energy-aware optimization space and service quality standards were improved with an evolutionary swarm intelligence algorithm, called the Scaling Optimization Algorithm."

"The results of the simulation of the proposed algorithm and its comparison with other algorithms reviewed in this paper hinted at the superiority of this algorithm in reducing energy consumption and improving the service quality standards."

Introduction

"LTE is a standard for high-speed wireless communication for mobile devices and data terminals based on Global System for Mobile Communications (GSM)/EDGE/ Universal Mobile Telecommunications System (UMTS)/HSPA technologies."

"The obtained results in such a condition may indicate high overlap in radio frequency with the communication channels using Wi-Fi and Create Load Balance Between SDN (CLBBSDN) and 5G networks [95, 96]."

"LTE is a 4G wireless communication standard developed by 3GPP or 3GPP to provide the speed ten times faster than the speed of Third Generation (3G) technologies for mobile devices such as smartphones, tablets, notebooks, and wireless hotspots."

"The availability of radio at different frequencies and the use of the 3GPP structure have made LTE one of the most important network services [97, 98]."

"Based on the preliminary explanations presented in this section, there should be an energy-aware system that has no problem regarding energy consumption, when it is connected to the 5G, LTE, LTE-A and LTE-U network."

Related Work

"To solve the problem of energy-aware locating with D2D communication in smartphones, a variety of different heuristic and meta-heuristic methods have been proposed by researchers in recent years."

"In [99–101], the authors investigated energy-based source localization methods through D2D communications in cellular networks aiming at maximizing the energy efficiency of the investigated D2D links."

"In [102], the researchers addressed D2D communications in unlicensed spectra and used network coding technique in bi-directional communications to propose an effective MAC scheme based on an Adaptive Cooperative NC-Based MAC (ACNC-MAC) protocol for the D2D data exchange."

"Channel allocation and power control were taken into account for improving the network throughput by reducing interference in D2D communications [103]."

"Improving the spectrum efficiency and maximizing the energy efficiency in D2D communications have attracted the attention of many researchers [7, 104, 105]. For example in [106], the subcarrier assignment and power allocation as a joint optimization problem were considered."

The Proposed Algorithm

"Based on the research gap identified in reviewing the existing literature, a new approach is proposed for locating D2D communications on smartphones according to DPT and MPT protocols along with distortion rate in which energy-aware optimization and service quality standards are improved using the SOA."

"Transmitter and receiver structures are created and the channel in LTE-U is chosen and estimated so that a transmitter signal is placed in the communication channels, which may suffer from noise, distortion rate, and other parameters that are comprehensively investigated in this section."

"The parametric signal modeling method to detect and reduce noise, distortion rate, and other factors is the approximation of least squares."

"The result of the formulation for energy in the optimal tuning mode for improving localization will be as follows: In general, the different steps of the proposed method are as follows: An LTE-A context is first created in D2D (IoT) communications along with a series of objects that are smartphones."

Simulation Results

"To localize D2D communications on smartphones, it is necessary to review the relevant literature at the time of simulation such as D2D communications at the time of simulation, the technologies required to connect nodes to it, data routing operations, energy consumption measurement, and quality of service standards, along with the most important goal which is locating."

"Improving localization and then transferring smartphone data like D2D communications requires a structure from the IoT. As the IoT does not have defined dimensions in the real world (although sometimes and in some places, its dimensions are defined), there is a need in the world of simulation to define at least one of its dimension for the initial deployment of nodes or equipment that are connected to this network."

"When data streaming occurs across the D2D communications for nodes connected to LTE-A technology, the OFDM communication channel is used with some modulations, including BPSK, QPSK, 16QAM or 64QAM."

Conclusion

"IoT, as a technology based on Internet protocols, has permeable segments that result in unsolved challenges."

"One of the most important challenges of IoT is the issue of multimedia data routing, which has various effects on its subsequent segments, including routing (volume of traffic consumed), energy consumption, service quality standards, reliability, and other relevant issues."

"The default routing protocols of this research are the DPT and MPT routing protocols, which are regarded as a type of dynamic routing protocol for multimedia data transfer."

"Distortion rate was considered and an evolutionary swarm intelligence algorithm, called as the SOA, was utilized to guarantee the energy-aware routing and transmission of multimedia data, service quality standards, and the primary purpose of the current study, i.e., localization."

"The optimum routing is required to stream data when transmitting and receiving them on the D2D communication channels on smartphones."

Deep Reinforcement Learning Mechanism for Deadline-Aware Cache Placement in Device-to-Device Mobile Edge Networks [107]

This is a machine-generated summary of:

Somesula, Manoj Kumar; Mothku, Sai Krishna; Kotte, Anusha: Deep reinforcement learning mechanism for deadline-aware cache placement in device-to-device mobile edge networks [107]

Published in: Wireless Networks (2022)

Link to original: https://doi.org/10.1007/s11276-022-03135-1

If you want to cite the papers, please refer to the original.

For technical reasons we could not place the page where the original quote is coming from.

Abstract-Summary

"Caching the most likely to be requested content at the mobile devices in a cooperative manner can facilitate direct content delivery without fetching content from the remote content server and thus alleviate the user-perceived latency, reduce the burden on backhaul and minimize the duplicated content transmissions."

"To content popularity, it is also essential to consider the users' dynamic behaviour for real-time applications, which can further improve communication between user devices, leading to efficient content service time."

"We study an essential issue: caching content by taking advantage of user mobility and the randomness of user interaction time."

"We consider a realistic scenario in a cooperative caching problem with user devices moving at various velocities."

"We formulate the cache placement problems as maximization of saved delay with capacity and deadline constraints by considering the contact duration and inter-contact time among the user devices."

Introduction
"Mobile edge caching (MEC) is a prominent technique that utilizes the edge nodes (i.e., base stations and mobile devices) as caching nodes to bring the contents near users, thus alleviating the core network burden and enhancing user Quality of Experience (QoE) [86]."

"To providing content near to users, caching at mobile devices even reduces network latency [87]."

"The user's mobility, short contact time, and inter-contact time make it challenging for mobile devices to cache content proactively."

"The contributions of this work are as follows: Design an Integer linear programming problem for cooperative content caching problem: maximization of saved download delay subject to constraints, namely cache capacity and deadline of the content in D2D enabled mobile edge networks."

"Design a cooperative caching scheme based on deep reinforcement learning mechanism, DDPG to speed up the learning rate and enhance content placement at mobile users to maximize the saved delay."

Related Work
"Li and others [88] have investigated a distributed caching mechanism to select appropriate user nodes and allocate content at selected nodes in cellular networks using D2D communications."

"Zhao and others [89] have presented user mobility and interest prediction based cache placement and cache-aware replacement mechanisms to maximize the utility in IoT networks."

"Sun and others [92] have investigated a caching mechanism in a D2D environment by considering the user mobility to minimize the network latency."

"Jiang and others [93] have formulated the content caching in D2D networks as multi-armed bandit problem and presented two Q-learning based multi agent learning mechanisms."

"This inspires us to utilise machine learning methodologies to solve the exciting problem of designing an efficient content placement mechanism in a cache-enabled D2D network while considering the contact duration and inter-contact time."

System Model and Problem Formulation
"The user may get the desired content from more than one mobile users who are in the communication range of the user."

"Neighbour user If the user requested content is not available at the local storage, then the content can be obtained from the nearby users in the communication range based on the availability of the content at neighbouring users."

"Base station If the requested content does not exist with any users in the communication range, the content would be served from the associated base station by fetching the content from the central server through a backhaul link."

"The user may request the content based on the content popularity."

"The user retrieves the content from the base station, which is associated with the requesting user."

"If the requesting user obtains l segments from the users in contact, then the remaining content segments will be retrieved from the associated base station."

Deep Reinforcement Learning Mechanism

"We designed a reinforcement learning mechanism to solve the proposed problem where the agent learns an optimal caching scheme to cache the content at the user devices in the mobile edge network."

"We present deep reinforcement learning mechanism to select the more preferable mobile devices based on the highest reward."

"The action represents a set of mobile devices over which the content is cached to maximize the saved delay."

"Agent performs an action based on the transition probability from one state to the next state."

"The maximum reward (i.e., reinforcement signal) obtained from all potential actions is used to update a Q-learning agent's policy, independent of the agent's policy."

"We apply DRL to address the presented optimization problem by characterizing the state, action, and reward."

"Compared to traditional RL algorithms, DDPG can solve optimization problems with huge state and action spaces with high dimensional concerns."

Performance Evaluation

"In scenario 1, the number of contents is 600, the deadline considered is 800 s, the Zipf parameter is 0.6, and the cache capacity ranges from 0.2 GB and the number of user devices varies from 10 to 50 with step size 10."

"In scenario 3, the number of users is 20, the deadline considered is 800 s, the Zipf parameter is 0.6, the cache capacity is 0.2 GB, and the number of contents varies from 200 to 1000 with step size 200."

"The number of contents is 600, the deadline considered is 800 s, the Zipf parameter is 0.6, the cache capacity is 0.2 GB and the number of user devices varies from 10 to 50 with step size 10."

"GCP considers the mobility of the users into the caching decisions; hence it achieves a better acceleration ratio, but the lack of learning mechanism it caches the redundant copies of the content at the users leads to a lower acceleration ratio than DQN and DDPGCP."

Conclusion

"The cooperative cache placement problem has been analyzed in device-to-device mobile edge networks by placing the content to maximize the saved delay with deadline and capacity constraints."

"Since the proposed problem is NP-hard, we designed a caching scheme for large-sized D2D enabled mobile edge networks by integrating a deep reinforcement learning based deep deterministic policy gradient mechanism to enhance the long term reward and speed up the learning process."

"It has been demonstrated that exploiting user cooperation, content deadlines, and randomness of user interaction information results in a considerable performance gain in D2D enabled mobile edge networks."

"The simulation results show that the proposed cooperative cache placement improves up to 23%, 24% and 25% on cache hit rate, acceleration ratio and offload ratio compared with RC, MPC, GCP and DQN, respectively."

Radio Resource Management: Approaches and Implementations from 4G to 5G and Beyond [108]

This is a machine-generated summary of:

Akhtar, Tafseer; Tselios, Christos; Politis, Ilias: Radio resource management: approaches and implementations from 4G to 5G and beyond [108]

Published in: Wireless Networks (2020)

Link to original: https://doi.org/10.1007/s11276-020-02479-w

If you want to cite the papers, please refer to the original.

For technical reasons we could not place the page where the original quote is coming from.

Abstract-Summary

"Radio resource management (RRM) schemes that are evolved during LTE/LTE-A network environment period will remain relevant for 5G, however, these schemes must become more intelligent and adaptive for future as features and requirements of network and users will be diverse and highly demanding."

"A comprehensive view is provided upon various aspects of RRM, its challenges and existing schemes."

"The existing RRM schemes are presented with their respective architecture which has significant impact on the approaches."

"Study of legacy and state of the art RRM schemes is presented with their features and inefficiencies in the modern telecommunication era of heterogeneous, ultra-dense, very low latency and highly reliable mobile network."

"The need of RRM solution is critical and this paper aim is to outline the challenges, existing solutions and directions for research to find and develop smarter and more adaptive schemes for future."

Introduction

"Radio resource management is key for satisfying the standard requirements of 5G environment, as it demands efficient management of interference and spectrum while considering fairness and Quality of Service (QoS) requirements as well."

"Energy and spectrum management are other key aspects of RRM and more efficient approaches are required to handle these challenging issues in 5G environment."

"Heterogeneous Networks (HetNets) are typically composed of various small cells and macro cell which create a hyper dense environment that can help in satisfying the diverse Quality of Experience (QoE) requirements of 5G system [109]."

"Highly dense HetNets will become key network in 5G environment and small cells which are essential part of HetNets will play key role in efficient implementation and realization of 5G environment."

"The existing RRM schemes will not work with evolving 5G environment, as the challenges of next generation communication technologies are different and more complex in nature."

Overview of 5G System Performance Challenges

"The technological requirements are massive where network should be scalable and self-sustainable, also 5G system will include the existing RATs (i.e., WLAN, Bluetooth, 3G, 4G, etc.) for mm Wave backhaul operations [110]."

"Evolved radio access technologies (RATs) backbone with ultra dense networks (UDNs) and futuristic applications with elaborate use cases require far more sophisticated RRM schemes for 5G with massive traffic volume [111, 112]."

"This challenge will require smart extensions from traditional cellular technology, where humongous number of machine type communications (MTCs) will also become the extended part of cellular network, which will need seamless integration with existing cellular RATs."

"Smart city, e-Health, traffic management requires ultra-low latency and very high reliability as human lives will depend upon the decision made through these systems [113]."

"In UDNs environment, smart functionality such as minimization of signaling, synchronization, node sleep mode, low link distance can help in bringing down the energy consumption, which is very critical for 5G and future network [114]."

Radio Resource Management

"There are few main functions of RRM in LTE/LTE-A cellular technologies such as packet scheduling, resource allocation, link adaptation and control, radio admission and removal and handover management."

"The scheduling decisions depend upon various parameters such as channel state information (CSI), QoS requirements of UEs, channel quality indicator (CQI), interference, etc. Another important RRM function is link adaptation which is managed at MAC layer for improving user throughput at targeted block error rate (BER) [115]."

"Radio admission control entity belongs to layer-3 in LTE/LTE-A, that decides if a radio bearer will become part of the network or not, this decision depends upon the radio bearer requirement of resources and availability."

"User handover and mobility management is done at radio resource control (RRC) layer, in LTE/LTE-A the decision of handover is made at eNB level and signal measurement happens at UE level with only hard handover support [116]."

Evolution from Small Cells to Femtocells

"Femtocells in the indoor environment can significantly reduce the power consumption with better QoS and improve the coverage due to proximity with users."

"Self-optimization, auto-configuration capabilities of femtocell with existing small cell environment make them highly efficient plug and play device, also automatic integration ability of femtocell with end users enable them for high scale deployment [117]."

"Femtocell network modeling requires evolution in traditional cellular model, a simple scenario involves one base station which is connected to mobile users and other small cells (downlink), also users (uplink) will act as interferer [118, 119]."

"Many high-level formulation models, such as game theory can be used for better resource allocation, distribution of channel gain and power control to improve the efficiency of overall cellular model scenario [120]."

"Selection of suitable access control model can be key for any cellular model involving femtocell."

Issues and Challenges of RRM in HetNets

"The HetNets environment is more challenging due to dense deployment of small cells as RRM approach should be adaptive and scalable."

"Fairness is mainly categorized into local and global fairness [121], local fairness is linked with the scheduling of packets among various UEs and global fairness is associated with the fairness of resource allocation between backhaul, access and direct links in multi-hop network."

"In small cell environment, if the resource requirement at cell level is satisfied, the allocation is called 'globally fair', whereas if the minimum required data rate of every UE is ensured with maximization of resource allocation is called 'max-min fair'."

"The allocation of resources to UEs are done for each TTI, if computation time required for executing RRM algorithm is more that single TTI (i.e., permissible computation time) then this approach is not feasible to use in LTE/LTE-A system."

RRM Schemes in Heterogeneous LTE/LTE-A Networks with Small Cells

"It is one of the simplest approaches to allocate the resources (PRBs) to UEs, on the basis of CSI and interference information."

"The issue of co-tier interference is still a major concern and it is addressed by designing centralized scheme for two tier femtocell network using proportion fair scheduling [122], where UEs are aware of their surrounding interfering eNBs with their respective signal power."

"In this scheme PRBs are allocated to UEs by central entity (resource manager) based on the CQI and interference information of every UE channel."

"Is co-channel allocation where every UE share the same spectrum which leads to better resource utilization, but it suffers from high interference."

"Various stochastic resource allocation is effective in interference management and resource utilization as well as in low complexity implementation, however, QoS requirements considerations are missing in these approaches."

RRM Schemes with Relay Nodes in LTE/LTE-A

"Two different scheduling techniques are used for PRBs allocation, in backhaul sub frame eNB uses buffer-aware scheduling for every RN, then proportional fairness scheduling is used by eNB and RNs for accessing sub frame to allocate PRBs among MUEs and RUEs."

"A genetic algorithm is proposed for resource allocation in LTE-A multi-hop environment where the main goal of scheme is throughput and fairness maximization with load balancing between RNs and their respective eNBs."

"Maximization of logarithmic average rate by every UEs is the problem with many constrains such as guaranteed stability if packets queue is related to RNs and eNB, fix average and instantaneous power threshold, limited number of active packets, one PRB to every UE, no concurrent transmission between access and backhaul link for same sub frame, etc. [123]."

"This approach is efficient for multi-hop LTE-A network with high spectral efficiency, local and global fairness, better intracell and inter-RN interference management."

Radio Interference Management (RIM) Schemes

"The recursive core is achieved using merge algorithm which has low complexity with better QoS performance in UDNs however, cross-tier interference management in multi RATs environment costs high CSI overheads."

"Learning based adaptive power control (LeAP) is also suggested which uses the neighboring cell interference information to reduce multi-tier intercell interference [124], learning algorithm uses the transmission power control variable sets to optimize the overall throughput and latency of the network [125]."

"Use of fast heuristic technique can help in the convergence of optimal solution for improving QoS. A fully decentralized heuristic approach with self-organization is suggested with interference and power control management where transmission power is dynamically controlled according to QoS requirements of UEs [126]."

"Current and instantaneous CSI at transmitter (CSIT) is used for MIMO based interference management schemes [127–129], MIMO Broadcast Channel (BC) which uses multi-antenna can achieve higher degree of freedom (DoF) and low delay with CSI feedback."

Spectrum-Efficient Management (SEM) Schemes

"Spectrum resource scheduling uses the available spectrum and assign it to every user efficiently by meeting their QoS requirements such as low latency, high data rates, longer life of battery, reliability, etc. In future, it is expected that a scheduler and RB (i.e., smallest resource radio unit) will be assigned to every UEs in the shared channel of available spectrum where every RB has 12 OFDM sub-carriers at adjacent position with 15 kHz spacing between sub-carriers."

"CR utility's main functions are sensing, managing, sharing and mobilizing the spectrum according to primary and secondary user's QoS requirement without interference."

"DSA can be deployed using exclusive or opportunistic CR approach such as exclusive spectrum access where spectrum bands are exclusively allocated to a mobile network for managing the interference efficiently, however, it suffers from low utilization of spectrum [130]."

Energy-Efficient Management (EEM) Schemes

"Relay nodes can be used for creating multiple connections between source and destination, with many links different path fading, diversity gain, spectral efficiency combinations are possible which can help in reduction of data transmission and energy consumption [131, 132]."

"Adaptive modulation and coding (AMC) for different transmission resources can be used for link adaptation and MAC aspects in multi-user environment with hybrid multi-access technique to improve energy efficiency [133]."

"Ultra-dense small cell and mmWave communication is used while analyzing backhaul network energy efficiency in 5G, two different architectures are used to model traffic for backhaul, the outcome indicates distributed solutions are better compared to centralized solutions for overall spectral efficiency [134]."

"The energy consumption of one bulb in VLC system is comparable to RF-based transmission of equal data density for analysis, however, VLC system is more sensitive towards weather, also information is carried by light intensity (power) which need to positive and real-value signal."

Hybrid Resource Management (HRM) Schemes

"Secondary service provider (SSP) is introduced for harvesting and accessing the spectrum of SUs using CR ability of infrastructure, SU cooperate with CR routers of network and use harvested bands which increase reusability of available spectrum and capacity [135]."

"Channel access aware (CAA) scheme for user association is proposed for better resource allocation and interference management which can also achieve high spectral efficiency and traffic load balancing for down-link at BSs."

"H-CRANs can improve spectral and energy efficiency using different scenarios such as coordinated multi-point transmission, cloud based cooperative RRM, self-organizing network (SON) and high scale cooperative antennae [136]."

"Bio-inspired energy and channel (BEACH) management method is introduced for better channel utilization and energy efficiency in wireless multi-radio network [137]."

"The algorithm uses bio-behaviors which are suitable to optimize the network resource allocation which in time improve the spectrum and energy efficiency of RAN systems [138, 139]."

5G Wireless Network Design

"Multi path effect is created due to reception of signal by antenna through various paths [140], power delay profiles (PDP) can help in analyzing multi path effects for mm-wave environment."

"Wider width beam antennas have high accuracy for received signal estimation whereas shorter width beam antennas have spatial directivity advantages, therefore best combination of width can help the antennas to achieve high SNR as well as low delay spread [141]."

"In MIMO enabled mm-wave system, obtaining information from every antenna element is difficult and RF beam forming requirements for overcoming poor links limit the system, use of narrow beam by transceiver can be helpful in this scenario, as data transmission happen only between best transceiver beam pairs [142]."

"Massive MIMO, beam forming, small cell, centralization of architecture can help in efficient FD realization, whereas intelligent scheduling, better power control and rate selection can lead to high gain in capacity through FD communication."

Architectural Paradigm Shift for 5G and Slicing Implementation

"New resource allocation mechanism is proposed that uses the various relationship exist with different players including network slice tenants and providers of network infrastructure [143]."

"Network slicing logical architecture has radio access plane, multiple RATs with heterogeneous network and efficient cooperation mechanism for 5G. WiFi access point and small cell deployment can also improve the demands of increasing traffic in 5G [144]."

"Slicing strategies are encoded into binary sequences and these sequences are used for genetic optimization, which enable efficient inter-slice resource management by using mobile network operator decisions and requests from tenants."

"SDN decouples the control and user plane at gateway in CN environment, control signaling can be reduced using intelligent control functions even in large distributed nodes network however, ultra high density of network and devices with higher mobility has management challenges when network-slicing is done."

Conclusion

"The architectural shift that is needed to move towards 5G and beyond is transforming the future network into hyper dense HetNets, and to satisfy the features and QoS requirement of 5G wireless network radio resource management schemes will remain extremely critical."

"The problem of resource management is multi-dimensional and has multiple challenges such as interference management, resource utilization, QoS requirement, fairness and energy management, therefore the solutions should be adaptive to change according to the user's and network demands."

"This paper indicates the RRM schemes are essentially including the small cell environment in their approach however, designing a fair, energy and resource efficient solution with interference mitigation in hyperdense environment is very challenging issue."

"It can be concluded that the future of RRM depends upon the deployment environment and this area of research will remain interesting and challenging as the features and requirements will remain highly demanding for 5G and beyond."

Remote Radio Head Scheduling in LTE-Advanced Networks [145]

This is a machine-generated summary of:

Ramdev, Mandeep Singh; Bajaj, Rohit; Sidhu, Jagpreet: Remote Radio Head Scheduling in LTE-Advanced Networks [145]

Published in: Wireless Personal Communications (2021)

Link to original: https://doi.org/10.1007/s11277-021-08916-z

If you want to cite the papers, please refer to the original.

For technical reasons we could not place the page where the original quote is coming from.

Abstract-Summary

"The emerging services and key enhancements such as Further Enhancement of Downlink Multiple-Input Multiple-Output (MIMO), Heterogeneous Networks, and Carrier Aggregation (CA) in LTE-A has improved performance of LTE-A networks."

"Scheduling optimization still remains one of the biggest challenges in high speed data transmission network."

"Scheduling in LTE-A networks are performed at various levels; User Equipment (UE), Serving Gateway (SGW), Air Interface and eNodeB. Remote Radio Head (RRH) is an extremely specialized device installed at antenna of eNodeB for optical to electrical signal conversion, amplification of signals and Uplink and Downlink Scheduling."

"This paper proposes a soft computing based scheduler for RRH."

"The proposed algorithm aims to improve performance of scheduling."

"From experimental results, it is observed that proposed model succeeds to achieve significantly better performance as compared to state-of-art algorithms."

Introduction

"LTE improves user experience through increases in bandwidth efficiency and reduced latency."

"The support for multimedia applications; (1) Video Conferencing, (2) Audio Conferencing, (3) HD Video Streaming has made LTE-A most significant for users and service providers [146]."

"Goals, LTE-A was expected to achieve; (1) Increased Data Throughput, (2) Decreased Latency, (3) Improved Flexibility of Spectrum Allocation, (4) Increased Reliability of Data Transmission and (5) Increase in Communication Efficiency."

"Extensive literature analysis indicates less significant on schedulers is performed in LTE networks, however some significant work is performed in RRH for cloud based systems."

"VoIP and browsing requires high data rate with low transmission loss and low latency in contrast to Video Streaming Dynamic environment need soft computing based schedulers which have a proven their significance [147] and scheduling can be termed as most significant problem in LTE."

Literature

"Monghal and others [148] introduced a new scheduling algorithm based on PF algorithm that tried to balance coverage of network as well as cell throughput."

"This algorithm actually utilize adaptive scheduling which able to dynamically adjust fairness index based upon inputs received."

"The results could have been better in terms of throughput if soft computing based scheduling algorithms could have been employed instead of generic algorithms."

"Content aware strategies can be categorized roughly into following subclasses; (1) Quality driven scheduling,, (2) Proxy driven radio resource allocation strategy and (3) Client driven approach quality driven scheduling approach focuses on scheduling strategies for video streaming traffic."

"PF is the most researched algorithms in scheduling of LTE based networks."

"The simplicity and adaptability of PF algorithm made it most significant algorithms to employ in scheduling LTE based networks."

The Proposed Optimization Algorithm

"The simulations are run using Round Robin Scheduler as well as Proportional Fair Scheduler at various Transmission Time Intervals (TTI) with comparative analysis on results."

"Real Time Traffic has rigorous throughput requirements and is given more priority over Non-Real."

"The Non-Real Time traffic queue consists of packets which have data from delay sensitive applications."

"The main advantage of using PF scheduler is that it prioritizes real time applications over others and consequently guaranteeing a bounded delay to data packets hence maximizing system throughput."

"Scheduling Module schedules traffic based upon the scheduling algorithm used."

"As soon as the input reaches the RRH, it is sent into the default traffic classifier for classifying the traffic into real-time and non-real time traffic for an efficient scheduling."

"This section will analyze, through simulation, performance of soft computing based Proportional Fair Algorithm with comparative analysis between PF and RR scheduler on identical scenario."

Results and Discussion

"The results clearly indicated that PF scheduler has better Average Spectral Efficiency in contrast to RR Scheduler."

"In Rank 2 distribution it can be clearly seen that Rank 2 in RR scheduling is significantly high which clearly indicates priority of MIMO over the Rank 2 in PF Scheduling."

"While comparing the fairness index of RR and PF scheduling, the fairness index of RR Scheduling is 0.706048 in contrast PF Scheduling is 0.711594 which is a marginal 0.78% increase in final value at 10 TTI."

"The peak throughput achieved is 7.10 Mb/s using RR scheduler and 8.05 Mb/s using PF scheduler."

"A marginal 3.71% increase can be analyzed in Average UE Spectral efficiency while employing PF Scheduler."

"This indicated that PF scheduler has more consistency as there is less deviation from peak to average throughput of PF scheduler as compared to RR Scheduler."

Conclusion

"This paper presented a classification for better understanding of downlink scheduling in LTE-A networks."

"It presented a comparative analysis of generic and soft computing based scheduler employed on RRH."

"The results indicate that better scheduling algorithms deployment on RRH has attained significant output from LTE-A. Results were compared with existing work and are found significantly better."

"It has also been proved in the present work that soft-computing based schedulers are significantly better than generic scheduler."

Estimation of Distribution Algorithm for Joint Resource Management in D2D Communication [149]

This is a machine-generated summary of:

Ahmad, Mushtaq; Naeem, Muhammad; Iqbal, Muhammad: Estimation of Distribution Algorithm for Joint Resource Management in D2D Communication [149]

Published in: Wireless Personal Communications (2019)

Link to original: https://doi.org/10.1007/s11277-019-06459-y

If you want to cite the papers, please refer to the original.

For technical reasons we could not place the page where the original quote is coming from.

Abstract-Summary

"Use of same resource of cellular users by D2D pairs creates interference problem among users."

"Simultaneous resource management and mode selection in emerging cellular networks having D2D capabilities can surely improve overall system throughput."

"This work addresses the problem of overall throughput maximization of emerging cellular networks while ensuring the power and interference constraints."

"This paper applies evolutionary technique, like estimation of distribution algorithm which has the potential to address the complex problems having combinatorial nature such as joint resource management strategy in D2D. Near optimal solution convergence is achieved by the algorithm with minimal number of iterations."

Introduction

"An interference avoidance based resource allocation scheme using Particle Swarm Optimization (PSO) technique has been demonstrated in [150] in order to enhance overall spectral efficiency by allocating the same frequency resource of one Cellular user to more than one D2D pair."

"The literature review unveils that there is no evolutionary algorithm related research work which attempts to maximize the overall system throughput in cellular network having D2D features and simultaneously controls the admission of the users, assigns modes (to be in cellular or D2D mode) and allocates power complying with QoS and interference constraints."

"This work fills the missing link and attempts to enhance the throughput of the Cellular System having D2D capabilities using EDA with thresholds (statistical based evolutionary algorithm) while jointly considering features of access control, selection of mode and power allocation of users."

System Model and Problem Formulation

"This paper devises a technique for joint access control and selection of mode in order to maximize the overall throughput of the system and the number of admitted users complying with the power and minimum rate constraint requirements."

"The constraint C1 ensures compliance of threshold of minimum rate for each admitted user."

"Users in violations of constraint C1 are not allowed to transmit."

"Power constraints C2, C3 and C4, limit the transmit powers for D2D, cellular uplink and cellular down link modes respectively."

"C6 allocates binary mode selection and user admission variables."

"Constraints C7 ensures that if a user is not selected, its respective power becomes zero."

"Constraints, C8 and C9 ensure quality of uplink and down link data rates of cellular users."

Proposed Approach to a Solution

"An EDA optimizes an objective function based on the statistics of the candidate solutions of population."

"This algorithm has advantage in the sense that storage of candidate solutions are not required rather statistical characteristics are preserved while proceeding ahead in iterations."

"EDA being iterative algorithm, calculates population using probability of previous generation."

"As we move ahead in iterations, the solution converges to the best population."

"In EDA with a threshold algorithm, the probability of the typical candidate is shaved off in case the probability of that candidate exceeds the already defined upper or lower limit."

"The population of candidate solution contains number of expected solutions, which converge to the optimal solution as the algorithm proceeds."

Simulation Results

"EDA with threshold provides better utility values and convergence to the best utility value is attained with less number of iterations."

"As EDA attains the best utility value with less number of iteration proving the fact that this algorithm is computationally more efficient compared to classic EDA and GA."

"As the number of users are increased, EDA with Threshold provides better utility value (better performance) compared to the other two considered algorithms."

"Spectral efficiency affects can be seen more prominently as the number of users become more larger. The simulation results confirm that EDA with threshold provides better utility values hence an excellent means to jointly control admission of users and selection of modes with more spectral efficiency."

Conclusion

"A convergent solution to address the problem of joint admission control, mode selection and power allocation has been proposed in this paper."

"We used EDA algorithm with threshold to solve the problem and compared the performance with classic EDA and GA."

"Results confirm the superiority of EDA with threshold over the other two considered algorithm in terms of yielding better utility values, computationally more efficient, spectrally more efficient and more efficient in terms of selection of users."

Deep Reinforcement Learning Aided Cell Outage Compensation Framework in 5G Cloud Radio Access Networks [151]

This is a machine-generated summary of:

Yu, Peng; Yang, Xiao; Zhou, Fanqin; Li, Hao; Feng, Lei; Li, Wenjing; Qiu, Xuesong: Deep Reinforcement Learning Aided Cell Outage Compensation Framework in 5G Cloud Radio Access Networks [151]

Published in: Mobile Networks and Applications (2020)

Link to original: https://doi.org/10.1007/s11036-020-01574-8

If you want to cite the papers, please refer to the original.

For technical reasons we could not place the page where the original quote is coming from.

Abstract-Summary

"The cell outage compensation (COC) problem for RRH in 5G C-RAN is very important."

"Although deep reinforcement learning (DRL) has been applied to many scenarios related to the self-organizing network (SON), there are fewer applications for cell outage compensation."

"Aiming at the cell outage scenario in C-RAN with the goal of maximizing the energy efficiency, connectivity of RRH while meeting service quality demands of each compensation user, a framework based on DRL is presented to solve it."

"Compensation users are allocated to adjacent RRHs by using the K-means clustering algorithm."

"DQN is used to find the antenna downtilt and the power allocated to compensation users."

"It verifies the efficiency of the DRL-based framework and its effectiveness in meeting user requirements and handling cell outage compensation."

Introduction

"The method to maximize the overall performance of the network when the failure occurs is a challenging problem for future 5G C-RAN networks."

"To obtain optimized solutions, network management functions such as SON use cases requiring quantities of computations for parameter adjustments, which are tough tasks, and machine learning algorithms are efficient for them [152], especially in 5G heterogeneous networks [153]."

"A DRL-based framework for RRH COC in 5G C-RAN is presented, which maximizes the connectivity of RRHs while ensuring that the service quality of each compensation user is satisfied."

"Our main contribution is proposing a DRL-based framework for COC in 5G C-RAN for the first time."

"This paper applies DRL to COC of 5G C-RAN networks for the first time."

"A staged DRL framework is proposed to make COC problems more easier to be resolved, which is user connection with K-means algorithms in the first stage, and use deep Q-learning for power and tilt jointly optimization in the second stage."

Related Work

"This algorithm considers just one parameter which is reference signal power, and the application scenario is LTE which does not apply to 5G C-RAN scenarios."

"In paper [154], the authors proposed an automatic and self-organized RL-based approach for COC to automatically adjust downlink transmission power levels and antenna tilt value to fill the coverage and capacity gap."

"In paper [155], the authors propose a new COC method, which adds a new self-healing radio to each base station in the cloud-radio access network (C-RAN)."

"In paper [156], the authors proposed an enhanced immune-genetic algorithm named adaptive chaos immune-genetic algorithm (ACIGA) to minimize public safety network (PSN) system outage probability, which compensates the C-RAN-based PSN using both cooperative transmission and power adjustment."

"In paper [157], the authors present a novel DRL-based framework for power-efficient resource allocation in cloud RANs to minimize power consumption and meeting demands of wireless users."

System Model

"That mostly COC compensation models aim at maximizing user throughput for outage/interrupt BSs or regional coverage, few considering BS energy efficiency and coverage gap together."

"To maximize the connectivity and energy efficiency of active BSs while ensuring that the demand of each compensation user is satisfied, the optimization problem is formulated as follows: In this model, the optimization goal is in [12], where the decision variables are the user connection relationship x, antenna downtilt φ and BS transmission power allocation π as these parameters are all easier to be adjusted with SON entity in network management system."

"Constraint [14] ensures that each user must be served by just one BS."

"Constraint [16] ensures that every user is served by an active BS."

DRL-Based COC Framework

"To reduce the action space size in the framework, we first group compensation users in clusters by using the K-means clustering algorithm and allocate users to adjacent BSs."

"According to the result, compensation users in the same cluster are allocated to the nearest adjacent base station, and the connection relationship between BSs and users is determined."

"DRL agent will make an action decision a_t based on the state s_t, adjust the antenna downtilt φ_i of each BS i and allocate transmission power π_{ij} from every BS i to each compensation user j. The state space, action space and reward function of the DRL-based framework are defined as follows: State Space: (1) The connection relationship x_{ij} between BS i and user j; (2) The antenna downtilt φ_i of each BS i; (3) Allocated transmission power π_{ij} from BS i to compensation user j. Action Space: (1) Changes of the BS antenna downtilt."

Simulation Results

"The compensation users are classified according to the positions of adjacent base stations and allocated to five adjacent base stations using the K-means clustering algorithm."

"One is to use the K-means clustering algorithm, the other is to use the way that compensation users are allocated to the nearest BSs with shortest distance (Shortest-Dis)."

"Compared with using the way that compensation users are served by the shortest distance BSs, the convergence speed is faster and the reward is higher while using the K-means clustering algorithm."

"It means DRL methods here achieve better optimization targets with worse SINR (still above acceptable level), and may scarify a little interference influence, which takes on the tradeoff between efficiency and user QoS. From the above analysis, it can be got that DRL with K-means clustering is the best solution comparing to other connections and intelligent algorithms, and it shows the tradeoff between the optimization targets and SINR."

Conclusion and Future Work

"The DRL-based framework is proposed to maximize the BS connectivity, energy efficiency while meeting the demand of each compensation user solves the COC problem in 5G C-RAN."

"A well-trained network can quickly resolve COC problems, and its result is closed to the optimal value comparing to traditional and enhanced genetic algorithms."

"As this paper adopts a relatively simple simulation scenario to evaluate the efficiency of the proposed algorithm, next complex scenarios such as heterogeneous C-RAN networks will be considered."

"Besides, we will improve the proposed DRL-based COC algorithm to find a better training model with a quicker convergence rate and a higher accuracy algorithm."

Emperor Based Resource Allocation for D2D Communication and QoF Based Routing over Cellular V2X in Urban Environment (ERA-D²Q) [158]

This is a machine-generated summary of:

Alghamdi, Saleh A.: Emperor based resource allocation for D2D communication and QoF based routing over cellular V2X in urban environment (ERA-D²Q) [158]

Published in: Wireless Networks (2020)

Link to original: https://doi.org/10.1007/s11276-020-02273-8

If you want to cite the papers, please refer to the original.

For technical reasons we could not place the page where the original quote is coming from.

Abstract-Summary

"Achieving less interference in terms of resource allocation and a higher quality of forwarding (QoF) during routing is a major problem owing to the highly mobile environment used in such communications."

"To overcome this bottleneck, the present study focuses on emperor-based resource allocation for device-to-device (D2D) communications and QoF routing (ERA-D²Q) in C-V2X. Our ERA-D²Q offers three sub-sequential processes: (1) coalition-based routing, (2) D2D communications and SignRank oriented pedestrian selection, and (3) emperor based resource allocation."

"To achieve QoF routing, ERA-D^{2}Q introduces coalition game theory (CGT), which elects an optimum relay and reduces the amount of time for a packet transmission."

"The best pedestrian sends a relay request to the 5G base station (BS) to discover a destination for D2D communications."

"To optimize the resource blocks assigned to the D2D communications, an emperor penguin colony algorithm is established for the proposed ERA-D^{2}Q, which is employed in an OMNET++ simulator."

"The evaluation results prove that our ERA-D^{2}Q enhances the PDR and throughput by up to 35%, and decreases the ATD by up to 48%, the jitter by up to 33%, and the MoS by 40% compared to the existing methods QFRG, CF, SMRS, SRA, and DTDMD."

Introduction

"To attain a better transmission rate during D2D communications, a proper resource allocation is required [159]."

"Allocating resources for D2D communications remains crucial for V2X because such communications suffer under a high density of vehicles [160]."

"Resource allocation for D2D communications in C-V2X allowing the ability to handle such a high density and effectively avoid a communication overhead has been developed."

"Interference-free resource allocation is important for obtaining a lossless transmission during D2D communications."

"Studies, we found that many problems have yet to be resolved in succeeding processes such as routing, D2D communications, and resource allocation."

"Achieving a high transmission rate in D2D communications is a major issue owing to the lack of an optimum device discovery."

"To attain an optimum resource block allocation for D2D communications, an emperor penguin colony (EPC) algorithm is applied for the pedestrians."

Related Studies

"This section discusses previous state-of-the-art studies related to routing, D2D communications, and resource allocation in a vehicular environment."

"Further analysis is required during the selection of a relay in D2D communications."

"Xujie and others [161] proposed resource allocation schemes based on the optimization algorithm for D2D communications in a cellular network."

"A Hungarian method was utilized to effectively allocate resources to the D2D communications."

"Juhyun and others [162] offered a resource allocation scheme for cooperative D2D communications in cellular networks with multiple D2D pairs."

"The average achievable rate of a D2D pair was investigated to allocate proper resources to the D2D communications."

"In the studies described in [162, 163], more time is required to provide a proper resource allocation to each D2D pair."

"The ILA scheme allocates resources based on the preset transmission rate requirement of each D2D pair."

Problem Statement

"Transmission losses are increased while routing the packets between the source and destination owing to an absence of mobility oriented metrics (speed) in the relay selection."

"Durgesh and others [164] described a stochastic model for mobility aware relay selection (SMRS) in 5G D2D communications."

"The link capacity limitation is high, leading to a packet loss during D2D communications."

"D2D communications are established based solely on the distance constraints, leading to a reduction in the quality of the multimedia data."

"Pimmy and others [165] suggested a sector-based resource allocation (SRA) scheme for device-to-device communications with better QoE and QoS constraints."

"Frequent resource allocations are required owing to an improper selection of cellular users for resource sharing."

System Model

"This section describes the ERA-D^{2}Q functionalities, such as routing, D2D communications, and resource allocation, when using our proposed algorithms."

"Our proposed ERA-D^{2}Q contributes to three sequential processes, QoF routing, high data-rate D2D communications, and an optimized resource block allocation."

"We select an optimum pedestrian to communicate with the 5G BS because the D2D device does not have the ability to communicate directly with it."

"To provide proper communications, we propose the use of a SignRank algorithm based pedestrian selection."

"The BS receives a relay request from the pedestrian, and then selects the optimum relay node to establish D2D communications with the destination device."

"Data Size: This metric also plays a vital role in providing the proper resource blocks to the D2D communications and avoids wasted resources and shortages by estimating the size of the data to be transmitted during D2D communications."

Simulation Results and Discussions

"To evaluate the significance of our ERA-D^{2}Q framework, succeeding metrics have been considered: Packet delivery ratio (PDR) The PDR is the ratio of packets received at the destination vehicle with respect to the packets transmitted from the source vehicle."

"The comparison results between the PDR and the number of vehicles shows that our ERA-D^{2}Q approach performs better than the other existing methods."

"The ratio of successful packets received at the destination vehicle is high in our approach because we utilize a proper method for both routing and D2D relay selection."

"Under this situation as well, our ERA-D^2Q approach achieves a much smaller increase in the ATD owing to the connectivity strength and ETD metrics during the forwarder selection."

"Our ERA-D^2Q method increases the PDR by up to 35% in comparison to the existing QFRG, CF, and SMRS methods."

Conclusion

"Three sequential processes are applied, namely, routing, D2D communications, and pedestrian selection and resource allocation."

"D2D communications are applied under two scenarios, within the coverage of the discoverer and out of the coverage of the discoverer."

"A pedestrian is selected though the SignRank algorithm to achieve a better communication between the BS and D2D devices."

"The BS selects an optimum path through the MIT2F algorithm, which provides a lossless transmission during D2D communications."

"The pedestrian implements the EPC algorithm, which optimizes the resource block allocated to the D2D pair."

Intensive Benchmarking of D2D Communication over 5G Cellular Networks: Prototype, Integrated Features, Challenges, and Main Applications [166]

This is a machine-generated summary of:

Hussein, Hanan H.; Elsayed, Hussein A.; Abd El-Kader, Sherine M.: Intensive Benchmarking of D2D communication over 5G cellular networks: prototype, integrated features, challenges, and main applications [166]

Published in: Wireless Networks (2019)

Link to original: https://doi.org/10.1007/s11276-019-02131-2

If you want to cite the papers, please refer to the original.

For technical reasons we could not place the page where the original quote is coming from.

Abstract-Summary

"The evolving Fifth generation (5G) cellular wireless networks are envisioned to provide higher data rates, lower end-to-end latency, and lower energy consumption for devices."

"In order to achieve 5G requirements, a lot of new technologies are needed to operate in 5G. Device-to-device communication (D2D) is one of the key technologies provided to enhance 5G performance."

"This paper provides an intensive benchmarking of the integration of D2D communication into cellular network focusing on the potential advantages, different recent prototypes, classifications, and applications for D2D technology."

Introduction

"5G wireless cellular network deployment is the main key technology of the next wireless cellular network evolution."

"All these previous technologies are outside of the scope of this paper except D2D. D2D Communication is one of the competent technologies for 5G. In previous mobile generations (i.e. 1G–4G), D2D technology did not take much attention; however, it is predicted that applying such a technology will play a vital role in enhancing the era of wireless cellular communication."

"To a traditional wireless cellular network, D2D technology can provide more power savings due to the close distance among connected devices."

"One of the main advantage is its flexibility; so that D2D can be integrated with different technologies to enhance the network's performance such as vehicle-to-vehicle (V2V) technology, mmW technology, Internet of Things (IoT), mode selection, and Artificial intelligence (AI)."

D2D Technology Main Advantages

"From the definition, in the case of devices that near in distance, using D2D technology reduces latency between devices and consumed power."

"Another advantage of applying D2D technology is that it can take some load off of the cellular network in a local area such as big shopping centers or stadiums."

"D2D technology can be an efficient solution in case of emergency networks (an urgent communication network can be established within a short time to replace the damaged infrastructure)."

"Using D2D communications beside cellular communications could raise the spectrum utilization and network capacity."

D2D Main Classification

"This mode can be classified into two types; namely, full control mode and autonomous mode."

"When devices are managed by a cellular network or operator, the mode is called controlled mode; otherwise it is called autonomous mode."

"Autonomous mode means that DUs participate in the control."

"The D2D communication mode is classified into three types."

"The first one is dedicated mode which allows the D2D pair to exchange data directly without involving eNB."

"Shared mode is the second type in which both DUs and CUs can share the same RB; also known as underlay mode."

"The third type is a cellular mode as in the traditional cellular communication where all users transmit and receive data through the core network without applying direct communication among users, which is known as overlay mode."

D2D Scenarios or Applications

"Using such D2D technology can enhance the network performance especially delay reduce between users to meet the real-time voice constrains."

"Local data service can be provided using D2D communication when two col-located users or devices, within D2D coverage area, want to exchange data."

"One type of video streaming that can offload the traffic from wireless networks is Point-to-Point (P2P) video streaming."

"P2P video streaming promotes all users to share their videos with others through direct connection."

"D2D technology can provide more performance improvement than normal cellular networks between devices that desire to share video streaming."

"Overall M2M network delay can be less because of direct routing of D2D traffic."

"Simulation results proved that EH-based D2D protocol performs better nodes life time due to energy efficiency for cluster heads and relay nodes; and also offered extended coverage area."

D2D Prototypes and Experiments

"The procedure of this system to achieve its target is as follows: (1) Dataspotting system activates the content sharing system via the cellular operator in case of crowded areas or rush hours. (2) It gathers all DUs' locations and lists their available contents. (3) The size of each data spot is assigned based on DUs' locations and contents."

"WiFi Direct (WFD) is considered as the radio access technique for D2D connectivity since it provides a lager data rate up to 300 Mbps, and extended coverage range up to 70 m. The main experiment and the results obtained in [167] are as follows: To prototype the proposed model, a video application is shared among users that close in distance."

D2D Integrated Features

"The simulation results demonstrate that CFRM can enhance the stability of the cellular network and increase the network throughput for IoT by exploiting D2D. Mode selection is the problem of choosing whether two users should communicate through a direct link (D2D) or via eNB [168] in order to optimize network performance."

"The optimal mode selection depends on the performance measure to enhance network output such as sum rate, transmit power, energy consumption, and system capacity; and on the state information available when making the decision such as physical distance, the channel quality of the links, and interference level."

"Results proved that the proposed algorithm has better performance in terms of network throughput and D2D throughput, and network QoS. Another RL algorithm was investigated that based power control mechanism for underlying D2D connectivity in cellular networks [169]."

D2D Challenges

"One of the main challenges that face cooperative D2D networking is energy consumption problem; because of evolved DUs in such networks are battery-powered devices."

"Priori process means initiating D2D discovery before the DUs start to communicate by sharing certain information between devices."

"Interference between DUs and CUs is one of the main challenges that face D2D technology."

"In the case of in-band shared resource communication, interference among cellular and D2D devices can exist."

"Radio resource management optimization issue indicates how to optimally allocate frequency resources to overall DUs so as to optimize some performance parameters such as SINR, latency, power consumption, reliability, QoS, interference, channel utilization, energy efficiency ...etc."

"Because of the nature of D2D networks, DUs become less secure than CUs under eNB control."

"A D2D connection between two DUs in a cellular network is proposed in [170]."

Conclusion

"An intensive overview is provided about D2D communication technology with the existence of infrastructure cellular network."

"The D2D communication technology underlying mobile network provides numerous benefits such as raising overloaded cellular network efficiency to reduce latency and minimize consumed energy."

"Besides that, the paper highlighted the vital challenges that D2D technology faces such as handover, energy consumption, interference, resource allocation, and security."

"Most researches and prototypes are still in the initial study stages of D2D technology performance in case of simple scenarios or under certain conditions."

Decoupling Uncertainty Quantification from Robust Design Optimization [171]

This is a machine-generated summary of:

Chatterjee, Tanmoy; Chowdhury, Rajib; Ramu, Palaniappan: Decoupling uncertainty quantification from robust design optimization [171]

Published in: Structural and Multidisciplinary Optimization (2019)

Link to original: https://doi.org/10.1007/s00158-018-2167-0

If you want to cite the papers, please refer to the original.

For technical reasons we could not place the page where the original quote is coming from.

Abstract-Summary

"In order to deal with such issues, an efficient approximation-based generalized RDO framework has been proposed."

"Since RDO formulation comprises of statistical terms of the performance functions, the proposed framework deals with approximation of those statistical quantities, rather than the performance functions."

"The proposed framework allows transformation of the RDO problem to an equivalent deterministic one."

"Unlike traditional surrogate-assisted RDO, the proposed framework yields desirable results in significantly less number of functional evaluations."

"The generality of the proposed methodology allows any surrogate models to be employed within this framework, provided it is capable of capturing the functional non-linearity."

Introduction

"The solution of an RDO problem will draw in uncertainty quantification of the response and its integration within an optimization routine."

"The accuracy of surrogate models can prove to be a vital factor in stochastic optimization framework, such as RDO."

"In order to address the prominent drawbacks as discussed above, the novelty of this paper lies in the following: A simplistic yet efficient approximation-based framework has been proposed in order to assist RDO framework to yield optimal solutions in considerably lesser number of functional evaluations."

"These models have been integrated within RDO framework to yield optimal solutions in a cost-effective manner (as illustrated in the first point)."

"The generality of the proposed RDO framework allows it to be applied to any of the available surrogate models, which is capable of accurately approximating the response statistics."

Surrogate-Assisted Robust Design Optimization

"Other related improved RDO approaches can be found in [172–174]."

"In a broad sense, surrogate models can be integrated within RDO setup by employing the following approaches."

"It involves building the surrogate model at the onset before the optimization commences and subsequently utilized for predicting the performance functions at every iteration within the optimization routine."

"It involves building a surrogate model for the performance functions at each optimization iteration."

"The approximation accuracy of a HF-RDO approach is considerably high so as to treat non-linear problems."

"In all the above frameworks, large number of simulations have to be performed for evaluating the response statistics of the performance functions within each optimization iteration."

"In order to avoid large number of simulations at each iteration within the optimization routine, a novel framework for surrogate-assisted RDO has been proposed in the following section."

Proposed Surrogate-Assisted Robust Design Optimization Framework

"The fundamental concept underlying the proposed RDO framework is to approximate the response statistics, rather than the response quantities itself."

"It can be observed from steps 2, 7, and 8 of algorithm 1 that n_{s1} represents the number of training points for building the meta-models in approximating the response statistics (i.e., the mean and SD of response)."

"It is also clear from steps 5 and 6 of algorithm 1 that n_{s2} represents the number of actual response evaluation in the proposed RDO framework."

"The efficiency of the proposed RDO framework has been clearly illustrated as there are no actual response evaluations and Monte Carlo simulation (MCS) within the optimization cycle, evident from step 9."

"The approximation framework as illustrated in steps 1–8 of algorithm 1 renders generality to the proposed RDO approach so that any surrogate model can be employed; however, they should be robust enough to capture the variation of response statistics accurately, to avoid convergence to a false optima."

Proposed Surrogate Models

"Compressive sampling aims to determine the set of most sparse coefficients α by solving an optimization problem of the following form Where, s = {0, 1} and positive diagonal weight matrix W is a measure of sparsity of α and $||\psi\alpha - g||_2$ is the measure of the accuracy of the approximation model in determining the realizations."

"Such a regularization may be achieved by determining a solution α with minimum number of non-zero terms [175] which can be formulated as an optimization problem of the following form Where the semi-norm $||\alpha||_0 = \{c: \alpha_c \neq 0\}$ represent the number of non-zero components of α."

"Since the ℓ_1-norm is convex, the optimization problem P_1 accepts a unique solution which conforms to the unique solution to problem P_0 for sufficiently sparse α [175]."

"OMP and BP algorithms, respectively, as illustrated above, impart p-adaptivity to the proposed models."

Numerical Validation

"In order to access the performance of the proposed surrogate-assisted RDO frameworks, they have been utilized for solving few numerical examples in this section."

"For the RDO formulation of the above problem, each of the random variables have been considered to be normally distributed with coefficient of variation of 5%."

"It can be observed that the optimal solutions as obtained by PM1-RDO and PM2-RDO almost match with that of MCS-based solutions and other surrogate models."

"Considering the complexity of approximation, it is worth acknowledging the robustness of the proposed RDO frameworks which perform a single approximation of the response statistics and accurately map the overall functional behavior, eventually leading to the true optima."

"The proposed RDO frameworks have been utilized to solve two FE-based practical problems."

Summary and Conclusions

"A novel and efficient surrogate-assisted RDO framework has been proposed."

"The highlights in relation to the proposed RDO framework have been enumerated as follows: The proposed surrogate-assisted RDO methodology is a generalized framework, in which any of the existing and/or improvised surrogate models can be integrated in a straightforward manner."

"Large number of simulations can be avoided to approximate the response statistics, making the proposed RDO framework inherently suitable for large-scale problems."

"The proposed methodology may not prove to be suitable for solving problems in which the design variables are not random in nature and/or the random variables are less in number."

"The PCE bases utilized make the proposed models generalized for stochastic computations and reduce computational effort."

"The study has paved way for the proposed surrogate-assisted RDO frameworks to be applied to further complex large-scale applications."

In-Band Device to Device (D2D) Communication and Device Discovery: A Survey [176]

This is a machine-generated summary of:

Hayat, Omar; Ngah, Razali; Zahedi, Yasser: In-Band Device to Device (D2D) Communication and Device Discovery: A Survey [176]

Published in: Wireless Personal Communications (2019)

Link to original: https://doi.org/10.1007/s11277-019-06173-9

If you want to cite the papers, please refer to the original.

For technical reasons we could not place the page where the original quote is coming from.

Abstract-Summary

"Device to device (D2D) communication is one of the potentials to achieve the established standards for 5G. It represents a direct communication between two devices located in the vicinity of each other."

"In order to initiate the D2D communication, device discovery is needed, and it is a major design issue for D2D communication."

"To achieve a potential discovery, there are some requirements such as energy efficient discovery capability, supporting a large number of devices and proximity discovery in the network assisted underlay D2D communication network."

"D2D discovery techniques are evaluated, its main challenges are analyzed and potential solutions are also suggested."

Introduction

"A Device-to-Device (D2D) communication permits two proximity devices to communicate directly to each other without passing through a Base Transceiver System (BTS)."

"In terms of the ISM band, however, the spectrum is unlicensed which helps in deploying D2D communication without any alteration to present communication systems but several issues e.g., interference, quality of service and delay etc., are available and more are highlighted in [177]."

"Based on these technologies, Flash-LinQ established an effectual method for synchronization, device discovery, and resource management for D2D communication in cellular networks."

"IEEE 802.X group has completed much work on proximate device discovery and D2D communication [178]."

"For the case of in-band D2D communication, allocated cellular spectrum is shared between cellular and D2D users, while for out-band communication, an unlicensed spectrum, e.g., 2.4 GHz band, is utilised for communicating."

"One of the primary issues for establishing a D2D communication is the device discovery."

Classification of D2D Communication

"D2D communication, which is used based on unlicensed spectrum is called Out-Band D2D (OBD) communication."

"The inspiration behind utilizing OBD communication is to dispose of the interference issue amongst D2D and cellular networks."

"The main challenge in OBD communication is to manage the interference between D2D users and cellular users."

"One of the technologies in 5G is IBD D2D communication, which can dramatically improve the data rate, manage the interference between cellular users (CU) and D2D users (DU) and so on, to meet the standards of 5G. The inspiration for picking IBD communication is typically the ability of control over licensed spectrum [179]."

"IBD communication is categorized into underlay and overlay D2D communication."

"In overlay D2D communication, dedicated resources are used for DU and CU [180]."

"IBD communication can enhance the spectrum efficiency of the cellular network by reusing the resources (based on underlay communication)."

D2D Implementation Scenarios

"Even though D2D communication is not developed yet [181], it is as of now being concentrated by the 3GPP standardization authority."

"3GPP currently has decided that D2D communication in LTE focuses on public safety networks [182]."

"Possible D2D implementation scenarios for D2D communication are given below."

"In order to achieve a proper communication, devices in the coverage area can act as a relay [183]."

"They devices require the communication between each other, and this is done based on communication through the BTS and then there is a need to communicate via BTS."

D2D Challenges

"Current work can be categorized into two approaches, distributed D2D, and centralized D2D. In the centralized D2D approach a specific unit in the cellular system, for example, data network gateway and mobility management unit, distinguishes that it might be better for two DUs to setup a communication."

"In distributed D2D approach the DUs communicate their identities occasionally so that different DUs might know about its presence and chooses whether it should begin a D2D communication with it."

"A practical implementation D2D communication challenges in heterogeneous networks are mainly device discovery, inter cell interference, effective medium access control, and resource allocations [182]."

"D2D communication starts once devices in the vicinity of each other are discovered."

"Device discovery for in-band D2D communication is a major and first challenge that needs to be explored vitally [184, 185]."

Device Discovery

"Another autonomous device discovery method is proposed which is based on routing protocol as stochastic, Bio inspired and hieratical [186]."

"In this proposed method, neighboring devices discovery is based on the calculated SINR using received beacon signal."

"Using an innovative functionality of Wi-Fi, direct base device discovery method is proposed in [187], and it presents an experimental assessment that depicts the performance in practical situations."

"A self-governing ProSe server base device discovery scheme is proposed in [188]."

"An energy efficient discovery mechanism is proposed in [189], where the device discovery process takes the benefit of proximity services explicitly where in DU triggers their D2D abilities."

"Radio channel values based device discovery process is proposed in [190]."

"A link base device discovery method is proposed in [191]."

"Device beaconing method based discovery and enabled services based scheme are proposed in [192]."

Possible Way of Discovery in Cellular Networks

"The network generates D2D discovery signal and illuminates both source and destination devices."

"The benefit of the packet based method is that full information identified associated with device discovery."

"The benefit of packet based discovery is that all the information in the packet is self-contained and has a minimum network dependence In the signature base device discovery, utilizing a signature includes the transmission of the special waveform on the physical layer."

"The information that can be incorporated into the signature is commonly a much smaller complex than packet based device discovery."

"The lack of signature-based discovery is that only the signature cannot carry full device information."

"The beacon base device discovery process is sequenced as follows: (1)."

"One of proposed method is femto cell GPS base device discovery which will help full for above parameters and also discover the device when out of coverage using relay devices information."

Future Work

"This can be more enhanced to afford, more efficient device discovery, accurate throughput, power efficiency, minimum delay in D2D communications of the underlay cellular networks."

"There is actual slight work to illuminate the essential architecture in demand to the provision of device discovery in D2D communications in the underlay cellular networks."

"It is stimulating to auxiliary consideration on the ability of the present centralized cellular network to maintain D2D communication such as device discovery, connection setup, network registration, intervention control, resources allocation, security."

"Software defined based networks design can also help for device discovery in D2D communication."

"We rely on a performance assessment by means of the current network simulators such as NS3, OPNET, OMNET or an experimental assessment can help in enlightening together an actual presentation and new challenges like device discovery of D2D communications in the underlay cellular networks."

Conclusion

"We are providing a wide study of the existing literature on device discovery in D2D communications in underlay cellular networks."

"The key issue in underlay D2D communication is device discovery, power control mechanism, interference controlling between D2D users and cellular users."

"In Overlay D2D communication, interference is not a major issue because D2D users and cellular users do not overlap."

"Our survey presented that D2D communication in in-band underlay cellular networks is under developing and device discovery is the first important phase for the initiation of D2D communication."

A Comparative Review of Recent Advances in Decoding Algorithms for Low-Density Parity-Check (LDPC) Codes and Their Applications [193]

This is a machine-generated summary of:

Roberts, Michaelraj Kingston; Anguraj, Parthibaraj: A Comparative Review of Recent Advances in Decoding Algorithms for Low-Density Parity-Check (LDPC) Codes and Their Applications [193]

Published in: Archives of Computational Methods in Engineering (2020)
Link to original: https://doi.org/10.1007/s11831-020-09466-6

If you want to cite the papers, please refer to the original.

For technical reasons we could not place the page where the original quote is coming from.

Abstract-Summary

"In the domain of wireless communication systems, Error Control Coding (ECC) schemes are one of the widely relied upon or responsible methodology for securing the integrity and authenticity of the data transmission process."

"Recent developments in the wireless communication based technologies have witnessed the pliable nature of low-density parity-check (LDPC) codes and their contributions which cannot be overstated."

"As of now, the decoding schemes based on LDPC codes have emerged as one of the most promising and effective coding scheme for addressing several key problems of reliable data communication."

"A thorough investigation and comparison is carried out on various LDPC decoding algorithms based on their performance, similarities, scalability, numerical stability and feasibility for hardware realization."

Introduction

"According to their error correcting mechanisms, decoding algorithms for LDPC codes can be categorized into two main classes, i.e., soft-decision [194] and hard-decision [195] based approaches."

"Since the decoding algorithms for LDPC codes were on the edge of this survey's relevance, only few studies on LDPC decoding were referred."

"In Fang and others [196] have proposed an exclusive survey on the design and analysis of protograph based LDPC codes."

"In this survey only few LDPC based decoding algorithms were discussed as the survey was focused exclusively towards the encoding design of binary protograph based LDPC codes."

"In this survey a detailed discussion on the code construction techniques along with the design guidelines for LDPC encoders and decoders were provided."

"The holistic view of recent surveys on LDPC based decoding algorithms and its applications reveal that there exists a substantial gap between the amount of recent research advancements and their coverage in survey papers [196–203]."

Importance of LDPC Based Decoding Algorithms

"Recent studies on the research challenges and issues related to LDPC codes, it is evident that there still exists a wide demand and need for the selection of computationally efficient decoding algorithm and architectures which are expected to be permanent [204]."

"For the past few years, research works on LDPC codes have been carried out in the scientific community by the coding researchers and theorists throughout the world."

"A dedicated research team from IIT Madras has been working in the area of coding theory, and they have introduced many new decoding schemes related to LDPC in the past few years."

"To these research works, at national level many research ideas based on LDPC codes were published by leading scientific journals of various domains [205–207]."

"It is pretty evident that still research ideas based on LDPC codes are widely studied and explored in INDIA."

Overview of Some Popular Hard Decision Based Weighted Bit Flipping LDPC Decoding Algorithms

"Chen [208] introduced a new adaptive-weighted multi bit-flipping based decoding algorithm using the concept of ordered statistics."

"Tiwari and others [209] in their inspiring work proposed a novel hybrid weighted bit flipping based decoding algorithm (HWBF) for LDPC codes."

"The performance of this hybrid weighted bit flipping algorithm was reported to be superior in terms of coding gain improvement, iteration count and decoding time reduction when compared with other popular bit flipping algorithms."

"A new type of multi-thresholding based bit flipping (MTBF) algorithm for decoding structured LDPC codes was proposed by Liu and others [210]."

"Elsanadily and others [211] designed a modified bit flipping based decoding algorithm for generalized LDPC codes over binary channels."

"Liu and others [212] introduced a hard decision bit-flipping decoder based on the concept of adaptive bit-local thresholding and demonstrated it with two novel algorithms."

"The proposed decoder design is based on a multi-threshold bit flipping algorithm."

Review of Iterative Optimization Based Bit Flipping Algorithms

"View point several outstanding research proposals were formulated for hard decision based LDPC decoding algorithms using the concept of Gradient descent based bit flipping (GDBF)."

"The decoding steps of this improved GDBF algorithm is given below Even though, this improved version of GDBF has better decoding stability than GDBF, in some of the cases if the local optimum traps the search point the escape process fails and the decoding loop of the search point remains trapped."

"The decoding steps of this algorithm are given below. Also, for improving the decoding efficiency for multi bit flipping mode of operation, a new adaptive inversion threshold is proposed in this work."

"A brief review highlighting the latest advancements of hard decision based decoding algorithms for LDPC codes based on Gradient descent bit flipping (GDBF) methodology is presented."

Optimization Method Based Soft Decision Decoding Algorithms for LDPC Codes

"Wei and others [213] in their outstanding research work proposed a reduced complexity Linear Programming (LP) decoding algorithm for LDPC codes using the concept of ADMM method."

"Through simulation results it is evident that the improved ADMM based decoding algorithm can save around 20% of the decoding time while demonstrating good error rate performances."

"This algorithm development was focused with an objective to improve the error correcting performance of the conventional ADMM based decoding algorithm through a simplified computational approach."

"Through experimental simulations, the decoding performance of this algorithm is shown to achieve better decoding performance than the conventional ADMM based decoding algorithm for LDPC codes."

"The feasibility analysis of this scheme indicates that the complexity requirements of this algorithm to perform processing and computation in every single decoding iteration is much lesser when compared to the existing ADMM based decoding schemes [213–217]."

Conclusion and Future Work

"For the past few decades, the development of error control coding techniques based on LDPC decoding algorithms has emerged as an active area of research and source of inspiration for several outstanding researchers worldwide."

"Based on the meta-analysis on the current developments, it is quite interesting to observe that the hard decision based decoding algorithms have emerged as a successful technique for solving many real world problems."

"ADMM based decoding approaches for LDPC codes using LP process have demonstrated their popularity and effectiveness in tackling wide range of data transmission problems with their simple, yet powerful decoding framework."

"The future development process of the research works on LDPC decoding algorithms should evaluate the use of different techniques based on optimization, linear programming and parameter estimation to motivate new and promising solutions."

Radar Cross Section Minimization for Step Structures Using Topology Optimization [218]

This is a machine-generated summary of:

Chen, Fei; Zhu, Jihong; Zhang, Weihong: Radar cross section minimization for step structures using topology optimization [218]

Published in: Structural and Multidisciplinary Optimization (2022)

Link to original: https://doi.org/10.1007/s00158-021-03110-y

If you want to cite the papers, please refer to the original.

For technical reasons we could not place the page where the original quote is coming from.

Abstract-Summary

"As one of the important indicators of this performance, radar cross section (RCS) has been extensively studied."

"A significant limitation on the RCS of flight vehicles lies in the fact that tiny gaps will transform into steps during structural deformation, where the local RCS will be significantly affected."

"This study investigates a topology optimization method to minimize the RCS of step structures when subjected to external loads."

"The conventional method of moments is utilized to calculate the RCS."

"To implement the gradient-based optimization, the sensitivity of RCS with respect to the design variables is derived analytically by the adjoint method."

"Together with a structural volume limitation, the average RCS of target incident angles will be set as the object to obtain the structural topology."

Introduction

"Hassan et al. [219] utilized the density-based topology optimization to design patch antennas with a gradient-based optimization algorithm."

"As a classical electromagnetic scattering problem, few studies have attempted to investigate the topology optimization on RCS, not only because of the own calculation complexity of RCS, but also the complication of analytical sensitivity analysis of the object with respect to the design variables under various numerical algorithms."

"Haji-Ahmadi et al. [220] performed the topology optimization to design the unit cell of meta-surface as the RCS reducer by the binary particle swarm optimization."

"This investigation aims to minimize the RCS of step structures subjected to external loads through topology optimization."

Problem Formulation and Computational Methodologies

"When performing mechanical FEA on aerospace structures, shell elements are usually utilized to discretize the thin-walled model to reduce the calculation scale."

"The facets required for the RCS calculation of steps in the thickness direction cannot be formed by the location of nodes extracted from FEA solvers."

"To verify the accuracy of the simplified model by shell elements and ASEs in mechanical performance analysis and RCS calculation, comparisons of the solid model and simplified model are performed in ANSYS and FEKO, respectively."

"The accuracy of the model simplified by shell elements and ASEs can be ensured in mechanical behavior and RCS calculation."

"The method of moments (MOM) [221], finite-difference time-domain (FDTD) method [222], and MLFMA [223] are three mainstream ways to calculate RCS of steps."

"To deal with this difference, we divide one quadrilateral mechanical element into two triangular elements here for the RCS calculation."

The Formulation and Implementation of Topology Optimization

"The RCS minimization problem for the step structures subjected to external loads can be written as here: s is the number of incident angles that need to be considered for RCS minimization."

"To perform the gradient-based topology optimization, the sensitivities of the object and constraints with respect to the design variables need to be provided."

"That the sensitivities of structural compliance and volume fraction commonly exist in other articles dealing with topology optimization problems [224, 225]."

Numerical Results

"The compliance constraint of the structure must be considered in the RCS minimization design."

"To further verify the effect of the proposed method on the overall RCS reduction, the RCS minimization designs for different numbers of target angles are tested."

"Compared with the stiffness design, the average RCS of target angles is reduced by 8.10%, 11.80%, and 18.74%, respectively, which demonstrates that the proposed method is effective in RCS reduction."

"It is apparent that the overall RCS of RCS minimization designs is lower than that of the stiffness design, which reveals that the RCS minimization optimization for target angles is effective for simultaneously reducing the overall RCS."

"Same as the previous example, to verify the effect of the number of target angles on reducing the overall RCS, the RCS minimization designs with different numbers of target angles are studied here with the same stiffness constraint."

Conclusions

"This paper proposed an efficient topology optimization approach for the RCS reduction of step structures."

"The main findings of this paper are drawn as follows: By changing the load-carrying path of the structure, the proposed method can make the relatively large deformation area avoid the step domain, thus reducing the RCS of the functional area."

"This method can achieve a considerable RCS reduction of target angles by a lower stiffness loss."

"With the gradual relaxation of the compliance constraint, the changes of step height in the two examples present different trends, which indicates that the step height is not the only factor affecting the RCS of the step structure."

"Considering the efficiency of optimization, it is only necessary to minimize the average RCS of representative angles."

"Besides, the proposed method is expected to reduce the RCS of practical structures with multiple functional domains and steps after the improvement of computational power."

A Computationally Efficient Algorithm to Estimate the Parameters of a Two-Dimensional Chirp Model with the Product Term [226]

This is a machine-generated summary of:

Shukla, Abhinek; Grover, Rhythm; Kundu, Debasis; Mitra, Amit: A computationally efficient algorithm to estimate the parameters of a two-dimensional chirp model with the product term [226]

Published in: Multidimensional Systems and Signal Processing (2023)

Link to original: https://doi.org/10.1007/s11045-023-00879-7

If you want to cite the papers, please refer to the original.

For technical reasons we could not place the page where the original quote is coming from.

Abstract-Summary

"Several methods have been proposed for parameter estimation of these models."

"We consider the problem of parameter estimation of two dimensional chirp models and propose a computationally efficient estimator and establish asymptotic theoretical properties of the proposed estimators."

"The proposed estimators are observed to have the same rates of convergence as the least squares estimators."

"Further, the proposed estimators of chirp rate parameters are shown to be asymptotically optimal."

Introduction

"The parameter estimation problem for two dimensional (2D) chirp models is encountered in many real-life applications such as 2D-homomorphic signal processing, magnetic resonance imaging (MRI), optical imaging, interferometric synthetic aperture radar (INSAR), modeling non-homogeneous patterns in the texture image captured by a camera due to perspective or orientation (see e.g., [227–229] and the references cited therein)."

"Cubic phase function (CPF) proposed in O'Shea [230], was extended in Zhang et al. [231] for similar 2D chirp signal modeling."

"A finite step computationally efficient procedure for a similar 2D chirp signal model proposed in Lahiri and others [232] was proved asymptotically equivalent to LSEs."

"The proposed algorithm is motivated by the fact that a 2D chirp model with five non-linear parameters can be viewed through a number of 1D chirp models with two non-linear parameters, and hence computational complexity of estimators for the 2D models can be reduced."

Estimation Methodology

"This paper addresses the problem of parameter estimation of a 2D chirp signal model defined as follows: Here, y(m, n) is the observed real valued signal and X(m, n) is the additive noise term."

"Our methodology is developed by estimating parameters of these 1D chirps based on a particular column (or row) vector of data matrix, rather than estimating the whole 2D chirp parameters based on the full data matrix."

"This procedure reduces computational burden to estimate model parameters drastically."

Theoretical Results

"We first state the model assumptions required to derive the theoretical asymptotic properties explicitly."

"Under these general assumptions, we have derived strong consistency and asymptotic normality of the estimators."

"The obtained results are stated in the following theorems."

Simulation Results

"We compare the performance of the proposed estimators with the asymptotically optimal estimators such as LSEs, and ALSEs, and fast but sub-optimal 2D-multilag-HAF estimators."

"The second part shows the lower computational cost of the proposed estimators as compared to the LSEs."

"The third part exemplifies the ability of the proposed estimators to extract original gray-scale texture from one contaminated with noise and reproduce the original texture."

"Further, as sample size increases, the gap between the MSEs of the proposed estimators and LSEs decreases."

"The computational advantage of the proposed estimators over the conventional LSEs is quite significant."

"We analyze this data using the proposed estimator, the optimal LSEs, ALSEs, and also with the 2D-multilag-HAF method."

"That we have used true values as initial guess for obtaining optimal estimators, LSEs and ALSEs because of the computational complexity and grid-search method for obtaining 2D-multilag-HAF estimators and proposed estimators."

Conclusion

"The paper proposes a computationally efficient estimators with the same convergence rate as the LSEs or ALSEs."

"The key idea is to develop a strategy by disintegrating the 2D model into several 1D chirp models and then design an optimal estimation method to obtain estimates of the 2D model parameters."

"The proposed estimators are not only asymptotically unbiased but also have an asymptotic normal distribution and same rate of convergence as that of the LSEs."

"Extensive numerical simulations firmly support the theoretical results and also unveil the gigantic gap between time required for obtaining proposed estimates and the LSEs."

"Further, our proposed methodology can be adapted for parameter estimation of more general signal models by utilizing a sequential procedure similar to Grover and others [233]."

Watermarking Techniques for the Security of Medical Images and Image Sequences [234]

This is a machine-generated summary of:

Alquhayz, Hani; Raza, Basit: Watermarking Techniques for the Security of Medical Images and Image Sequences [234]

Published in: Arabian Journal for Science and Engineering (2021)

Link to original: https://doi.org/10.1007/s13369-021-06254-7

If you want to cite the papers, please refer to the original.

For technical reasons we could not place the page where the original quote is coming from.

Abstract-Summary "The secure storage and exchange of such images and image sequences present challenges for hospital management."

"We design a systematic dual-medium watermarking model that focuses on the security of these images and sequences."

"Watermark bits (Wbs) are embedded in the transformed and temporal domains of images and image sequences, respectively, instead of in the spatial domain."

"The images are transformed using integer wavelet transforms, and motion vectors (MVs) are estimated from the image sequences to transform them into the temporal domain using seven block-matching algorithms."

"The best-performing MVs during frame reconstruction are reconstructed using a genetic algorithm to improve imperceptibility in the watermark-embedded image sequences."

"Wbs are embedded in regions of image sequences with significant movements."

Introduction

"A prospective solution that ensures the security and integrity of medical images is watermarking, i.e., the embedding of secret information in the original data (the cover medium)."

"In medical image sequences, the region of interest is extremely small because Wbs can only be embedded in a small region."

"The basic watermarking technique for images involves embedding Wbs in the spatial domain of a cover image, and various methods have been proposed in this regard [235–238]."

"A user-defined dual watermarking technique is proposed for medical images and image sequences, where Wbs can be embedded in images and image sequences to authenticate originality."

"Embedding Wbs in image sequence regions of minimum or no movement allows more bits per MV to be embedded compared with state-of-the-art methods, thereby minimizing the differences in embedded and actual MVs."

Literature Review

"Watermarking in the spatial domain is less secure, as the modification of pixel intensity can produce salt-and-pepper noise in embedded images [239]; it is noteworthy that all the pixels are independent and the embedded information can be detected."

"Watermark embedding in the transformed domain increases the imperceptibility of digital images because every single Wb is distributed in neighboring pixels instead of a single pixel [240]."

"To spatial [241] and frequency [242–244] domains, the temporal domain is viable for embedding Wbs in image sequences."

"Embedding secret information in image sequences by changing the magnitude of its MVs is the simplest and most basic technique."

"Imperceptibility is improved by embedding watermark data in the MVs of smooth regions in consecutive frames, which are non-salient."

"The Wbs were embedded in the transformed domain of the images and the temporal domain of the image sequences."

Background Knowledge
"We converted the medical images into the transform domain using IWT and image sequences into the temporal domain by estimating the MVs using a BMA."

"Three-step search (TSS) The basic aim of all BMA techniques other than ES is to reduce the computational complexity of the algorithms."

"Li [245] proposed a center-biased search pattern by adding eight more points around the center of the search window in the first step, as well as reduced the computational cost by introducing a halfway search stop."

"Four step search (4SS) 4SS is an improved version of NTS, in which the center-biased searching pattern and halfway search stop are retained; it further reduces the computational cost with the addition of a fourth step [246]."

"ARP is implemented in the first step to reduce the initial search area, followed by SDSP in the remaining steps."

Proposed Approach for Wb Embedding
"The Wbs were embedded in one sub-band at a time, depending on the pixel values in the decision band."

"Modulus 3 of the updated LL became 0, satisfying condition c. Wbs were embedded at different locations in all four sub-bands."

"The following methods can be used to embed Wbs in image sequences: embedding in the spatial domain of each frame or selected frames, transforming frames into another domain, and using the temporal domain."

"At the receiver end, the Wbs were extracted under the following conditions: Db value "0" indicates Wb was embedded in LSB of HH Db value "1" indicates Wb was embedded in LSB of LH Db value "2" indicates Wb was embedded in LSB of HL The extracted watermark bits were compared with the actual watermark to authenticate the medical image."

"We embedded the Wbs in the temporal domain of image sequences/videos."

Experiments and Results
"We not only increased the embedding capacity of the cover image, but also successfully achieved higher imperceptibility."

"The watermark embedding capacity of the proposed method is equivalent to that of the sub-band of the cover images, and it differs for different image sequences."

"To evaluate watermark embedding in image sequences, PSNR values were computed before and after embedding the Wbs."

"Although the method by Wu [247] yielded higher PSNR values compared with the methods or Bhuiyan and Chang, the embedding capacity afforded was extremely low."

"The PSNR values indicate the change in the frames reconstructed using the actual and embedded MVs."

"The embedding capacity was set to 25 BpF for all image sequences to compare the proposed method with the other methods."

"The minimum degradation in the embedded image sequences indicated that the proposed model yielded the highest imperceptibility."

Conclusion and Future Studies

"A user-defined watermark embedding model to ensure the security of medical images and image sequences was proposed herein."

"In the image watermarking process, a medical cover image was decomposed into four sub-bands by applying IWT."

"Embedding Wbs in such MVs, which reconstruct degraded frames, can cause a reduction in imperceptibility."

"We utilized a GA to select MVs, thereby enabling the reconstruction of the original frames with minimum change in the PSNR even after embedding the Wbs."

"We improved the security of medical images and image sequences by embedding watermarks with higher imperceptibility."

"The capacity of the dual-medium watermark embedding model can be increased by embedding more Wbs per pixel in medical images."

"The dual-medium watermark embedding system can be used to embed the encryption key in one cover medium and the encrypted medical image in another."

Design and Implementation of a Smart Elderly Positioning Management System Based on Wireless Communication Network [248]

This is a machine-generated summary of:

Yu, Tiantian; Wang, Ruping: Design and implementation of a smart elderly positioning management system based on wireless communication network [248]

Published in: EURASIP Journal on Wireless Communications and Networking (2021)

Link to original: https://doi.org/10.1186/s13638-021-02021-2

For technical reasons we could not place the page where the original quote is coming from.

Abstract-Summary

"Wireless communication refers to long-distance transmission communication between multiple nodes without propagation through conductors or cables, and wireless communication can be carried out by radio and electromagnetic waves, etc. Judging from the current situation, the home space positioning design of the medical and elderly care communities is not perfect, which affects the traditional home care."

"Wireless communication can give wisdom, coordination, and development space in the smart elderly care space positioning design, allowing the elderly to interact, contact, and know their precise location and behavior."

"This article mainly studies the design and implementation of smart pension space positioning under BIM technology."

"The basic technology and communication technology of personnel deployment based on BIM technology are introduced, the development strategy of wireless communications node is studied, the installation of equipment radio personnel and communication equipment is designed, and the system management platform for the design and development of staff placement in pension institutions is provided."

"The experimental results of this paper show that the smart pension space positioning under the BIM technology solves the management loopholes in nursing homes, shortage of nursing staff, inefficiency, lack of services, and family conflicts."

"The development of wireless communication technology promotes the development of intelligent pension space."

Introduction

"The implementation of BIM technology also includes management changes, which can exchange information throughout the entire life cycle of the building to support the integration of design, construction, and management."

"This paper designs an intelligent pension management system based on BIM technology, with information technology applied to pension institutions to improve the efficiency of staff management and transaction management in pension institutions, reduce labor costs, and effectively alleviate the "large gap of nursing staff."

"This paper mainly studies the realization of the smart pension space positioning system under BIM technology and the application of BIM in smart space pension management."

"Tell about the design and implementation of a smart pension space positioning system based on BIM technology, and make a series of analysis of the safety of the elderly to ensure that they can be located and monitored at any time and any place."

Method/Experiment

"Taking BIM technology as the research background and taking the elderly as the research object, the spatial design and spatial positioning of the residential houses are carried out."

"This article adopts wireless radio frequency positioning technology for the design of the positioning system, which is a positioning technology consisting of three parts: a radio frequency tag, a radio frequency reader, and a software system, which transmits information through radio frequency signals."

"The function of the positioning system is judged by the evacuation of the elderly in the retirement space, and the effect of the space design is judged by the length of the evacuation time [249]."

BIM Technology Smart Pension Space Positioning Method

"BIM technology can make the community building information accurate to the component level and can show the whole community environment with three-dimensional model, which makes the data more accurately express the reality."

"Physical motor function Although the physical function of the elderly aged 65–75 years gradually declines, they can basically take care of themselves."

"After 85 years old, the elderly's physical function will accelerate decline, and they need continuous care."

"Psychological changes brought about by changes in social roles. After retirement, many old people gradually become leisurely from a busy and regular life, which makes a lot of free time for the old people unable to have fun."

"Psychological changes brought about by physiological changes. The elderly's physical function is gradually declining, and there are certain obstacles in their activities."

"The special needs of the elderly for the community environment. The physical and psychological characteristics of the elderly determine their way of life and mentality."

Experiment of Smart Pension Spatial Positioning

"Based on the special theme of the elderly, it provides a space for the elderly to realize their social needs through life, learning, entertainment, sports, and other activities."

"For the physical and mental conditions of the elderly, the environment of the space used by the elderly activities should be constructed with more familiar, comfortable and humane interior decoration materials, and non-reflective materials should be selected in the sanatorium."

"When designing a comfortable and happy indoor environment for the elderly, the use of color is also crucial."

"Comfort, peace and health are the best aging environment pursued by most elderly people."

"According to the principle of aging adaptability of elderly care center in indoor environment, the design of space environment of elderly care center is based on many low-saturation color choices, such as milky white, beige, light gray, and green."

"The living environment of the elderly is very important for them."

Results and Discussion

"In the position detection experiment, the CC2430 node was used as a gateway to connect with the computer to form a coordinated entrance."

"The reference nodes are arranged in a 10 m square, and the blind nodes are placed in different test positions."

"Since the installation result of the engine CC2431 has an analysis of 0.25 m and there is a reduction result, the installation result is a limited combination of coordinate X and Y data close to the actual position."

"Due to the influence of the environment on the building, the detection error is slightly larger than that of the open environment, but Kalman filtering can still obtain a better estimate of the trajectory, reflecting the change of the actual moving target movement position."

"When we select the maturity level of 11 elements of a certain evaluated object, we can find out the corresponding scores from the above scoring table, and then, we can get the BIM score of this object after accumulation."

Conclusions

"The system that affects the quality of old-age care in smart elderly communities is divided into two levels: standard level and factor level."

"Entropy weight method is adopted: Through comparison, the weight of the influencing factors is clarified, and the TOPSIS evaluation method of the impact of BIM on the quality of care of the elderly in the smart community is clarified."

"The TOPSIS method was used to evaluate the impact of BIM on the quality of care for the elderly in smart communities."

"We demonstrated the impact of BIM on improving the quality of service for the elderly in smart communities, analyzed the relative proximity of each factor system to have a corresponding impact, and found that BIM has a particularly significant impact on the living environment of the elderly."

Adaptive Image Enhancement and Dynamic-Template-Matching-Based Edge Extraction Method for Diamond Roller On-Machine Profile Measurement [250]

This is a machine-generated summary of:

Li, Jing; Lu, Yushun; Shen, Nanyan; Pu, Junhao; Ma, Zehui: Adaptive image enhancement and dynamic-template-matching-based edge extraction method for diamond roller on-machine profile measurement [250]

Published in: The International Journal of Advanced Manufacturing Technology (2022)

Link to original: https://doi.org/10.1007/s00170-022-09113-3

If you want to cite the papers, please refer to the original.

For technical reasons we could not place the page where the original quote is coming from.

Abstract-Summary

"The complex and harsh environment faced by diamond roller on-machine profile measurement brings challenges to roller image processing technology."

"This paper focuses on the original image enhancement and edge extraction, which has a vital impact on the measurement results of the roller profile in the complex environment."

"The key research is the multi-scale Retinex (MSR) improved by the fast Deriche filter and the saturation correction method with gating mechanism, and the image enhancement effect with strong edge-preserving ability is realized through parameter optimization."

"It lays the foundation for realizing high-precision diamond roller on-machine profile measurement."

Introduction

"The measurement method of diamond roller profile is similar to that of ordinary grinding wheel, which can be divided into contact measurement and non-contact measurement."

"Many Canny edge detection algorithms combined with image enhancement are proposed recently, which improve the performance of Canny algorithm in edge connectivity [251, 252]."

"Compared with these methods, Retinex theory has a better performance in image enhancement, which can achieve a balance among dynamic range compression, edge enhancement, and color constancy, and also enhance the different kinds of images of multiple features [253, 254]."

"All these interference areas on image greatly affect the edge detection of roller, which are required to be removed to obtain the roller profile for improving the edge detection quality."

"An adaptive image enhancement processing method and a dynamic-template-matching-based edge extraction method are explored in this paper, which can adapt to the complex environment to obtain accurate roller edges and lay the solid foundation for high-precision on-machine measurement of roller profile dimensions."

Method

"The axes of lens and light source are coaxial, which ensures the images of clear and sharp edges and thus the measurement accuracy."

"The upper computer unit provides the functions such as light source control, camera control, image acquisition, camera calibration, image processing, profile evaluation, and the communication with computer numerical control (CNC)."

"For the on-machine measurement of diamond roller, it is quite difficult to obtain high-quality images due to the complex machining environment under the influence of ambient light, coolant, and wear debris."

"Image enhancement and ROI sub-pixel edge extraction are the key to ensure the profile measurement accuracy."

"This paper focuses on image processing methods to obtain high-quality image of the ROI profile, which can effectively reduce the impact of harsh environment on the edge detection of roller."

Improved Adaptive Image Enhancement Based on MSR

"The red–green–blue (RGB) color gamut is converted to luminance-bandwidth-chrominance (YUV) color gamut, which directly enhances the gray-scale brightness Y component and better maintains the color of the original image, and also reduces the computation."

"If the Y component of enhanced gray-scale brightness is directly converted to the RGB space, the image may suffer from a certain degree of saturation loss."

"Taking the edge and information complexity of the image into consideration, the fitness function of optimization problem is constructed by combining edge intensity and gray information entropy."

"Then the optimization problem is solved by intelligent algorithm to obtain the optimal parameters for the enhancement of a certain roller image."

"Then the Tenengrad gradient of the enhanced image obtained by the current parameters after Sobel filtering is calculated according to the edge points."

"The image enhancement under YUV color gamut only considers the change of its gray-scale brightness."

Edge Extraction Based on Dynamic Template Matching

"To ensure the quality of extracted edge, an adaptive threshold method is used to segment the image and dynamic template matching is utilized to extract roller profile, before sub-pixel edge extraction is performed by polynomial interpolation."

"In order to effectively eliminate the influence of wear debris and coolant attached to roller profile, a sub-pixel edge extraction method based on dynamic template matching is proposed."

"When the bounding rectangle area of the maximum profile in the segmented image is smaller than that in the template image, the image is eliminated; otherwise, the shape matching method based on Hu moment [255] is used to calculate the matching degree between the image and the template."

"A high-quality roller profile image without interference areas that can be finally used for edge extraction is acquired."

Experiment and Discussion

"The roller profile on-machine inspection software developed by Visual Studio using the proposed image enhancement and edge extraction methods is installed on industrial computer typed Advantech TPC-1581WP-433BE under Windows 7 operating system."

"MSR and several image enhancement methods based on its improvement [256–262] are compared with the proposed method."

"The image enhancement effect of this method is most conducive to extracting the edge of the diamond roller in all methods."

"The enhancement effect of the roller region and edge of L2LP [261] and SuPeR [262] shown in subfigures (e) and (f) are close to that of the proposed method and the profile edges are sharper, but the proposed method can better suppress background noise."

"The proposed method can filter out most of the background noise outside the roller profile and obtain the most uniform image brightness, the clearest image quality, and the best enhancement effect in general."

Conclusion

"In order to overcome the influence of the complex environment of on-machine measurement on the image quality of the diamond roller profile, this paper focuses on the improved MSR and dynamic NCC template matching to achieve image enhancement with strong edge-preserving ability and obtain a high-quality roller profile without interference areas for edge extraction, which helps ensure high-precision on-machine measurement of roller profile."

"In the improved MSR method, a novel model is proposed to describe the edge quality of the image."

"It indicates that this method has great application value and its image imaging effect is most conducive to extracting the edge of the diamond roller among the comparison methods."

"The experimental results of the ROI edge extraction show that the effective edge extraction of different diamond rollers dressed under various working conditions is realized, and the extracted edges are smoother than the traditional NCC method."

RANSAC Algorithm for Instantaneous Frequency Estimation and Reconstruction of Frequency-Modulated Undersampled Signals [263]

This is a machine-generated summary of:

Djurović, Igor: RANSAC algorithm for instantaneous frequency estimation and reconstruction of frequency-modulated undersampled signals [263]

Published in: EURASIP Journal on Advances in Signal Processing (2021)

Link to original: https://doi.org/10.1186/s13634-021-00726-6

If you want to cite the papers, please refer to the original.

For technical reasons we could not place the page where the original quote is coming from.

Abstract-Summary

"Proposed matching pursuit and greedy techniques are inefficient for signals with several phase parameters since they require a search over multidimensional space."

"An alternative is proposed here based on the random samples consensus algorithm (RANSAC) applied to the instantaneous frequency (IF) estimates obtained from the time-frequency (TF) representation of recordings (undersampled or signal with missing samples)."

"The proposed technique is tested against third- and fifth-order polynomial phase signals (PPS) and also for signals corrupted by noise."

Introduction

"Similar holds for parametric estimation where signals can be described with a significantly smaller number of parameters than the number of samples required by sampling theorem [264, 265]."

"The RANSAC algorithm is applied to IF estimates calculated from the TF representations of undersampled signals."

"It means that if we have a parametric model of signal with a small number of parameters (for example quadratic phase signal or other signals with one or two parameters), it can be more efficiently processed with matching pursuit or greedy algorithm strategies."

"For signals with more parameters, for example, PPS of order three or higher, it would be better to employ the RANSAC algorithm."

"The RANSAC algorithm complexity is not so dependent on the number of signal parameters and it represents an excellent alternative to existing strategies in the field."

Signal Model and TF Representation

"The STFT can be written as where T{ } is an operator applied to modulated signal samples $x(n + k)w_h(k)\exp(-j\omega n)$."

"The IF as a crucial feature of the FM signals can be estimated maximizing STFT: This IF estimator is biased with the bias increasing as window width does."

"The final estimate is selected by simple criterion function, i.e., matching correlator or ML penalty function, giving robust technique to determine optimal window width in the STFT and final algorithm output."

"Our goal is a precise estimation of phase parameters of signal $\{a_0, a_1, \ldots, a_M\}$ and consequently signal reconstruction based on available signal samples $x(n_i)$, $i = 1, \ldots, K$. In the next section, we present the RANSAC-based procedure for the estimation of the parameters of undersampled PPSs."

RANSAC for Signal Reconstruction

"The random samples selection should be repeated multiple times, and the most accurate among reconstructed signals/IF estimates is selected based on an appropriate criterion."

"In the case of undersampled signals, refinement is even more important since it can improve relatively inaccurate estimates without outliers what cannot be done for IF estimate samples with outliers."

"It should be noted that in the rough stage it is performed only on M randomly selected samples of the IF estimate, while in the fine stage it is performed on all available K signal phase samples."

"The signal filtering is commonly performed in the refinement stage of the algorithm in both the QML and QML-RANSAC to reduce noise-like effects from the estimate and from this point of view it could be useful: However, it can introduce additional errors distorting dechirped signal $x'(n_i)$ used in the further procedure."

Number of Trials in the RANSAC Algorithm

"The main finding from the experiment is that the percentage of outliers in IF estimates is smaller than in the time domain, i.e., the percentage of correctly estimated IF samples is larger than K/N ratio in almost all cases of interest."

"It can be seen that only for extremely narrow window $h = 8$ and/or for small ratio K/N bellow K/N = 10%, percentage of outliers in the IF estimate is larger than the number of missing samples in the original signal."

"For a known percentage of IF estimate outliers (as demonstrated previously it can be approximated or at least overestimated with the number of missing samples), Λ can be significantly reduced, i.e., the number of trials can be selected in dependence on the percentage of IF estimate outliers (or percentage of missing samples) while in the matching pursuit techniques it is constant."

Results and Discussion

"It can be seen a relatively large number of outliers for K/N = 7.8% without refinement while in the case of the O'Shea refinement percentage of outliers is moderate (quantified below)."

"For K/N = 8%, there are no outliers in the IF estimation in the case of application of the refinement but there is 7% of outliers when the refinement is not applied."

"For K/N = 6.6%, the percentage of outliers in the case of application of the refinement is only 4% while in the case without refinement it is 21%, for K/N = 5.3% the proposed technique with refinement has only 11% outliers while without refinement it has almost 4 times more, i.e., 43%."

"It can be seen that for K/N = 12.5%, the proposed RANSAC procedure with implemented refinement achieves almost perfect results with minor errors at interval borders."

Multicomponent Signals

"Consider a multicomponent signal with Q FM components: The goal is to estimate signal parameters of all components for an undersampled signal."

"The procedure is relatively simple: we are estimating the strongest component firstly by the previously described procedure and after estimation of its parameters the component is removed from the mixture and estimation is performed for weaker components."

"As we have already claimed, it can work for signals of different amplitudes where the STFT maximization can be used to estimate parameters of the strongest component."

"It is possible to estimate intersecting components in the TF plane with the same amplitude but with more trials in the RANSAC procedure than in the case of mono-component signals."

"Assume that the percentage of outliers in the IF estimation is e. In the worst-case scenario, the remaining 1 − e percentage of correct estimates is uniformly distributed on Q signal components (1 − e)/Q. The RANSAC requires that all M samples belong to the same component number."

Conclusion

"The QML-RANSAC technique is applied to the parameter estimation of undersampled FM signals."

"The proposed technique overcomes the main problem in the matching pursuit and greedy strategies generalization to signals with more parameters by employing the RANSAC algorithm."

"It is shown that without employing refinement of the phase coefficients mediocre results are obtained while with the O'Shea refinement results are significantly improved reducing the number of outliers in the final estimation for several orders of magnitude."

"Improvement in the multicomponent undersampled signals estimation and reconstruction will be also considered in our future research."

Bibliography

1. Pasandideh, F., Cesen, F. E. R., Pereira, P. H. M., Rothenberg, C. E., & de Freitas, E. P. (2023). An improved particle swarm optimization algorithm for UAV base station placement. *Wireless Personal Communications*. https://doi.org/10.1007/s11277-023-10334-2
2. Akram, T., Awais, M., Naqvi, R., Ahmed, A., & Naeem, M. (2020). Multicriteria UAV base stations placement for disaster management. *IEEE Systems Journal, 14*(3), 3475–3482.
3. Zhong, X., Huo, Y., Dong, X., & Liang, Z. (2020). Qos-compliant 3-d deployment optimization strategy for UAV base stations. *IEEE Systems Journal, 15*, 1795–1803.
4. Wang, L., Zhang, H., Guo, S., & Yuan, D. (2022). Deployment and association of multiple UAVs in UAV-assisted cellular networks with the knowledge of statistical user position. *IEEE Transactions on Wireless Communications, 21*(8), 6553–6567.
5. Qiu, C., Wei, Z., Yuan, X., Feng, Z., & Zhang, P. (2020). Multiple UAV-mounted base station placement and user association with joint fronthaul and backhaul optimization. *IEEE Transactions on Communications, 68*(9), 5864–5877.
6. Malathy, S., Jayarajan, P., Nour Hindia, M. H. D., Tilwari, V., Dimyati, K., Noordin, K. A., & Amiri, I. S. (2021). Routing constraints in the device-to-device communication for beyond IoT 5G networks: A review. *Wireless Networks*. https://doi.org/10.1007/s11276-021-02641-y

7. Yang, Z.-Y., & Kuo, Y.-W. (2017). Efficient resource allocation algorithm for overlay D2D communication. *Computer Networks, 124*, 61–71.
8. Esmat, H. H., Elmesalawy, M. M., & Ibrahim, I. I. (2018). Uplink resource allocation and power control for D2D communications underlaying multi-cell mobile networks. *AEU-International Journal of Electronics and Communications, 93*, 163–171.
9. Lin, Z., Huang, L., Zhao, Y., Du, X., & Guizani, M. (2017). P2P-based resource allocation with coalitional game for D2D networks. *Pervasive and Mobile Computing, 42*, 487–497.
10. Gong, W., Li, G., & Li, B. (2018). System utility based resource allocation for D2D multicast communication in software-defined cellular networks. *AEU-International Journal of Electronics and Communications, 96*, 138–143.
11. Bai, F., Sadagopan, N., Krishnamachari, B., & Helmy, A. (2004). Modeling path duration distributions in MANETs and their impact on reactive routing protocols. *IEEE Journal on Selected Areas in Communications, 22*(7), 1357–1373. https://doi.org/10.1109/JSAC.2004.829353
12. Muchtar, F., Abdullah, A. H., Hassan, S., Khader, A. T., & Zamli, K. Z. (2019). Energy conservation of content routing through wireless broadcast control in NDN based MANET: A review. *Journal of Network and Computer Applications, 131*, 109–132.
13. Chithaluru, P., Tiwari, R., & Kumar, K. (2019). AREOR–Adaptive ranking based energy efficient opportunistic routing scheme in Wireless Sensor Network. *Computer Networks, 162*, 106863.
14. Bello-Salau, H., Aibinu, A. M., Wang, Z., Onumanyi, A. J., Onwuka, E. N., & Dukiya, J. J. (2019). An optimized routing algorithm for vehicle ad-hoc networks. *Engineering Science and Technology, an International Journal, 22*(3), 754–766.
15. Al-Dhief, F. T., Sabri, N., Fouad, S., Latiff, N. M. A., & Albader, M. A. A. (2017). A review of forest fire surveillance technologies: Mobile ad-hoc network routing protocols perspective. *Journal of King Saud University-Computer and Information Sciences, 31*(2019), 135–146.
16. Li, M., Zhang, L., Li, V. O., Shan, X., & Ren, Y. (2005). An energy-aware multipath routing protocol for mobile ad hoc networks. *ACM SIGCOMM Asia, 5*, 10–12.
17. Ladas, A., Deepak, G. C., Pavlatos, N., & Politis, C. (2018). A selective multipath routing protocol for ubiquitous networks. *Ad Hoc Networks, 77*, 95–107.
18. Dubey, R., Mishra, P. K., & Pandey, S. (2020). Mixed uplink, downlink channel allocation and power allocation schemes for 5G networks. *Wireless Personal Communications*. https://doi.org/10.1007/s11277-020-07148-x
19. Lin, S., Yusheng, J., Ping, W., & Liu, F. (2013). Resource allocation using particle swarm optimization for D2D communication underlay of cellular networks. In *2013 IEEE Wireless Communications and Networking Conference* (pp. 129–133).
20. Islam, M. T., Taha, A.-E. M., Akl, S., & Choudhury, S. (2015). A local search algorithm for resource allocation for underlaying device-to-device communications. In *2015 IEEE Global Communications Conference* (pp. 1–6).
21. Yu, G., Xu, L., Feng, D., Yin, R., Li, G. Y., & Jiang, Y. (2014). Joint mode selection and resource allocation for device-to-device communications. *IEEE Transactions on Communications, 62*(11), 3814–3824.
22. Wang, F., Xu, C., Song, L., Zhao, Q., Wang, X., & Han, Z. (2013). Energy-aware resource allocation for device-to-device underlay communication. In *2013 IEEE International Conference on Communications* (pp. 6076–6080).
23. Algedir, A., & Refai, H. H. (2017). Adaptive D2D resources allocation underlaying (2-tier) heterogeneous cellular networks. In *2017 IEEE 28th Annual International Symposium on Personal, Indoor, and Mobile Radio Communications (PIMRC), Montreal, QC* (pp. 1–6).
24. Ali, M., Qaisar, S., Naeem, M., & Mumtaz, S. (2016). Energy efficient resource allocation in D2D-assisted heterogeneous networks with relays. *IEEE Access, 4*, 4902–4911.
25. Yang, C., Xu, X., Han, J., Rehman, W., & Tao, X. (2014). GA based optimal resource allocation and user matching in device to device underlaying network. In *2014 IEEE Wireless Communications and Networking Conference Workshops* (pp. 242–247).

26. Kashyap, K., Pathak, S., & Yadav, N. S. (2021). Optimization of spreading code using modified differential evolution for wireless communication. *Wireless Personal Communications*. https://doi.org/10.1007/s11277-021-08949-4
27. Mandi, M. V., Hari-Bhat, K. N., & Murali, R. (2015). Performance evaluation of DS-CDMA system using chaotic binary sequences. *Computer Applications: An International Journal, 2*(1), 37–52. https://doi.org/10.5121/caij.2015.2104
28. Okagbue, H., Adamu, M., & Anake, T. (2019). Differential evolution in wireless communications: A review. *International Journal of Online and Biomedical Engineering (iJOE), 15*(11), 29. https://doi.org/10.3991/ijoe.v15i11.10651
29. Chiu, C., Tong, Y., & Cheng, Y. (2019). Comparison of self-adaptive dynamic differential evolution and particle swarm optimization for smart antennas in wireless communication. *International Journal of Communication Systems, 32*(9), e3941. https://doi.org/10.1002/dac.3941
30. Chi, R., Li, Z., Chi, X., Qu, Z., & Tu, H. (2021). Reactive power optimization of power system based on improved differential evolution algorithm. *Mathematical Problems in Engineering, 2021*, 1–19. https://doi.org/10.1155/2021/6690924
31. Mathew, M., Wisner, B., Ridwan, S., McCarthy, M., Bartoli, I., & Kontsos, A. (2020). A bio-inspired frequency-based approach for tailorable and scalable speckle pattern generation. *Experimental Mechanics*. https://doi.org/10.1007/s11340-020-00631-3
32. Pan, B., Lu, Z., & Xie, H. (2010). Mean intensity gradient: An effective global parameter for quality assessment of the speckle patterns used in digital image correlation. *Optics and Lasers in Engineering, 48*(4), 469–477.
33. Sharma, K., & Pandey, G. P. (2020). Pattern correction in planar antenna array using multi-objective particle swarm optimization and continuous excitation. *Iranian Journal of Science and Technology, Transactions of Electrical Engineering*. https://doi.org/10.1007/s40998-020-00378-6
34. Khan, S. U., Rahim, M. K. A., Aminu-Baba, M., & Murad, N. A. (2017). Correction of failure in linear antenna arrays with greedy sparseness constrained optimization technique. *PLoS One, 12*(12), 1–18.
35. Patidar, H., & Mahanti, G. K. (2017). Failure correction of linear antenna array by changing length and spacing of failed elements. *Progress in Electromagnetics Research M, 61*, 75–84.
36. Muralidharan, R., Vallavaraj, A., Mahanti, G., & Patidar, H. (2017). Qpso for failure correction of linear array of mutually coupled parallel dipole antennas with desired side lobe level and return loss. *Journal of King Saud University, Engineering Sciences, 29*(2), 112–117.
37. Guney, K., Durmus, A., & Basbug, S. (2014). Antenna array synthesis and failure correction using differential search algorithm. *International Journal of Antennas and Propagation, 2014*, 276754. https://doi.org/10.1155/2014/276754
38. Grewal, N., Rattan, M., & Patterh, M. (2014). A linear antenna array failure correction with null steering using firefly algorithm. *Defence Science Journal, 64*(2), 136–142.
39. Muralidaran, R., Vallavaraj, A., Patidar, H., & Mahanti, G. K. (2014). Failure correction of linear array antenna with multiple null placement using cuckoo search algorithm. *Ictact Journal on Communication Technology, 05*(01), 877–881.
40. Allen, D. (2014). Prognostic optimization of phased array antenna for self-healing. In *Annual Conference of the Prognostics and Health Management Society* (Vol. 6).
41. Acharya, O. P., Patnaik, A., & Sinha, S. N. (2014). Limits of compensation in a failed antenna array. *International Journal of RF and Microwave Computer-Aided Engineering, 24*(6), 635–645.
42. Bhargav, A., & Gupta, N. (2013). Multiobjective genetic optimization of nonuniform linear array with low sidelobes and beamwidth. *IEEE Antennas and Wireless Propagation Letters, 12*, 1547–1549.
43. Robinson, J., & Rahmat-Samii, Y. (2004). Particle swarm optimization in electromagnetics. *IEEE Transactions on Antennas and Propagation, 52*(2), 397–407.

44. Rodriguez, J. A., Ares, F., Moreno, E., & Franceschetti, G. (2000). Genetic algorithm procedure for linear array failure correction. *Electronics Letters, 36*(2), 196–198.
45. Beng-Kiong, Y. (1999). Array failure correction with a genetic algorithm. *IEEE Transactions on Antennas and Propagation, 47*(5), 823–828.
46. Redvik, J. (1999). Simulated annealing optimization applied to antenna arrays with failed elements. In *IEEE Antennas and Propagation Society International Symposium. 1999 Digest. Held in conjunction with: USNC/URSI National Radio Science Meeting (Cat. No.99CH37010)* (Vol. 1, pp. 458–461).
47. Lozano, M. V., Rodriguez, J. A., & Ares, F. (1999). Recalculating linear array antennas to compensate for failed elements while maintaining fixed nulls. In *IEEE Antennas and Propagation Society International Symposium. 1999 Digest. Held in conjunction with: USNC/URSI National Radio Science Meeting (Cat. No. 99CH37010)* (Vol. 3, pp. 2048–2051).
48. Das, A., Mandal, D., Ghoshal, S., & Kar, R. (2018). Concentric circular antenna array synthesis for side lobe suppression using moth flame optimization. *AEU International Journal of Electronics and Communications, 86*, 177–184.
49. Ravipudi, J. L., & Neebha, M. (2018). Synthesis of linear antenna arrays using jaya, self-adaptive jaya and chaotic jaya algorithms. *AEU International Journal of Electronics and Communications, 92*, 63.
50. Das, A., Mandal, D., Ghoshal, S., & Kar, R. (2017). An efficient side lobe reduction technique considering mutual coupling effect in linear array antenna using bat algorithm. *Swarm and Evolutionary Computation, 35*, 26–40.
51. Li, X., & Yin, M. (2012). Optimal synthesis of linear antenna array with composite differential evolution algorithm. *Scientia Iranica, 19*(6), 1780–1787.
52. Saxena, P., & Kothari, A. (2016). Ant lion optimization algorithm to control side lobe level and null depths in linear antenna arrays. *AEU International Journal of Electronics and Communications, 70*(9), 1339–1349.
53. Pappula, L., & Ghosh, D. (2014). Linear antenna array synthesis using cat swarm optimization. *AEU International Journal of Electronics and Communications, 68*(6), 540–549.
54. Goswami, B., & Mandal, D. (2013). A genetic algorithm for the level control of nulls and side lobes in linear antenna arrays. *Journal of King Saud University, Computer and Information Sciences, 25*(2), 117–126.
55. Mandal, D., Ansari, M. A. I., Kar, R., & Ghoshal, S. (2012). Non-uniform concentric circular antenna array design using ipso technique for side lobe reduction. *Procedia Technology, 6*, 856–863. 2nd International Conference on Communication, Computing & Security [ICCCS-2012].
56. Aboul-Seoud, A. K., Mahmoud, A. K., & Hafez, A. E.-D. S. (2010). A sidelobelevel reduction (sll) for planar array antennas with 30 DB attenuators weight precision. *Aerospace Science and Technology, 14*(5), 316–320.
57. Grewal, A. R. M., Singh, N., & Manjeet Singh, P. (2017). A non-uniform circular antenna array failure correction using fire fly algorithm. *Wireless Personal Communications, 97*, 845–858.
58. Khan, S. U., Rahim, M. K. A., & Ali, L. (2018). Correction of array failure using grey wolf optimizer hybridized with an interior point algorithm. *Frontiers of Information Technology & Electronic Engineering, 19*(9), 1191–1202.
59. Singh Grewal, N., Rattan, M., & Singh Patterh, M. (2017). A linear antenna array failure correction using improved bat algorithm. *International Journal of RF and Microwave Computer-Aided Engineering, 27*(7), e21119.
60. Das, A., Mandal, D., & Kar, R. (2020). An optimal compact time-modulated circular antenna array synthesis using krill herd optimization. *Annals of Telecommunications*. https://doi.org/10.1007/s12243-020-00827-7
61. Ram, G., Mandal, D., Kar, R., & Ghoshal, S. P. (2015). Circular and concentric circular antenna array synthesis using cat swarm optimization. *IETE Technical Review, 32*, 204–217.

62. Haupt, R. L., & Werner, D. H. (2007). *Genetic algorithms in electromagnetics*. IEEE Press and Willey-Interscience.
63. Haupt, R. L. (1995). An introduction to genetic algorithm for electromagnetics. *IEEE Antennas and Propagation Magazine, 37*(2), 7–15.
64. Panduro, M., Mendez, A., Dominguez, R., & Romero, G. (2006). Design of non-uniform antenna arrays for side lobe reduction using the method of genetic algorithm. *International Journal of Electronics and Communications, 60*(7), 713–717.
65. Kennedy, J., & Eberhart, R. (1995). Particle swarm optimization. In *Proceedings of the IEEE International Conference on Neural Network, Perth* (pp. 1942–1848).
66. Boeringer, D. W., & Werner, D. H. (2004). Particle swarm optimization versus genetic algorithm for phased array synthesis. *IEEE Transactions on Antennas and Propagation, 52*(3), 771–779.
67. Sahib, M., Najjar, Y., Dib, N., & Khodier, M. (2008). Design of non-uniform circular antenna arrays using the particle swarm optimization. *Journal of Electrical Engineering, 59*(4), 216–220.
68. Ingber, L. (1993). Simulated annealing: Practice versus theory. *Mathematical and Computer Modelling, 18*(11), 29–57.
69. Rattan, M., Patterh, M. S., & Sohi, B. S. (2009). Optimization of circular antenna arrays of isotropic radiators using simulated annealing. *International Journal of Microwave and Wireless Technologies, 1*(05), 441–446.
70. Singh, U., & Kamal, T. S. (2011). Design of non-uniform circular antenna arrays using biogeography-based optimization. *Journal of Microwave Antenna Propagation, IET, 5*(11), 1365–1370.
71. Sharaqa, A., & Dib, N. (2013). Circular antenna array synthesis using firefly algorithm. *International Journal of RF and MicrowaveComputer-Aided Engineering, 24*(2), 139–146.
72. Das, A., Mandal, D., Ghoshal, S. P., & Kar, R. (2018). Moth flame optimization-based design of linear and circular antenna array for sidelobe reduction. *International Journal of Numerical Modelling, 32*(1), 1–15.
73. Das, A., Mandal, D., Ghoshal, S. P., & Kar, R. (2018). A heuristic approach to design linear and circular antenna array for side lobe reduction. *Iranian Journal of Science & Technology Transactions of Electrical Engineering, 43*(1), 67–76.
74. Gandomi, A. H., & Alavi, A. H. (2012). Krill herd: A new bio-inspired optimization algorithm. *Communications in Nonlinear Science and Numerical Simulation, 17*, 4831–4845.
75. Gandomi, A. H., & Alavi, A. H. (2016). An introduction of krill herd algorithm for engineering optimization. *Journal of Civil Engineering and Management, 22*(3), 302–310.
76. Sharma, P., & Dhubkarya, D. C. (2020). An energy efficient OFDM–MIMO systems for multimedia data transmission based on hybrid fuzzy approach. *Wireless Personal Communications*. https://doi.org/10.1007/s11277-020-07109-4
77. Zhou, L., Zhu, C., Ruby, R., Wang, X., Ji, X., Wang, S., et al. (2017). QoS-aware energy-efficient resource allocation in OFDM-based heterogeneous cellular networks. *International Journal of Communication Systems, 30*(2), e2931.
78. Ren, Z., Chen, S., Hu, B., & Ma, W. (2014). Energy-efficient resource allocation in downlink OFDM wireless systems with proportional rate constraints. *IEEE Transactions on Vehicular Technology, 63*(5), 2139–2150.
79. Ge, X., Huang, X., Wang, Y., Chen, M., Li, Q., Han, T., et al. (2014). Energy-efficiency optimization for MIMO–OFDM mobile multimedia communication systems with QoS constraints. *IEEE Transactions on Vehicular Technology, 63*(5), 2127–2138.
80. Chen, X., Hwang, J.-N., Lee, C.-N., & Chen, S.-I. (2015). A near optimal QoE-driven power allocation scheme for scalable video transmissions over MIMO systems. *IEEE Journal of Selected Topics in Signal Processing, 9*(1), 76–88.
81. Eraslan, E., & Daneshrad, B. (2017). Low-complexity link adaptation for energy efficiency maximization in MIMO–OFDM systems. *IEEE Transactions on Wireless Communications, 16*(8), 5102–5114.

82. Guo, R., Zhu, P., Li, J., Wang, D., & You, X. (2019). Energy optimization algorithms for MIMO–OFDM based downlink C-RAN system. *IEEE Access, 7*, 17927–17934.
83. He, J., Li, B., Deng, L., Tang, M., Gan, L., Fu, S., et al. (2016). Experimental demonstration of bidirectional OFDM/OQAM-MIMO signal over a multicore fiber system. *IEEE Photonics Journal, 8*(5), 1–8.
84. Krishna, C. M., Somasekhar, B., & Swaroop, V. S. (2017). A study on PAPR reduction and channel estimation in MIMO–OFDM system using wavelet transform. In *2017 International Conference on Energy, Communication, Data Analytics and Soft Computing (ICECDS)* (pp. 1707–1711). IEEE.
85. Somesula, M. K., Kotte, A., Annadanam, S. C., & Mothku, S. K. (2022). Deadline-aware cache placement scheme using fuzzy reinforcement learning in device-to-device mobile edge networks. *Mobile Networks and Applications*. https://doi.org/10.1007/s11036-022-02010-9
86. Yao, J., Han, T., & Ansari, N. (2019). On mobile edge caching. *IEEE Communications Surveys & Tutorials, 21*(3), 2525–2553.
87. Yu, S., Dab, B., Movahedi, Z., Langar, R., & Wang, L. (2019). A socially-aware hybrid computation offloading framework for multi-access edge computing. *IEEE Transactions on Mobile Computing, 19*(6), 1247–1259.
88. Li, J., Liu, M., Lu, J., Shu, F., Zhang, Y., Bayat, S., & Jayakody, D. N. K. (2019). On social-aware content caching for d2d-enabled cellular networks with matching theory. *IEEE Internet of Things Journal, 6*(1), 297–310.
89. Zhao, D., Wang, H., Shao, K., & Zhu, Y. (2016). Deep reinforcement learning with experience replay based on SARSA. In *2016 IEEE Symposium Series on Computational Intelligence (SSCI)* (pp. 1–6). IEEE.
90. Zhang, S., Quan, W., Li, J., Shi, W., Yang, P., & Shen, X. (2018). Air-ground integrated vehicular network slicing with content pushing and caching. *IEEE Journal on Selected Areas in Communications, 36*(9), 2114–2127.
91. Zhang, W., Wu, D., Yang, W., & Cai, Y. (2019). Caching on the move: A user interest-driven caching strategy for d2d content sharing. *IEEE Transactions on Vehicular Technology, 68*(3), 2958–2971.
92. Sun, R., Wang, Y., Lyu, L., Cheng, N., Zhang, S., Yang, T., & Shen, X. (2020). Delay-oriented caching strategies in d2d mobile networks. *IEEE Transactions on Vehicular Technology, 69*(8), 8529–8541.
93. Jiang, W., Feng, G., Qin, S., Yum, T. S. P., & Cao, G. (2019). Multi-agent reinforcement learning for efficient content caching in mobile d2d networks. *IEEE Transactions on Wireless Communications, 18*(3), 1610–1622.
94. Rostami, A. S., Mohanna, F., & Keshavarz, H. (2021). Presenting an optimal energy-aware locating structure using the Internet of Things and device-to-device communications on smartphones. *Wireless Personal Communications*. https://doi.org/10.1007/s11277-021-08114-x
95. Tavakkolai, H., Yadollahi, N., Yadollahi, M., Hosseinabadi, A. R., Rezaei, P., & Kardgar, M. (2015). Sensor selection wireless multimedia sensor network using gravitational search algorithm. *Indian Journal of Science and Technology, 8*, 1–6.
96. Bozorgi, S. M., Rostami, A. S., Hosseinabadi, A. R., & Balas, V. E. (2017). A new clustering protocol based on renewable energy and multi-hop routing for energy harvesting-wireless sensor networks. *Computers & Electrical Engineering, 64*, 233–247.
97. Sangaiah, A. K., Sadeghilalimi, M., Hosseinabadi, A. R., & Zhang, W. (2019). Energy consumption in point-coverage wireless sensor networks via Bat algorithm. *IEEE Access, 7*, 180258–180269.
98. Jeon, J., Niu, H., Li, Q., Papathanassiou, A., & Wu, G. (2015). LTE with listen-before-talk in unlicensed spectrum. In *IEEE International Conference on Communication Workshop (ICCW)* (pp. 2320–2324).
99. Hoang, T. D., Le, L. B., & Le-Ngoc, T. (2016). Energy-efficient resource allocation for D2D communications in cellular networks. *IEEE Transactions on Vehicular Technology, 65*, 6972–6986.

100. Tang, R., Zhao, J., Qu, H., & Zhang, Z. (2016). Energy-efficient resource allocation for 5G full-duplex enabled device-to-device communication. In *IEEE Globecom Workshops (GC Wkshps)* (pp. 1–7).
101. Wu, D., Wang, J., Hu, R. Q., Cai, Y., & Zhou, L. (2014). Energy-efficient resource sharing for mobile device-to-device multimedia communications. *IEEE Transactions on Vehicular Technology, 63*, 2093–2103.
102. Datsika, E., Antonopoulos, A., Zorba, N., & Verikoukis, C. (2017). Cross-network performance analysis of network coding aided cooperative out band D2D communications. *IEEE Transactions on Wireless Communications, 16*, 3176–3188.
103. Xu, J., Guo, C., & Zhang, H. (2018). Joint channel allocation and power control based on PSO for cellular networks with D2D communications. *Computer Networks, 133*, 104–119.
104. Swain, S. N., Thakur, R., & Murthy, C. S. R. (2017). Design and stochastic geometric analysis of an efficient Q-Learning based physical resource block allocation scheme to maximize the spectral efficiency of Device-to-Device overlaid cellular networks. *Computer Networks, 119*, 71–85.
105. Javadpour, A., Wang, G., Rezaei, S., & Li, K.-C. (2020). Detecting straggler MapReduce tasks in big data processing infrastructure by neural network. *The Journal of Supercomputing, 76*, 6969–6993.
106. Peng, B., Hu, C., Peng, T., Yang, Y., & Wang, W. (2013). A resource allocation scheme for D2D multicast with QoS protection in OFDMA-based systems. In *24th International Symposium on Personal Indoor and Mobile Radio Communications (PIMRC)* (pp. 12383–12387).
107. Somesula, M. K., Mothku, S. K., & Kotte, A. (2022). Deep reinforcement learning mechanism for deadline-aware cache placement in device-to-device mobile edge networks. *Wireless Networks*. https://doi.org/10.1007/s11276-022-03135-1
108. Akhtar, T., Tselios, C., & Politis, I. (2020). Radio resource management: Approaches and implementations from 4G to 5G and beyond. *Wireless Networks*. https://doi.org/10.1007/s11276-020-02479-w
109. Zhang, N., Cheng, N., Gamage, A. T., Zhang, K., Mark, J. W., & Shen, X. (2015). Cloud assisted HetNets toward 5G wireless networks. *IEEE Communications Magazine, 53*(6), 59–65.
110. Hur, S., Kim, T., Love, D. J., Krogmeier, J. V., Thomas, T. A., & Ghosh, A. (2013). Millimeter wave beamforming for wireless backhaul and access in smallcell networks. *IEEE Transactions on Communications, 61*(10), 4391–4403.
111. Osseiran, A., et al. (2014). Scenarios for 5G mobile and wireless communications: The vision of the METIS project. *IEEE Communications Magazine, 52*(5), 26–35.
112. Rappaport, T. S., et al. (2013). Millimeter wave mobile communications for 5G cellular: It will work! *IEEE Access, 1*, 335–345.
113. Department of Communications. (2013). *South Africa connect: Creating opportunities, ensuring inclusion South Africa's broadband policy*. South African Government.
114. Khan, S., & Mauri, J. L. (2014). *Green networking and communications: ICT for sustainability*. CRC Press.
115. Dahlman, E., Parkvall, S., Skold, J., & Beming, P. (2008). *3G evolution HSPA and LTE for mobile broadband*. Academic Press.
116. Evolved Universal Terrestrial Radio Access (E-UTRA) and Evolved Universal Terrestrial Radio Access Network (E-UTRAN). (2013). Overall description: Stage 2, Sophia-Antipolis, France, TS 36.300. Retrieved from http://www.3gpp.org/DynaReport/36.300.htm .
117. Chandrasekhar, V., & Andrews, J. G. (2009). Uplink capacity and interference avoidance for two-tier femtocell networks. *IEEE Transactions on Wireless Communications, 8*(7), 3498–3509.
118. Shi, Z., Reed, M. C., & Zhao, M. (2010). On uplink interference scenarios in two-tier macro and femto co-existing UMTS networks. *EURASIP Journal on Wireless Communications and Networking, 2010*(240745), 1–8.

119. Garcia, L. G. U., Kovacs, I. Z., Pedersen, K., Costa, G. W. O., & Mogensen, P. (2012). Autonomous component carrier selection for 4G femtocells—A fresh look at an old problem. *IEEE Journal on Selected Areas in Communications, 30*, 525–537.
120. Kang, X., Zhang, R., & Motani, M. (2012). Price-based resource allocation for spectrum-sharing femtocell networks: A stackelberg game approach. *IEEE Journal on Selected Areas in Communications*. https://doi.org/10.1109/GLOCOM.2011.6133874
121. Ma, Z., Xiang, W., Long, H., & Wang, W. (2011). Proportional fair resource partition for LTE-advanced networks with type I relay nodes. In *Proceedings of IEEE ICC: Kyoto, Japan, June* (pp. 1–5).
122. Chandrasekhar, V., & Andrews, J. (2009). Spectrum allocation in tiered cellular networks. *IEEE Transactions on Communications, 57*(10), 3059–3068.
123. Jiu, H., Liang, B., Li, J., & Yang, X. (2013). Dynamic joint resource optimization for LTE-advanced relay networks. *Proceedings of IEEE Transactions Wireless Communications, 12*(11), 5668–5678.
124. Deb, S., & Monogloudis, P. (2015). Learning based uplink interference management in 4G LTE cellular systems. *IEEE/ACM Transactions Network, 23*(2), 398–411.
125. Deb, S., Monogioudis, P., Miernik, J., & Seymour, J. P. (2014). Algorithms for enhanced inter-cell interference coordination (eICIC) in LTE HetNets. *IEEE/ACM Transactions Network, 22*(1), 137–150.
126. Xu, X., Kutrolli, G., & Mathar, R. (2013). Dynamic downlink power control strategies for LTE femtocells. In *Proceedings of IEEE 7th International Conference Next Gener. Mobile Apps Serv. Technology, Prague, Czech Republic, September 25–27* (pp. 181–186).
127. Lejosne, Y., Slock, D., & Yuan-Wu, Y. (2014). Net degrees of freedom of decomposition schemes for the MIMO IC with delayed CSIT. In *Proceedings of International Symposium Information Theory (ISIT'14), June 29/July 4* (pp. 1742–1746).
128. Lejosne, Y., Slock, D., & Yuan-Wu, Y. (2013). Net degrees of freedom of recent schemes for the MISO BC with delayed CSIT and finite coherence time. In *Proceedings of Wireless Communications Network Conference (WCNC'13), April 7–10* (pp. 3040–3045).
129. Lejosne, Y., Slock, D., & Yuan-Wu, Y. (2013). Space time interference alignment scheme for the MISO BC and IS with delayed CSIT and finite coherence time. In *Proceedings of International Conference Acoust, Speech, Signal Processing (ICASSP'13), May 26–31* (pp. 4868–4872).
130. Singh, B., Koufos, K., & Tirkkonen, O. (2014). Co-primary inter-operator spectrum sharing using repeated games. In *Proceedings of IEEE International Conference Communications System (ICCS'14), November 19–21* (pp. 67–71).
131. Li, G. Y., et al. (2011). Energy-efficient wireless communications: Tutorial, survey, and open issues. In *IEEE Wireless Communication* (pp. 28–35).
132. Yang, Y., et al. (2009). Relay technologies for WiMAX and LTE-advanced mobile systems. *IEEE Communications Magazine, 47*(10), 100–105.
133. Mantzoukas, K. P., Sagkriotis, S. E., & Panagopoulos, A. D. (2014). On the design of energy-efficient wireless access networks: A cross layer approach. In S. Khan & J. L. Mauri (Eds.), *Green networking and communications ICT for sustainability* (pp. 49–61). CRC Press.
134. Ge, X., Cheng, H., Guizani, M., & Han, T. (2014). 5G wireless Backhaul networks: Challenges and research advances. *IEEE Network*, 6–12.
135. Pan, M., Zhang, C., Li, P., & Fang, Y. (2012). Spectrum harvesting and sharing in multi-hop CRNs under uncertain spectrum supply. *IEEE Journal on Selected Areas in Communications, 30*(2), 369–378.
136. Imran, A., Zoha, A., & Abu-Dayya, A. (2014). Challenges in 5G: How to empower SON with big data for enabling 5G. *IEEE Network, 28*(6), 27–33.
137. Chih-Lin, I., Rowell, C., Han, S., Xu, Z., Li, G., & Pan, Z. (2014). Toward green and soft: A 5G perspective. *IEEE Communications Magazine, 52*(2), 66–73.

138. Olwal, T. O., Djouani, K., Kogeda, O. P., & vanWyk, B. J. (2012). Joint queueperturbed and weakly coupled power control for wireless backbone networks. *International Journal of Applied Mathematics and Computer Science, 22*(3), 749–764.
139. Olwal, T. O. (2010). *Decentralised dynamic power control for wireless backbone mesh networks*. Ph.D. dissertation, Dept. computer Sci., Univ. Paris-EST, Creteil, Paris, France and Dept. Elect. Eng., Tshwane Univ. Technology.
140. Rappaport, T. S. (1996). *Wireless communications: Principles and practice*. Prentice-Hall.
141. Rappaport, T. S., Ben-Dor, E., Murdock, J. N., & Qiao, Y. (2012). 38 GHz and 60 GHz angle-dependent propagation for cellular & peer-to-peer wireless communications. In *Proceedings of IEEE International Conference Communications* (pp. 4568–4573).
142. Vook, F. W., Ghosh, A., & Thomas, T. A. (2014). MIMO and beamforming solutions for 5G technology. In *Proceedings of IEEE Microwave Symposium (IMS)* (pp. 1–4).
143. Bega, D., Gramaglia, M., Banchs, A., Sciancalepore, V., Samdanis, K., & Costa-Perez, X. (2017). Optimising 5G infrastructure markets: The business of network slicing. In *IEEE INFOCOM 2017—IEEE Conference on Computer Communications, Atlanta, GA* (pp. 1–9).
144. Zhang, H., et al. (2015). Coexistence of Wi-Fi and heterogeneous small cell networks sharing unlicensed spectrum. *IEEE Communications Magazine, 53*(3), 158–164.
145. Ramdev, M. S., Bajaj, R., & Sidhu, J. (2021). Remote radio head scheduling in LTE-advanced networks. *Wireless Personal Communications*. https://doi.org/10.1007/s11277-021-08916-z
146. Holma, H., & Toskala, A. (2009). *LTE for UMTS OFDMA and SC-FDMA based radio access*. Wiley.
147. Sivaraj, N., & Palanisamy, P. (2016). Soft computing based power control for interference mitigation in LTE femtocell networks. *Procedia Computer Science, 79*, 93–99.
148. Monghal, G., Pedersen, K. I., Kovacs, I. Z., & Mogensen, P. E. (2008). QoS oriented time and frequency domain packet schedulers for the UTRAN long term evolution. In *VTC Spring 2008-IEEE Vehicular Technology Conference* (pp. 2532–2536). IEEE.
149. Ahmad, M., Naeem, M., & Iqbal, M. (2019). Estimation of distribution algorithm for joint resource management in D2D communication. *Wireless Personal Communications*. https://doi.org/10.1007/s11277-019-06459-y
150. Sun, S., & Shin, Y. (2014). Resource allocation for D2D communication using particle swarm optimization in LTE networks. In *2014 International Conference on Information and Communication Technology Convergence (ICTC)* (pp. 371–376). IEEE.
151. Yu, P., Yang, X., Zhou, F., Li, H., Feng, L., Li, W., & Qiu, X. (2020). Deep reinforcement learning aided cell outage compensation framework in 5G cloud radio access networks. *Mobile Networks and Applications*. https://doi.org/10.1007/s11036-020-01574-8
152. Ayoubi, S., Limam, N., Salahuddin, M. A., Shahriar, N., Boutaba, R., Estrada-Solano, F., & Caicedo, O. M. (2018). Machine learning for cognitive network management. *IEEE Communication Magazine, 56*(1), 158–165.
153. Sun, S., Kadoch, M., & Ran, T. (2015). Adaptive SON and cognitive smart LPN for 5G heterogeneous networks. *Mobile Network and Applications, 20*(6), 745–755.
154. Moysen, J., & Giupponi, L. (2014). A reinforcement learning based solution for self-healing in LTE networks. In *2014 IEEE 80th Vehicular Technology Conference (VTC2014-Fall), Vancouver, BC* (pp. 1–6).
155. Selim, M., Kamal, A. E., Elsayed, K., Abdel-Atty, H. M., & Alnuem, M. (2016). Fronthaul cell outage compensation for 5G networks. *IEEE Communications Magazine, 54*(8), 169–175.
156. Yin, M., Li, W., Feng, L., Yu, P., & Qiu, X. (2017). Multi-cell cooperative outage compensation in cloud-RANs based 5G public safety network. *IEEE Access, 5*, 17309–17321.
157. Xu, Z., Wang, Y., Tang, J., Wang, J., & Gursoy, M. C. (2017). A deep reinforcement learning based framework for power-efficient resource allocation in cloud RANs. In *IEEE International Conference on Communications (ICC), Paris, 2017* (pp. 1–6).
158. Alghamdi, S. A. (2020). Emperor based resource allocation for D2D communication and QoF based routing over cellular V2X in urban environment (ERA-D^{2}Q). *Wireless Networks*. https://doi.org/10.1007/s11276-020-02273-8

159. Bonjorn, N., Foukalas, F., & Pop, P. (2018). Enhanced 5G V2X services using sidelink device-to-device communications. In *2018 17th Annual Mediterranean Ad Hoc Networking Workshop (Med-Hoc-Net)*.
160. Sun, W., Strom, E. G., Brannstrom, F., Sou, K. C., & Sui, Y. (2018). Radio resource management for D2D-based V2V communication. *IEEE Transactions on Vehicular Technology, 65*(8), 6636–6650.
161. Li, X., Zhou, L., Chen, X., Qi, A., Li, C., & Yanli, X. (2018). Resource allocation schemes based on intelligent optimization algorithms for D2D communications underlaying cellular networks. *Mobile Information Systems, 2018*, 10.
162. Lee, J., & Lee, J. H. (2019). Performance analysis and resource allocation for cooperative D2D communication in cellular networks with multiple D2D pairs. *IEEE Communications Letters, 23*(5), 909–912.
163. Liang, L., Li, G. Y., & Xu, W. (2017). Resource allocation for D2D-enabled vehicular communications. *IEEE Transactions on Communications, 65*(7), 3186–3197.
164. Singh, D., & Ghosh, S. C. (2019). Mobility aware relay selection in 5G 2D communication using stochastic model. *IEEE Transactions on Vehicular Technology, 68*, 1–12.
165. Gandotra, P., Jha, R. K., & Jain, S. (2018). Sector based radio resource allocation algorithm for better quality of service and experience in device to device to communication. *IEEE Transactions on Vehicular Technology, 67*(7), 5750–5765.
166. Hussein, H. H., Elsayed, H. A., & Abd El-Kader, S. M. (2019). Intensive Benchmarking of D2D communication over 5G cellular networks: Prototype, integrated features, challenges, and main applications. *Wireless Networks*. https://doi.org/10.1007/s11276-019-02131-2
167. Devos, M., Ometov, A., Mäkitalo, N., Aaltonen, T., Andreev, S., & Koucheryavy, Y. (2016). D2D communications for mobile devices: Technology overview and prototype implementation. In *2016 8th International Congress on Ultra Modern Telecommunications and Control Systems and Workshops (ICUMT)* (pp. 124–129). IEEE.
168. Liu, Y. (2016). Optimal mode selection in D2D-enabled multibase station systems. *IEEE Communications Letters, 20*(3), 470–473.
169. Nie, S., Fan, Z., Zhao, M., Gu, X., & Zhang, L. (2016). Q-learning based power control algorithm for D2D communication. In *2016 IEEE 27th Annual International Symposium on Personal, Indoor, and Mobile Radio Communications (PIMRC)* (pp. 1–6). IEEE.
170. Sedidi, R., & Kumar, A. (2016). Key exchange protocols for secure device-to-device (D2D) communication in 5G. In *2016 Wireless Days (WD)* (pp. 1–6). IEEE.
171. Chatterjee, T., Chowdhury, R., & Ramu, P. (2019). Decoupling uncertainty quantification from robust design optimization. *Structural and Multidisciplinary Optimization*. https://doi.org/10.1007/s00158-018-2167-0
172. Messac, A. (1996). Physical programming: Effective optimization for computational design. *AIAA Journal, 34*, 149–158.
173. Chen, W., Wiecek, M., & Zhang, J. (1999). Quality utility—A compromise programming approach to robust design. *Journal of Mechanical Design—The ASME, 121*, 179–187.
174. Chen, W., Sahai, A., Messac, A., & Sundararaj, G. (2000). Exploration of the effectiveness of physical programming in robust design. *Journal of Mechanical Design—The ASME, 122*, 155–163.
175. Doostan, A., & Owhadi, H. (2011). A non-adapted sparse approximation of PDEs with stochastic inputs. *Journal of Computational Physics, 230*, 3015–3034.
176. Hayat, O., Ngah, R., & Zahedi, Y. (2019). In-band device to device (D2D) communication and device discovery: A survey. *Wireless Personal Communications*. https://doi.org/10.1007/s11277-019-06173-9
177. Agiwal, M., Roy, A., & Saxena, N. (2016). Next generation 5G wireless networks: A comprehensive survey. *IEEE Communications Surveys & Tutorials, 18*(3), 1617–1655.
178. Lien, S.-Y., Chien, C.-C., Tseng, F.-M., & Ho, T.-C. (2016). 3GPP device-to-device communications for beyond 4G cellular networks. *IEEE Communications Magazine, 54*, 29–35.

179. Shen, Y., Jiang, C., Quek, T. Q., & Ren, Y. (2016). Device-to-device-assisted communications in cellular networks: An energy efficient approach in downlink video sharing scenario. *IEEE Transactions on Wireless Communications, 15*, 1575–1587.
180. Araniti, G., Raschellà, A., Orsino, A., Militano, L., & Condoluci, M. (2017). Device-to-device communications over 5G systems: Standardization, challenges and open issues. In *5G mobile communications* (pp. 337–360). Springer.
181. Orange, J. S.-B., Armada, A. G., Evans, B., Galis, A., & Karl, H. (2016). White Paper for Research Beyond 5G (Final Edit), Networld (NW) 2020, version 1.0 (pp. 1–43).
182. Lin, X., Andrews, J. G., Ghosh, A., & Ratasuk, R. (2014). An overview of 3GPP device-to-device proximity services. *IEEE Communications Magazine, 52*(4), 40–48. https://doi.org/10.1109/MCOM.2014.6807945
183. Tehrani, M. N., Uysal, M., & Yanikomeroglu, H. (2014). Device-to-device communication in 5G cellular networks: Challenges, solutions, and future directions. *IEEE Communications Magazine, 52*(5), 86–92.
184. Hayat, O., Ngah, R., & Zahedi, Y. (2017). Cooperative device-to-device discovery model for multiuser and OFDMA network base neighbour discovery in in-band 5G cellular networks (in English). *Wireless Personal Communications, 97*(3), 4681–4695.
185. Hayat, O., Ngah, R., & Zahedi, Y. (2018). Device discovery for D2D communication in in-band cellular networks using sphere decoder like (SDL) algorithm. *EURASIP Journal on Wireless Communications and Networking*. https://doi.org/10.1186/s13638-018-1083-8
186. Kaufman, B., Aazhang, B., & Lilleberg, J. (2009). Interference aware link discovery for device to device communication. In *Forty-third Asilomar Conference on, Signals, Systems and Computers* (pp. 297–301).
187. Camps-Mur, D., Garcia-Saavedra, A., & Serrano, P. (2013). Device-to-device communications with Wi-Fi Direct: Overview and experimentation. *IEEE Wireless Communications, 20*, 96–104.
188. Sharmila, K., Mohan, V., Ramesh, C., & Munda, S. P. (2016). Proximity services based device-to-device framework design for direct discovery. In *2016 2nd International Conference on Advances in Electrical, Electronics, Information, Communication and Bio-Informatics (AEEICB)* (pp. 499–502).
189. Prasad, A., Kunz, A., Velev, G., Samdanis, K., & Song, J. (2014). Energy-efficient D2D discovery for proximity services in 3GPP LTE-advanced networks: ProSe discovery mechanisms. *IEEE Vehicular Technology Magazine, 9*, 40–50.
190. Lee, W., Kim, J., & Choi, S. W. (2016). New D2D peer discovery scheme based on spatial correlation of wireless channel. *IEEE Transactions on Vehicular Technology, 65*(12), 10120–10125.
191. Bello, O., & Zeadally, S. (2016). Intelligent device-to-device communication in the Internet of Things. *IEEE Systems Journal, 10*(3), 1172–1182.
192. Doppler, K., Ribeiro, C. B., & Kneckt, J. (2011). Advances in D2D communications: Energy efficient service and device discovery radio. In *2011 2nd International Conference on Wireless Communication, Vehicular Technology, Information Theory and Aerospace & Electronic Systems Technology (Wireless VITAE)* (pp. 1–6).
193. Roberts, M. K., & Anguraj, P. (2020). A comparative review of recent advances in decoding algorithms for low-density parity-check (LDPC) codes and their applications. *Archives of Computational Methods in Engineering*. https://doi.org/10.1007/s11831-020-09466-6
194. Roberts, M. K., Mohanram, S. S., & Shanmugasundaram, N. (2019). An improved low complex offset min-sum based decoding algorithm for LDPC codes. *Mobile Networks and Applications, 24*(6), 1848–1852.
195. Chen, Y., Cui, H., Lin, J., & Wang, Z. (2020). Fine-grained bit-flipping decoding for LDPC codes. *IEEE Transactions on Circuits and Systems II: Express Briefs, 67*(5), 896–900.
196. Fang, Y., Bi, G., Guan, Y. L., & Lau, F. C. M. (2015). A survey on protograph LDPC codes and their applications. *IEEE Communication Surveys and Tutorials, 17*(4), 1989–2016.

197. Bonello, N., Chen, S., & Hanzo, L. (2011). Low-density parity-check codes and their rateless relatives. *IEEE Communication Surveys and Tutorials, 13*(1), 3–26.
198. Wang, Z., Cui, Z., & Sha, J. (2011). VLSI design for low-density parity-check code decoding. *IEEE Circuits and Systems Magazine, 11*(1), 52–69.
199. Ankan, E., Hassan, N. U., Lentmaier, M., Montorsi, G., & Sayir, J. (2015). Challenges and some new directions in channel coding. *Journal of the Communications Network, 17*(4), 328–338.
200. Andrade, J., Falcao, G., Silva, V., & Sousa, L. (2016). A survey on programmable LDPC decoders. *IEEE Access, 4*, 6704–6718.
201. Hailes, P., Xu, L., Maunder, R. G., Al-Hashimi, B. M., & Hanzo, L. (2016). A survey of FPGA-based LDPC decoders. *IEEE Communication Surveys and Tutorials, 18*(2), 1098–1122.
202. Fang, Y., Han, G., Cai, G., Lau, F. C. M., Chen, P., & Guan, Y. L. (2018). Design guidelines of low-density parity-check codes for magnetic recording systems. *IEEE Communication Surveys and Tutorials, 20*(2), 1574–1606.
203. Shao, S., Hailes, P., Wang, T.-Y., Wu, J.-Y., Maunder, R. G., Al-Hashimi, B. M., & Hanzo, L. (2019). Survey of turbo, LDPC, and polar decoder ASIC implementations. *IEEE Communication Surveys and Tutorials, 21*(3), 2309–2333.
204. Awais, M., Razzaq, A., Ahmed, A., & Masera, G. (2019). LDPC check node implementation using reversible logic. *IET Circuits, Devices & Systems, 13*(4), 443–455.
205. Gupta, R. S., Kamal, T. S., & Singh, P. S. (2019). Concatenated LDPC-TCM codes for better performance of OFDM-FSO system using gamma-gamma fading model. *Wireless Personal Communications, 106*(4), 2247–2260.
206. Singh, J., Gupta, M., & Bhullar, J. S. (2019). Construction of girth-8 (3, L)-QC-LDPC codes of smallest CPM size using column multipliers. *Designs, Codes and Cryptography, 88*(1), 1–9.
207. Aswathy, G. P., Nair, G. K., & Ahamed, T. P. I. (2019). Parallel concatenated Gallager codes for reliable data transmission in cognitive radio networks. *Physical Communication, 37*(12), 1–14.
208. Chen, T.-C. (2013). Adaptive-weighted multibit-flipping decoding of low-density parity-check codes based on ordered statistics. *IET Communications, 7*(14), 1517–1521.
209. Tiwari, H. D., Tiwari, H. D., & Lee, K.-Y. (2014). Hybrid weighted bit flipping low density parity check decoding. *Digital Signal Processing, 28*, 82–92.
210. Liu, Y., Niu, X., & Zhang, M. (2015). Multi-threshold bit flipping algorithm for decoding structured LDPC codes. *IEEE Communications Letters, 19*(2), 127–130.
211. Elsanadily, S., Mahran, A., & Elghandour, O. (2017). Two-side state-aided bit-flipping decoding of generalized low density parity check codes. *IEEE Communications Letters, 21*(10), 2122–2125.
212. Liu, Y., & Zhang, M. (2019). Hard-decision bit-flipping decoder based on adaptive bit-local threshold for LDPC codes. *IEEE Communications Letters, 23*(5), 789–792.
213. Wei, H., Jiao, X., & Mu, J. (2015). Reduced-complexity linear programming decoding based on ADMM for LDPC codes. *IEEE Communications Letters, 19*(6), 909–912.
214. Jiao, X., Wei, H., Mu, J., & Chen, C. (2015). Improved ADMM penalized decoder for irregular low-density parity-check codes. *IEEE Communications Letters, 19*(6), 913–916.
215. Debbabi, I., Gal, B. L., Khouja, N., Tlili, F., & Jego, C. (2016). Fast converging ADMM-penalized algorithm for LDPC decoding. *IEEE Communications Letters, 20*(4), 648–651.
216. Liu, X., & Draper, S. C. (2016). The ADMM penalized decoder for LDPC codes. *IEEE Transactions on Information Theory, 62*(6), 2966–2984.
217. Jiao, X., & Mu, J. (2016). Lowering the error floor of ADMM penalized decoder for LDPC codes. *China Communications, 13*(8), 127–135.
218. Chen, F., Zhu, J., & Zhang, W. (2022). Radar cross section minimization for step structures using topology optimization. *Structural and Multidisciplinary Optimization*. https://doi.org/10.1007/s00158-021-03110-y

219. Hassan, E., Wadbro, E., & Berggren, M. (2014). Topology optimization of metallic antennas. *IEEE Transactions on Antennas and Propagation, 62*(5), 2488–2500.
220. Haji-Ahmadi, M.-J., Nayyeri, V., Soleimani, M., & Ramahi, O. M. (2017). Pixelated checkerboard metasurface for ultra-wideband radar cross section reduction. *Scientific Reports, 7*(1), 1–12.
221. Harrington, R. F. (1993). *Field computation by moment methods*. Wiley-IEEE Press.
222. Yee, K. (1966). Numerical solution of initial boundary value problems involving Maxwell's equations in isotropic media. *IEEE Transactions on Antennas and Propagation, 9*(1), 5–6.
223. Song, J. M., Lu, C. C., & Chew, W. C. (1997). Multilevel fast multipole algorithm for electromagnetic scattering by large complex objects. *IEEE Transactions on Antennas and Propagation, 45*(10), 1488–1493.
224. Sigmund, O. (2001). A 99 line topology optimization code written in matlab. *Structural and Multidisciplinary Optimization, 21*(2), 120–127.
225. Bendsøe, M. P. (1989). Optimal shape design as a material distribution problem. *Structural Optimization, 1*(4), 193–202.
226. Shukla, A., Grover, R., Kundu, D., & Mitra, A. (2023). A computationally efficient algorithm to estimate the parameters of a two-dimensional chirp model with the product term. *Multidimensional Systems and Signal Processing*. https://doi.org/10.1007/s11045-023-00879-7
227. Francos, J. M., & Friedlander, B. (1995). The polynomial phase difference operator for modeling of nonhomogeneous images. In *Proceedings, International Conference on Image Processing* (Vol. 2, pp. 276–279). IEEE.
228. Francos, J. M., & Friedlander, B. (1998). Two-dimensional polynomial phase signals: Parameter estimation and bounds. *Multidimensional Systems and Signal Processing, 9*(2), 173–205.
229. Francos, J. M., & Friedlander, B. (1999). Parameter estimation of 2-D random amplitude polynomial-phase signals. *IEEE Transactions on Signal Processing, 47*(7), 1795–1810.
230. O'Shea, P. (2002). A new technique for instantaneous frequency rate estimation. *IEEE Signal Processing Letters, 9*(8), 251–252.
231. Zhang, K., Wang, S., & Cao, F. (2008). Product cubic phase function algorithm for estimating the instantaneous frequency rate of multicomponent two-dimensional Chirp signals. In *2008 Congress on Image and Signal Processing* (Vol. 5, pp. 498–502). IEEE.
232. Lahiri, A., Kundu, D., & Mitra, A. (2013). Efficient algorithm for estimating the parameters of two dimensional Chirp signal. *Sankhya B, 75*(1), 65–89.
233. Grover, R., Kundu, D., & Mitra, A. (2021). An efficient methodology to estimate the parameters of a two-dimensional Chirp signal model. *Multidimensional Systems and Signal Processing, 32*(1), 49–75.
234. Alquhayz, H., & Raza, B. (2021). Watermarking techniques for the security of medical images and image sequences. *Arabian Journal for Science and Engineering*. https://doi.org/10.1007/s13369-021-06254-7
235. Pizzolante, R., Carpentieri, B., & Castiglione, A. A secure low complexity approach for compression and transmission of 3-D medical images. In *Eighth International Conference on Broadband and Wireless Computing, Communication and Applications, October 2013* (pp. 387–392). IEEE.
236. Chitra, K., & Venkatesan, V. P. Spatial domain watermarking technique: An introspective study. In *ICIA-16: Proceedings of the International Conference on Informatics and Analytics, August 2016 Article No.: 50* (pp. 1–6). https://doi.org/10.1145/2980258.2980363
237. Muyco, S. D., & Hernandez, A. A. Least significant bit hash algorithm for digital image watermarking authentication. In *ICCAI '19: Proceedings of the 2019 5th International Conference on Computing and Artificial Intelligence, April 2019* (pp. 150–154). https://doi.org/10.1145/3330482.3330523
238. Zeki, A., Abubakar, A., & Chiroma, H. (2016). An intermediate significant bit (ISB) watermarking technique using neural networks. *Springerplus, 5*(1), 868.

239. Begum, M., & Uddin, M. S. (2020). Digital image watermarking techniques: A review. *Information, 11*(2), 110.
240. Shaik, A., & Thanikaiselvan, V. (2018). Comparative analysis of integer wavelet transforms in reversible data hiding using threshold based histogram modification. *Journal of King Saud University Computer and Information Sciences, 33*(7), 878–889.
241. Soleimanpour-Moghadam, M., & Talebi, S. (2013). A novel technique for steganography method based on improved genetic algorithm optimization in spatial domain. *Iranian Journal of Electrical and Electronic Engineering, 9*(2), 67–75.
242. Cedillo-Hernandez, A., Cedillo-Hernandez, M., Garcia-Vazquez, M., Nakano-Miyatake, M., Perez-Meana, H., & Ramirez-Acosta, A. (2014). Transcoding resilient video watermarking scheme based on spatio-temporal HVS and DCT. *Signal Processing, 97*, 40–54.
243. Gaobo, Y., Xingming, S., & Xiaojing, W. (2006). A genetic algorithm based video watermarking in the DWT domain. In *2006 International Conference on Computational Intelligence and Security, Guangzhou, China* (pp. 1209–1212). IEEE.
244. Himeur, Y., & Boukabou, A. (2018). A robust and secure key-frames based video watermarking system using chaotic encryption. *Multimedia Tools and Applications, 77*(7), 8603–8627.
245. Li, R., Zeng, B., & Liou, M. L. (1994). A new three-step search algorithm for block motion estimation. *IEEE Transactions on Circuits and Systems for Video Technology, 4*(4), 438–442.
246. Po, L.-M., & Ma, W.-C. (1996). A novel four-step search algorithm for fast block motion estimation. *IEEE Transactions on Circuits and Systems for Video Technology, 6*(3), 313–317.
247. Wu, D.-C., & Tsai, W.-H. (2003). A steganographic method for images by pixel-value differencing. *Pattern Recognition Letters, 24*(9–10), 1613–1626.
248. Yu, T., & Wang, R. (2021). Design and implementation of a smart elderly positioning management system based on wireless communication network. *EURASIP Journal on Wireless Communications and Networking*. https://doi.org/10.1186/s13638-021-02021-2
249. Haitao, F. (2015). Application of BIM technology in HVAC engineering. *Intelligent Buildings City Information, 000*(012), 89–91.
250. Li, J., Lu, Y., Shen, N., Pu, J., & Ma, Z. (2022). Adaptive image enhancement and dynamic-template-matching-based edge extraction method for diamond roller on-machine profile measurement. *The International Journal of Advanced Manufacturing Technology*. https://doi.org/10.1007/s00170-022-09113-3
251. Yu, Z. Y., Wang, J., & Lu, G. D. (2019). Optimized self-adapting contrast enhancement algorithm for wafer contour extraction. *Multimedia Tools and Applications, 78*(22), 32087–32108. https://doi.org/10.1007/s11042-019-08019-9
252. Wang, D. Y., Tang, C., Shi-Ju, E., Gao, C. F., & Ge, B. Z. (2021). Image edge detection based on guided filter Retinex and adaptive Canny. *Optics and Precision Engineering, 29*(2), 443–451. https://doi.org/10.37188/OPE.20212902.0443
253. Liu, S. X., Long, W., He, L., Li, Y. Y., & Ding, W. (2021). Retinex-based fast algorithm for low-light image enhancement. *Entropy, 23*(6), 746. https://doi.org/10.3390/e23060746
254. Hari, U., Bevi, A. R., & Ramachandran, B. (2021). Performance analysis of Retinex based algorithms for enhancement of low light images. *Journal of Physics Conference Series, 1964*(6), 062046. https://doi.org/10.1088/1742-6596/1964/6/062046
255. Qi, M. X., & Sun, W. (2017). Screw target matching algorithm based on image pyramid structure and Hu moment. *Modular Machine Tool & Automatic Manufacturing Technique, 9*, 19–22. https://doi.org/10.13462/j.cnki.mmtamt.2017.09.006
256. Rahman, Z., Jobson, D. J., & Woodell, G. A. (1996). Multi-scale retinex for color image enhancement. In *Proceedings of 3rd IEEE International Conference on Image Processing, Lausanne, Switzerland* (pp. 1003–1006). https://doi.org/10.1109/ICIP.1996.560995
257. Jobson, D. J., Rahman, Z., & Woodell, G. A. (1997). A multiscale retinex for bridging the gap between color images and the human observation of scenes. *IEEE Transactions on Image Processing, 6*(7), 965–976. https://doi.org/10.1109/83.597272
258. Petro, A. B., Sbert, C., & Morel, J. M. (2014). Multiscale retinex. *Image Processing On Line, 4*(18), 71–88. https://doi.org/10.5201/ipol.2014.107

259. Ahn, H., Keum, B., Kim, D., & Lee, H. S. (2013). Adaptive local tone mapping based on retinex for high dynamic range images. In *2013 IEEE International Conference on Consumer Electronics (ICCE), Las Vegas, NV, USA* (pp. 153–156). https://doi.org/10.1109/ICCE.2013.6486837
260. Guo, X. J., Li, Y., & Ling, H. B. (2017). LIME: Low-light image enhancement via illumination map estimation. *IEEE Transactions on Image Processing, 26*(2), 982–993. https://doi.org/10.1109/TIP.2016.2639450
261. Fu, G., Duan, L., & Xiao, C. X. (2019). A hybrid L2−LP variational model for single low-light image enhancement with bright channel prior. In *2019 IEEE International Conference on Image Processing (ICIP), Taipei, Taiwan* (pp. 1925–1929). https://doi.org/10.1109/ICIP.2019.8803197
262. Lecca, M., & Messelodi, S. (2019). SuPeR: Milano Retinex implementation exploiting a regular image grid. *Journal of the Optical Society of America. A, Optics, Image Science, and Vision, 36*(8), 1423–1432. https://doi.org/10.1364/JOSAA.36.001423
263. Djurović, I. (2021). RANSAC algorithm for instantaneous frequency estimation and reconstruction of frequency-modulated undersampled signals. *EURASIP Journal on Advances in Signal Processing*. https://doi.org/10.1186/s13634-021-00726-6
264. Stanković, S., Orović, I., & Stanković, L. J. (2017). Polynomial Fourier domain as a domain of signal sparsity. *Signal Processing, 130*, 243–253.
265. Djurović, I., & Simeunović, M. (2018). Review of the quasi maximum likelihood estimator for polynomial phase signals. *Digital Signal Processing, 72*, 59–74.

MIX
Papier aus verantwortungsvollen Quellen
Paper from responsible sources
FSC® C105338

If you have any concerns about our products,
you can contact us on
ProductSafety@springernature.com

In case Publisher is established outside the EU,
the EU authorized representative is:
Springer Nature Customer Service Center GmbH
Europaplatz 3, 69115 Heidelberg, Germany

Printed by Libri Plureos GmbH
in Hamburg, Germany